# Principles of Precambrian Geology

*Dedicated to the memory of
William Roy Goodwin,
1919–1944.*

# Principles of Precambrian Geology

**Alan M Goodwin**
*Department of Geology*
*University of Toronto*
*Toronto*
*Canada*

**ACADEMIC PRESS**
*Harcourt Brace and Company Publishers*
London   San Diego   New York   Boston   Sydney   Tokyo   Toronto

ACADEMIC PRESS LIMITED
24–28 Oval Road
London NW1 7DX

*U.S. Edition published by*
ACADEMIC PRESS, INC.
San Diego, CA 92101

This book is printed on acid-free paper

Copyright © 1996 by
ACADEMIC PRESS LIMITED

*All rights reserved*
No part of this book may be reproduced in any form
or by any means, electronic or mechanical including
photocopying, recording or any information storage and
retrieval system without permission from the publishers.

A catalogue record for this book is available
from the British Library

ISBN 0-12-289770-6

Typeset by Photographics
Printed in Great Britain by Butler and Tanner Ltd, Frome and London.

# Contents

1 **Distribution and tectonic setting of Precambrian crust**   1
1.1 Introduction   1
1.2 Global distribution   5
   1.2.1 Areal proportions   5
   1.2.2 Paleomagnetism and continental reconstructions   7
1.3 Radiometric dating   8
1.4 Orogenies and tectonic cycles   9
   1.4.1 Introduction   9
   1.4.2 Tectonic framework by craton   9
1.5 Precambrian classification scheme   21
1.6 Geologic setting by craton   23
   1.6.1 Cathaysian Craton   23
   1.6.2 Siberian Craton   24
   1.6.3 East European Craton   25
   1.6.4 Greenland Shield (North American Craton)   31
   1.6.5 North American Craton (less Greenland Shield)   33
   1.6.6 South American Craton   36
   1.6.7 African Craton   38
   1.6.8 Indian Craton   44
   1.6.9 Australian Craton   46
   1.6.10 Antarctic Craton   49

2 **Archean crust**   51
2.1 Introduction   51
   2.1.1. Distribution   51
   2.1.2. Salient characteristics   51
2.2 Cathaysian Craton   54
   2.2.1 Sino–Korean Craton   54
   2.2.2 Dabie Uplift   57
2.3 Siberian Craton   57
   2.3.1 Aldan Shield   57
   2.3.2 Anabar Shield   58
   2.3.3 Pericratonic mobile belts   58
   2.3.4 Buried basement   59
   2.3.5 Median massifs   59
2.4 East European Craton   59
   2.4.1 Ukrainian Shield   59
   2.4.2 Baltic Shield   60
2.5 Greenland Shield (North American Craton)   62
   2.5.1 Introduction   62
   2.5.2 Southern West Greenland   62
   2.5.3 Scottish Shield Fragment (North Atlantic Craton)   65
2.6 North American Craton (less Greenland Shield)   66
   2.6.1 Superior Province   66
   2.6.2 Slave Province   72
   2.6.3 Nain Province   75
   2.6.4 Churchill Province   75
   2.6.5 Minnesota River Valley Inlier   76
   2.6.6 Archean gneiss terrains in Wisconsin and Michigan   77
   2.6.7 Wyoming Uplift   79
2.7 South American Craton   81
   2.7.1 Guiana Shield   81
   2.7.2 Central Brazil Shield   81
2.8 African Craton: Southern Africa   88
   2.8.1 Kaapvaal Craton   88
   2.8.2 Limpopo Mobile Belt   98
   2.8.3 Zimbabwe Craton   100
2.9 African Craton: Central Africa   103
2.10 African Craton: West African Craton and Trans-Saharan Mobile Belt   104
   2.10.1 Man Shield   104
   2.10.2 Reguibat Shield   106
   2.10.3 Tuareg Shield   106
2.11 Indian Craton   106
   2.11.1 Dharwar Craton   106
   2.11.2 Granulite Domain (South Indian Highlands)   110
   2.11.3 Eastern Ghats Belt   111
   2.11.4 Singhbhum Craton of Singhbhum–Orissa   111
   2.11.5 Rajasthan and Bundelkhand blocks (Aravalli Craton)   113
   2.11.6 Bhandara Craton   114
   2.11.7 Mineral deposits   114
2.12 Australian Craton   114
   2.12.1 Pilbara Block   114

|   |   |
|---|---|
| 2.12.2 Hamersley Basin | 116 |
| 2.12.3 Yilgarn Block | 118 |
| 2.12.4 Others | 120 |
| 2.13 Antarctic Craton | 121 |
|     2.13.1 Napier Complex of Enderby Land | 121 |

## 3 Early Proterozoic crust — 123

| | |
|---|---|
| 3.1 Introduction | 123 |
|   3.1.1 Distribution | 123 |
|   3.1.2 Salient characteristics | 123 |
| 3.2 Cathaysian Craton | 127 |
|   3.2.1 Sino–Korean Craton | 127 |
| 3.3 Siberian Craton | 128 |
|   3.3.1 Crustal regeneration | 128 |
|   3.3.2 Fold belts | 129 |
| 3.4 East European Craton | 129 |
|   3.4.1 Baltic Shield | 129 |
|   3.4.2 Ukrainian Shield and Voronezh Massif | 133 |
|   3.4.3 Buried basement | 134 |
| 3.5 Greenland Shield (North American Craton) | 134 |
|   3.5.1 Nagssugtoqidian Mobile Belt | 134 |
|   3.5.2 Rinkian Mobile Belt | 134 |
|   3.5.3 Ketilidian Mobile Belt | 135 |
|   3.5.4 Scottish Shield Fragment (North Atlantic Craton) | 135 |
| 3.6 North American Craton (less Greenland Shield) | 135 |
|   3.6.1 Introduction | 135 |
|   3.6.2 Circum-Superior fold belts | 136 |
|   3.6.3 Kapuskasing Structural Zone | 146 |
|   3.6.4 Western Churchill Province | 146 |
|   3.6.5 Eastern Churchill Province | 148 |
|   3.6.6 Wopmay Orogen | 149 |
|   3.6.7 Western USA | 150 |
| 3.7 South American Craton | 151 |
|   3.7.1 Maroni–Itacaiunas Mobile Belt Amazonian Craton | 151 |
|   3.7.2 Goias Massif, Tocantins Province, Central Brazil Shield | 152 |
|   3.7.3 São Francisco Craton, Atlantic Shield | 152 |
| 3.8 African Craton: Southern Africa | 153 |
|   3.8.1 Introduction | 153 |
|   3.8.2 Transvaal Supergroup Basin Complex | 153 |
|   3.8.3 The Bushveld Complex | 156 |
|   3.8.4 The Waterberg–Soutpansberg–Matsap–Umkondo Basins | 158 |
|   3.8.5 Magondi Mobile Belt | 158 |
|   3.8.6 Kheis Belt | 159 |
| 3.9 African Craton: Central (Equatorial) Africa | 159 |
|   3.9.1 Ubendian (Ruzizian) Belt | 159 |
|   3.9.2 Usagaran Belt | 159 |
|   3.9.3 Ruwenzori Fold Belt | 160 |
|   3.9.4 Kimezian Assemblage | 161 |
|   3.9.5 Francevillian Supergroup | 161 |
|   3.9.6 Angolan Craton | 162 |
|   3.9.7 Kunene Anorthosite Complex | 162 |
|   3.9.8 Bangweulu Block (Zambian Craton) | 162 |
| 3.10 African Craton: Northwest Africa | 163 |
|   3.10.1 West African Craton | 163 |
|   3.10.2 Tuareg Shield | 164 |
|   3.10.3 Benin Nigeria Shield | 164 |
|   3.10.4 Anti-Atlas Domain | 164 |
| 3.11 African Craton: Northeast Africa and Arabia | 165 |
|   3.11.1 Uweinat Inlier | 165 |
|   3.11.2 Arabian–Nubian Shield | 166 |
| 3.12 Indian Craton | 166 |
|   3.12.1 Copperbelt Thrust Zone | 166 |
|   3.12.2 Aravalli–Delhi Belt | 166 |
|   3.12.3 Bhandara Craton | 167 |
|   3.12.4 Others | 168 |
| 3.13 Australian Craton | 168 |
|   3.13.1 West Australian Shield (Craton) | 168 |
|   3.13.2 North Australian Craton | 171 |
|   3.13.3 South Australia | 173 |
|   3.13.4 Barramundi Orogeny | 175 |
| 3.14 Antarctic Craton | 176 |

## 4 Mid-Proterozoic crust — 177

| | |
|---|---|
| 4.1 Introduction | 177 |
|   4.1.1 Distribution | 177 |
|   4.1.2 Salient characteristics | 177 |
| 4.2 Cathaysian Craton | 180 |
|   4.2.1 Introduction | 180 |
|   4.2.2 Lithostratigraphy | 180 |
| 4.3 Siberian Craton | 182 |
|   4.3.1 Introduction | 182 |
|   4.3.2 Akitkan and related groups | 182 |
|   4.3.3 Burzyanian Group (Early Riphean) | 182 |
|   4.3.4 Yurmatinian Group (Mid-Riphean) | 185 |
| 4.4 East European Craton | 185 |
|   4.4.1 Baltic Shield | 185 |

| | | | | |
|---|---|---|---|---|
| 4.4.2 Ukrainian Shield | 188 | 5.4 East European Craton | 223 |
| 4.4.3 Buried East European Craton | 188 | 5.4.1 Interior platform and pericratonic downwarps | 223 |
| 4.4.4 Southern Urals | 188 | 5.4.2 Baltic Shield | 224 |
| 4.5 Greenland Shield (North American Craton) | 189 | 5.4.3 Median massifs | 225 |
| 4.5.1 Gardar Province | 189 | 5.5 Greenland Shield (North American Craton) | 229 |
| 4.5.2 East Greenland Fold Belt | 190 | 5.5.1 East Greenland | 229 |
| 4.6 North American Craton (less Greenland Shield) | 191 | 5.5.2 Northern Greenland | 229 |
| 4.6.1 Paleohelikian Subera | 191 | 5.6 North American Craton (less Greenland Shield) | 229 |
| 4.6.2 Elsonian Disturbance (Orogeny) | 196 | 5.6.1 Arctic Province | 230 |
| 4.6.3 Neohelikian Subera | 197 | 5.6.2 Cordilleran Orogen | 230 |
| 4.7 South American Craton | 204 | 5.6.3 Appalachian Orogen | 233 |
| 4.7.1 Amazonian Craton: Guiana and Central Brazil Shields (less Tocantins Province) | 204 | 5.7 South American Craton | 234 |
| | | 5.7.1 Tocantins Province | 234 |
| 4.7.2 São Francisco Craton | 205 | 5.7.2 São Francisco Craton | 234 |
| 4.7.3 Tocantins Province (Central Brazil Shield) | 206 | 5.7.3 Atlantic Shield | 235 |
| 4.8 African Craton: Southern Africa | 206 | 5.8 African Craton: Southern Africa | 236 |
| 4.8.1 Namaqua–Natal Mobile Belt | 206 | 5.8.1 Introduction | 236 |
| 4.8.2 Koras–Sinclair–Ghanzi Rifts | 209 | 5.8.2 Damara Province | 236 |
| 4.9 African Craton: Central Africa | 209 | 5.8.3 Gariep Belt | 239 |
| 4.9.1 Kibaran Belt | 209 | 5.8.4 Malmesbury Group (Saldanian Belt) | 239 |
| 4.9.2 Irumide Belt | 210 | 5.8.5 Nama Group | 239 |
| 4.10 African Craton: Northern Africa | 212 | 5.8.6 Zambezi Belt | 239 |
| 4.10.1 Darfur and Eastern Tchad Basement | 212 | 5.9 African Craton: Central Africa | 240 |
| | | 5.9.1 Introduction | 240 |
| 4.10.2 Tuareg Shield | 212 | 5.9.2 Circum-Congo Basin Sequences | 241 |
| 4.11 Indian Craton | 212 | 5.9.3 Mozambique Belt | 245 |
| 4.11.1 Cuddapah–Kaladgi–Godavari Region | 212 | 5.9.4 Madagascar (Malagasy) | 247 |
| | | 5.9.5 Seychelles Islands | 247 |
| 4.11.2 Eastern Ghats Belt | 213 | 5.9.6 Central African Belt (Cameroon–Central African Republic–Southern Sudan) | 248 |
| 4.12 Australian Craton | 213 | |
| 4.12.1 Carpentarian Division | 214 | |
| 4.12.2 Musgravian Division | 216 | 5.10 African Craton: Northwest Africa | 248 |
| 4.13 Antarctic Craton | 218 | 5.10.1 Taoudeni Basin | 248 |
| | | 5.10.2 Anti-Atlas Domain | 249 |
| | | 5.10.3 Trans-Saharan Mobile Belt, Tuareg and Benin–Nigeria Shields | 249 |
| 5 Late Proterozoic crust | 219 | 5.11 African Craton: Northeast Africa | 251 |
| 5.1 Introduction | 219 | 5.11.1 Arabian–Nubian Shield | 251 |
| 5.1.1 Distribution | 219 | 5.12 Indian Craton | 254 |
| 5.1.2 Salient characteristics | 219 | 5.12.1 Vindhyan Basin | 254 |
| 5.2 Cathaysian Craton | 221 | 5.12.2 Sri Lanka | 254 |
| 5.3 Siberian Craton | 222 | 5.13 Australian Craton | 256 |
| 5.3.1 Karatavian (late Riphean) Division | 222 | 5.13.1 Adelaide Geosyncline | 256 |
| | | 5.13.2 Officer, Amadeus, Ngalia and Georgina Basins | 258 |
| 5.3.2 Kudashian (latest Riphean) Division | 223 | 5.13.3 Kimberley Region | 258 |
| | | 5.13.4 Tasmania | 258 |
| 5.3.3 Vendian (latest Proterozoic) Division | 223 | 5.14 Antarctic Craton | 258 |

## 6 Evolution of the continental crust — 261

- 6.1 Introduction — 261
- 6.2 Endogenous processes and products — 261
  - 6.2.1 Archean heat flow and geothermal gradients — 261
  - 6.2.2 The nature of Archean ocean crust — 262
  - 6.2.3 Granitoid associations — 262
  - 6.2.4 Growth rate of continental crust — 264
  - 6.2.5 Composition of continental crust — 264
  - 6.2.6 High grade metamorphic terrains — 265
  - 6.2.7 Mafic dyke swarms — 265
  - 6.2.8 The expanding earth theory — 266
- 6.3 Exogenous processes and products — 266
  - 6.3.1 Sea water composition — 266
  - 6.3.2 Uraniferous conglomerates — 267
  - 6.3.3 Banded iron formation (BIF) — 267
  - 6.3.4 Redbeds and paleosols — 269
  - 6.3.5 Glaciogenic and evaporite deposits — 269
  - 6.3.6 Oxidation state — 270
  - 6.3.7 Biogenesis — 270
  - 6.3.8 Sedimentation — 271
  - 6.3.9 Phosphorites — 273
  - 6.3.10 Mineral deposits — 273
- 6.4 Summary crustal development by stage — 274
  - 6.4.1 Hadean (formative) stage (4.6–3.9 Ga) — 274
  - 6.4.2 Archean stage (3.9–2.5 Ga) — 275
  - 6.4.3 Early Proterozoic stage (2.5–1.8 Ga) — 276
  - 6.4.4 Mid Proterozoic stage (1.8–1.0 Ga) — 276
  - 6.4.5 Late Proterozoic stage (1.0–0.57 Ga) — 277
- 6.5 Preferred model for the evolution of the continental crust — 277
  - 6.5.1 Introduction — 277
  - 6.5.2 Concluding statement — 278

**References** — 281

**Index** — 319

# Preface

*Principles of Precambrian Geology* represents an updated abridgment of my 1991 book *Precambrian Geology: The Dynamic Evolution of the Continental Crust*. The new book covers the same ground and reaches the same broad conclusions as its parent, but in reduced, less densely illustrated format. The aim of the new book remains the same: to provide a modern, comprehensive statement on the nature and evolution of Earth's Precambrian crust. A meaningful Precambrian perspective is achieved by precisely delineating appropriate geologic elements in terms of time, space and nature with due regard to presently unresolved aspects, all within the context of Earth's stage-by-stage crustal evolution. As previously considered (1991), the main focus is placed upon geologic inventory rather than tectonic models. The resulting broad synthesis provides a suitable framework for assessing various Earth dynamic-biospheric hypotheses including the well-known plate-tectonic paradigm according to which Earth's crust presently moves, interacts and regenerates, and the less well-known but broadly inclusive Gaia hypothesis which postulates that the physico–chemical conditions of Earth's lithosphere–atmosphere–hydrosphere are favourably influenced by the presence of life itself. The present book is designed, in brief, to serve alike the needs of student, teacher, explorationist, and general examiner of Earth's continental crust. Such a broad comprehensive coverage is particularly relevant in this age of increasing specialization.

Abridgment has involved reductions in text (from 666 to 325 pages), illustrations, including figures and tables, (from 198 to 115) and references (from 1721 to 1609, the last involving deleting 754 deletions and 642 superceeding additions). The required text reduction has been achieved mainly at the expense of geologic–geochronologic details and relationships, so important to fabric richness but not to broad synthesis. Deleted data have been replaced where possible by summary statements, pertinent references and increased reliance on retained illustrations. However, anyone interested in fuller details is urged to refer to the parent (1991) book.

Updating the text has involved incorporating the substance of 642 selected, all-English language articles published in 50 regular journals and 61 books and memoirs. The new articles bear the following four key relationships, with article percentages in brackets:
1 by publication year, 1992–94 (67), 1991–92 (22) and 1954–90 (11);
2 by publication medium, Precambrian Research (30), Geology (12), Canadian Journal of Earth Sciences (11), Books/Memoirs (10), Economic Geology (7), Earth and Planetary Science Letters (4), Nature (2), and 44 other journals (24);
3 by Precambrian age, Archean (33), early (18)-, mid (15)-, and late (17)- Proterozoic, and undivided (17);
4 by geographic distribution, North America (34), Africa (19), Australia (9), India (7), Fennoscandia (6), other Preambrian cratons (7), and undivided (18).

To prevent possible misinterpretation and confusion, the term Precambrian 'platform', used in the 1991 book and following long-established Russian practise, has been replaced with the term Precambrian 'craton', in accord with common world usage. Following the AGI Glossary, craton refers to a part of the Earth's crust which has attained stability, and which has been little deformed for a prolonged period. Precambrian cratons, which centre existing continents or pre-existing but now-dismembered supercontinents (e.g. Gondwanaland), include both exposed basement shields and platforms, the latter, where present, composed of flat-lying or gently consolidated mainly sedimentary strata, such platforms representing integral parts of the designated cratons. Precambrian cratons were developed in a myriad of sizes, ages, and durations, whether now wholly, partly or non-existent. For identification purposes throughout the text, particular Precambrian cratons are designated adjectivally as required; thus, 'composite craton' and 'component craton' refer respectively to the large predominating Precambrian craton of an existing continent, and to the smaller contained cratons therein; and 'Eocambrian (end-Precambrian) craton' refers to the final stable Precambrian craton that existed at the close of Precambrian time, 570 Ma ago, whether now wholly extant, partly dismembered, or non-existent.

The term 'terrain' is retained as a general designation of a particular rock unit or association. The term 'terrane' is selectively restricted to a fault-bounded body of rock of regional extent characterized by a geologic history different from that of contiguous terranes, such as to indicate significant juxtaposing plate motions as outlined by Coney et al (1980). 'Suspect terrane' or 'possible terrane' are useful terms where tectonic relationships are equivocal. Also, the term 'domain' may be usefully applied to geologic areas of distinct characteristics from one another without tectonic implications. The term 'granitoid–greenstone' association is preferred to the less precise and, in view of prevailing granitoid terminology, lithologically misleading 'granite–greenstone' alternative. 'Gondwanaland' is retained for the pre-Mesozoic southern supercontinent rather than the admittedly neater 'Gondwana' (land of Gonds), in deference to the earlier named Gondwana Supergroup, an essentially terrigenous sequence of sediments laid down in peninsular India.

The updating-abridgment process has brought to light the many rapid advances currently underway in Precambrian studies. Thus the impact of geochronology, notably U–Pb zircon dating and related isotopic systematics has added tremendous precision to rock unit correlations both surficial and intrusive, to unravelling complex crustal inheritance relationships, and to establishing fundamental petrogenic developments in presently exposed but once-deeper-seated rock units. One successful example is provided by ongoing research in the fabulous gold-endowed Witwatersrand Basin, South Africa, involving skilfully coordinated field, isotopic and geophysical studies. The plate-tectonic process, with growing application especially to later Precambrian studies, is now increasingly leavened in earlier Precambrian studies by recognition of a markedly different, probably plume-dominated early Earth. However, little is known about the timing and nature of the intervening process transitions across Precambrian time.

Needless to say, any synthesis of such a vast topic as Precambrian geology is bound to be selective, with many important aspects short-changed. However, the purpose of this book will have been achieved if it provides the reader with meaningful perspective and stimulates interest and curiosity in this fascinating branch of Earth science. After all, reaching out to the stars, the current space age challenge, is more than matched by reaching back across Precambrian time to the actual birth and infancy of our planet. Such is the justification and rationale for perusing this book.

I acknowledge the great value and stimulation provided by comments, criticisms, discussion and references provided by Hans Hoffman, Warren Hamilton, Murray Frarey, Norman Evenson and Walter Mooney, which have resulted in substantial improvements to the text. Thanks are due to Subash Shanbhag, University of Toronto, for skilful drafting services. I am grateful to Academic Press personnel for shepherding the text through to publication. Once again, I acknowledge my great debt to Precambrian geologists of the world – wherever and whenever – upon whose shoulders this work securely rests.

# Chapter 1

# Distribution and Tectonic Setting of Precambrian Crust

## 1.1 INTRODUCTION

The bulk of Earth's Precambrian crust is located in nine Precambrian cratons—large, subcircular to oblong, tectonically stable continental entities composed of Precambrian rocks of diverse types and ages, which dominate the main continents—Asia, Europe, Greenland, North America, South America, Africa, India, Australia and Antarctica (Figs 1-1, 1-2; Table 1-1). The nine Precambrian cratons, together with rare neighbouring island microcontinents, comprise both (1) exposed shields, also called craton, block, uplift, rise, belt, nucleus, ridge, etc.; and (2) buried (i.e. sub-Phanerozoic) basement and cover. Additional Precambrian crust lies in numerous median massifs (inliers), scattered within long, linear, pericratonic Phanerozoic mobile belts, and in certain peripheral and isolated oceanic environments. Some intercratonic clusters of median massifs may, in fact, represent now-fragmented former Precambrian cratons, e.g. Kazakhstan region situated between the East European and Siberian cratons.

Asia, lying east of the Urals and in present context north of the Indian Subcontinent, contains two independent Precambrian cratons: Cathaysian in the southeast and Siberian in the north. The (1) *Cathaysian Craton* includes three component cratons—Sino–Korean, Tarim and Yangtze—each characterized by restricted Precambrian exposure and correspondingly widespread buried basement. Additional pericratonic Precambrian crust lies in numerous median massifs enclosed in closely compressed Caledonian, Variscan and Tanshanian fold belts. The (2) *Siberian Craton* to the north comprises the Aldan (-Stanovoy) and Anabar shields, Olenek Uplift, four restricted peripheral Precambrian fold belts in the south and west, and extensive buried interior basement and platform cover. The encircling Phanerozoic fold belts contain six main Precambrian massifs, of which Kolyma–Omolon is by far the largest. The neighbouring East Arctic Shelf also contains some presently ill-defined, buried Precambrian crust.

Europe is cored by the (3) *East European (Russian) Craton*, comprising the comparatively large Baltic (Fennoscandian) Shield and much smaller Ukrainian Shield, the slightly buried Voronezh Uplift and Volga–Kama Anteclise to the east, with intervening aulacogen-induced troughs, and the deeply buried interior basement including the Moscow (–Baltic) and Caspian syneclises. The small Scottish Shield Fragment, a rifted piece of the pre-drift North Atlantic Craton, occupies parts of Scotland and Ireland; related Precambrian crust underlies much of central–southern England. At least 10 Precambrian median massifs are contained in the Variscan–Hercynian fold belts including Armorican, Central, Bohemian, Vosges–Black Forest, Iberian and Uralian massifs. Substantial but presently ill-defined, buried Precambrian crust lies in the neighbouring West Arctic Shelf.

The *Greenland Precambrian Shield*, part of the North American Craton, occupies all of this island-continent but the East Greenland Caledonides and North Greenland Fold Belt, both of which have some Precambrian inliers. Exposed shield rock is effectively restricted to the narrow, ice-free coastlines.

(4) The *North American Craton*, less Greenland, includes (a) in the northern part, the unusually large Canadian Shield with buried extensions beneath the Hudson, Arctic, Interior and St Lawrence lowlands; (b) the comparatively small Wyoming Uplift to the west; and (c) substantial buried basement in south-central (midcontinent) USA. Numerous Precambrian inliers crop out in the enclosing Phanerozoic fold belts.

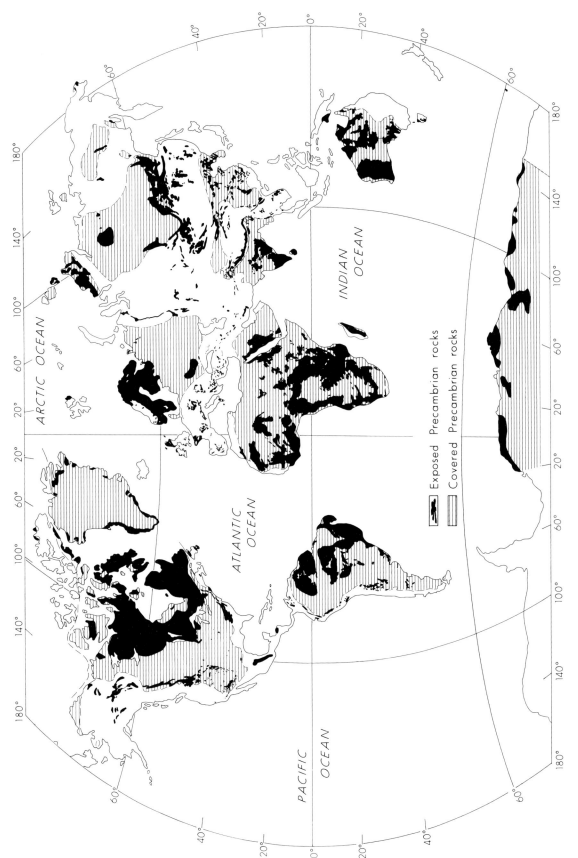

**Fig. 1-1.** Global Precambrian sketch-map showing the distribution of exposed and buried (sub-Phanerozoic) Precambrian crust within the conventionally defined continents. Data plotted on National Geographic Society base-map 'The World', National Geographic Magazine (Washington, December 1981).

*Table 1-1.* Precambrian composite cratons (Roman numerals), component cratons, shields, blocks, belts, etc. (Arabic numerals), and neighbouring median massifs, inliers, etc. (lower case letters) by continent.

*Asia* (excluding India)

I    Cathaysian Craton
  (1) Sino-Korean Craton, (2) Tarim Craton, (3) Yangtze Craton
    (a) Himalayan Massifs, (b) Pamirs, (c) South Tien Shan Massif

II   Siberian Craton
  (1) Aldan Shield, (2) Anabar Shield, (3) Olenek Uplift, (4) Baikal Belt, (5) East Sayan Belt, (6) Stanovoy Belt, (7) Yenisei Ridge, (8) Turukhansk Uplift
    (a) Kolyma-Omolon Massif, (b) Taigonos Block, (c) Okhotsk Massif, (d) Altai-Sayan Massif, (e) Taymyr Belt, (f) East Arctic Shelf

*Europe*

III  East European (Russian) Craton
  (1) Baltic Shield, (2) Ukrainian Shield, (3) Voronezh Uplift, (4) Timan-Pechora Extension, (5) Volga-Kama Anteclise, (6) Caspian Syneclise, (7) Moscow (-Baltic) Syneclise
    (a) Uralian Inliers, Variscan massifs including (b) Armorican Massif, (c) Massif Central, (d) Bohemian Massif, (e) British Precambrian including Scottish Shield Fragment, (f) West Arctic Shelf

*North America-Greenland*

IV  Greenland Shield
  (1) Archaean Block, (2) Nagssugtoqidian, Rinkian and Ketilidian belts, and buried extensions
    (a) Caledonide massifs, (b) North Greenland massifs

IV  North America Craton (excl. Greenland Shield)
  (1) Canadian Shield and buried extensions, (2) Wyoming uplift, (3) Central (US) Belt (buried) and buried extensions of Grenville Belt
    Inliers in the (a) Cordilleran, (b) Ouachitan, (c) Appalachian and (d) Innuitian fold belts

*South America*

V   South American (-Patagonian) Craton
  (1) Guiana Shield, (2) Central Brazil Shield, (3) Atlantic Shield, and buried extensions
    (a) Cordilleran inliers

*Africa*

VI  African (-Arabian) Craton
  Kalahari Craton (southern Africa): (1) Kaapvaal Craton, (2) Zimbabwe Craton, (3) Limpopo Belt, (4) Namaqua-Natal Belt, (5) Rehobothian Domain, (5) Koras-Sinclair troughs, (7) Nama Basin
  Congo Craton (central Africa): (1) Kasai-Angolan Craton, (2) Chaillu Craton, (3) Gabon Craton, (4) Bouca Craton, (5) Bomu-Kibalian Craton, (6) Tanzania Craton, (7) Zambian Craton (Bangweulu Block), (8) Madagascar Craton, (9) Ubendian-Ruzizian Belt, (10) Ruwenzori (Buganda-Toro) Belt, (11) Kibalian Belt, (12) Irumide Belt, (13) Lurio Belt
  West African Craton (northwestern Africa): (1) West African Craton including Reguibat and Man shields, Taoudeni and Volta basins, Gourma Aulacogen, and Rockelides, Marampa and Kasila belts, (2) Tuareg Shield, (3) Benin Nigeria Shield
  East Saharan Craton (north-central Africa)
  Northeastern Africa: (1) Tibesti, Uweinat and Tchad inliers, (2) Arabain-Nubian Shield
  Mobile belts: (1) Damara-Katanga-Zambezi, (2) Central African (Cameroon-West Nile), (3) Mozambique, (4) West Congo, (5) Trans-Saharan (Pharusian-Dahomeyan), (6) Kaoko, (7) Gariepian, (8) Saldanian

*India* (subcontinent)

VII  Indian Craton
  (1) Dravidian Shield including Western Dharwar Craton, Eastern Dharwar Craton and Southern Highlands Granulite Terrain, (2) Bhandara Craton, (3) Singhbhum Craton, (4) Aravalli Craton including the Bundelkhand Complex, (5) Eastern Ghats Belt, (6) Sri Lanka Craton

*Australia*

VIII Australian Craton
  (1) West Australian Shield including Pilbara and Yilgarn blocks, Capricorn Orogen and Bangemall Basin, (2) North Australian Craton with adjoining Northeast Orogens and buried extensions, (3) Central Australian mobile belts, (4) Gawler-Nullarbor Block, (5) Curnamona Craton
    (a) Tasmanian inliers

*Antarctica*

IX  Antarctic Craton
  (1) East Antarctic Metamorphic Shield
    (a) Transantarctic Mountains inliers

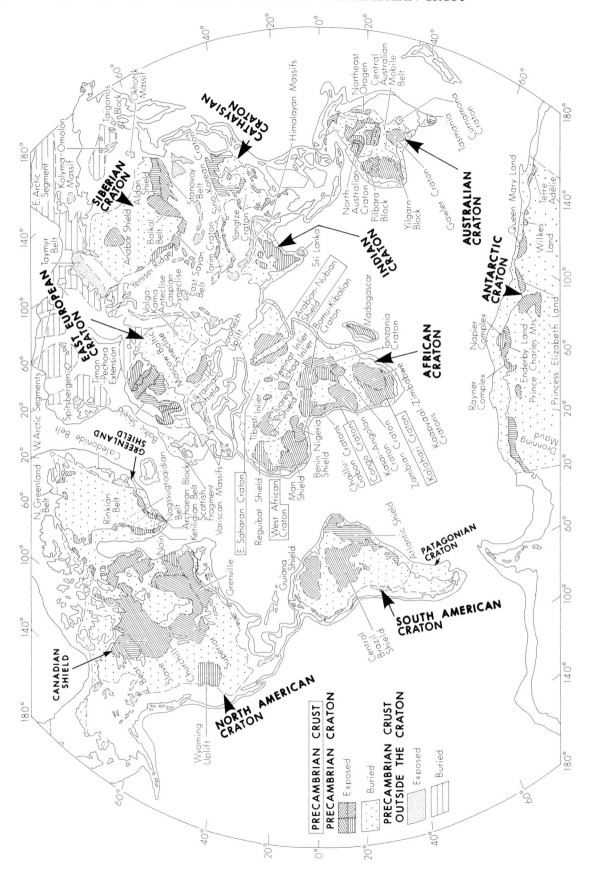

(5) The *South American Craton* comprises the Guiana, Central Brazil and Atlantic shields, with buried basement beneath (a) the intervening Amazon, Parnaiba, and Paraná basins; (b) the Sub-Andean Foredeep to the west; and (c) the Atlantic margin deposits to the east. The adjoining, south-tapering, Phanerozoic-dominated Patagonian Craton to the south is integrated here for convenience. Precambrian inliers are scattered along the length of the Andean Chain to the west, including the substantial Arequipa–Cuzco Massif exposed on the Pacific coast at 15–20°S.

(6) The *African (–Arabian) Craton* occupies all of the continent with the exception of the restricted Cape, Mauritanide and Atlas fold belts, located respectively at the south, northwest and north margins. This unusually large craton is conveniently divided by an orthogonal system of late Precambrian (Pan-African) mobile belts, into five parts: (a) Kalahari (southern); (b) Congo (equatorial); (c) West African (northwestern); (d) East Saharan (north–central) cratons; and (e) Arabian–Nubian (northeastern) Shield. The system of Pan-African belts includes from west to east, the N-trending (a) Pharusian–Dahomeyan; (b) West Congo–Gariep; and (c) Mozambique and, from south to north, the ENE-trending (d) Saldanian; (e) Damara–Katanga–Zambezi; (f) Central African; and (g) an unnamed northern African belt. Two large central subsidence basins—Taoudeni and Congo—are located in the West African and Congo cratons respectively. The southerly Kalahari Craton is largely obscured in the northwest by the Kalahari desert. The rifted Madagascar and Seychelle Islands are included in the African Craton.

(7) The *Indian Craton* with Sri Lanka includes exposed Precambrian terrains in the southern (Dravidian), eastern (Eastern Ghats), northeastern (Chotanagpur–Singhbhum) and northwestern (Aravalli) blocks, with substantial buried extensions beneath the west–central Deccan Traps and the unusually extensive Ganges–Indus flood plain which encroaches northward on the Precambrian inlier-charged Himalayan Fold Belt, products of extensive Cenozoic subduction–collision events.

(8) The *Australian Craton* underlies all of the continent and adjoining shelf except for the Tasman Fold Belt in the east. The craton extends northward beneath the Arafura Sea to incorporate a south–central embayment in Papua-New Guinea. The craton is conveniently divided by the Central Australian Mobile Belt network into the North Australian, West Australian (Yilgarn and Pilbara blocks), Gawler (–Nullarbor) and Curnamona cratons and Northeast orogens, each with specific subdivisions.

Finally, (9) the *Antarctic Craton* is dominated by the East Antarctic Metamorphic Shield, which adjoins the Transantarctic Mountains (Fold Belt) which itself contains numerous Precambrian inliers. The great bulk of the Precambrian Craton lies beneath the continental ice sheet, with exposures mainly restricted to the coastline including those in Dronning Maud Land and Enderby Land, Prince Charles Mountains, Princess Elizabeth, Queen Mary and Wilkes lands and Terre Adélie.

Figures 1-1, 1-2 and Table 1-1 provide useful basic references for the rest of the book, including this chapter. In existing plate tectonic terms (Lowman, 1992, fig. 2), the Cathaysian, Siberian and East European cratons all lie in the Eurasian Plate, the North American, South American, Antarctic and African (–Arabian) cratons in plates of the same names, and the Indian and Australian cratons both in the Indian–Australian Plate.

## 1.2 GLOBAL DISTRIBUTION

### 1.2.1 AREAL PROPORTIONS

Methodology

Areas of exposed and buried (i.e. sub-Phanerozoic) Precambrian crust were calculated by planimeter surveys of sub-Phanerozoic geologic maps, prepared as follows for each continent. The 1 : 10 000 000 scale UNESCO World Atlas maps (1976) were used as the base for each continent. Precambrian age destinations in these maps were modified as required in response to more recent information. Continental margins were extended in all cases to include continental shelves according to

**Fig. 1-2.** Global Precambrian sketch-map showing the distribution of Precambrian cratons including exposed shields (cratons, fold-belts, blocks, etc.) and buried (sub-Phanerozoic) platforms, together with Precambrian median massifs (inliers) within the continents as bounded by the continental slopes. Craton divisions are as listed in Table 1-1. Certain Cordilleran and Appalachian allochthonous thrust-slices with contained Precambrian liners overlying the North American Craton margins as well as Kalahari Desert cover in southern Africa, as illustrated in Fig. 1-1, are omitted for basement clarity. Data plotted on National Geographic Society base-map, "The World", National Geographic Magazine (Washington, 1981).

the National Geographic Society base map, *The World* (published in Washington DC, 1981). Sub-Phanerozoic geologic contacts were drawn according to the best available controls.

The resulting calculated areas of Precambrian rocks by continent compare closely with those provided both by Poldervaart (1955) and, for the relevant continents, by Hurley and Rand (1969). According to Poldervaart's results, the total measured continental crust is $148 \times 10^6$ km$^2$, of which 72% ($105 \times 10^6$ km$^2$) is Precambrian continental crust. According to the new calculations prepared for this book (Table 1-2), the total measured Precambrian crust is $106 \times 10^6$ km$^2$. However, Cogley (1984) has shown that the continental crust may be more extensive than previously considered. Thus recent marine geophysical surveys have revealed that continental crust probably underlies several depressed peripheral and isolated oceanic plateaus, connecting ridges and continental rises, and even portions of some abyssal plains, all beyond the continental margins. Cogley's revised estimate of continental crust is $210.4 \times 10^6$ km$^2$. Some portion of this 'new-found' continental crust may be Precambrian and would be thereby excluded from the measurements used in this text.

These problems notwithstanding, the global Precambrian geologic map prepared for the present survey provides a satisfactory basis for this study of Precambrian crust. Figures 1-1, 1-2, 2-1, 3-1, 4-1, 5-1 and Tables 1-1, 1-2, 2-1, 3-1, 4-1, 5-1 are direct outgrowths of this global Precambrian map.

*Table 1-2.* Areal proportions of exposed and buried (sub-Phanerozoic) Precambrian crust by continent and era/eon. Exposed Precambrian crust includes all tectonically uplifted domains which, as a result of non-deposition or erosion, are free of consolidated Phanerozoic cover (e.g. Canadian Shield) whereas buried Precambrian crust includes buried extensions beneath consolidated Phanerozoic cover (e.g. Central Belt, USA and Officer Basin, Australia).

| | | | Precambrian era/eon (%) | | | |
| | | | Proterozoic era | | | Archean eon |
| Continent | Area ($10^3$ km$^2$) | Percentage of total | Late (0.6–1.0 Ga) | Mid (1.0–1.7 Ga) | Early (1.7–2.5 Ga) | (>2.5 Ga) |
|---|---|---|---|---|---|---|
| *Exposed crust only* | | | | | | |
| Asia[1] | (2 670) | 9 | 25 | 42 | 11 | 22 |
| Europe | (1 595) | 5 | 35 | 17 | 28 | 20 |
| North America[2] | (5 969) | 20 | 10 | 23 | 37 | 30 |
| South America[3] | (5 366) | 18 | 33 | 36 | 15 | 16 |
| Africa[4] | (10 684) | 35 | 54 | 8 | 18 | 20 |
| India | (847) | 3 | 22 | 30 | 6 | 42 |
| Australia | (2 329) | 7 | 12 | 41 | 28 | 19 |
| Antarctica | (845) | 3 | 37 | 38 | 5 | 20 |
| Total | (30 305) | 100 | 33 | 23 | 22 | 22 |
| *Exposed plus buried crust* | | | | | | |
| Asia[1] | (8 033) | 9 | 46 | 30 | 21 | 3 |
| Europe | (9 507) | 8 | 45 | 11 | 20 | 24 |
| North America[2] | (19 470) | 20 | 4 | 30 | 49 | 17 |
| South America[3] | (18 419) | 13 | 52 | 33 | 10 | 5 |
| Africa[4] | (28 381) | 29 | 75 | 6 | 7 | 12 |
| India | (3 837) | 4 | 47 | 15 | 2 | 36 |
| Australia | (7 657) | 8 | 15 | 55 | 20 | 10 |
| Antarctica | (10 632) | 9 | 37 | 38 | 5 | 20 |
| Total | (105 936) | 100 | 43 | 22 | 21 | 14 |

[1]Excluding India
[2]Including Greenland
[3]Including Patagonia
[4]Including Arabia and Madagascar

## Results

The calculated areas of exposed and buried Precambrian crust as preserved in Earth's eight continents are listed in Table 1-2. The area of exposed Precambrian crust is 30 305 000 km$^2$. The proportions by continent, as listed, reveal that Africa contains most exposed Precambrian crust; that the four Atlantic continents—Europe, the two Americas and Africa—collectively contain 78% of the total; and that, assembled by former supercontinents, Laurasia (Asia, Europe, North America) and Gondwanaland (South America, Africa, India, Australia, Antarctica) respectively represent 34% and 66% of the total; or Laurasia:Gondwanaland = 1 : 2.

The area of total (exposed plus buried) Precambrian crust is 105 936 000 km$^2$, with the proportions by continent as listed. Thus Africa again dominates, though in reduced proportion. The same four Atlantic continents together account for 72% of the total. The former supercontinents of Laurasia and Gondwanaland respectively account for 37% and 63% of the total; or Laurasia:Gondwanaland = 1 : 1.7.

The area of exposed Precambrian crust (30 305 000 km$^2$) represents 29% of the total Precambrian crust (105 936 000 km$^2$). The total Precambrian crust, in turn, constitutes 50% of Cogley's estimated continental crust (210 405 000 km$^2$) (Cogley 1984), which itself forms 41% of Earth's total surface (513 183 000 km$^2$). In brief, the preserved areas of (1) exposed and (2) total (exposed plus buried) Precambrian crust, products of 87% of geologic time (0.6–4.6 Ga) and the focus of this text, respectively account for (1) 6% and (2) 21% of Earth's present total surface, encompassing continental and oceanic crust. Any consideration of Precambrian crustal volumes (i.e. including crustal depth) is fraught with uncertainties. Whereas the present Precambrian-bearing continental crust is three to four times as thick as the intervening oceanic crust, thereby moving in the direction of substantially higher overall Precambrian crustal proportions, the downward projection of Precambrian terrains within the continents despite increasing geophysical deep crustal probing (Durrheim and Mooney 1991, Klemperer 1992) is highly conjectural. Accordingly, all Precambrian crustal proportions used herein are area based.

### 1.2.2 PALEOMAGNETISM AND CONTINENTAL RECONSTRUCTIONS

According to plate tectonics the continents are in general motion one to the other, the pattern of the Wilson cycle allowing for recurring fission and fusion of supercontinents. The present distribution pattern of the continents, then, is ephemeral. How were they distributed in the past? This questions is of fundamental importance to Precambrian geologists, as it is to all students of the continental crust. Indeed it would be difficult to overemphasize the significance of such palaeoreconstructions in interpreting the dynamic evolution of the continental crust. This warrants careful consideration of both the paleomagnetic methods and their current limitations. (Key references: Piper et al 1973, McElhinney and McWilliams 1977, Dunlop 1981, Irving and McGlynn 1981, McWilliams 1981, Piper 1983, 1991, 1992, Onstott et al 1984, Tarling 1985, Etheridge and Wyborn 1988, Butler 1992, Park 1992, Duncan and Turcotte 1994, Onstott 1994).

What do paleomagnetic studies reveal about the motions of cratons in the past and their paleoreconstructions? Was the tempo of assembly and dispersal episodic and essentially random, or cyclic and systematic on a global scale (Dickinson 1993, Duncan and Turcotte 1994)? Are orogenic belts mainly intra- or inter-cratonic? In brief, the available paleomagnetic record is markedly sporadic and the interpretations conflicting and controversial (Onstott 1994). This is so, even in the case of late Proterozoic, the most favoured Precambrian era for such interpretations (van der Voo and Meert 1991, Dalziel 1992, Powell et al 1993).

In view of the widely contradictory interpretations, it must be concluded that, despite the great potential and need, any Precambrian paleoreconstruction is to be used with extreme caution and with careful consideration of all assumptions used and existing limitations. Only through multidisciplinary approach will the debate over the origin of mobile belts and presence, absence, periodicity or longevity of supercontinents be resolved (Onstott 1994). Successful Precambrian application demands unusually precise paleomagnetic measurements in concert with equally precise geochronologic–geologic studies (Buchan et al 1994). Continuing effort is required to bring this potentially decisive technique to fruition.

Available geologic–paleomagnetic evidence demonstrates the existence of the supercontinent Pangea (0.6–0.2 Ga) and points to the presence of the earlier supercontinent of Rodinia (Vendia) (1.0–0.6 Ga) (Bond et al 1984, Hoffman 1988, 1991, McMenamin and McMenamin 1990, Dalziel 1992a, b, Brookfield 1993, Powell et al 1993). Evidence for earlier supercontinents, including those proposed for the mid-Proterozoic (1.8–1.0 Ga, Piper et al 1973) and even longer (2.7–0.6 Ga, Piper 1983) is less certain (Duncan and Turcotte 1994).

Based mainly on non-paleomagnetic data and therefore largely unverifiable, the inferred Rodinian supercontinent formed by 1.3–1.0 Ga and broke-up by 625–555 Ma when the Pacific Ocean opened. The western margin of Laurentia was adjacent to the eastern margin of Australia–Antarctica; either parts of South America and Baltica or only South America were adjacent to the eastern margin of Laurentia. Many other partial Precambrian paleoreconstructions have been proposed, based on local geologic relationships, and on 'apparent polar wander paths' (APWP). However, all suffer from at least some problems of verification (Tarling 1985).

## 1.3 RADIOMETRIC DATING

A tectonic classification system of Precambrian rocks is used in this book, based on the recognition of successive tectonic imprints on Precambrian terrains of substantial size.

By way of background, large regions of Precambrian rocks, such as shields, are commonly divided into recognizable and distinctive structural domains or provinces. Recognition of structural provinces (Gill 1949, Wilson 1949, Holmes 1951) is based largely on differences in overall structural trends and style of folding, which are products of the last major orogeny to affect the rocks in a particular province. Boundaries of provinces and subprovinces are typically drawn along major unconformities or orogenic fronts, where structures on one side are typically truncated by younger structures on the other side. However, the structural criteria used for defining the boundaries indicate only the relative ages of the structures on either side (Stockwell 1961, 1982).

Typically, each of the component structural provinces of a Precambrian shield has been involved in more than one period of orogeny, and each contains rocks of several or many ages. The last major orogeny not only imposed its presently distinctive structural style but characteristically imposed comparative tectonic stability or cratonization. Thus the assembled provinces of a shield bear witness to the progressive cycle-by-cycle tectonic cratonization of that shield, a process called 'continental accretion'.

The advent of radiometric (isotopic) dating, first stated explicitly by Holmes and Lawson in 1927, led to the recognition that individual structural provinces carry a characteristic radiometric imprint which is also the product of the last major orogeny to affect the province. This recognition has revolutionized Precambrian geology which, in the practical absence of index fossils, faced irresolvable problems of regional Precambrian correlation. The resulting chronologic framework is fundamental to understanding the dynamic evolution of the continental crust.

The historical development and detailed methodology of radiometric dating are fully described in a number of books and publications (Dalrymple and Langphere 1969, Doe 1970, Faure and Powell 1972, York and Farquhar 1972, Harper 1973, Moorbath 1976, Faure 1977, 1986, O'Nions et al 1979, Froude et al 1983). Established methods, extensively used in the 1950s–1960s and early 1970s, include K–Ar, Rb–Sr, and U–Pb methods either on single minerals or, where appropriate, on whole-rock samples. New developments in high precision, solid-source mass spectrometry have made possible the exploitation of the alpha decay of $^{147}$Sm to $^{143}$Nd, and of $^{176}$Lu to $^{176}$Hf as a rock dating and petrogenetic investigative method (O'Nions et al 1979, Patchett et al 1981).

The introduction of the concordia diagram (Wetherill 1956) ultimately resulted in U–Pb zircon dating becoming the most useful of the isotopic systems for determining stratigraphic ages. Subsequent enhancements included the development of low contamination techniques (Krogh 1973), the use of air abrasion (Krogh 1982), and the introduction of the $^{205}$Pb spike (Krogh and Davis 1975).

The recent application of the sensitive highmass resolution ion-microprobe (SHRIMP) to the analysis of single grains, thereby avoiding painstaking mineral separation, gives great promise, and the ion-microprobe is now the state-of-the-art instrument for the analysis of diminishing volume. Indeed, the ability of the SHRIMP to measure isotopic heterogeneities within individual zircon crystals on the scale of a few tens of micrometres has transformed the study of the age and history of polymetamorphic rocks and removed some of the ambiguities involved in the interpretation of conventional single- and multi-grain zircon analyses (Compston et al 1984).

Direct and precise dating, expressed in millions ($10^6$) of years (Ma) or billions ($10^9$) of years (Ga) measured backward from the present, can clearly yield fundamental information on the timing and duration of many Precambrian events and processes.

## 1.4 OROGENIES AND TECTONIC CYCLES

### 1.4.1 INTRODUCTION

Each of the nine Precambrian cratons attained tectonic stability in successive stages. In general, the stages correspond to tectonic cycles, each marked by a culminating orogeny. Each cycle preferentially affected a particular part of the composite craton, commonly a linear fold belt or more irregular terrain. Thus cratonization was achieved in stages, the successively stabilized parts referred to as orogenic provinces or blocks. The final craton is a patchwork mosaic of structural provinces, each with characteristic age or limited range of ages, that has been assembled by processes collectively called 'continental accretion', inclusive of additions upon (epi), within (intra-), around (peri-) and under (sub-) the existing continental crust.

Most of the nine Precambrian cratons bear the imprint of three Archean (> 2.5 Ga) and three Proterozoic (2.5–0.57 Ga) tectonic cycles. Generally speaking, all nine cratons comprise an older Precambrian (> 1.8 Ga) basement, with varied younger Precambrian peri- and intracratonic accretions (fold belts) and more or less Precambrian platform cover. Thus the resulting tectonic classification reflects progressive craton stabilization.

The role of plate-tectonics in Precambrian cratonization requires constant appraisal. Geological processes have changed greatly as Earth has progressively lost heat, and actualistic plate-tectonic models become progressively less applicable to successively older rocks. Thus the evidence for operation of modern plate-tectonic processes is strong in later Proterozoic (< 1.3 Ga) crust, but weak-to-absent in Archean (> 2.5 Ga) crust. In this regard the possible role of other models, including voluminous magmatism, remains controversial and largely unexplored.

As will be appreciated, Precambrian cratons of the world stand at different levels of understanding, depending on a variety of factors, including bedrock exposure, access, economic exploration incentives, population density and, especially, intensity, duration and calibre of scientific studies. These factors, together with the virtual absence of Precambrian index fossils, the immensity of Precambrian time, the comparative scarcity of the Precambrian sedimentary record and the complementary dominance of hitherto largely intractable gneissic terrains, compound the inherent Precambrian imprecision (Harland et al 1982, Wright 1985). As a result, any current worldwide Precambrian classification system is at best a useful contemporary approximation, to be improved upon and eventually replaced, though not in the foreseeable future, by a more rigorous chronostratigraphic system such as that used in the Phanerozoic.

To avoid unnecessary duplication, references cited elsewhere in the book are used sparingly in the remainder of this chapter.

### 1.4.2 TECTONIC FRAMEWORK BY CRATON

#### Cathaysian Craton

Four main orogenies are widely recognized in older Precambrian (> 1.85 Ga) rocks of the Sino–Korean Craton: Qianxi, Fupingian, Wutaian and Luliangian; and three in younger Precambrian rocks—Dongan, Sibaoan and Jinningian (Yangtze) (Figs 1-2, 1-3a). The Qianxi Orogeny, responsible for widespread granulite facies metamorphism, provides the mid–late Archean boundary. The Fupingian Orogeny, involving both widespread granitoid emplacement and amphibolite–granulite facies metamorphism, marks the important Archean–Proterozoic boundary. Granulite-grade mafic enclaves at Tsaozhuang, Hebei Province, dated at 3.5 Ga, may reflect a still earlier tectonic cycle marking the early–mid Archean boundary.

Early Proterozoic (2.5–1.85 Ga) deposition on the embryonic craton is widely represented by the Wuati and Hutuo groups. The Wutaian Orogeny at 2.2 Ga, involving widespread granitoid emplacement, effectively cratonized the Sino–Korean basement, site of subsequent Hutuo sedimentation. Additional local cratonization occurred during two phases of the Luliangian (Zhongtian) Orogeny at 1.8–1.7 Ga. This completed consolidation of the Sino–Korean basement. The Tarim Craton to the west experienced a similar, though not identical, history. Subsequently both the Sino–Korean and Tarim basements underwent taphrogenic rupture with development of mainly NE- and subordinate

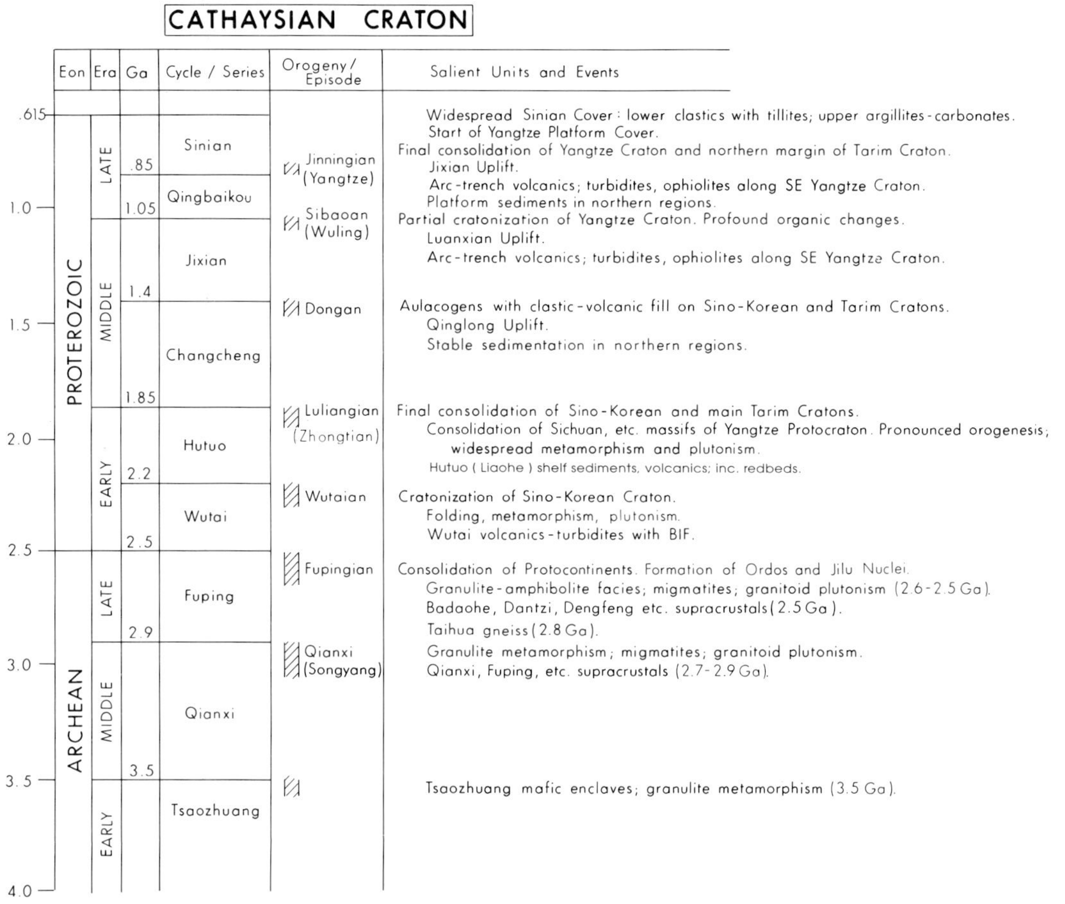

**Fig. 1-3a.** Summary chrono-stratigraphic development of Precambrian crust of the Cathaysian Craton. Salient crustal units and events are arranged in relation to internal orogenies and resulting tectonic cycles.

NW-trending aulacogens ranging in age from 1.85 to 0.8 Ga. The Tarim Craton was essentially consolidated by Sibaoan time (1.0 Ga).

Following earlier Proterozoic growth stages, the Yangtze Craton to the south was finally consolidated during the Jinningian Orogeny (0.85 Ga), to initiate widespread Sinian platform cover. The Yangtze and Sino–Korean cratons, based on dating of eclogites, collided and consolidated at about 209 Ma to complete the composite craton (Ames et al 1993).

## Siberian Craton

Two main earlier Precambrian (> 1.7 Ga) orogenies—Aldanian and Stanovoyan—effectively stabilized this large composite craton (Figs 1-2, 1-3b). Of these, the older Aldanian Orogeny (2.6 Ga), marking the Archean–Proterozoic boundary, itself achieved major cratonization. Still older Archean ages have been determined locally in gneisses of East Sayan Province (3.2 Ga), Anabar Shield (2.9 Ga) and nearby Omolon Massif (3.4 Ga), thereby demonstrating widespread early sialic crust. The Omolonian Orogeny (3.4 Ga) provides a tentative Katarchean–Archean boundary.

During early Proterozoic time (2.6–1.65 Ga), epicratonic rocks filled local troughs and grabens on the Aldan Shield. The very large latitudinal Stanovoy Fault, marking the Stanovoy–Aldan boundary, developed during this time (1.95 Ga). The Stanovoyan Orogeny (19.5–1.8 Ga) effectively completed craton consolidation.

Following Stanovoyan consolidation, the Akitkan taphrogenic 'system', characterized by numerous aulacogens, marks the early–late Proterozoic (Riphean) boundary.

During Riphean (1.65–0.65 Ga) time, the interior Siberian platform rythmically subsided, resulting in four successive psammite–carbonate–pelite cycles, the respective chronostratigraphic divisions named Bourzianian, Yurmatinian, Karatavian and Kudash.

## Siberian Craton table

| Eon/Era | Ga | Cycle/Series | Orogeny/Episode | Salient Units and Events |
|---|---|---|---|---|
| LATE PROTEROZOIC / RIPHEAN | .57 / .56 / .65 / .68 | Vendian / Kudash | Baikalian | Ultramafic-alkaline intrusions; kimberlites. Tillites; carbonate-pelite sequences; flysch and mixtites; Ediacara-type metazoa. Final consolidation of full platform with addition of peripheral massifs and fold belts. |
| | 1.0 | Karatavian | | Local granitoids. |
| | 1.40 | Yurmatinian | | Major subsidence and vast platform cover (Platform stage): cyclically alternating psammite-carbonate-pelite deposits. Thick deposits in peripheral geosynclines, sites of future fold belts. |
| | 1.65 | Bourzianian | | Ulkan Laccolith (alkaline granitoid) (1.65 Ga). Aulacogen stage: clastics and bimodal volcanic fill. Craton border rifting with development of major pericratonic fold belts |
| EARLY PROTEROZOIC | 1.95 | Akitkan | Stanovoyan | Final consolidation of main craton: thick, extensive crust. Stanovoy Fault (1.9 Ga); granitoids; retrogression. Extensive reworking and metamorphism of belt network across the platform; local granulite metamorphism. |
| | 2.6 | Oudokan | | Epicratonic rifting with bimodal volcanic red-bed fill; including Oudokan-Kodar and Ulkan Troughs (to 1.9 Ga). |
| ARCHEAN | 2.9 | Subgan | Aldanian | Epi-Archean Platform Consolidation: Granitoid intrusion and granulite facies metamorphism (2.6-2.5 Ga). Gneiss terranes and greenstone belts (2.96 Ga). eg. Aldan and Stanovoy Domains. Anabar Gneiss. Aldan mafic gneiss (3.2 Ga). Gneisses of Onotsk Graben, Sayan Province (3.2 Ga). Sayan greenstone belts (+3.2 Ga). |
| KATARCHEAN | 3.2 / 3.4 | Aldan | Omolonian | Granulite facies metamorphism (3.4 Ga) (Omolon Massif). Aldan basement gneiss (3.4 Ga). |

**Fig. 1-3b.** Summary chrono-stratigraphic development of Precambrian crust of the Siberian Craton. Salient crustal units and events are arranged in relation to internal orogenies and resulting tectonic cycles.

This was succeeded by the important tillite-bearing Vendian Cycle (0.65–0.57 Ga). Concurrent pericratonic rifting resulted in major trough (geosynclinal) accumulations. These were peripherally accreted during the Baikalian Orogeny (~0.57 Ga) in the form of the Baikal–Patom, East Sayan, Yenisei and Turukhansk fold belts, to complete the composite craton.

### East European Craton

Three principal orogenies recorded in the northwestern part (*Baltic Shield*) of this composite craton, namely Lopian (Karelian), Svecofennian and Baikalian, provide a threefold division into Archean eon and early and late Proterozoic eras (Figs 1-2, 1-3c(i)). The late Proterozoic era itself is punctuated by the Hallandian (–Gothian) and Sveconorwegian events.

In the Baltic Shield, the easternmost Kola–Karelian basement (Svecokarelian or Pre-Svecofennian Domain) which is composed of high grade gneiss and granitoid–greenstone terrains, is the product of older Saamian and younger Lopian cycles with essential stabilization by 2.6 Ga. Ensuing early Proterozoic (2.6–1.8 Ga) additions comprise Karelian–Kalevian cratonic cover transitional westward to the Svecofennian assemblage (1.93–1.87 Ga), a major juvenile magmatic-supracrustal accretion. This marks substantial consolidation of the shield. Subsequent events include widespread anorogenic plutonism notably rapakivi granites (1.7–1.5 Ga) and crustal reworking to form, successively westward, the Transscandinavian (Smaland–Varmland) (1.8–1.6 Ga) and Southwest Scandinavian (to 0.57 Ga) belts.

In the Ukrainian Shield to the south (Fig. 1-3c(ii)), Archean events and products include Auly magmatism, Konka greenstones and Dnieper Complex, culminating in the Karelian Orogeny (2.7–2.5 Ga) to achieve effective consolidation. Subsequent early Proterozoic events are characterized

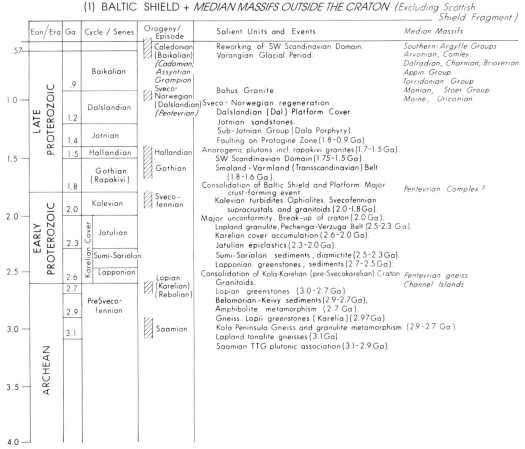

**Fig. 1-3c(i).** Summary chrono-stratigraphic development of Precambrian crust of the East European Craton—Baltic Shield and median massifs outside the craton. Salient crustal units and events are arranged in relation to internal orogenies and resulting tectonic cycles.

by widespread crustal rifting with major Banded Iron Formation (BIF) accumulation (2.3 Ga) in major N-trending geosynclinal troughs (Krivoy Rog). Intermittent magmatism–metamorphism to 1.4 Ga consolidated the shield.

In Riphean–Vendian times (1.65–0.57 Ga) the interior basement of the enlarging composite craton underwent wholesale taphrogenesis involving rifting and subsidence (Aulacogen Stage) followed by widespread cyclic sedimentation–volcanism (Platform Stage). The resulting chronostratigraphic divisions match those in the Siberian Craton cover to the east. Pericratonic accretions representing collisional products during the Baikalian–Salairian Orogeny (0.6–0.5 Ga) completed the parent craton.

Corresponding events affected nine main median massifs located in Phanerozoic belts to the west (Caledonides), south (Hercynian) and east (Uralian). (Fig. 1–3c(i)).

### Greenland Shield (North American Craton)

The Archaean Block, magnificently exposed in southern West Greenland (Fig. 1-2) has provided an unusually complete record of early Archean events, including development of Amîtsoq gneisses (~3.8 Ga), and still older Isua-Ikasia supracrustal rocks (Fig. 1-3d(i)).

Recorded mid-Archean events include emplacement of Ameralik Dykes, accumulation of Malene supracrustal rocks, and intrusion of both anorthosite–gabbro layered complexes, and Nûk gneisses (3.0 Ga). This was followed by varied deformations and intrusions culminating with the Qôrqut Granite at 2.6–2.5 Ga to complete the main history of the Archaean Block.

The comparatively narrow Nagssugtoqidian Mobile Belt to the north bears the imprint of the orogeny of that name at 1.8 Ga. Still farther north,

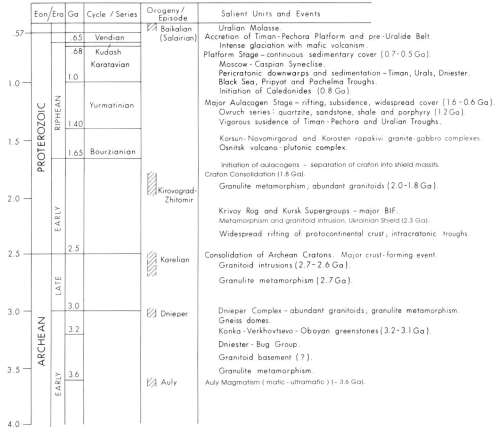

**Fig. 1-3c(ii).** Summary chrono-stratigraphic development of Precambrian crust of the East European Craton—Ukrainian Shield and interior platform. Salient crustal units and events are arranged in relation to internal orogenies and resulting tectonic cycles.

the exceptionally broad and complex Rinkian Mobile Belt with substantial Archean infrastructure, was metamorphosed to higher grade including granulite facies at ~1.8 Ga.

The similar Ketilidian Mobile Belt to the south was uniquely intruded by Gardar nepheline syenites at 1.3–1.1 Ga.

In the East Greenland Caledonian Belt, Krummedal sediments, intensely deformed during the Carolinidian Orogeny at 1.2–0.9 Ga, mark the mid to late Proterozoic transition. Eleanore Bay and Hagen Fjord sediments, including prominent tillites, subsequently accumulated prior to the Caledonide Orogeny (~0.5 Ga).

The Scottish Shield Fragment, a rifted part of the original North Atlantic Craton, includes the Lewisian Complex (2.9 Ga), Badcallian granulite facies metamorphism (2.7 Ga), Laxfordian deformation–metamorphism–intrusion (1.8–1.4 Ga), and Grampian (Assyntian, Cadomian) orogenic events (0.6–0.5 Ga).

### North American Craton (less Greenland Shield)

The Canadian Shield, including the easternmost Nutak segment, Labrador, illustrates with unusual clarity the cyclic development of Precambrian crust (Figs 1-2, 1-3d(ii)). Six main orogenies are recognized marking respective eon/era boundaries: (1) Uivakian Orogeny in Labrador, Mortonian event in the Minnesota Valley inlier, and related Slave Province event producing the Acasta gneiss (3.96 Ga), Earth's oldest known intact rock; (2) in Superior Province, the Wanipigowan and related orogenies with a closing date of 2.9 Ga; (3) Kenoran (Algoman) Orogeny, at 2.65–2.55 Ga, mainly in Superior Province, which marks the Archean–Proterozoic boundary; (4) Hudsonian (Penokean)

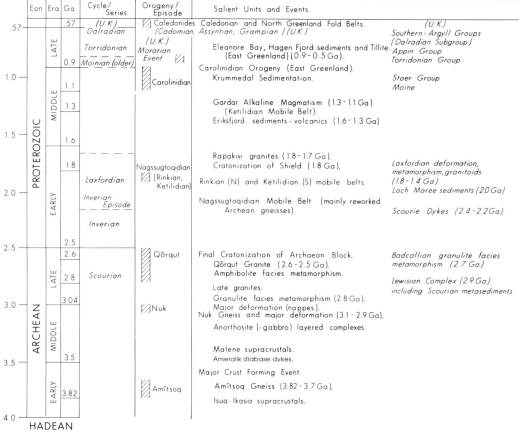

**Fig. 1-3d(i).** Summary chrono-stratigraphic development of Precambrian crust of the North American Craton—Greenland Shield and Scottish Shield Fragment. Salient crustal units and events are arranged in relation to internal orogenies and resulting tectonic cycles.

Orogeny, mainly in Churchill, Bear and Southern provinces, closing at about 1.85 Ga; (5) Elsonian Disturbance, mainly in Eastern Canada, at 1.4 Ga; and (6) the Grenvillian Orogeny, mainly expressed in the Grenville Province and closing at 1.0 Ga. Additionally, the Avalonian Orogeny, Newfoundland, closed about 620 Ma ago, very close to the Precambrian–Cambrian boundary. Precambrian rocks in the mainly buried southern–western region (Fig. 1-2) correspond closely to those of the nearby Canadian Shield. The almost totally buried midcontinent Central Belt comprises at least two parallel ENE-trending orogenic belts formed respectively at 1.78–1.69 and 1.68–~1.47 Ga. Both the Central Belt and parts of the eastward adjoining Grenville Belt experienced widespread anorogenic (granite–anorthosite) intrusions at 1.48–1.38 Ga.

The Working Group on the Precambrian Rocks of USA and Mexico has proposed the following subdivisions of Precambrian time (Harrison and Peterman 1980, Reed et al 1993a): the Archean–Proterozoic boundary is set at 2500 Ma; the Archean eon is divided at 3.3 and 2.9 Ga into early, mid and late Archean eras; the Proterozoic eon is divided at 1.6 and 0.9 Ga into early, mid and late Proterozoic eras. Pre 3900 Ma time is referred to as pre-Archean (elsewhere called Hadean) time.

Hoffman (1989) interprets development of the North American craton as outlined above in terms of Archean microcontinents welded by successive Proterozoic collisional orogens.

## South American Craton

The oldest established Archean events (3.5–3.2 Ga) are from the Imataca Complex in Guiana Shield, Goias Massif in Central Brazil Shield, and Boa Vista gneiss and Jequié Complex in Atlantic Shield,

## NORTH AMERICAN CRATON[1] LESS GREENLAND SHIELD

| Eon | Era | Ga | Orogeny/Episode | Salient Units and Events | Nutak Segment (Nain Province) |
|---|---|---|---|---|---|
| | HADRYNIAN | .57 .62 1.0 | Avalonian (Franklinian) | Iapetus: Atlantic-Arctic-Pacific (?) opening. Local Craton Accretions. Windermere, Rapitan, Grand Canyon, Avalon, Ocoee and Great Smoky sediments including diamictites and local BIF-volcanics. Pericratonic Rifting - Accumulation in Arctic, Cordilleran and Appalachian Belts. | |
| PROTEROZOIC | NEO-HELIKIAN | 1.0 1.4 | Grenvillian (East Kootenay, Racklan) | Full Craton Accretion (Grenville Belt). Grenville Front Tectonic Zone (1.2-1.0 Ga). Midcontinent - pericratonic rifting and volcanism: Keweenawan, Seal Lake, Coppermine, Muskox (c.1.1 Ga). Grenville Supergroup (1.3-1.1 Ga). Belt-Purcell, Wernecke, etc. Supergroups (1.4-0.9 Ga). | |
| | PALEO-HELIKIAN | 1.5 1.6 1.8 | Elsonian Mazatzal Central Plains (Labrador) | Anorogenic Intrusion and Craton Rifting (1.5-1.4 Ga). Anorthosites, gabbros, rapakivi granites, rhyolites. Central Orogenic Belts (Craton Accretion). Southern Province (USA) (1.7-1.5 Ga). Northern Province (USA) (1.8-1.7 Ga). | |
| | APHEBIAN | 2.0 | Hudsonian (Penokean Wopmay) | Epicratonic rifting: Athabasca, Dubawnt, Martin, Sioux, Quartzite redbeds (1.8-1.5 Ga). Major Cratonization: deformation, metamorphism, plutonism. Trans-Hudson, Wopmay, Kapuskasing, Penokean fold belts (1.9-1.8 Ga). Circum-Superior BIF-bearing fold belts (c. 2.0 Ga). Sudbury Irruptive (1.84 Ga). Nipissing Diabase (2.15 Ga). Huronian Supergroup (2.4-2.1 Ga). | Major deformation and metamorphism |
| ARCHEAN | LATE | 2.5 2.6 2.7 2.9 | Kenoran (Algoman, Fiordian) | Cratonization of Archean Provinces. Major crust forming events (~2.6 GA). Granitoid plutonism (2.76-2.65 Ga). Greenstones: Slave Province - 2.68-2.65 Ga. Superior Province [mainly 2.76-2.70 Ga; also 2.85-2.80 Ga; and 3.0-2.9 Ga.] Churchill Province: Kaminak - 2.7 Ga. Prince Albert - 2.9 Ga. | Major deformation, metamorphism and plutonism. Granulite metamorphism. |
| | MIDDLE | 3.0 3.1 3.4 | Wanipigowan (Laurentian, Hopedalian) | Granulite amphibolite metamorphism; tonalitic gneiss, granite (3.1 Ga). Slave Province basement gneiss (3.15 Ga). Beartooth supracrustals (Wyoming) (3.3 Ga). | Major Reactivation of gneiss (3.1 Ga). Anorthosite-gabbro complexes. Upernavik supracrustals. Saglek diabase dikes. |
| | EARLY | 3.5 3.9 | Uivakian (Mortonian) | Morton (-Michigan) Gneiss (tonalitic) (+3.4 Ga). Pre-Morton supracrustals (Minnesota). Acasta gneiss (3.96 Ga), Slave Province | Major crust forming event. Uivak Gneiss: deformation, metamorphism (3.5 Ga). Pre-Uivak supracrustals. |
| | Pre-Archean (Hadean) | 4.0 | | | |

[1] - excluding Greenland Shield but including Nutak Segment of North Atlantic Craton.

**Fig. 1-3d(ii).** Summary chrono-stratigraphic development of Precambrian crust of the North American Craton excluding Greenland Shield. Salient crustal units and events are arranged in relation to internal orogenies and resulting tectonic cycles.

culminating in the Gurian Orogeny (3.0 Ga) (Figs 1-2, 1-3e).

The Jequié Orogeny (2.6 Ga), which provides the Archean–Proterozoic boundary, is broadly expressed in late Archean greenstone belts and gneissic terrains.

The Transamazonian Orogeny (2.1–1.9 Ga), marking the early–mid Proterozoic boundary effected widespread cratonization.

Unusually extensive epicratonic cover and cratogenic plutons of the Guiana and Central Brazil shields (e.g. Uatuma Volcano Plutonic Complex and Roraima–Goritore Formation) characterize early mid-Proterozoic time (1.9–1.4 Ga). The Jari–Balsino episode (1.6–1.4 Ga) resulted in development of the Rio Negro–Jurueno and concurrent San Ignacio (Bolivia) belts, culminating in the Rondonian Orogeny at ~1.0 Ga. This completed cratonization of these two large shields and intervening terrains to form the Amazonian Craton. In the Atlantic Shield to the east similar activities in the São Francisco Province produced the Espinhaço and Uruaçuanos (Uruaçu) fold systems.

Finally, the Brasíliano Orogeny (700–480 Ma) resulted in widely interspersed, mainly N-trending high grade mobile belts across the Atlantic and Central Brazil shields to complete the composite craton.

## African Craton

### Southern Africa

The classification illustrated (Figs 1-2, 1-3f(i)) is that of the IUGS Subcommission on Stratigraphy (Tankard et al 1982, Plumb and James 1986). Some of the boundaries are highly diachronous. An alternative revised South African time scale is provided

| Eon | Era | Ga | Orogeny/Episode | Salient Units and Events |
|---|---|---|---|---|
| | | .57 | | Final Cratonization |
| | LATE | .7 | Braziliano | Major granitoid plutonism, granulite metamorphism (0.7-0.5 Ga) Paraguai-Araguai, Brasilia, Don Feliciano, Ribeira Belts. Barborema, Mantiqueira Provinces. San Luis, Luis Alves and Rio de la Plata cratonic fragments in Brasiliano Belt. |
| PROTEROZOIC | MIDDLE | 1.1 | Rondonian (Uruaçuano Espinhaço) Sunsas | Final Cratonization of Amazonian Craton Sunsas-Aguapei (Rondonian) Belts (1.0 Ga) Espinhaço, Uruaçuanos fold systems, Chapada Diamantino cover (1.75-1.2 Ga) |
| | | 1.4 | Jari-Balsino San Ignacio (Parguazan, Madeira) | San Ignacio Belt (~1.3 Ga) Rio Negro-Jurueno Belt (1.7-1.4 Ga) (Amazonian Craton). Rifting and Anorogenic Activities: rapakivi granites, platform sediments. Roraima-Goritore red beds (1.7-1.6 Ga). San Ignacio Belt protoliths. |
| | | 1.7 | | |
| | | 1.9 | Trans-amazonian (Minas) | Taphrogenic Rifting; Uatuma Volcanoplutonic Complex (1.9-1.7 Ga). Consolidation of Amazonian Craton and São Francisco Province. Major granitoid plutonism (2.1-1.9 Ga). Maroni-Itacaiunas Mobile Belt. Minas (BIF), Mirante, Serrinha, and Jacobina (Au, U) Belts. (~2.2 Ga). Pastora, Vila Nova, Amapa, etc. (BIF, Mn) greenstones (2.25 Ga). |
| | EARLY | 2.1 | | |
| | | 2.2 | | |
| ARCHEAN | LATE | 2.6 | Jequié (Aroan, Rio das Velhas) | Consolidation of Archean cratons in Guiana, Central Brazil, Atlantic Shields. Rio das Velhas (Mn, Au) greenstones (São Francisco Province). (2.8 Ga) |
| | | 2.8 | | Salobo greenstones, Serra dos Carajás BIF, Inaja Group (Guaporé Shield) (2.75 Ga). Xingu and Pakairama Nuclei (Amazonian Craton). Imataca migmatites (2.7 Ga). |
| | | 3.0 | | |
| | MIDDLE | 3.3 | Gurian | Goias greenstones, ultramafic massifs, granulites (Tocantins Province) (~3.2 Ga). Jequié Complex (São Francisco Province) (3.2 Ga). |
| | | 3.5 | | Boa Vista gneiss (São Francisco Province) (~3.4 Ga). Imataca Complex, BIF (Guiana Shield) (3.5-3.2 Ga). |
| | EARLY | | | |

Fig. 1-3e. Summary chrono-stratigraphic development of Precambrian crust of the South American Craton. Salient crustal units and events are arranged in relation to internal orogenies and resulting tectonic cycles.

by Johnson et al (1989). The validity of the tectonic cycle concept is appraised by Cooper (1990).

The oldest reliable date (3.6 Ga) is from the Ancient Gneiss Complex, Kaapvaal Craton, followed by episodic granitoid intrusion, deformation and metamorphism to 3.4 Ga. Deposition of Swaziland and Sebakwian greenstones and sediments was initiated at 3.5 Ga. Widespread tonalite plutonism, with deformation and later high level granite intrusion, led to consolidation of the component Kaapvaal Craton by 3.0 Ga, marking the mid–late Archean boundary.

Late Archean through early Proterozoic (3.0–1.6 Ga) epicratonic sedimentation–volcanism, represented successively by Pongola, Witwatersrand, Ventersdorp, Transvaal and Waterberg–Matsap–Soutpansberg supergroups, was punctuated by late Archean (2.8–2.5 Ga) granitoid intrusions, and by mafic magma injections, notably the Bushveld Complex and associates at 2.0 Ga.

To the north, in Zimbabwe, late Archean events included Bulawayan–Shamvaian greenstone development. Widespread granulite facies metamorphism, migmatization and renewed granitoid plutonism, which resulted in consolidation of the Zimbabwe Craton and Limpopo Belt by 2.5 Ga. This was followed immediately by late granitoid plutonism and, locally, Great Dyke emplacement (2.46 Ga) in the newly consolidated Kaapvaal–Limpopo–Zimbabwe composite.

In the early Proterozoic, the Kheis and Magondi belts evolved in equivalent positions along the western flanks of the Kaapvaal and Zimbabwe cratons respectively. To the south and west, the Namaqua–Natal Belt was accreted at 1.2–1.0 Ga to consolidate effectively the entire Kalahari (Southern Africa) Craton. A number of concurrent rift basins (Koras–Sinclair–Ghanzi) developed along the western and northern margins of the Kalahari Craton.

The Damara Mobile Belt and associates, part of

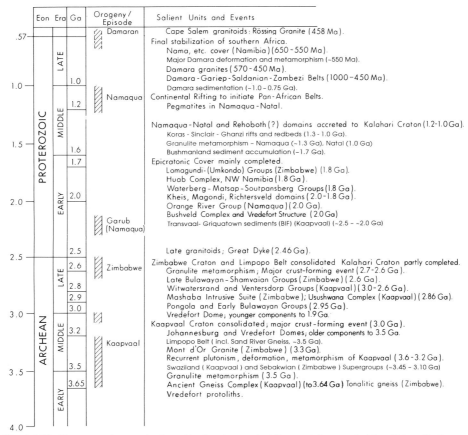

**Fig. 1-3f(i).** Summary chrono-stratigraphic development of Precambrian crust of the African Craton—southern Africa. Salient crustal units and events are arranged in relation to internal orogenies and resulting tectonic cycles.

the late Proterozoic (550 Ma) Pan-African chain effectively welded the Kalahari Craton to central-northern counterparts (see below).

Central and northern Africa

The classification used follows that of Cahen et al (1984) (Fig. 1-3f(ii)). The Archean–Proterozoic boundary is placed at 2.5 Ga. Archean time is divided into three eras, Archean I, II and III; and Proterozoic time into four eras, Proterozoic I–IV. The boundaries chosen correspond to comparatively widespread tectonothermal events, selected by Cahen et al (1984) as representing chronologic 'milestones' in the evolution of African crust.

Medium to high grade metamorphism, dated at 3.5–3.4 Ga, is locally recorded in eastern, central and northern parts. Pre 2.9 Ga old greenstones are identified in north–central (Bandas and Dekoa), northeastern (Ganguan, Kibalian and Nyanzian) and northwestern (Loko and Liberian) parts. Granulite metamorphism (e.g. Watian orogeny), widely dated at ~2.9 Ga in central, southeastern, northwestern and northeastern parts, provides the Archean II–III boundary.

Late Archean (III) time is characterized by development of greenstones and gneiss–migmatite terrains. The Liberian Orogeny (2.7–2.55 Ga) of West Africa and correlatives in central–northern Africa effectively consolidated the cratons of this vast region to terminate Archean time.

Following widespread supracrustal accumulation, including that of Birrimian greenstones, the main Eburnean Orogeny (~2.1 Ga) strongly imprinted the west African, central–east Saharan and central (equatorial) African domains. This was followed by widespread post-orogenic granitoid plutonism to 1.75 Ga, and dwindling anorogenic activities to 1.55 Ga. The net result was widespread

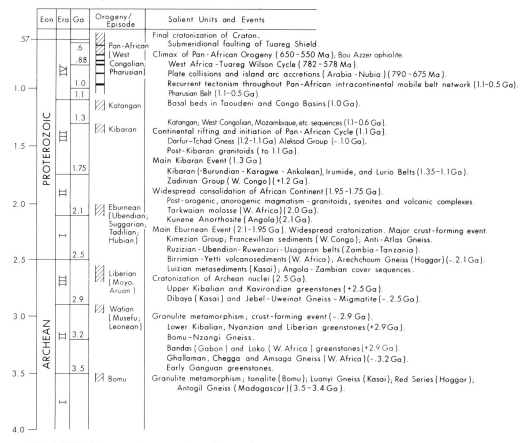

**Fig. 1-3f(ii).** Summary chrono-stratigraphic development of Precambrian crust of the African Craton—central-northern Africa. Salient crustal units and events are arranged in relation to internal orogenies and resulting tectonic cycles.

early Proterozoic (I and II) basement stabilization in central and northern Africa.

Later Proterozoic time (III) is marked in central Africa especially by the main Kibaran event at 1.4–1.3 Ga, followed by late-stage Kibaran activities to 1.0 Ga. These and related events stabilized central and eastern Africa.

By 1.05 Ga initial Pan-African cycle (Proterozoic IV) sedimentation occurred in a continental network of mobile belts and enclosed mega-basins (e.g. Congo Basin) across the enlarging craton. Six main events (1050, 950, 860, 785, 685 and 600 Ma) are recorded throughout Africa, marking successive tectonic stages, including accretion of the ophiolite-rich Arabian–Nubian magmatic arc (790–675 Ma). The last two events, at 685 and 600 Ma, are particularly widespread, marking the climax of Pan-African events and effectively consolidating the African Craton.

### Indian Craton

The oldest dated rocks (~3.4 Ga) lie in the Aravalli, Singhbhum and Dharwar cratons of northern and southern India. A major crust-forming event (Orogeny I), culminating at 3.0 Ga, stabilized the Peninsular Gneiss of south–central India, which forms basement to both the late Archean Dharwar supracrustal belts and platform sediments of the South Indian Highlands (Granulite Domain) (Figs 1-2, 1-3g).

The Archean–Proterozoic boundary, at 2.5 Ga, is marked by another major crust-forming event (Orogeny II) characterized by high grade metamorphism (charnockites), to consolidate the Dravidian Shield (southern India).

Early Proterozoic activities included volcanism–sedimentation in the Aravalli and Singhbhum areas, initiation of Cuddapah Basin accumulation

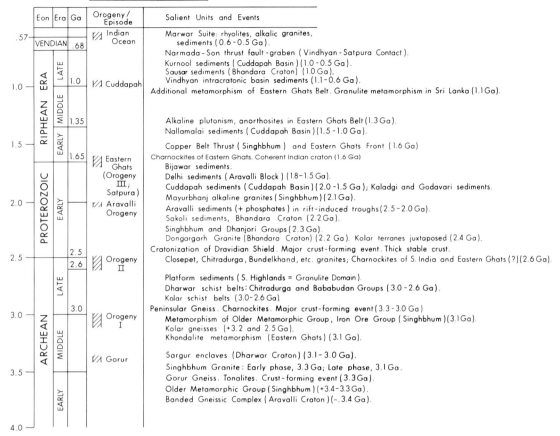

**Fig. 1-3g.** Summary chrono-stratigraphic development of Precambrian crust of the Indian Craton. Salient crustal units and events are arranged in relation to internal orogenies and resulting tectonic cycles.

(2.0 Ga) and terminal-era activities at 1.65 Ga involving Singhbhum (Copper Belt) Thrust and Eastern Ghats Orogeny (Orogeny III) to form a coherent Indian Craton.

Later Proterozoic (Riphean) time involved localized basinal rifting with platform accumulations in major continental repositories including Vindhyan and Cuddapah basins. The Eastern Ghats Belt experienced high grade metamorphism at 1.1–1.0 Ga, the possible time of its accretion to the Indian Shield. The classification scheme for post 1.65 Ga time comprises early, mid, and late Riphean terminating with the Vendian (to 0.57 Ga).

## Australian Craton

The Australian Craton has been broadly divided on the basis of progressive cratonization of component blocks (Plumb 1979, Rutland 1981). However, no existing classification is fully satisfactory in a rapidly developing chronometric assessment that reveals significant intercratonic diachroneity. The interim classification illustrated here is, in fact, a blend of tectonic and chronostratigraphic divisions with substantial adherence to the classification of Plumb and James (1986) (Figs 1-2, 1-3h).

The Archean–Proterozoic boundary is taken at 2.5 Ga, the time of final stabilization of the Yilgarn Block. However, craton diachroneity is illustrated by (1) the nearby Pilbara Block which was stabilized partly by 3.0 Ga and mainly by ~2.8 Ga; and (2) the Gawler Craton in South Australia which was finally stabilized by 2.3 Ga (Sleaford Orogen). A tentative internal Archean boundary at 3.0 Ga separates older 'Pilbarian' crust from younger 'Yilgarnian' crust. The latter includes initiation of the intercratonic Hamersley Basin at ~2.75 Ga with mega-BIF accumulation at ~2.5 Ga.

The Yilgarn Block has the current distinction of containing Earth's oldest dated crustal material (detrital zircons to 4.28 Ga), as well as the oldest known terrestrial anorthositic xenoliths (~3.7 Ga).

Proterozoic time was previously divided chrono-

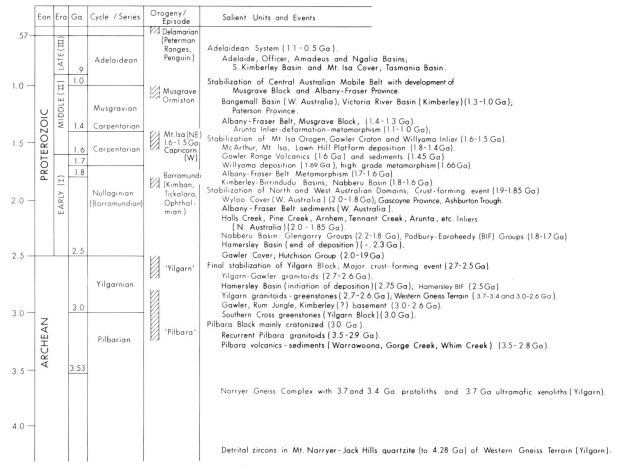

**Fig. 1-3h.** Summary chrono-stratigraphic development of Precambrian crust of the Australian Craton. Salient crustal units and events are arranged in relation to internal orogenies and resulting tectonic cycles.

stratigraphically into early, middle and later Proterozoic eras at 1.7 Ga and 1.0 Ga respectively, later revised to Proterozoic I, II and III, with internal boundaries at 1.6 Ga and 0.9 Ga respectively.

Early Proterozoic (I) time (Nullaginian (Barramundian) Cycle) encompasses the deposition, orogenesis, transitional tectonism and cratonization of much of the basement rocks of the North Australian Craton. The era also included (1) development of the Nabberu Basin (2.2–~1.7 Ga) in West Australia, together with termination of Hamersley Basin deposition (~2.3 Ga); and (2) orogenic activities in both the Musgrave–Arunta blocks of Central Australia and, following supracrustal accumulation, the Gawler region of South Australia (Kimban Orogeny, ~1.8 Ga). The early Proterozoic terrains of northern Australia and elsewhere were affected by the remarkably isochronous and tightly compressed Barramundi Orogeny at 1880–1850 Ma, the basis for a possibly forthcoming early Proterozoic tectonic division (2.5–1.8 Ga).

Mid-Proterozoic (II) time, from ~1.7–1.0 Ga, is separated at ~1.4 Ga into older Caprentarian and younger Musgravian divisions. The Mount Isa and Northeast orogenies at 1.5 Ga stabilized the eastern part of the North Australian Craton. Musgravian time is characterized by repeated orogenesis in the polymetamorphic Central Australian Mobile Belts, including the tectonically active Musgrave–Arunta blocks, Paterson Province and Albany–Fraser Province, all stabilized by the Musgrave event (~1.0 Ga).

Late Proterozoic (III) time (Adelaidean cycle), from 0.9 (–1.0) Ga to the base of the Cambrian and beyond, is characterized by a string of major tillite-bearing Adelaidean depositories initiated at ~1.1 Ga and later affected by the terminal Delamarian Orogeny (~0.5 Ga).

## Antarctic Craton

In East Antarctica, where bedrock exposure is extremely limited, access difficult and detailed studies rare, a number of high grade metamorphic events (at least six) have been locally identified, together with late Proterozoic–early Paleozoic greenschist facies metamorphism (Figs 1-2, 1-3i). Some of these events, notably those at ~900 Ma and ~500 Ma, are comparatively widespread throughout the craton. Napier Complex and adjoining Rayner Complex in Enderby Land, despite difficult access and limited exposure, are particularly well studied metamorphic complexes which represent major contributions to the elucidation of Precambrian geology.

An important granulite facies metamorphism at 2.5–2.4 Ga marks the Archean–Proterozoic boundary. Earlier Archean granulite facies events are recorded at 3.1–2.8 Ga and at 3930 Ma, the latter marking one of Earth's oldest dated rocks.

Successive granulite facies events, dated locally at ~1.6 Ga and 1.0–0.9 Ga respectively, serve to demarcate early, mid and late Proterozoic eras. The ~900 Ma old event is particularly widespread in East Antarctica. The Rayner Complex in East Antarctica forms part of an extensive Proterozoic mobile belt, product of an early Proterozoic (2.0–1.8 Ga) juvenile crust forming event.

Finally, a widespread ~500 Ma greenschist metamorphic event, coeval with the Ross Orogeny in the Transantarctic Mountains Fold Belt, completed the Antarctic Craton.

## 1.5 PRECAMBRIAN CLASSIFICATION SCHEME

The pattern of orogenic cycles and orogenies as presently known in the nine Precambrian cratons is summarized in Fig. 1-4. In most cratons, four prin-

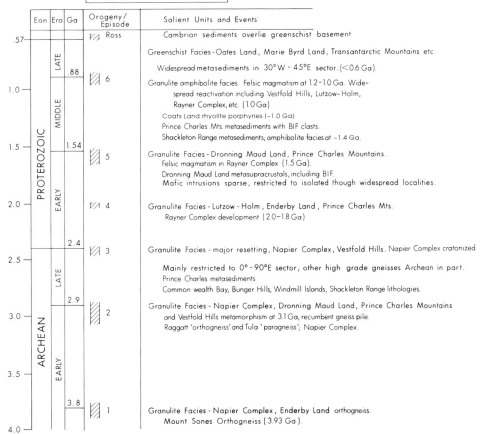

**Fig. 1-3i.** Summary chrono-stratigraphic development of Precambrian crust of the Antarctic Craton. Salient crustal units and events are arranged in relation to internal orogenies and resulting tectonic cycles.

22    DISTRIBUTION AND TECTONIC SETTING OF PRECAMBRIAN CRUST

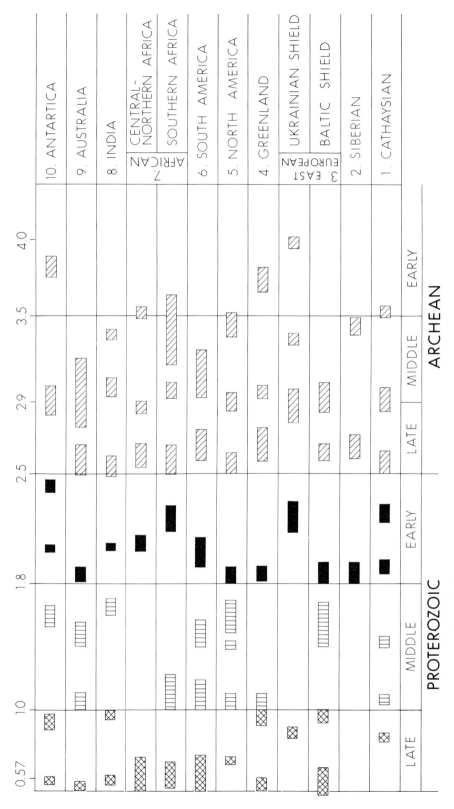

**Fig. 1-4.** Summary tectonic development of the nine Precambrian cratons and the resulting Precambrian classification scheme followed in this book. Earlier Archean subdivisions are tentative due to the paucity of critical geochronologic data.

cipal cycles, variably marked by culminating orogenies, terminate approximately at 2.5, 1.8, 1.0 and 0.6 Ga, the latter marking the generally accepted Precambrian–Cambrian boundary at 570 Ma. In the working scheme used in this text, the four dates delimit respectively the Archean–Proterozoic eon boundary and early, mid and late Proterozoic eras. In addition, the Archean eon is provisionally subdivided at 2.9 and 3.5 Ga into three component eras—early, middle and late.

Because a tectonic-based time-rock classification is used in this book, based on isotopic dating in structural provinces, rocks assigned to a particular era/eon typically include both newly formed (juvenile) crust and reworked (metamorphosed and recrystallized) older crust. Clearly, areal proportions of measured (i.e. preserved) Precambrian crust by era or eon do not reflect original growth rates of the continental crust. Rather, they reflect cumulative orogenic histories by era, the younger eras always gaining proportionately at the expense of the older. Indeed, given the pattern of wholesale recycling of the crust in successive cycles, it is remarkable that about 25% of the estimated original combined Archean plus early Proterozoic crust has, in fact, survived as such (see Chapter 6).

As elsewhere considered (Gostil 1960, Dearnley 1965, York and Farquhar 1972, Condie 1976a, 1989, Moorbath 1976, 1984, Cohen et al 1984, Dalziel 1992), the degree of both internal tectonic cyclicity by craton and of intercratonic, i.e. global, synchroneity remains equivocal. In the meantime, the present patterns are considered to be suggestive but by no means conclusive (Dalziel 1992) of some measure of global intra- and inter-craton consistency. Whatever the eventual outcome, current results provide an adequate tectonic-based classification for purposes of this text.

Indeed, the term 'Precambrian' is itself a suspect designation of pre-570 Ma Earth history as the Ediacaran megafossils in its uppermost part are more similar to those in overlying Cambrian strata than in underlying sequences (Hofmann 1986, Nisbet 1991). The concept of 'Precambrian' dates from a time when the early metazoan record was poorly known or disputed. Furthermore, the basis of Precambrian subdivision, whether time-stratigraphic, chronologic or other, is under dispute. It will be resolved only as more pertinent data—geologic, biologic and chronometric—come to hand. In the meantime, the well-entrenched term Precambrian is used, fully realizing its interim status.

As emphasized by Link et al (1993), the subdivision of Precambrian time and rocks has been a difficult and controversial matter. In the time scale proposed by the IUGS Subcommisison on Precambrian Stratigraphy (Plumb and James 1986), the eon/era boundaries are drawn to cut the minimum number of then-dated (1986) coherent rock sequences, and are therefore geochronologic (geologic time units) rather than chronostratigraphic (time–stratigraphic units), as prevails in the Phanerozoic (i.e. post-570 Ma) time scale. Whereas there are clear-cut short-term advantages to the IUGS chronologic time scale, it is considered here that the long-term advantages to an eventual chronostratigraphic Precambrian time scale, so cogently argued by Crook (1989) and Cloud (1987), greatly outweigh the inevitable intervening uncertainty and confusion in classification. In this regard, the basis for a satisfactory chronostratigraphic time scale will continue to be sought in areas of phenomenologic study such as (1) orogenic cycles and orogenies as possibly related to recurring supercontinent fusion–fission (Hartnady 1993, Duncan and Turcotte 1994); and (2) biomolecular paleontology (fossil molecules) dealing with the molecular record of ancient life (Nowlan 1993). These and other fields of phenomenologic study have enormous potential for critical contributions to many relevant fields of science, including Precambrian classification.

Two significant differences emerge between the classification used in this text and the IUGS subdivision of Precambrian time (Plumb and James 1986): (1) the early–mid Proterozoic boundary is here placed at 1.8 Ga rather than at 1.6 Ga; and (2) the mid–late Proterozoic boundary is placed at 1.0 Ga rather than at 0.9 Ga.

It now remains in this chapter to establish the general geologic setting by Precambrian craton which provides the basic framework for a systematic consideration of the geologic record by era–eon (Chapters 2–5), this culminating in a Precambrian synthesis (Chapter 6).

## 1.6 GEOLOGIC SETTING BY CRATON

Building on the above tectonic overview by craton, the geologic settings of the nine Precambrian cratons are briefly summarized consecutively below.

### 1.6.1 CATHAYSIAN CRATON

The Cathaysian Craton with its three component cratons—Sino–Korean, Tarim and Yangtze—forms a westward tapering, deeply indented triangle 5000 km long, up to 2200 km wide, and 4 000 000 km² in area (Figs 1-3a, 1-5a). (Key references: Cheng et al 1982, Zhang et al 1984, Sun and Lu 1985, Yang et al 1986.)

The *Sino–Korean craton*, 1 704 000 km² in area, includes most of north China, the southern part of northeast China, Bohai Bay (Sea), the northwestern Yellow Sea Region, and the Gueonggi Massif of North Korea.

Archean rocks, mainly gneiss, migmatite and massive tonalitic plutons with metasupracrustal intercalations (Qianxi and Fuping groups), underlie the deeply subsided Ordos and Ji-Lu nuclei in the centre–east, with exposures largely restricted to the upturned edges especially to the north. Archean rocks were widely affected by the Fupingian Orogeny at 2.5 Ga.

Early Proterozoic rocks of the Wutai Group, mainly volcanic-derived schists and gneisses with some intercalated dolomitic marble and ferruginous sediments, are common in the eastern and west–central parts of the craton where affected by the Wutai Orogeny (2.2 Ga).

A system of 1.85 Ga aulacogens which developed along the craton margins contain up to 10 km thick sequences (Hutuo Group) of unmetamorphosed sediments including redbeds, carbonates, tillites, manganese–phosphate-bearing strata, and alkalic volcanic rocks.

The Luliangian Orogeny (~1.86 Ga) completed craton consolidation, the resulting crystalline basement overlain by local mid to late Proterozoic strata and more widespread Phanerozoic cover.

The comparatively inaccessible and little exposed *Tarim Craton*, forms a lozenge-shaped block some 1600 km by up to 600 km or 717 000 km² in area. The south–central part is interpreted, on slender evidence, to include a deeply buried Archean nucleus (South Tarim Nucleus). Precambrian basement, locally exposed at the craton peripheries, comprises Archean gneiss and early to mid Proterozoic schist, quartzite and fossiliferous marble. Sinian (850–615 Ma) volcanic rocks and tillites are conformably overlain by Phanerozoic strata.

The *Yangtze Craton* forms a highly irregular, W-enlarging ellipse, 1600 km by 300–800 km or 1 560 000 km² in area. Archean rocks have not been identified in this craton. Early Proterozoic basement rocks, concentrated at the northern perimeter and in the interior Central Sichuan Massif, were affected by the Luliangian orogeny (1.85 Ga). Extensive mid–late Proterozoic cover including thick Sinian (0.85–0.61 Ga) molasse–tillites–carbonates cover the central part of the craton, while ophiolite-bearing magmatic arc–turbidite accretions occur at the southeastern and western margins. The craton was fully consolidated by 850–800 Ma (Jinningian Orogeny). It was moved to its present site *vis-à-vis* the other two component cratons in the Triassic (209 Ma).

## 1.6.2 SIBERIAN CRATON

The fault-bounded Siberian Craton forms an irregular polygon some 2500 km across and $4.4 \times 10^6$ km² in area with the broad Verkhoyansk re-entrant to the northeast and narrow Baikal indentation to the south (Figs 1-3b, 1-5b). The craton includes two prominent Archean–early Proterosoic, anteclise-centred shields—Aldan(–Stanovoy) in the southeast and Anabar in the north. Additional Precambrian exposures lie in peripheral fold belts to the south and west, as well as in more distant median massifs, notably Taymyr to the north and Okhotsk to the east. The interior basement of the craton is overlain by variable thicknesses of Riphean–Vendian (1.65–0.56 Ga) and Phanerozoic platform cover. The main elements of the buried interior basement are the Tunguska and Vilyuy syneclises with intervening Central Siberian Anteclise. (Key references: Nalivkin 1970, Salop 1983, Khain 1985, Sokolov and Fedonkin 1990, Rundqist and Mitrofanov 1992.)

### Aldan Shield and Stanovoy Fold Belt

The *Aldan Shield*, an elongated trapezoid 1200 km by 300–400 km, or 360 000 km² in area, is composed of predominantly older Archean (3.4–3.2 Ga) high grade gneiss–migmatite with interspersed late Archean (3.0–2.5 Ga) granitoid–greenstone belts. The dominant fold direction is to the north–northwest. Occasional early to mid-Proterozoic epiclastic–volcanic-filled troughs lie in this basement. Gently dipping Riphean–Vendian strata cover the northward shelving Aldan Anteclise.

The equivalent sized *Stanovoy Fold Belt* (Ridge) adjoining across the Stanovoy Fault to the south contains similar Archean assemblages (Aldan–Stanovoy Complex) which, however, have experienced major early Proterozoic tectonic reworking. Large early Proterozoic gabbro–anorthosite plutons are situated along the Stanovoy Fault. The main part (~60%) of the Stanovoy Belt is formed of Mesozoic–Cenozoic granitoid plutons, in sharp contrast to their virtual absence in the Aldan Shield to the north.

### Anabar Shield

The Anabar Shield (Block, Uplift) in the north, a triangular block, 300–400 km on the side, is mainly composed of NNW-trending, high-grade Archean gneiss–migmatite with intercalated greenstones, BIF and marble. These basement rocks were selectively mylonitized and retrogressed at 1.9 Ga along NNW-trending tectonic zones. This crystalline basement is unconformably overlain by peripherally shelving Riphean–Cambrian–Ordovician cover.

### Peripheral fold belts

The Baikal Fold Belt, a 1200 km-long, NNE-trending hooked fold system, located west of the Aldan Shield comprises a southeastward early Proterozoic sedimentary transition from thin (to 1500 m), gently folded sandstone–conglomerate platform facies in the north and west, to very thick (to 6000 m), isoclinally folded, northward-verging, slope-and-rise (eugeoclinal) volcano–turbidite–carbonate facies, at prevailing amphibolite facies metamorphism, in the south and east.

The corresponding East Sayan, Yenisei and Turukhansk fold belts to the west and north constitute similar disrupted Archean–early Proterozoic zones displaying prominent cratonward structural vergence. Accompanying late Proterozoic–Phanerozoic platform cover dips gently basinward from these uplifted basement cores.

### Interior basement and platform cover

The interior basement of the Siberian Craton, as known, is composed mainly of Archean rocks, partly reworked in early Proterozoic time (1.9 Ga) to form NNW-trending, alternating Archean–early Proterozoic complexes of uncertain relative proportions. The basement surface shows substantial relief, forming a number of large depressions including the huge Triassic trap-filled Tunguska Syneclise in the northwest and the Mesozioc Vilyui Syneclise in the east, the latter a northeastern extension of the Chara–Lena Trough.

Elsewhere, the Riphean–Vendian–Phanerozoic platform cover dips gently in homoclinal fashion from the craton margins to the interior.

### Taymyr Fold Belt

This markedly arcuate fold belt, tectonically divorced from the Siberian Craton by the intervening Mesozoic–Cenozoic Khatanga Trough, includes the E-trending Taymyr Belt proper in the south and the laterally continuous N-trending Severnaya Zemlya island archipelago in the north (see Fig. 1-1).

The Taymyr Belt contains the curvilinear Kara Massif, about 1200 km by 450 km, composed of an Archean to early Proterozoic crystalline basement with infolded Riphean cover. Late Riphean–Vendian molasse covers the southern edge of the Kara Massif.

## 1.6.3 EAST EUROPEAN CRATON

The East European (Russian) Craton (Fenno-Sarmatia) occupies much of European Russia (i.e. west of the Ural Mountains) and Scandinavia (Fig 1-3c(i,ii), 1-5c(i)–(iii)). The fault bounded craton forms an iregular pentagon 2500–3000 km across and 5 350 000 km$^2$ in area. The craton includes two prominent shields: the larger, rectangular Baltic (Fennoscandian) Shield in the northwest, and the smaller, curvilinear Ukrainian Shield in the southwest. Additional positive elements include, in the east, the slightly buried Voronezh Anteclise (Uplift), and in the west, the Belorussian (Byellorussian) Anteclise. Elsewhere, the Riphean–Vendian–Phanerozoic cover is generally 2–4 km thick, but locally attains 20 km thick. (Key references: Khain 1985, Gaál and Gorbatschev 1987, Rundqvst and Mitrofanov 1992, Gorbatschev, 1993.)

### Baltic Shield

The Baltic Shield, which includes practically the entire Scandinavian Peninsula, the Finnish–Russian border zone of Karelia, and the Kola Peninsula, forms a rectangle about 2000 km long (northeast to southwest) by 1600 km wide, or 2.2 × 10$^6$ km$^2$ in area (Figs 1-5c(i), 2-5, inset).

The Baltic Shield (Franke 1993, Gorbatschev and Bogdanovo 1993) is tectonically divided into six major, westward younging major provinces: (1) *Kola Peninsula Province*, in the northeast, is underlain mainly by complexly deformed, high-grade Archean gneiss and amphibolites with local fault-bounded slivers of early Proterozoic meta-supracrustal cover. (2) The adjoining narrow, sinuous *Belomorian Province* and on-strike *Lapland*

# GEOLOGIC SETTING BY CRATON

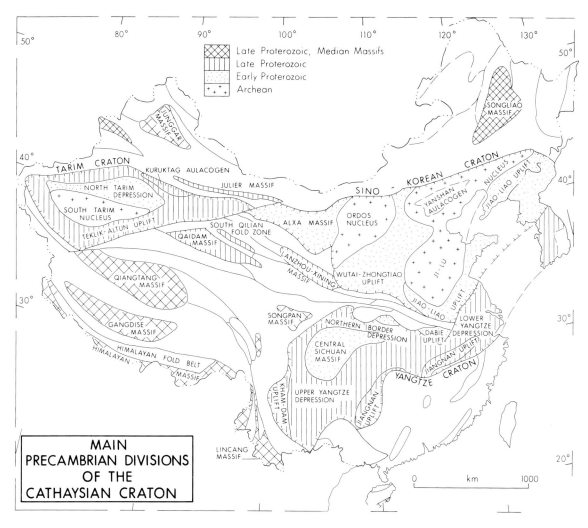

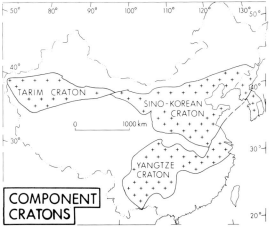

**Fig. 1-5a.** Main geologic outline and divisions of the Cathaysian Craton showing craton outline, main geologic features, and relevant political and geographic divisions (adapted in part from Atlas of Palaeogeography of China, 1985, Map 141).

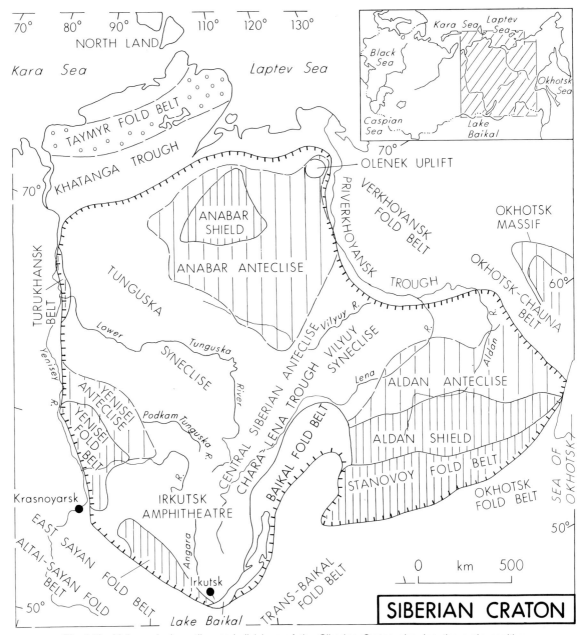

**Fig. 1-5b.** Main geologic outline and divisions of the Siberian Craton showing the main positive elements (exposed shields, fold-belts and adjoining anteclises) and negative elements (buried syneclises and troughs) (adapted from Salop 1977, Fig. 6 and Shatzki and Bogdanoff 1961, Fig. 1).

*Granulite Belt* are variably composed of Archean medium to high grade metapelites, amphibolites, BIF and granitoid gneiss–charnockites, disposed in complexly deformed, W-verging nappes. (3) To the west, the 400 km-wide rectangular *Karelian Province* comprises Archean granitoid–greenstone basement with considerable early Proterozoic Karelian–Kalevian cover, locally pierced by mantled gneiss domes. (4) Incorporating the uppermost Kalevian strata and transitional westward across an E-verging overthrust fault zone, which incorporates the Outokumpu suture, lies the broad (up to 1000 km) early Proterozoic *Svecofennian Province* (Domain). Svecofennian rocks are characterized by extensively granitized arc-type felsic metavolcanics–flysch assemblages and unusually large calc-alkalic granitoid intrusions. (5) To the west, the N-trending, 20–150 km-wide *Transscandinavian Belt* is composed mainly of ~ 1.75 Ga granitoid plutons. (6) It is succeeded westward by the *Southwest Scandinavian*

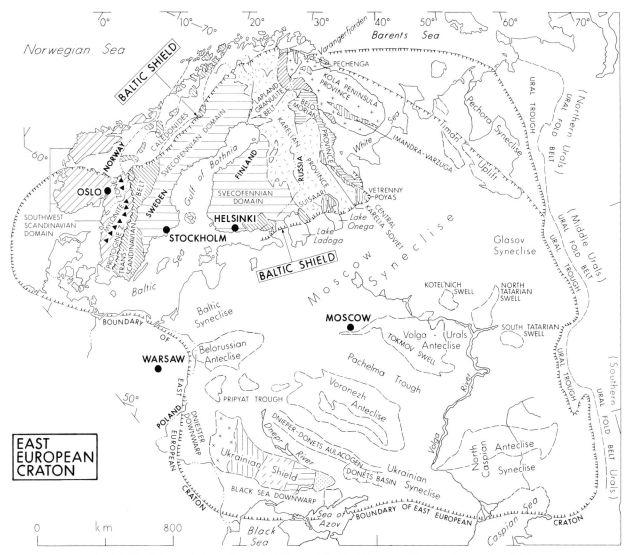

**Fig. 1-5c(i).** Main geologic outline and divisions of the East European Craton—Main craton divisions, Baltic Shield subdivisions, and Uralian inliers (based on Khain 1985, Fig. 2, Gaál and Gorbatschev 1987, Fig. 2, and Shatzki and Bogdanoff 1959, Fig. 1).

*Domain* (Sveconorwegian Province), a 500 km-wide, unusually complex metasupracrustal–granitoid assemble of diverse ages but dominated by 1.75–1.50 Ga Gothian rocks variably reworked during the Hallandian (1.5–1.4 Ga), Dalslandian (Grenvillian, Sveconorwegian) (1.25–0.9 Ga), and Caledonian (0.6–0.4 Ga) orogenies.

## Ukrainian Shield

To the south, the smaller, generally poorly exposed, gently curvilinear Ukrainian Shield (Fig. 1-5c(ii)) 1000 km by 100–320 km or 200 000 km² in area, is crossed in the eastern part by the renowned 'big bend' of the Dneiper River (Salop 1983, Shcherbak et al 1984, Khain 1985).

The shield contains five principal Archean granitoid–greenstone-rich blocks, that are mutually separated by narrow meridional BIF-bearing early Proterozoic synclinoria. Of these, the 70 km-wide Krivoy Rog–Kremenchug arenite–BIF–carbonate–iron ore-rich synclinorium has been traced by drilling for more than 220 km and by magnetic anomalies for a total distance of 1000 km, reaching the Kursk Magnetic Anomaly of the Voronezh Massif (Anteclies) to the northeast.

The northeast boundary of the Ukrainian Shield is marked by SE-trending faults of the unusually

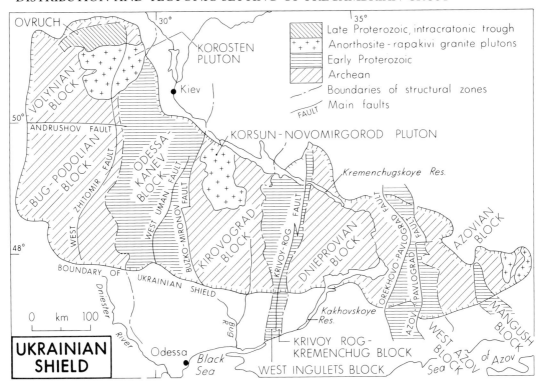

**Fig. 1-5c(ii).** Main geologic outline and divisions of East European Craton—Ukrainian Shield subdivisions (from Khain 1985, Fig. 6 and published with permission of the author).

deep Dneiper–Donets Aulacogen. To the south and west respectively, the crystalline basement plunges smoothly beneath the Black Sea and Dniester downwarps.

### Voronezh Uplift

The Voronezh Uplift (Anteclise), located 300 km to the northeast across the Dnieper–Donets Aulacogen, mostly lies under shallow cover but locally outcrops on the Don River around Kursk and Starvy Oskol. It has been extensively explored by drilling and opened by quarries at the Kursk Magnetic Anomaly, site of extensive BIF with rich iron ore deposits.

Still further northeast, and separated by the major NNW-trending Pachelma Trough (Aulacogen), lies the Volga–Urals (Kama) Anteclise and other moderately to slightly buried positive craton elements.

### Inner structure of the basement

The buried basement has been explored by magnetic and gravity surveys and by several thousand boreholes. It comprises some 23 relatively small, isometric-elongated high-grade Archean blocks (massifs) with intervening early Proterozoic medium-grade fold belts which represent either retrogressed Archean crust or juvenile early Proterozoic crustal additions (Gorbatsev 1993) (Fig. 1-5c(iii)). Later Proterozoic rapakivi granites and metamorphic overprinting (Gothian–Dalslandian) are prevalent in the southwestern and western parts.

Following final cratonization at 1.8 Ga the crystalline basement was variably buried beneath Riphean–Vendian platform cover (Fig. 1-5c(i)). Basement fragmentation and subsidence began in early Riphean time with development of a remarkable conjugate system of NE- and NW-trending extensional troughs (Aulocogen Stage), some of which later developed by progressive enlargement into major craton-wide depressions (Platform Stage). The principal positive and negative elements of the buried basement are as noted above.

### Peripheral belts and median massifs

*The Timan–Pechora Extension*, a 400 km-broad NW-trending marginal zone to the northeast is separated from the East European Craton proper by the Timan fault, a prominent NW-trending, NE-dipping thrust zone which can be traced northwest-

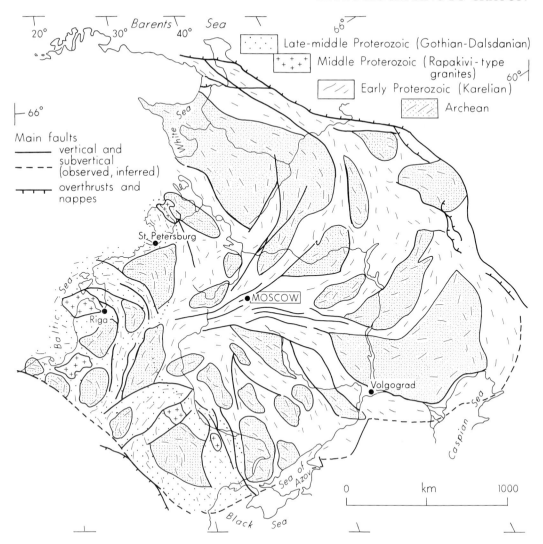

**Fig. 1-5c(iii).** Main geologic outline and divisions of East European Craton—Interior basement geology beneath the platform cover (based on Khain 1985, Fig. 8).

ward along the Murmansk coast of the Kola Peninsula to Varangerfjorden, Norway, where intensely folded mid-Riphean–Vendian sediments are overthrust southward on to the older Baltic Shield.

*The Uralian Fold Belt*, a long string of low-lying Precambrian massifs to the east, is separated from the East European Craton by the Ural Trough (cis-Uralian Foredeep). The western Uralian zone consists of a system of large uplifts and troughs, including the Bashkir Uraltau composed of W-verging thrust slices of mildly metamorphosed, mid–upper Riphean sandstone–limestone–conglomerate sequences up to 10 km thick. The eastern Uralian zone, which is composed of Riphean–Paleozoic platform sediments, has generally subsided in relation to the western zone resulting in deeply buried and rarely exposed rock sequences.

Precambrian-bearing median massifs, present in both adjoining Variscan (Hercynide) and Caledonide fold belts are considered in Chapter 5 (Figs 1-2, 1-3c(i)).

### 1.6.4 GREENLAND SHIELD (NORTH AMERICAN CRATON)

Greenland is an unusually large island-continent with a surface area of $2.2 \times 10^6$ km², of which about 80% is covered by Inland Ice (Figs 1-3d, 1-5d(i)). The ice-free marginal rim is usually from 8 to 40 km wide but locally up to 250 km wide, with the bedrock magnificently displayed in a bare mountainous region penetrated by long-steep-walled fjords.

By far the largest part of the island is composed

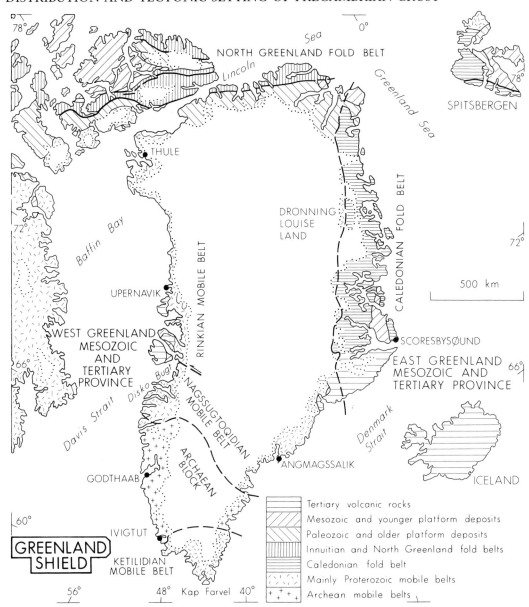

**Fig. 1-5d(i).** Main geologic outline and divisions of the North American Craton—Greenland Shield divisions (from Escher and Watt 1976, Fig. 1).

of Precambrian crystalline rocks of the Greenland Shield. This is flanked to the north and east by the North Greenland and Caledonian fold belts respectively. Mesozoic and Tertiary cover sequences occur locally. (Key references: Escher and Watt 1976, Moorbath et al 1981, Hamilton et al 1983, Kinny 1986, Kalsbeek et al 1993a, b, Nutman et al 1993.)

Precambrian provinces

Four major structural provinces are recognized in the Greenland Shield. The Archaean Block, 300–700 km wide, is flanked by the late Archean to early Proterozoic Nagssugtoqidian and Rinkian belts to the north and Ketilidian Belt to the south (Fig. 1–5d(i)).

The Archaean Block is mainly composed of complexly deformed, high-grade Amitsoq orthogneiss dated to 3820 Ma. The gneisses contain still older metasupracrustal units (Akilia and Isua metasediments), as well as younger metasupracrustal units, anorthosites, diabase dikes, and late granites (2.5 Ga). The structural pattern is highly varied and complex.

The Nagssugtoqidian Belt to the north, about 300 km wide, comprises Archean gneisses, largely reworked in early Proterozoic time (~1.8 Ga), but containing sizeable bodies of unreworked Archean crust, together with thin early Proterozoic metasupracrustal infolds. Amphibolite facies prevails with local granulite facies. A pronounced regional planar fabric prevails.

The much broader and less well-known Rinkian Belt to the north, with its own distinctive structural style, is also composed of widespread basement gneisses with local metasupracrustal cover, collectively folded and metamorphosed at upper greenschist–amphibolite facies. Isotopic ages are common in the 1870–1650 Ma range. However, large masses of Archean crust are known to be present.

To the south of the main Archaean Block, the narrow Ketilidian Belt is characterized by large granitoid plutons, 1.85–1.75 Ga, including rapakivi and other late granites with variable early Proterozoic metasupracrustal cover, all affected by Ketilidian Orogeny (1.8 Ga). At the southern tip of Greenland a variety of dykes and central complexes of nepheline syenites and other silica-undersaturated rocks constitute the 1.3–1.1 Ga Gardar Province.

In east Greenland, Krummedal metasediments bear the imprint of the Carolinidian Orogeny (~1.0 Ga), and late Proterozoic Eleanore Bay and Hagen Fjord metasediments of the Caledonian Orogeny.

Spitsbergen Archipelago, to the northeast, includes restricted patches of late Proterozioc metasupracrustal rocks, including prominent Varangian-age tilloids of the Hecla Hoek Geosyncline.

### Scottish Shield Fragment (North Atlantic Craton)

This small, yet historically important crustal fragment (Figs 1-2, 1-3d(i)) is considered here along with Greenland as rifted segments of the pre-Mesozoic North Atlantic Craton. (Key references: Watson 1975a, Park and Tarney 1987, Craig 1991, Park et al 1994.)

The Lewisian Complex (2.9 Ga), exposed in the northwest Highlands, Inner and Outer Hebrides, and as far south as Ireland, is dominated by tonalitic-granodioritic gneiss which contain small masses of older Scourian metasediments and of layered mafic–ultramafic metamorphites. Granulite facies metamorphism of age 2.7 Ga prevails.

Lewisian rocks were later affected by emplacement of Scourie dolerites and norites (2.4–2.2 Ga), Loch Maree volcanism–sedimentation (2.0 Ga), and Laxfordian metamorphism–deformation (1.9– 1.8 and 1.6–1.5 Ga). Accumulation of Torridonian sandstones at 975 and 790 Ma was followed by widespread uplift and erosion marked by an unconformity. Finally, Dalradian sediments were deposited some time between 800–700 Ma and ~400 Ma.

## 1.6.5 NORTH AMERICAN CRATON (LESS GREENLAND SHIELD)

The North American Craton (less Greenland Shield) forms a large ovoid fault-bounded crystalline mass about 5000 km in diameter and $17 \times 10^6$ km$^2$ in area. This craton is encircled by Phanerozoic fold belts (Innuitian, Cordilleran, Sierra Madre, Ouachitan and Appalachian) but for the rifted northeastern margin facing Greenland (Figs 1-3(ii), 1-5d (ii)). However, including Greenland, the Phanerozoic encirclement is completed by the Caledonian and North Greenland mobile belts. About one-third of the craton is dominated by the uniquely large Canadian Shield. Buried basement extensions in Canada and USA have been extensively explored by drilling and survey techniques. Numerous Precambrian inliers lie within the surrounding Phanerozoic fold belts. (Key references: Stockwell 1982, Hoffman 1989, 1990, Mooney and Braile 1989, Card 1990, Reed et al 1993a, b, Lewry et al 1994.)

### Canadian Shield

The Canadian Shield is a large orthogonal craton 3000 km in diameter and $5.5 \times 10^6$ km$^2$ in area; at its centre lies the 1000 km-wide Hudson Bay Lowlands. The shield is bounded by Phanerozoic sedimentary onlap, rare fold belts and oceanic crust.

The Canadian Shield is divided into seven structural (tectonic) provinces each of distinctive tectonic imprint (Fig. 1-5d(ii), inset). Of these, two (Superior and Slave) are mainly Archean with Kenoran (2.5 Ga) imprint; three (Churchill, Bear and Southern) are mainly Aphebian with Hudsonian (1.8 Ga) imprint; and two (Grenville and Nain) are mainly Helikian with Grenvillian (1.0 Ga) imprint/influence. Rare Hadrynian rocks are practically confined to the shield margins. The small easternmost Nutak domain represents a rifted segment of the Archaean Block in Greenland.

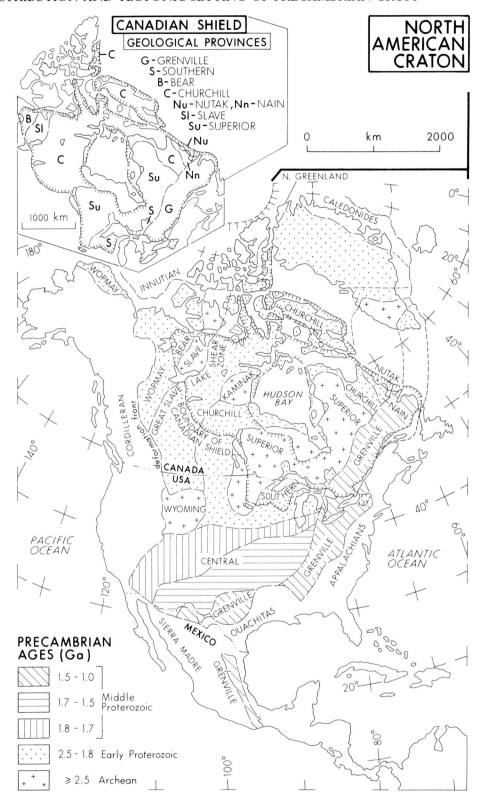

**Fig. 1-5d(ii).** Main geologic outline and divisions of the North American Craton with Greenland Shield in pre-drift position (Hoffman 1989, and published with permission of the author).

### Archean provinces (>2.5 Ga)

*Superior Province*, the dominant Archean entity, forms a large deeply indented ovoid, 2500 km by 700–1000 km or $1.6 \times 10^6$ km$^2$ in area. The province is about equally divided into (1) the northeastern high-grade gneiss–migmatite-rich Ungava Belt (Domain) with unusually large granulitic terrains; and (2) the southern–western low–medium grade metavolcanic–metasedimentary–gneiss–pluton-rich part characterized by E-trending, alternating granitoid–greenstone and metasedimentary–gneiss–pluton superbelts (subprovinces). The volcanic-rich greenstone belts are commonly 2.7–2.8 Ga but range to 3.0 Ga, whereas the gneiss terrains, including granulites range to at least 3.35 Ga. Granulite-rich crust (Pikwitonei) forms the northwestern boundary (Nelson Front) of Superior Province and underlies rare small interior domains. The narrow, irregular, NE-trending, 600 km-long Kapuskasing Structural Zone transects the province from Lake Superior vicinity to James Bay (Hudson Bay).

*Slave Province*, a smaller ovoid craton 1000 km to the northwest and 800 km by 400 km or 225 000 km$^2$ in area, comprises prevailing granitoid gneiss (2.6–3.96 Ga) with associated, N-trending, 2.7 Ga metasedimentary–metavolcanic (greenstone) belts.

*Other* smaller Archean domains (e.g. Kaminak) are present in the intervening Churchill Province.

### Aphebian provinces (2.5–1.8 Ga)

*Churchill Province*: Dominant tectonic trends in this large ($2.1 \times 10^6$ km$^2$) structurally varied province are concave to the south about Hudson Bay. Predominant gneiss–migmatite terrains include proportionately high reworked Archean basement infrastructure with variable Aphebian cover. Northern and western granulite-rich parts have been alternatively reclassified as Archean and renamed Rae and Hearne provinces (Fig. 3-2). The Trans-Hudson orogen, a prominent NE-trending zone in the southwest, is characterized by juvenile Aphebian volcanic–turbidite–granitoid accretions including discrete greenstone belts. Eastern Churchill Province likewise includes substantial reworked Archean infrastructure, also alternatively reclassified as Archean, and sutured to adjoining Archean provinces by New Quebec, Torngat, Foxe and Dorset orogens respectively (Fig. 3-2).

A garland of BIF-rich Aphebian foreland fold-and-thrust belts (2.0 Ga), including Labrador Trough and Belcher Belt, practically encircles Superior craton. Helikian quartzite–redbed-rich epicratonic structures (1.7–1.5 Ga) are widely distributed across the province, including Athabasca, Dubawnt, Bathurst and Coppermine basins.

*Bear Province*: The eastern boundary of this northwesternmost province is a high-angle unconformity on basement rocks of Slave province. The overlying Aphebian metasedimentary–igneous rocks mainly represent 1.9–1.8 Ga juvenile crustal accretions collectively imprinted by the Wopmay (Hudsonian) orogeny (~1.8 Ga). These rocks extend westward beneath Phanerozoic cover to the edge of the Cordillera.

*Southern Province*: This small but complex province is rich in Aphebian rocks but includes both Archean basement inliers and rift-induced Helikian (1140–1120 Ma) volcanic rich cover. Restricted older Aphebian (2.5–2.2 Ga) off-craton thickening, tillite-bearing, quartzitic wedges are separated from more extensive younger Aphebian (2.2–1.8 Ga) sequences comprising thinner BIF-rich shelf facies and S-thickening, also BIF-rich, volcano–turbidite facies. The unique Ni-rich Sudbury Igneous Complex (Irruptive) (1.9–1.8 Ga) is of possible meteorite impact origin. The 250 km-wide, ENE-trending Penokean Fold Belt south of Lake Superior bears the imprint of the Penokean (Hudsonian) Orogeny (1.9–1.8 Ga).

### Helikian provinces (1.8–1.0 Ga)

*Nain Province*: This small, mainly N-trending province to the northeast, is subdivided into three gneissic subprovinces of distinctive structural trends, two of which contain large, discordant anorogenic anorthosite–adamellite intrusions.

Two small adjoining Archean-rich segments on the Atlantic coast constitute the *Nutak domain*, a rifted correlate of the Archaean Block in Greenland.

*Grenville Province (Belt)*: This NE-trending province is 2000 km long and 300–600 km wide. The subsurface extension is at least as long again, reaching to Texas and even Mexico. The 4000 km long Grenville Front, on the northwest, one of Earth's great structural discontinuities, marks the junction of the belt with a variety of older Precambrian terrains. The Grenville Province represents an eroded orogenic belt distinctive for its widespread high grade metamorphism, complex deep level structures and abundant anorthosites. A northwest tectonic transport pattern is widely developed. Grenville rocks carry a pronounced Grenvillian (1.0 Ga) imprint, attributed to collisional tectonics.

### Hadrynian domains (1.0–0.57 Ga)

Limited Hadrynian supracrustal repositories are mainly restricted to shield peripheries including, in the Arctic region, Coppermine River area, Brock and Minto inliers, and Franklinian Fold Belt and the Lake Superior Basin in the south.

### Wyoming Uplift

This comparatively small (750 km diameter), subcircular upthrust, located at the Churchill–Central Cordilleran junction, is dominated by amphibolite-facies gneiss–migmatite–pluton terrains with local greenstone, amphibolite and ultramafic infolds, the latter including the renowned Stillwater Complex. NE-trending, open structures predominate, products of 2.7–2.6 Ga deformations. A complex E-trending shear zone, the Cheyenne Belt, separates this uplift from the Central Province to the south.

### Central Province (USA)

*Central Province*, a mainly buried ENE-trending belt, 3000 km by 1300 km, is characterized by older gneisses overlain by now widely infolded younger volcanic–turbidite-rich metasupracrustal sequences. Ages across this province range from 1.8–1.7 Ga in the northern part to 1.6–1.5 Ga in the southern–eastern parts; these are transected by the still younger NE-trending Grenville Belt (mainly by 1.1–1.0 Ga).

Both the Central and Grenville belts contain abundant mid-Proterozoic anorogenic complexes (1.5–1.4 Ga) including anorthosites, gabbros, bimodal volcanic suites, and predominating rapakivi granites.

### Median massifs

Numerous late Proterozoic–Cambrian (~900–450 Ma) inliers (massifs) are distributed in the adjoining Cordilleran, Appalachian and Innuitian fold belts. These now-deformed pericratonic quartzite–siltstone–carbonate sequences, many in the form of allochthonous slices, are remarkably uniform around the North American Craton, typically grading from thin sandstone shelf units to thicker off-shelf miogeoclinal sequences with, however, substantial volcano–turbidite facies present in the Appalachian massifs. Tillite–mafic volcanic–BIF associations characterize the base of the pericratonic sequences.

## 1.6.6 SOUTH AMERICAN CRATON

Precambrian rocks interlie much of the southward tapering South American Craton ($15 \times 10^6$ km$^2$) and form many small median massifs in the Andean Mobile Belt to the west (Figs 1-3e, 1-5e). (Key references: Gibbs and Barron 1983, Cordani et al 1985, Hasui and Almeida 1985, Teixeira 1985, Litherland et al 1989, Brito Neves and Cordani, 1991.)

The South American Craton includes the Guiana, Central Brazil and Atlantic shields, with buried extensions beneath (1) three large intervening basins; (2) the broad, continent-long Sub-Andean Foredeep to the west; and (3) narrow, discontinuous Atlantic coastal margin deposits to the east (Fig. 1-5e, inset 4). Rifted extensions occur across the Atlantic Ocean in Africa.

The Guiana Shield and all but the easternmost province (Tocantins) of the Central Brazil Shield form the Amazonian (Amazonic) Craton, a large coherent Archean–early Proterozoic basement entity with unusually widespread mid-Proterozoic epicratonic–anorogenic cover (Fig. 1-5e, inset 3). To the east, the Atlantic Shield plus Tocantins Province of the Central Brazil Shield, is characterized by long, sinuous, N–NE-trending, late Proterozoic (Brasíliano) granulite-rich mobile belts. These belts enclose or adjoin three older Precambrian interior massifs: (1) Goias Massif (Archean) in Tocantins Province in the west (Fig. 2-14); (2) parts of the São Francisco Craton (Archean to early Proterozoic) in the province of that name in the east; and (3) the largely buried mid-Proterozoic Rio Apa Massif (Craton) in the south (Fig. 1-5e, inset 3).

The Sub-Andean Foredeep, as provisionally known, is underlain by a group of continent-long, sinuous, ~1400 km broad, westward younging, mid to late Proterozoic (1.3–0.6 Ga) mobile belts. The sharp change in attitude of these largely buried belts, from N-trending in the south to NW-trending in the north, defines the axis of the Arica Elbow (Fig. 1-5e, inset 3), a fundamental tectonic feature of continental South America (Litherland et al 1985).

### Guiana and Central Brazil shields

The *Guiana Shield* (Fig. 1-5e, inset 1A, B) contains: (1) three comparatively small granitoid-rich Archean nuclei, Imataca, Pakairama and Xingu (northern segment); (2) in the centre–east the major NW–W, bifurcating, greenstone-bearing, early Pro-

# GEOLOGIC SETTING BY CRATON 37

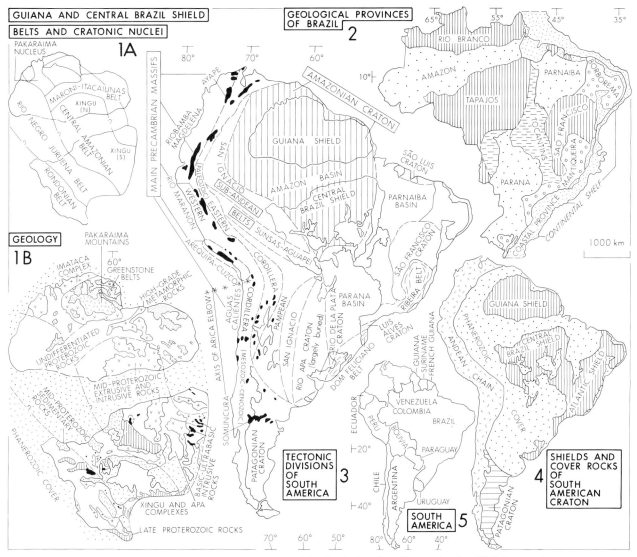

**Fig. 1-5e.** Main geologic outline and divisions of the South American Craton showing (1) Guiana and Central Brazil shields; (2) geologic provinces of Brazil; (3) tectonic divisions of South America; (4) shields and cover rocks of the platform; (5) political division of South America (adapted from Almeida et al 1981, Figs 1, 2, 3; Litherland et al 1985, Fig. 4; Gibbs and Barron 1983, Fig. 1; and Hasui and Almeida 1985, Fig. 2).

terozoic (2.2–1.8 Ga) Maroni–Itacaiunas Mobile Belt; and (3) in the west, part of the broad, NW-trending mid-Proterozoic (1.7–1.4 Ga) Rio Negro–Juruena Belt. Unusually extensive, rift-induced, mid-Proterozoic (1.85–1.55 Ga) volcanic–plutonic–sedimentary–anorogenic cover rests upon the Archean–early Proterozoic basement. That part of the Guiana Shield situated in Brazil is called the Rio Branco Province (Fig. 1-5e, inset 2).

The *Central Brazil Shield* (Fig. 1-5e, inset 1A,B), $2.4 \times 10^6$ km² in area comprises (1) the predominant Archean–Tapajos (Guapore) Province (Fig. 1-5e, inset 2), with its small southerly projection informally called the Bolivian shield area; and (2) to the east, the N-trending, bifurcating Tocantins Province.

The Tapajos (Guapore) Craton (Fig. 1-5e, inset 1A,B) comprises (1) the east–central Archean–mid Proterozoic Xingu nucleus (southern segment); (2) to the northeast, the small SE-trending extension of the early Proterozoic Maroni–Itacaiunas Mobile Belt, including the renowned Carajas region; (3) in the centre–west, the southeasterly extension of the granitoid-rich mid-Proterozoic (>1.4 Ga) Rio Negro–Juruena Mobile Belt; and (4) adjoining still farther to the southwest, part of the parallel (exterior), 1.3–1.0 Ga Rondonian (San Ignacio + Sunsas–Aguapei) Mobile Belt.

Tocantins Province (Fig. 2-14) comprises (1) two long, sinuous, S-trending, mobile belts—Paraguay–Araguaia on the west and Brasília on the east; (2) the small, intervening S–SE-trending Uruaçu Fold Belt; and (3) the centrally enclosed, Archean-rich Goias Massif.

### Atlantic Shield

The Atlantic Shield, an irregular, discontinuous linear complex, 4000 km by up to 800 km, forms much of the prominent Atlantic bight of South America. This tectonically complex shield, which is characterized by extensive Brasìliano (~600 Ma) overprint, comprises three structural provinces: São Francisco including Quadrilater Ferrifero in the centre–west, and Barborema and Mantiquiera in the coastal directions (Fig. 1-5e, inset 2).

Small adjoining basement fragments include São Luis Craton in the north and Rio de la Plata Craton in the south.

## 1.6.7 AFRICAN CRATON

In keeping with its great size, the African Craton includes diverse Precambrian units ranging from comparatively small Archean cratons, through now-restricted early to mid Proterozoic mobile belts and cover, to a plethora of late Proterozoic to early Paleozoic, polycyclic (Pan-African) mobile belts with their large accompanying interior basins (Figs 1-3f(i,ii), 1-5f(i,ii)). The roughly orthogonal (N–S and E–W) Pan-African belt system conveniently divides the composite craton into five component cratons: southern (Kalahari), central (Congo), northwestern (West African), north–central (East Saharan) and northeastern (Arabian–Nubian Shield). The main E-trending Pan-African belts are from south to north (1) Saldanian; (2) Damara–Katangan–Zambezi; and (3) Central African. The main N-trending Pan-African belts are, from west to east, (1) Rokelides (–Mauritanides); (2) Pharusian–Dahomeyan; (3) West Congo–Kaoko–Gariep; and (4) Mozambique. (Key references: Black 1980, Hunter 1981, Tankard et al 1982, Cahen et al 1984, Daly 1986a, b, Kröner et al 1987a, Pallister et al 1988, Porada 1989, Van Reenen et al 1992a.)

### Kalahari Craton (southern Africa)

The crudely hexagonal Kalahari Craton, about 2000 km across, comprises three main parts: (1) in the east, the Archean Kaapvaal and Zimbabwe cratons, with intervening polycyclic Limpopo Belt, and partly obscuring Archean to early Proterozoic epicratonic cover; (2) in the west and south, the early Proterozoic Magondi and Kheis belts, the flanking, mid-Proterozoic (to ~1.0 Ga) Namaqua–Natal Belt with the broadly coeval Koras–Sinclair–Ghanzi (Chobe) rift system; and (3) at the western margin, the small Eocambrian Nama Basin.

#### Archean domains

The *rectangular Kaapvaal Craton*, about 800 × 600 km, comprises an early- to mid-Archean (3.6–3.0 Ga) granitoid–greenstone basement with unconformably overlying late Archean to early Proterozoic (3.0–1.7 Ga) epicratonic basin cover. The Archean basement includes the Ancient Gneiss Complex and associates with ages to 3.64 Ga, and adjoining mafic–ultramafic-rich greenstones, notably the Barberton belt, with established ages to 3.44 Ga. The craton was partly stabilized by 3.0 Ga whereupon epicratonic cover accumulated, heralded by Pongola and basal Witwatersrand strata. Intermittent granitoid intrusions to 2.5 Ga completed Kaapvaal cratonization.

To the north, the *Zimbabwe Craton*, about 700 × 300 km, is formed of a similar granitoid–greenstone 'basement complex' with widespread platform cover. The gneiss–migmatite basement contains sparse earlier Archean (3.4 Ga) (Sebakwian) and more common later Archean (2.7 Ga) (Bulawayan–Shamvaian) greenstone belts. Widespread late Archean (~2.6 Ga) granitoid intrusions stabilized the craton.

A unique feature of this craton is the presence of the 2.5 Ga Great Dyke, a major NNE-trending linear graben-controlled mafic–ultramafic layered igneous complex.

The *Limpopo Belt*, a 150 km-wide intercratonic polycyclic complex, comprises medium to high grade gneisses with metasedimentary associates and layered mafic–ultramafic intrusions, including anorthosites. The growth history of the belt extends from ~3.5 Ga (Sand River Gneiss) to at least 2.5 Ga, the time of Great Dyke intrusion.

#### Proterozoic belts and basin

The early Proterozoic *Kheis, Magondi and Umkondo* belts, flanking the Kalahari craton on the

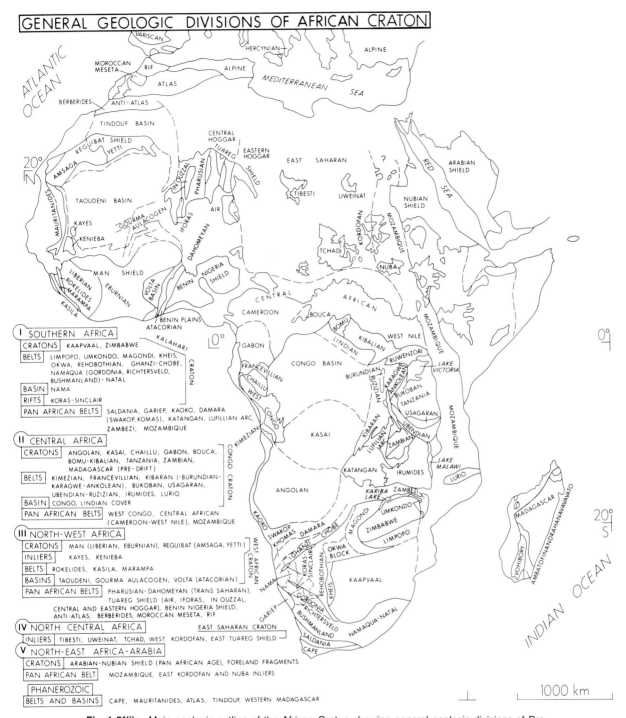

**Fig. 1-5f(i)a.** Main geologic outline of the African Craton showing general geologic divisions of Precambrian crust (adapted from Saggerson 1978, Fig. 1).

west and northeast, constitute thin-skinned, fold-and-thrust belts, involving 2.0 Ga metasedimentary–amphibolite assemblages with large-scale craton-directed vergence.

The mid-Proterozoic *Namaqua–Natal Belt* bordering the Kaapvaal Craton on the south and west, comprises polycyclically deformed, granulite-bearing gneiss–migmatite, representing reworked basement-cover assemblages with components dated to at least 2.0 Ga. Widespread metamorphic retrogression 1.3–1.0 Ga accompanied progressive late granitoid intrusions. The eastern Natal components

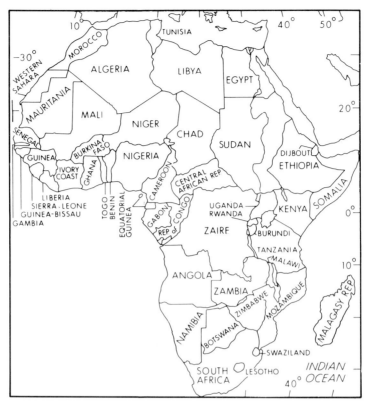

**Fig. 1-5f(i)b.** Political divisions of Africa.

were upthrust northward to partly overlie Archean basement during the terminal 1.0 Ga orogeny.

The arcuate (NW–NE-trending), 1.3–1.0 Ga, redbed–volcanic-filled *Koras–Sinclair–Ghanzi–Chobe* rift system developed concurrently to the west.

Finally, the small Eocambrian (650–550 Ma) *Nama Basin* at the western margin comprises eastward thinning, mixtite-bearing, quartzite–shale-carbonate epicratonic cover.

## Congo Craton (Central Africa)

The Congo Craton, a subcircular mass about 2500 km in diameter and $5.7 \times 10^6$ km$^2$ in area, comprises Archean cratons and early to mid Proterozoic fold belts now distributed about the late Proterozoic–Phanerozoic-filled Congo Basin, itself 1100 km in diameter. More distal circumjacent Pan-African foreland belts include Mozambique on the east, Damara–Katanga–Zambezi on the south, West Congo–Kimezian on the west and the provisionally defined Central African (North Equatorial) Belt on the north.

### Archean cratons

Archean crust is exposed in a ring of ancient cratons surrounding the Congo Basin, in clockwise succession from the south: the very large Kasai–Angolan composite, Chaillu, Gabon, Bouca (Yadé), Bomu–Kibalian, Tanzania and Zambian, the last (Bangweulu) with probable, but as yet unproven, Archean basement.

The cratons range in size from small round units some 500 km across (Chaillu) to very large, irregular masses up to 1800 km by 1300 km (Kasai–Angolan), and are partly buried by late Proterozoic–Phanerozoic cover. Most cratons are composed of dominant gneiss–migmatite dated to 3.4 Ga, with variable ~3.1 Ga and 2.7 Ga greenstone belts. Granulite-grade metamorphism at ~3.0 Ga is widespread. The complexly deformed metamorphites are commonly retrograded along zones of cataclasis. Late Archean (2.7 Ga) granitoid plutons are common.

## Proterozoic fold belts

Three restricted early Proterozoic fold belts adjoin the Tanzania Craton in the east: the NW–N-trending *Ubendian (–Ruzizian) Belt* on the southwest, the ENE-trending *Usagaran Belt* on the south, and the E-trending *Ruwenzori (Buganda–Toro) Belt* on the north. The belts, each from 500 to 1000 km long, and composed of quartzite, phyllite, slates, dolomitic marble and volcanic rocks, are tectonically interleaved with Archean basement gneiss, and intruded by 1.8 Ga Ubendian granitoid plutons.

On the west side of the Congo Basin, pre-West Congo (i.e. late Proterozoic) migmatites and gneisses of the ~2.1 Ga old *Kimezian Supergroup* are locally exposed along the internal zone of the West Congo Orogen.

On the east side of the basin, the now-deformed NNE–NE-trending, craton-verging mid-Proterozoic pelitic–arenaceous *Kibaran and Irumide belts* lie respectively between and to the south of the Kasai and Tanzania–Zambian massifs. Far to the southeast, a third parallel turbidite–ophiolite-bearing *Lurio Belt* forms a coeval remnant unit in the Pan-African Mozambique Belt. All three belts bear the imprint of the 1.3 Ga Kibaran orogeny, product of collisional tectonics.

## Congo Basin

The Congo (Zaire) Basin is a vast, subcircular area, some 1000–12 000 km in diameter. Subhorizontal Phanerozoic cover, 1000–1500 m thick conformably overlies late Proterozoic cover, itself 900–3000 m thick, which in turn unconformably overlies older Precambrian basement. Based on limited data, buried Precambrian basement terrains link with those exposed in the peripheral belts and cratons.

## Northwest Africa

The salient features of this vast region include (1) the West African Craton, a large ovoid domain 2300 km by 1700 km with its large, central Taoudeni Basin and smaller peripheral Volta Basin; and to the east, (2) the smaller Tuareg and Benin Nigeria shields; both enclosed in (3) the 1000 km-broad, 2500 km-long meridional polycyclic Pharusian–Dahomeyan or Trans-Saharan (650–550 Ma) Mobile Belt (Figs 1-3f(ii), 1-5f(i,ii)). The West African Craton itself is bounded by recent oceanic crust on the south, by the Pan-African Rokelide and Hercynian Mauritanide mobile belts on the southwest and west respectively, and northward, across the intervening late Proterozoic–Phanerozoic-filled Tindouf Basin, by the Hercynian Anti-Atlas Domain, itself containing a string of early Proterozoic inliers (Berberides).

## West African Craton

The West African Craton contains two older Precambrian (> 1.8 Ga) massifs, Man (Guinea, Leo, Liberian, Ivory Coast) Shield in the south and Reguibat Shield in the north.

*Man Shield*, 1700 km by 800 km, comprises two domains: (1) a smaller western Archean gneiss-rich domain (Liberian) with relic greenstone belts; and (2) a larger, eastern Birrimian-rich (2.1–2.0 Ga) greenstone–gneiss domain (Eburnean) which was deformed and metamorphosed during the Eburnean orogeny (2.1–1.95 Ga). Birrimian rocks are locally unconformably overlain by ~1.9 Ga Tarkwaian molasse.

*Reguibat Shield* in the north likewise includes an eastern, early Proterozoic (Yetti) domain composed of older (2.1–1.9 Ga) metasupracrustal assemblages widely intruded by Eburnean (1.97–1.76 Ga) granitoid plutons. Local unfolded Guelb el Habib molasse, itself intruded by 1.76 Ga plutons, is locally unconformably overlain by sediments of the Hank Supergroup (1.05 Ga), the oldest part of the adjoining Taoudeni Basin cover.

The vast *Taoudeni Basin* covers the central subsidence of the West African Craton. The buried Archean–early Proterozoic basement is unconformably overlain by flat-lying to little-deformed late Proterozoic (1.05 Ga) to Phanerozoic cover up to 1500 m thick. The smaller coeval Volta Basin on the southeastern flank is tectonically up-thrust westward on to the craton. The Gourma embayment in the east evolved as an aulacogen.

In the *Anti-Atlas Domain* to the north, Proterozoic basement inliers (Berberides) include the 787 Ma Bou Azzer ophiolitic suite which was obducted on to the craton at ~ 685 Ma.

## Trans-Saharan (Pharusian–Dahomeyan) Mobile Belt

This prominent meridional zone of Pan-African mobility is dominated by Pan-African plutonic–metamorphic components but incorporates crystalline basement rocks of the Tuareg and Benin Nigeria shields. The contact with the West African Craton takes the form of the Gourma–Dahomeyan Frontal Thrust, a west-verging upthrust.

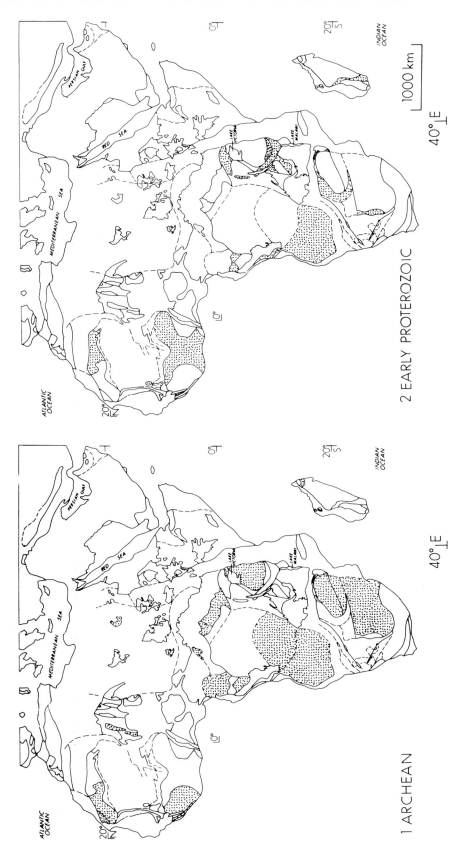

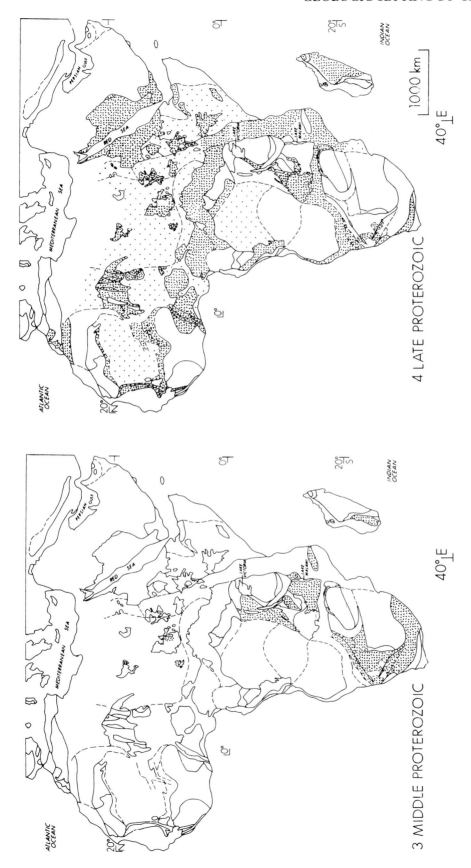

**Fig. 1-5f(ii)** Main geologic outline of the African Craton showing age designation of the geologic divisions by era/eon (adapted from Saggerson 1978, Fig. 1).

The *Tuareg Shield*, about 500 000 km² in area, comprises the central and eastern Hoggar regions together with Iforas and Air prolongations to the south. Crystalline basement rocks including high-grade metamorphites (3.0 Ga) are everywhere surrounded by Phanerozoic cover. Structurally, the shield is composed of huge submeridional slices formed by shield-wide wrench faults.

The pervasive Pharusian tectonothermal event between ~660 and ~604 Ma involved widespread granitoid intrusion and metamorphism. The easternmost Tuareg terrains are provisionally assigned to the East Saharan Craton.

In the *Benin Nigeria Shield* to the south, basement gneiss–migmatite, with dates to 3.0 Ga and even +3.5 Ga, are dispersed in widespread Pharusian (Dahomeyan) (~618 Ma) granitoid plutons.

### Northeast Africa and Arabia

Most of the vast northern fringe of Africa and adjoining Arabia is mantled by thick Phanerozoic sequences which form a generally undeformed cover to mainly deeply buried crystalline basement. However, the basement rocks are well exposed in (1) the easternmost Arabian–Nubian (Egypt–Sudan) Shield; (2) farther west, as isolated inliers—Uweinat, Tibesti, Tchad, Kordofan and Nuba Mountains; and (3) farther south and southwest where they merge with exposed basement rocks of Central Africa.

The easternmost *Arabian Shield* is composed of late Proterozoic (950–640 Ma) andesite–rhyolite-rich volcanic–plutonic complexes festooned with dismembered ophiolites, products of Pan-African arc-terrane crustal accretions.

The adjoining *Nubian Shield* to the west, likewise comprises dominant late Proterozoic volcanic–plutonic–turbidite complexes with tectonically dismembered ophiolite complexes marking sutures between previously separated crustal blocks.

### Pan-African belts (excluding Trans-Saharan) (Fig. 1-5f(i,ii))

The *Mozambique Belt*, a 5500 km by 600 km meridional polycyclic structure in eastern Africa, is the largest continuous Pan-African belt. It is characterized by: (1) a tectonic alternation of older Precambrian basement granulites with variable Irumide (1.1 Ga) and Pan-African (0.6 Ga) overprints, and late Proterozoic metasedimentary cover; (2) dispersed ophiolites; and (3) recumbent structures, thrusts, imbricates and isoclinal folds, all of prevailing westward vergence. These features are reasonably attributed to Pan-African (~615 Ma) continent–continent collisions (west Gondwana–east Gondwana) incorporating rocks of diverse ages strung out along the length of the continent.

The *Damara–Katanga–Zambezi* structure is an ENE-trending, 2000 km-long, continent-wide, platform-offshore turbidite belt association separating the Kalahari and Congo cratons. It incorporates the Damara orogen (1000–540 Ma) with coastal branches in the west and main Intracontinental branch to the east, the on-strike Katangan–Lufilian arc in the centre, and the Zambezi belt in the east. The Katangan supergroup of the Lufilian arc in the centre is stratigraphically divided into a lower, copper ore-rich arenite–argillite division (1.1–0.95 Ga) and an upper arenite–pelite–carbonate–mixtite–basalt division (0.95–0.6 Ga).

The *West Congo Belt* is a NNW-trending, 1300 km-long structure at the west margin of the Congo craton. It is divided into an eastern, external arenite–pelite zone, and a western internal turbidite–plutonite zone characterized by increasing deformation–metamorphism to the west. Despite the absence of identified ophiolites, the belt is interpreted in terms of early (~1.0 Ga) continental rifting, following by continent–continent (South America-central Africa) collision during Pan-African time (650–550 Ma).

The as yet poorly defined, 2000 km-long, latitudinal Central African (North Equatorial) Belt occupies a critical Congo–West African–East Saharan intercratonic position. The belt is composed of assorted lower grade metasupracrustal sequences, higher grade gneisses, and widespread granitoid plutons. Most belt components are considered to have been added as juvenile crustal accretions during the Pan-African cycle of events (830–550 Ma).

## 1.6.8 INDIAN CRATON

The Indian Craton (Figs 1-3g, 1-5g) forms a southward tapering triangular wedge about 2500 km on the side and $3.8 \times 10^6$ km² in area. Sri Lanka, offshore to the south, is a tectonic extension. The two southward converging faulted sides are bounded respectively by the Arabian Sea (west) and Bay of Bengal (east) of the Indian Ocean. Unusually thick Indo–Gangetic alluvium covers the northern and northwestern sectors, and Deccan Traps of Cretaceous–Olicogene age (60–65 Ma) cover the west–

# GEOLOGIC SETTING BY CRATON

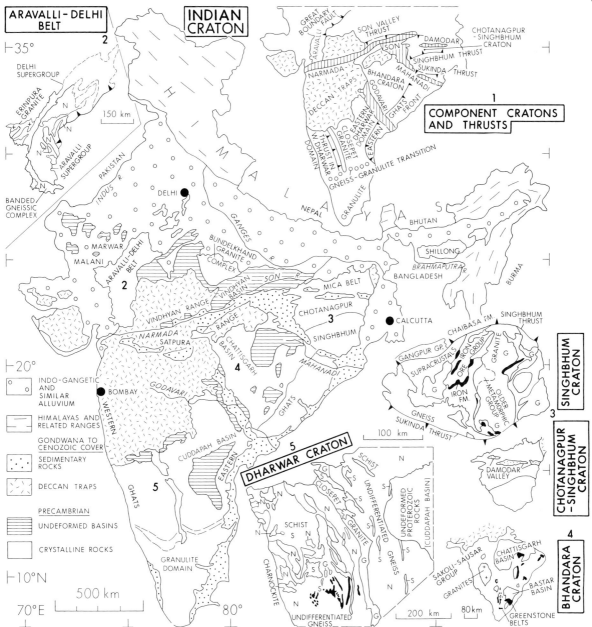

**Fig. 1-5g.** Main geologic outline and divisions of the Indian Craton showing main tectonic divisions; insets include (1) main cratons and thrusts, (2) Aravalli–Delhi Belt, (3) Chotanagpur–Singhbhum Craton, (4) Bhandara Craton, and (5) Dharwar Craton (adapted from Naqvi and Rogers 1987, Figs 1.1, 1.5, 2.1, 3.1, 5.2, 6.1, 7.1).

central sector. (Key references: Radhakrishna 1984, Taylor et al 1984, Chadwick et al 1986, Naqvi and Rogers 1987, Moorbath and Taylor 1988.)

Exposed Precambrian terrains, 847 000 km² in area or 22% of the total craton, are concentrated in the southern, eastern and north–central sectors, with local inliers in the Shillong region to the northeast. The craton is divided into seven parts by six linear structural joins (thrusts) (Fig. 1-5g, inset 1).

## Early Precambrian basement

The *Dharwar Craton* in the south (Fig. 1-5, inset 5) includes numerous submeridional Dharwar and Kolar schist belts (3.0–2.6 Ga), which unconformably overlie widespread gneiss–migmatite of the *Peninsular Gneiss Complex* (3.3–3.0 Ga), with enclosed Sargur enclaves (>3.0 Ga). Both Dharwar metasupracrustal rocks and Peninsular gneiss–

migmatites are cut by the 350 km-long, linear, meridional granite–migmatite assemblage known as the *Closepet granite* (2.6 Ga).

Dharwar metasupracrustal rocks are transitional southward by increase in metamorphic grade across the sinuous, E-trending, 30–60 km-wide Gneiss–Granulite Transition to the *Granulite Domain (South Indian Highlands)*. High grade facies metamorphism (charnockitization) occurred mainly at 2.6 Ga, involving both pre-existing tonalitic orthogneiss (3.4 Ga) and widespread platform-type metapelite–carbonate assemblages of uncertain 3.4–2.6 Ga age.

The NE–NNE-trending, 1200 km long by 100–200 km wide, *Eastern Ghats* coastal belt consists largely of subparallel, alternating layers of charnockites (high-grade, Mn-bearing metasedimentary complexes) and associated granitoid plutons (Fig. 5-17), products of repeated metamorphic events (3.1, 2.6, 1.3 Ga). All bear the imprint of high grade, deep-seated regional metamorphism.

In the northeast, the *Singhbhum (Iron Ore) Craton* (Fig. 1-5g, inset 3) includes an Older Metamorphic Group (OMG) composed of metapelite, quartzite, calcsilicates and amphibolites. Relict OMG patches, dated at ~3.35 Ga, are distributed within younger tonalite–trondhjemite gneiss. The Iron Ore Group (3.3–3.1 Ga) including important BIF, shales, phyllites and volcanic rocks is intruded by the Singhbhum and associated granites, with early and late phases dated respectively at 3.3 and 3.1 Ga. Thick sedimentary–volcanic accumulations, (2.3 Ga) in the Singhbhum region were variably deformed and intruded at 2.1–2.0, 1.6–1.5 and 0.9 Ga.

Similar gneiss–migmatite basement occurs in the *Bundelkhand Complex* in north–central India and in the Aravalli Range, Rajasthan, to the west (Fig. 1-5g, inset 2).

In the *Bhandara Craton* of central India (Fig. 1-5g, inset 4) extensive areas are underlain by gneisses with minor infolded schist belts. No certain Archean rock has been identified. Dongargarh Granite and nearby sediments are dated at 2.2 Ga. Younger basinal sequences include (1) the Sakoli Group (~2.2 Ga) of metapelites, BIF and amphibolites; and (2) the Sausar Group (~1.0 Ga) of psammitic, pelitic and calcareous sediments with notable manganese ore deposits.

### Late Precambrian basins

Numerous sedimentary basins, mainly of (mid-) late Proterozoic age, include the moderate sized *Cuddapah* in the southeast, *Chatisgarh* in the north-east and very large *Vindhyan* in the north–centre (Fig. 5-17). The larger basins are all characterized by aerially widespread, mature sandstone (orthoquartzite)–shale–carbonate sequences with thicknesses measured in kilometres. Deposition occurred along the coast of a tidal sea that included lagoonal, tidal flat and beach shoal elements under very stable conditions. Shallow marine redbeds, stromatolitic carbonates associated with phosphorite and magnesite, and glauconitic tidal flat deposits are widespread.

## 1.6.9 AUSTRALIAN CRATON

The Australian Craton, $7.6 \times 10^6$ km² in area, forms a varyingly indented rift-bound polygon with a northward horn crossing the Arafura Sea to include a south–central segment in New Guinea (Figs 1-2, 1-3h, 1-5h). Exposed Precambrian terrains, common in the western and coastal parts but scattered in the central desert region, amount to $2.3 \times 10^6$ km² or 30% of the total craton. (Key references: Plumb et al 1981, 1990, Rutland 1981, Bickle et al 1985, Hallberg 1986, Hickman 1983, Page 1988, Myers 1991, Drexel et al 1993.)

The six main craton subdivisions (Fig. 1-2) are: West Australian Shield including Pilbara and Yilgarn blocks; North Australian Craton; Northeast Orogen; Central Australian Mobile Belt; Gawler–Nullarbar Craton; and Curnamona Craton. Precambrian rocks therein are divided into basement domains and platform cover.

### Archean basement

The Pilbara and Yilgarn blocks dominate the West Australia Shield.

The smaller (60 000 km²) *Pilbara Block* to the north is underlain by dominant (60%) domal granitic batholiths (3.5–2.85 Ga) up to 100 km across which are separated by coeval low–medium grade metasupracrustal rocks of the Pilbara Supergroup. The BIF-rich Hamersley cover (see below) upon the southern flank of the Pilbara Craton ranges in age from late Archean to earliest Proterozoic (2.8–~2.3 Ga).

The much larger (650 000 km²) *Yilgarn Block* is divided into (1) the smaller, high grade, arcuate Western Gneiss Terrane in the west; and (2) the larger lower grade granitoid–greenstone provinces including the Eastern Goldfields province.

(1) The granulite facies gneisses contain numer-

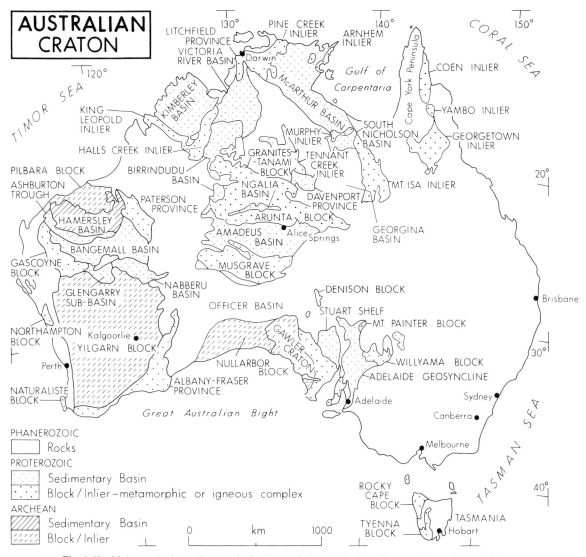

**Fig. 1-5h.** Main geologic outline and divisions of the Australian Craton (adapted from Wyborn 1988, Fig. 1).

ous metasedimentary enclaves characteristic of shelf facies paleoenvironments, as well as local inclusions of ~3.7 Ga gabbro–anorthosite, the oldest such coherent lithology so far identified on Earth. Detailed ion–microprobe analyses of clastic zircons in nearby quartzites yield ages as old as 4276 Ma.

(2) The lower grade provinces to the east are underlain by about 70% granitoid rocks (2.9–2.6 Ga) and 30% metasupracrustal rocks, the latter distributed in numerous interdomal greenstone belts (3.0–2.7 Ga) containing thick unicyclic to multicyclic mafic–felsic volcaniclastic sequences. Eastern Goldfields greenstone belts are famous for widespread Au and Ni mineralization.

Finally, the Gawler Craton in South Australia, an oval-shaped 800 km by 600 km domain centred upon the Gawler Ranges, contains scattered late Archean ages thereby pointing to original widespread Archean crust, now with pervasive Proterozoic tectono-magmatic overprint. A provisionally identified buried contiguous Archean block (Nullarbor) may extend ~600 km to the west.

Early Proterozoic domains

Early Proterozoic assemblages comprise: (1) a mosaic of linear belts (inliers) (2.1–1.9 Ga) including Halls Creek, Pine Creek, Tennant Creek and eight other units of North Australia; (2) the Middleback Ranges (1.8 Ga) and Willyama assemblage (1.7 Ga) of South Australia; and (3) the BIF-rich Nabberu cover sequence (2.0–1.6 Ga) upon the northern flank of the Yilgarn Block, West Australia.

(1) The North Australian belts (inliers) are uniformly characterized by early clastic–mafic volcanic to clastic–carbonate–BIF to late turbidite-rich stratigraphic sequences. Most strata are tightly to isoclinally folded and slaty cleavage or schistosity is ubiquitous. Greenschist facies metamorphism prevails. Mafic and felsic intrusions are widespread.

(2) In South Australia, early Proterozoic sedimentation is noted for its mature and uniform character over wide areas. In the Gawler Craton, iron ore has been mined in the Middleback Ranges (1.96–1.85 Ga) for more than half a century. In the nearby Curnamona Craton, the Willyama Supergroup comprises a largely structurally inverted, 7–9 km-thick arenite–pelite succession now in the form of high-grade gneisses. The succession contains the Broken Hill Pb–Zn–Ag deposit, originally one of Earth's largest known base metal sulphide deposits. Willyama deposition occurred at 1.69 Ga, and high-grade metamorphism at 1.66 Ga, placing it in the early–mid Proterozoic transition.

(3) In West Australia, the Hamersley and Nabberu basins, respectively flanking the Pilbara and Yilgarn blocks, are notable for major BIF development. The Mount Bruce Supergroup of the Hamersley Basin comprises, up-section, the mafic volcanic-rich Fortescue Group (~ 2.8 Ga), the thin-bedded, BIF-rich Hamersley Group (~ 2.5 Ga), and the clastic–carbonate Turee Creek Group (~2.4 Ga). These are unconformably overlain by the Wyloo Group (~2.0 Ga), which also fills the nearby Ashburton Trough. This trough, the Gascoyne Province and the Nabberu Basin together constitute the exposed elements of the Capricorn Orogen.

The Nabberu Basin, which also includes very large BIF but of granular oolitic (pelletal) or shallow-water type, contains at least two sedimentary cycles, an older cycle at 2.2–1.8 Ga and a younger cycle at ~1.8–1.7 Ga. Thus Nabberu BIF (~1.7 Ga) is some 800 Ma younger than the nearby Hamersley BIF (~2.5 Ga).

The Barramundi Orogeny (1880–1850 Ma)—a tightly isochronous, chemically coherent, dominantly felsic magmatic–volcanic event—affected much of northern–central Australia, where it induced widespread cratonization. Subsequent anorogenic bimodal magmatism (1.8–1.5 Ga) was less extensive.

### Mid-Proterozoic domains

Widespread bimodal anorogenic magmatism with concurrent trough-basin subsidence–deposition followed cratonization (by 1.85 Ga) of the North Australian Craton. The component trough-basins are characterized by gently dipping shelf sequences up to 5 km thick which are abruptly transitional to now-deformed, fault-bound troughs containing sequences up to 12 km thick. Stratigraphic sequences are rich in shallow-marine, intertidal, supratidal and fluviatile facies, including abundant evaporite–carbonate complexes. The basins host the classic stratabound Pb–Zn–Ag deposits of McArthur and Mount Isa (1.69 Ga), amongst other smaller Cu, Fe, U and Pb–Zn deposits. Large layered dolerite complexes were intruded at 1.7 Ga.

To the south, in the Musgrave and Arunta blocks and Albany–Froser Belt, major crustal dislocation was associated with mylonite development, granulite facies metamorphism (1.87–1.60 Ga) and felsic–mafic plutonism.

In South Australia, the Gawler Craton was the site of voluminous eruptions of mainly felsic with lesser bimodal Gawler Range Volcanics (1.6 Ga), along with synchronous sedimentation and accumulation of the slightly older, subjacent, major mineralized (Cu–U–Au) Olympic Dam breccia, with co-magmatic granitoid intrusions.

Finally, in western Australia, the Bangemall Basin (1.3–1.0 Ga) contains up to 8 km-thick arenite–pelite–carbonate sequences which unconformably overlie the core of the intercratonic (Yilgarn–Pilbara) Capricorn Orogen.

### Late Proterozoic platform cover

Following cratonization by 1.1–1.0 Ga, central–southern Australia experienced widespread subsidence with basin development. The principal depositories include the Adelaide Geosyncline of South Australia, Centralian Superbasin (Amadeus–Ngalia–Georgina basins) and Kimberley and Victoria River basins in the northwest.

The common Adelaidean succession is broadly divided into three groups, each separated by unconformities: a lower arenite–carbonate group; a middle diamictite with 'cap dolomite'–arenite–lutite–carbonate group; and an upper diamictite with 'cap dolomite'–lutites–arenites–carbonate group. The upper arenites contain the famous 'Ediacara' trace metazoan fossils. Correlations within and between basins are based on stromatolite zonation, diamictites and their 'cap dolomites', as well as magnetostratigraphy. Adelaidean deposition was more or less continuous from ~1.1 Ga to ~0.5 Ga. Broad subsidence became more important than rifting after Sturtian glaciation (676 Ma). Deposition ceased with the onset of the Delamarian Orogeny at ~ 500 Ma.

## 1.6.10 ANTARCTIC CRATON

The continent of Antarctica is geologically distinctive in being almost wholly covered by ice, essentially free of earthquake activity, and having retained a fixed location for the past 200 Ma or so. However, the glaciation of Antarctica is a quite recent condition, the continent having been ice-free for most of its geologic history. An additional noteworthy feature is the ultra high-grade metamorphic assemblage of the Napier Complex in Enderby Land (see below).

The Antarctic Craton (East Antarctic Metamor-

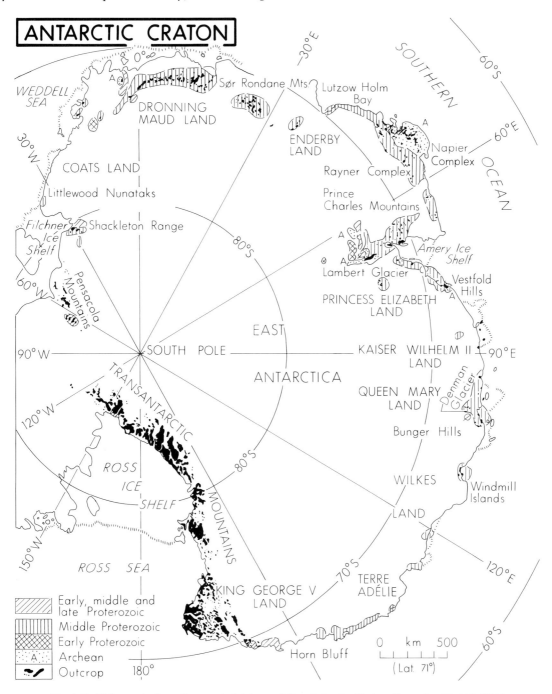

**Fig. 1-5i.** Main geologic outline and divisions of the Antarctic Craton (from James and Tingey 1983, Fig. 1).

phic Shield), $10.6 \times 10^6$ km² in area, is approximately bounded by longitudes 30°W and 150°E (Figs 1-3i, 1-5i). The sparse and scattered bedrock outcrops of east Antarctica consist predominantly of greenschist to granulite facies metamorphic rocks, ranging in age back to early Archean (3.93 Ga). (Key references: Ravich and Grikurov 1976, James and Black 1981, James and Tingey 1983, Yoshida and Kozaki 1983, Black et al 1986a,b, 1991, Black 1988, Grew et al 1988, Tingey 1991b, c.)

Rocks older than 3.0 Ga are rare, and a pre 3.0 Ga history has only been revealed for Enderby Land's Napier Complex. High grade metamorphism at 2.5 Ga has been documented from Napier Complex, Prince Charles Mountains and Vestfold Hills. Evidence for a 2.0–1.7 Ga tectonic episode is confined to sparse data from the southern Prince Charles Mountains and Terre Adélie/King George V Land. However, there is abundant evidence of widespread tectono-thermal activity between 1300 and 900 Ma, including granulite facies metamorphism. Thermal activity at about 500 Ma affected wide areas of the East Antarctic Shield. This activity is coeval with the Ross Orogeny in the Transantarctic Mountains and, in turn, with the 'Pan-African events' in Africa.

# Chapter 2

# Archean Crust

## 2.1 INTRODUCTION

### 2.1.1 DISTRIBUTION

The areas of exposed and buried (total) Archean crust is $15.5 \times 10^6$ km$^2$ (Table 2-1) or 15% of Earth's total preserved Precambrian crust (Table 1.2). Of this amount (Table 2-1), North America and Africa each account for 24% (Σ48%), Europe, India and Antarctica each for 10–12% (Σ34%) and Asia, South America and Australia each for 5–7% (Σ18%). The four Atlantic continents (Europe, the two Americas and Africa) account for 65% of the total. The three former Laurasian continents (Asia, Europe and North America) form 42% of the total, and those of former Gondwanaland the remaining 58%. In brief, the proportions of total preserved Archean crust are high in North America and Africa, and comparatively evenly distributed elsewhere (Fig. 2-1).

The area of exposed Archean crust is $6.6 \times 10^6$ km$^2$ (Table 2-1) or 22% of Earth's exposed Precambrian crust (Table 1-2). Of this amount, Africa and North America, in that order of abundance, together account for 59%, South America 13%, Asia 9% and the remaining four continents each 3–6%. The four Atlantic continents, as enumerated above, together account for 77% of the total, the former Laurasian continents 41%, and the former Gondwanaland continents 59%. Again, Africa and North America predominate; the proportion of exposed crust is moderately high in South America, while in Antarctica it is comparatively low, as largely dictated by the overwhelming continental ice sheet.

### 2.1.2 SALIENT CHARACTERISTICS (Table 2-2)

(1) Preserved Archean crust is distributed in some 36 typically subcircular to oblong exposed cratons with buried extensions, of which the largest is Superior Province in the Canadian Shield. The Archean cratons are broadly distributed across the continents with the largest concentrations, as stated above, in Africa and North American (Fig. 2-1; Table 2-1).

(2) Archean crust on average is composed of (a) 60% granitoid gneiss (50% orthogneiss and 10% paragneiss–migmatite); (b) 30% massive granitoid plutons; and (c) 10% greenstone belts. This translates to gneiss:plutons:greenstone = 6 : 3 : 1, or plutonites : metasupracrustals = 4 : 1. In terms of metamorphic grade, exposed Archean crust averages approximately 66% amphibolite-, 23% granulite- and 11% greenschist-facies (6 : 2 : 1). Plutonic rocks typically comprise an older, dominantly sodic tonalite–trondhjemite–granodiorite (TTG) suite (trondhjemitic differentiation series) with distinctive and strongly fractionated trace element patterns, and a younger calc-alkalic association dominated by potassic granites. The mean

*Table 2-1.* Distribution of preserved Archean crust by continent.

|  | Total crust (%) | Exposed crust (%) |
|---|---|---|
| Asia | 5.9 | 8.8 |
| Europe | 12.6 | 4.9 |
| North America | 23.8 | 26.7 |
| South America | 6.6 | 12.7 |
| Africa | 23.7 | 32.0 |
| India | 9.7 | 5.3 |
| Australia | 5.6 | 6.6 |
| Antarctica | 12.1 | 3.0 |
|  | 100.0 | 100.0 |

Total (exposed + buried) crust = 15 476 000 km$^2$
Exposed crust = 6 631 000 km$^2$ (43% of total)

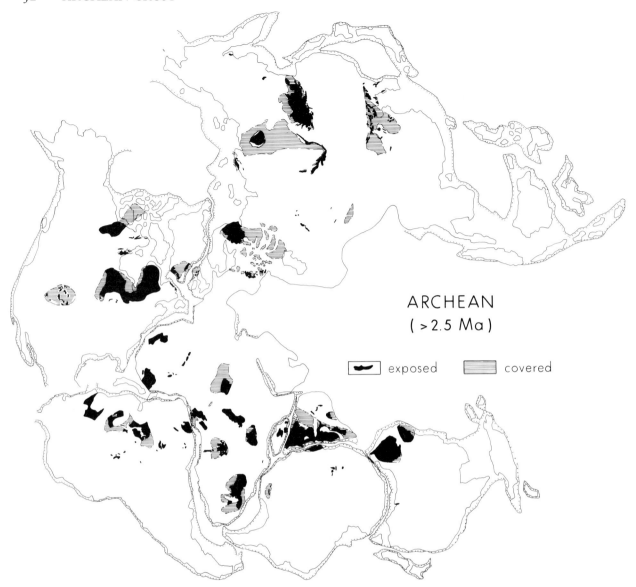

Fig. 2-1. Distribution of exposed and covered (i.e. sub-Phanerozoic) Archean crust as preserved in the pre-drift Pangean reconstruction of the continents (Briden et al 1971). Interior basement crust beneath late Proterozoic cover of the Siberian and East European platforms is included.

chemical composition of exposed Archean crust is that of Na-granodiorite. Bimodal (tonalite (–trondhjemite)–amphibolite) suites are characteristic. Layered mafic–ultramafic igneous intrusions, mainly gabbro–anorthosite, are present in small yet significant quantities.

(3) Medium to high grade gneiss–migmatite terrains, the predominant Archean crustal association, characteristically contain high grade metasupracrustal xenoliths, evidence of still older crust. These predominant older tonalitic gneisses typically display low Sr and Nd initial ratios, implying a temporally close mantle-derived petrogenesis. Also, their distinctively high La:Yb ratios indicate a comparatively deep partial melt mantle origin, involving a garnet–amphibole residue at mantle depths greater than 40 km. In contrast, the closely associated younger Archean igneous components, which take the form of calc-alkalic K-granites and granodiorites with characteristic Eu depletion and thereby higher-level crustal affinity, are indicated to have formed mainly by intracrustal melting.

The widespread presence, in the predominating medium to high grade terrains, of charnockites and other high grade metamorphic suites, typically including shallow-water sediment protoliths,

# INTRODUCTION

*Table 2-2.* Salient Archean characteristics.

| | |
|---|---|
| Distribution of preserved crust | Highest proportions in Africa and North America; elsewhere comparatively even. |
| Composition and metamorphism | Mainly amphibolite facies TTG suite gneiss-migmatite; with closely juxtaposed but subordinate (a) medium to high grade (granulite) gneiss and (b) medium to low grade (greenschist) granitoid-greenstone terrains. Mean composition that of Na-granodiorite. |
| | Common bimodal volcanic assemblages with widespread though subordinate peridotitic (high-T) komatiites and locally abundant andesites. |
| | Wacke-mudstone turbidites (deeper water) predominate; some fluviatile (shallower water) sediments; paucity of shelf facies, except in high grade gneiss and other restricted terrains. Common Algoma-type BIF. |
| | Notably high Ni and Cr and La:Yb; low Sr and REE; lack Eu anomalies. |
| | Layered gabbro-anorthosite bodies. |
| Deformation | Widespread earlier horizontal and later superimposed vertical deformations; highly varied and closely juxtaposed patterns. |
| | Local epicratonic basins, some extensive and richly mineralized, with subtabular clastic-bimodal volcanic fill. |
| Mineralization | Cu-Zn sulphides, Au–Ag; Au–U; Ni–Cr, asbestos; Fe–Mn; Ta–Li, Nb–Be. |
| Paleoenvironment | Reducing-low oxidation state. Dominant mantle-flux signature. |
| Paleobiology | Early chemical evolution leading to early widespread prokaryota. Microfossils and stromatolites. Cyanobacteria by 2.8 Ga. |
| Paleotectonics | Episodic massive juvenile TTG crustal accretions with diachronous growth and consolidation of cratons, especially at 3.8–3.5, 3.1–2.9 and 2.7–2.5 Ga. Major intracrustal K-granite plutonism and cratonization in latest Archean (2.7–2.5 Ga). |
| | Widespread horizontal and vertical microplate motions. Products of largely unestablished mantle-crust processes. |

implies (a) burial of primary surficial materials to deep crustal levels of 30–40 km and locally even 45–75 km; followed by (b) their net vertical uplift (rebound) to the present erosion surface astride a 'normal' crustal thickness of 30–45 km.

(4) Structural style in the gneiss terrains is characterized by recumbency and other subhorizontal attitudes and high intensity of deformation. Deformation patterns feature complex interference structures and commonly flattish folds, thereby implying a significant component of horizontal tectonic regimes in deeper crustal levels during the Archean. However, diapiric ascent of granitoid intrusions into dome cores, with accompanying steep deformations, is also very important.

(5) Closely juxtaposed with the medium to high grade gneiss terrains are the medium to low grade granitoid–greenstone terrains, a subordinate yet important Archean crustal component. Archean greenstone (schist) belts are of two broad ages—~3.5 and 2.7 Ga—but with intervening representatives. Greenstone stratigraphic sequences, commonly 7–17 km thick, are typically characterized by lower tholeiitic to komatiitic lava flows representing submarine lava plains and large shield volcanoes, and upper felsic pyroclastic concentrations with more or less clastic–chemical sedimentary associates. The proportions of andesites (intermediate composition) and of komatiites (ultramafic composition) in the greenstone belts vary widely with andesites locally predominating, whereas the overall mafic : felsic proportion is more uniform. Both multimodal and bimodal facies assemblages (i.e. with and without intermediate components respectively) may occur transitionally in the same greenstone belt. Some mafic sequences erupted upon felsic sequences. Some sequences contain two or more mafic–felsic successions. Associated, 10–15 km-thick, granitoid domes and composite batholiths, locally gneissic, are typically distributed in elliptical patterns involving networks of crumpled-synformal greenstones. Both granitoids and greenstones typically carry the geochemical signatures of juvenile crustal accretions, products of major crust-forming events. Structural style is dominated by vertical tectonic patterns, with, however, local recumbent patterns reflecting some horizontal tectonic movements. High-magnesium (peridotitic) komatiite lava flows, which are quantitatively restricted to the Archean, require estimated flow temperatures of about 1600°C, excessively high in modern terms and a principal support for the interpretation of higher than present heat flows in the Archean source environments. Unequivocal Archean ophiolites have not been recognized in any Archean greenstone belt.

(6) Archean sediments are typically of the turbidite facies, with greywacke-type volcaniclastics predominating. Rarer shelf facies, including orthoquartzites, silicified volcaniclastics, evaporites and

carbonates have been identified, especially in older Archean terrains. Chemical and biogenic sediments, notably BIF, are commonly intercalated. Archean sediments display highly variable trace-element abundances but, on average, have comparatively low levels of incompatible elements (Th, light rare earth elements (LREE)), low La : Yb ratios, high abundances of Cr and Ni and lack Eu anomalies. The common provenance is considered to be the Archean bimodal igneous suite, including volcanic–plutonic protoliths. Volcaniclastic sediments of this type, now predominantly micaceous gneiss, may form substantial (tens of kilometres wide) metasedimentary–paragneiss belts, typically intruded by peraluminous granites largely derived by partial melting of the micaceous rocks themselves.

(7) Epicratonic basins developed upon stabilized continental crust by 3.0 Ga in southern Africa, India, Antarctica and West Australia. Elsewhere, common platform cover followed widespread terminal Archean (~2.6 Ga) cratonization.

(8) Heat production in Archean times was probably two to three times the present rate as supported by the presence of peridotitic komatiites. However, geobarometry in the adjoining high grade gneiss terrains suggests moderately low geothermal gradients comparable to those of modern continental crust. This apparent paradox of closely juxtaposed 'hot–cold' environments points to the operation of some process of concentrated Archean heat loss involving unusually effective, heat-dispersing mantle convections (e.g. hotspots), represented by unusually long and/or unusually rapid mantle convection systems, directly alongside thermally insulated growing 'protocontinental' cores. This may reflect some precursor Archean plume-related, magma-dominated tectonic process.

(9) Archean mineral deposits feature volcanogenic base-metal (Cu, Zn, Pb, Au) massive sulphides; sedimentary iron and manganese; lode Au–Ag; Ni, Cu, Cr, Pt, V and asbestos of ultramafic–mafic association; and minor Ta–Li–Nb–Be of late pegmatoid association.

## 2.2. CATHAYSIAN CRATON

Archean basement rocks of the composite Cathaysian Craton are mainly exposed in the centre–east part of the component Sino–Korean Craton and locally in the adjoining Tarim Craton to the west (Figs 1-5a, 2-2).

### 2.2.1 SINO–KOREAN CRATON

Archean exposures in the Sino–Korean Craton are especially common (1) on both sides of the prominent, NNE-trending, sinistral Tan-Lu Fault, which crosses Bohai Bay; and (2) along the northern margins of the craton (Fig. 2-2; see also Sun et al 1992, fig. 1, adapted from Jahn 1990). Archean terrains contain, on average, 52% gneiss–migmatite, 39% massive granitoid plutons, and 9% metasupracrustal rocks. Gneiss–migmatites characteristically includes bimodal gneiss–amphibolite associations: migmatites are characteristically potassic. Interleaved and infolded volcanic-rich greenstone belts include variable dolomitic marble, mica schist, sillimanite gneiss, meta-arkose and BIF. Amphibolite facies metamorphism prevails except in the north where granulite facies predominates. Archean rocks are typically folded along two or more superimposed fold systems. A prominent stage of metamorphism–migmatization occurred at 2.5–2.6 Ga (Fupingian), marking the Archean–Proterozoic boundary. Earlier events are reported at 2.8 Ga (Qianxi) (Cheng et al 1984) and at 3.5 Ga at Tsaozhuang (Jahn et al 1987). A chronostratigraphic tradition that automatically places granulite facies terrains below those of amphibolite facies is now under critical review (Table 2-3). Stratigraphic relations in four typical areas are summarized below.

### East Qinling Range

Occasional late Archean exposures in central Henan Province include E-trending, granitoid–greenstone assemblages of the Dengfeng Group, and the tectonically juxtaposed high grade gneiss terrane of the Taihua Complex in the south (Zhang et al 1985) (Table 2-3, column 1). The Dengfeng Group comprises ultramafic–mafic–felsic volcanic–greywacke–BIF-bearing greenstone belts and enclosing granitoid–migmatitic gneiss complexes, in the ratio of greenstone : granitoid 1 : 4. Amphibolite facies prevails. Younger granitoid plutons include earlier Na-rich tonalites and later K-rich granites.

The adjoining upper amphibolite to granulite facies, complexly deformed Taihua Complex is dominated (~65%) by tonalitic gneisses with distributed older metasupracrustal patches.

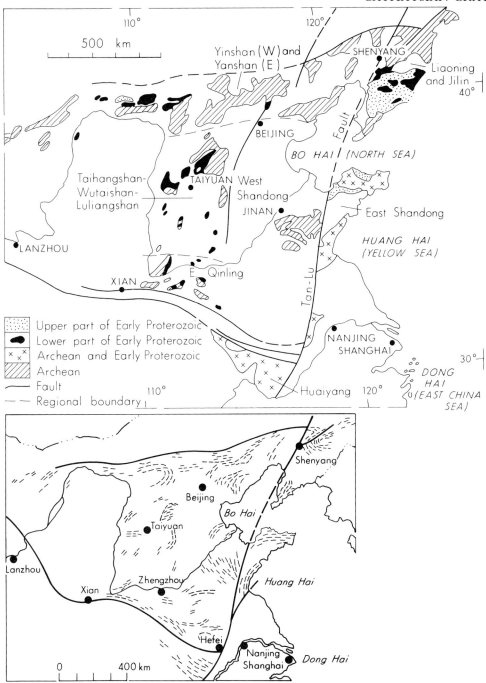

**Fig. 2-2.** Sketch maps showing the distribution of, and structural trends in, exposed Archean and early Proterozoic rocks in the Sino–Korean Craton, of the composite Cathaysian Craton. (Adapted from Cheng et al 1984, Fig. 1, and published with permission of the authors).

### Wutai–Taihang–Luliang District

Archean rocks in this central–western district (Table 2-3, column 2) belong to the Fuping Group (Liu et al 1985, Sun et al 1992) and the unconformably overlying Longquanguan Group (Yang et al 1986).

The older Fuping Group is composed of assorted gneiss, amphibolite and marble, with recognizable lower mafic volcanic components, all at prevailing granulite facies. The Longquanguan Group, of similar lithology, is less highly metamorphosed, being at greenschist–amphibolite facies.

*Table 2-3.* General stratigraphic arrangement in representative areas of the Sino-Korean Craton.

| Erathem | (1) Northern slope of East Qinling (south-central) | (2) Wutai-Taihang-Luliang (west-central) | (3) Yinshan (north central) | (4) Yanshan (East Hebei) (northeast) |
|---|---|---|---|---|
| Early Proterozoic | Songshan Group | Hutuo Group | Erdaowa (Majiadian) Group | Qinglonghe Group |
| | | ——————— Wutai Movement ——————— | | |
| ——— 2500 Ma ——— | | Wutai Group | Sanheming Group | Shuanshanzi Group |
| | | ——————— Tiepu Movement ——————— | | |
| Archean | Dengfeng Group Taihua Complex | Longquanguan Group Fuping Group (M) | Wulashan Group (M) | Badaohe Group |
| | | | ? | |
| | | | Jining Group (M) | Qianxi Group (M) |

(M) = with granulite facies rocks
From Cheng et al (1982), Table 2

The Lanzhishan Granite, intrusive into Fuping rocks, has been dated at 2560 Ma (Liu et al 1985), later supported by an inferred fuping depositional age of ~2.6 Ga (Sun et al 1992). These dates provide both a minimum Fuping age and a maximum age for the overlying early Proterozoic Wutai Group.

### Yinshan District

Archean rocks in the north central part (Table 2-3, column 3) comprise a tholeiite–pelite sequence (Jining Group) at high metamorphic grade, and a nearby tholeiite–greywacke–carbonate sequence (Wulashan Group) at prevailing amphibolite facies. Potassic granitoid plutons and related migmatites are common.

Archean granulites in the Datong–Huai area, 200 km west of Beijing, comprise a basement complex of felsic and mafic granulites in tectonic contact with a metapelite–metapsammite cover sequence. This juxtaposed association was affected by two prominent granulite facies events, dominantly at 2500 Ma and subordinately at 1800 Ma. Basement and cover are juxtaposed along a large, low-angle shear (detachment) zone interpreted to have played an important role in the overall uplift and cooling history of the north China granulite belt (Zhang et al 1994).

### Eastern Hebei District

The Yanshan District (Table 2-3, column 4), some 5000 km² in area, includes the Qianxi Group, currently the best studied Archean assemblage in China (Jahn and Zhang 1984a, b, Sun et al 1984, Zhai et al 1985, Jahn et al 1987). The overlying Badaohe Group is composed mainly of amphibolite and 'granulitites' (leptynite) of amphibolite facies representing intermediate–felsic volcaniclastic rocks, pelites and BIF.

The Qianxi Group itself is essentially composed of granulites of diverse compositions and their retrograded amphibolites. Migmatites are locally developed. Small ultramafic bodies are intercalated in the presumed lower stratigraphic part of the group, whereas metamorphosed BIF is well developed in the presumed upper stratigraphic parts. The igneous protoliths of the prevailing granulites are considered to belong to predominant tholeiitic and subordinate calc-alkalic volcanic series. The majority of the felsic granulites have compositions, including Rare Earth Elements (REE) abundances, corresponding to the Tonalite–Trordhjenite–Granodionite (TTG) igneous suite (Jahn and Zhang 1984a).

Numerous amphibolite blocks or enclaves within grey gneisses near the village of Tsaozhuang, eastern Hebei Province, have yielded a Sm–Nd date of 3470 Ma (Jahn et al 1987), interpreted to represent the time of protolithic basalt eruption and formation of a primitive mafic crust. Similar late Archean assemblages occur to the east near Bohai Bay in Jilin/Liaoning provinces (Jahn and Ernst 1990). Sensitive High Resolution Ion Microprobe (SHRIMP) analyses at two localities, respectively 160 km east of Beijing (Caozhuang) and 150 km south of Shenyang (Anshan area), provide dates of

3550 Ma or older from detrital zircons in metaquartzite, and 3804 Ma (protolith age) from zircons in sheared gneiss (Liu et al 1992).

## 2.2.2 DABIE UPLIFT

Within the rugged terrain of the Dabie mountains in the Huaiyang region (Fig. 2-2), the Dabie Group extends for about 400 km southeastward to the NNE-trending Tacheng–Lujiang (Tan–Lu) fault zone (Jahn 1990, Cheng Tingyu et al 1992, Xu Shutang et al 1992). The Dabie Group, which may be as much as 15 km thick, is composed of assorted gneisses, migmatites, marbles, BIF and amphibolites with minor serpentinite and eclogite. The presence of Archean crust was earlier suggested by U–Pb zircon dates of 3120 Ma and ~2500 Ma (Yang et al 1986). Protoliths of the common metamorphic rocks are considered to be Archean flysch with abundant volcaniclastic sediments (Wang et al 1989). Dabie metamorphic grade decreases from high amphibolite–granulite facies in the northwest, through common amphibolite facies in the centre, to low amphibolite–greenschist facies southeast of the Dabie Mountains. Eclogite, which is widespread in lenses or blocks up to 10 km long within the Dabie Group, contains diamond-bearing coesite and coesite pseudomorphs after both garnet and omphacite crystals.

However, recent studies in the Dabie Mountains document a two-phase cooling and exhumation history following Triassic (230–195 Ma) continental collision and metamorphism between the Sino–Korean and Yangtze component cratons. Blue schist through eclogite facies ages are now considered to record initial exhumation from that collision zone (Xu Shutong et al 1992, Eide et al 1994). The earlier suggested Archean crustal presence is placed in doubt.

## 2.3 SIBERIAN CRATON

Archean crust is exposed in three parts of the craton and vicinity (Fig. 1-5b): (1) the two main Archean-bearing shields: Aldan in the southeast and Anabar in the north; (2) as relict patches in the principal Proterozoic fold belts that partly frame the craton; and (3) median massifs in the encircling Phanerozoic fold belts. Substantial Archean crust also lies in the buried interior basement of the platform (Fig. 2-1).

## 2.3.1 ALDAN SHIELD

Aldan basement comprises dominant gneiss–migmatite with subordinate greenstone (–schist) belts and granitoid plutons. Basement gneisses are characteristically distributed in large domes or oval structures, separated by zones of compressed linear metasupracrustal (greenstone) belts.

The Archean assemblages are classically divided into two lithostratigraphic sequences: (1) an older, widespread, gneiss–migmatite-rich Aldan Supergroup; and (2) a younger, restricted, metasupracrustal-rich Subgan (Olondo) Group (Salop 1977, Kratz and Mitrofanov 1980).

(1) The Aldan Supergroup or Complex is rich in granulite-facies gneiss with considerable interleaved quartzite–marble–calcsilicate relic patches. The supergroup has been traditionally subdivided in three ascending parts: (a) in the central Aldan–Timpton block, a lower assemblage of thick basal quartzite overlain by pyroxene–amphibole gneiss; (b) to the east, unconformably overlying hypersthene–tonalitic gneiss, marble and calcsilicates with minor quartzite and mica gneiss; and (c) an upper sequence of rhythmically alternating garnet–biotite–gneiss with intercalated calcsilicates and graphitic gneiss.

Aldan assemblages show intricate interference patterns, products of polyphase deformation. Granulite facies metamorphism prevails with superimposed amphibolite facies (Sedova et al 1993). Maximum P–T conditions have been estimated at 9 kbar and 1000°C in the southern part of the shield, but only 5–7 kbar and 850°C in the central part.

Basement rocks in the central part of the Aldan shield are characterized by large fold ovals. The ovals, of which at least 12 have been identified, are from 80 to 350 km long (major axes). Gneiss domes, commonly located in the centre of the ovals, have granitoid cores and arc-shaped folds at the periphery which exhibit characteristic centripetal vergency. They are often isoclinally folded, the folds being complicated by small folds down to microplications. In plan, the ovals are often very complex and typically amoeboidal.

(2) The Subgan (Olondo) Group comprises predominant mafic-to-felsic metavolcanic rocks with minor intercalated quartzite, tufftitic schist and BIF, all at prevailing greenschist to amphibolite facies metamorphism. At least 30 such metasupracrustal (greenstone) belts are known in the Aldan Shield and adjoining Stanovoy Ridge (Kazansky and Mor-

alev 1981). Individual belts are 30–150 km long, 4–25 km wide, and 5–12 km in stratigraphic thickness. The characteristic trends are N–S, NW–SE and E–W, this reflecting intergneiss dome distribution.

Greenstone sequences are folded into narrow elongated synclinoria, partly complicated by thrust faults, and transformed into steeply dipping, lens-like monoclinal inliers. The basement relations of the greenstones have yet to be established.

Archean granitoid plutons of the Aldan Shield are broadly distributed in varied shapes and sizes. They are subdivided into two groups: (1) earlier plagio-granite, enderbite, hypersthene granite (charnockite) and associated pegmatites; and (2) later biotite–microcline granites, migmatites and pegmatites.

Similar Archean granitoid gneiss–greenstone terrains are present in the Stanovoy Ridge (fold belt) adjoining to the south. However, the Archean rocks of the Stanovoy Ridge were not only extensively metamorphosed in early Proterozoic time, but broadly intruded by Jurassic–Cretaceous granitic batholiths which now underlie about 60% of the exposed ridge. The boundary zone between the Aldan and Stanovoy blocks is characterized by the Stanovoy Fault, a system of deep fractures along which anorthositic gabbro, anorthosite and younger granitoid plutons were emplaced.

*Geochronology*: Central Aldan gneisses provide minimum zircon ages of 3.2 Ga and 3.4 Ga (Bibikova et al 1986), and within-grain zircon dates of 3.25 Ga and 3.35 Ga (Nutman et al 1992). Subgan volcanic rocks of the Olekma block provide provide zircon dates of ~3.0 Ga (Bibikova et al 1986, Nutman et al 1992). Numerous granitoid rocks in the shield have been dated at both 2.9–3.1 Ga (Neymarket et al 1993) and 2.5–2.6 Ga (Bibikova et al 1982, Bibikova and Krylov 1983, Bibikova 1984). The oldest age so far established for the adjoining Stanovoy Ridge is 2.8 Ga (Nutman et al 1992). It is noted that both Aldan Shield and Stanovoy Ridge basement rocks were partly reworked in the early Proterozoic at 1.9–2.0 Ga (Nutman et al 1992).

## 2.3.2 ANABAR SHIELD

Archean rocks of the Anabar Shield (Fig. 1-5b) comprise principally high grade gneiss, schist and migmatite which are preferentially folded on north–northwest axes and traversed by crush zones of similar trend. Fold ovals similar to those of the Aldan Shield are common (Salop 1983). Metamorphic grade is predominantly granulite facies. Metasedimentary relics are widespread. Granodioritic plutons form large elongated masses arranged parallel to the tectonic grain. Smaller mafic and ultramafic intrusions also occur.

Morphologically, the Anabar Shield forms a rocky plateau at 500–900 m above sea level. The main lithologic components are: (1) Archean granulites and meta-anorthosites of the Anabar Complex; and (2) the early Proterozoic Lamuyka Complex, at prevailing amphibolite facies, formed by tectonic reworking of Anabar rocks along NNW-striking, ENE-dipping belts of linear deep faulting, the belts 10–30 km wide and more than 200 km long.

(1) The common rock types in the Anabar Complex are hypersthene–plagioclase gneisses (enderbites), associated with metabasites, BIF, metapelites and marbles. The total estimated stratigraphic thickness is 15–20 km.

The Anabar Complex is deformed into narrow, linear, partly isoclinal, locally keel-shaped folds, and divided into segments by faults. Prevailing metamorphic conditions record P = 7–8 kbar and T = 780–850°C.

(2) The oldest components of the nearby early Proterozoic Lamuyka Complex are diaphthorites (retrograded secondary metamorphic rocks) composed of gneisses and amphibolites at prevailing amphipolite facies. Large elongated blocks of unaltered granulites, up to 10–15 km long, are occasionally preserved. The gneisses are typically tectonized, with common mylonites/blastomylonites. Migmatites are widespread, and porphyroblastic granodiorite and leucogranite bodies up to 10–12 km wide are encountered.

*Geochronology*: Anabar granulites provide zircon dates of 2.7 Ga, with a designated protolith age exceeding 3.2 Ga (Bibikova et al 1987). Accessory zircons from high-alumina gneiss yield concordia intercept age values of 2.9 Ga (Bibikova 1984). A later thermal event at ~1.9 Ga influenced development of the adjoining Lamuyka Complex (Bibikova et al 1987).

## 2.3.3 PERICRATONIC MOBILE BELTS

### Baikal Fold Belt

The hooked system of Baikal folds, concave to the south, extends southwestward for more than 1200 km, from the west end of Aldan Shield to the south end of Lake Baikal (Fig. 1-5b). Archean rocks exposed along the core of this fold belt are typified by the Muya Group, composed of assorted quartzite, metavolcanic rocks, BIF and marble.

Muya strata, which are intruded by as yet undated granitoid bodies, are overlain by early Proterozoic sediments, themselves cut by granitoid intrusions dated at 1900 Ma. In the Olekma–Vitim Mountainland of the Aldan Shield to the east, rocks closely resembling the Muya Group are cut by pegmatites reportedly dated at 2150–2540 Ma (Salop 1977).

### East Sayan Fold Belt

The East Sayan Fold Belt continues the pericratonic frame of the Siberian Craton extending northwestward for some 1200 km from the south end of Lake Baikal to the vicinity of Krasnoyarsk. Archean basement dated to 3.2 Ga (Bibikova 1984) is exposed in a number of blocks distributed along the length of the belt.

### Yenisei Fold Belt

Archean rocks form the major part of the Angara–Kansk Block, a median massif in the Yenisei Fold Belt. This polymetamorphic complex, known as the Kansk Group, is mainly composed of pyroxene-plagioclase gneiss and amphibolite with intercalated marble and calcsilicates. Granulite facies metamorphites are widely retrograded to amphibolite facies and, locally, to greenschist facies. Late Archean (~2.5 Ga) post-tectonic granitoid plutons are widely developed in the Yenisei Ridge (Salop 1977).

## 2.3.4. BURIED BASEMENT

The buried basement of the Siberian Craton is poorly known because of deep Riphean–Phanerozoic burial, sparse drill intersections and the blanketing magnetic influence of the widespread Siberian Traps. Based on the nature of exposed bedrock in the Aldan and Anabar shields and by analogy with the better known East European Craton to the west (see below) the buried basement is considered to comprise mainly (~75%) small Archean relic blocks (massifs) interspersed with early Proterozoic fold belts at least partly representing retrograded Archean crust and subordinate mid-Proterozoic trough infillings (Khain 1985).

## 2.3.5 MEDIAN MASSIFS

A number of Archean-bearing massifs lie in the surrounding Phanerozoic fold belts. The larger massifs lie to the east of the platform and include the Okhotsk, Taigonos, Omolon and Kolyma massifs (blocks) (Fig. 1-2). The latter three may form a continuous Kolyma–Omolon–Taigonos block extending for 1200 km from the Yana Fold Belt on the west to the Pacific coast on the east. The little-studied massifs typically comprise scattered Archean-bearing metamorphites with thick, widespread Riphean–Cambrian cover.

## 2.4 EAST EUROPEAN CRATON

Archean rocks of the East European Craton are well exposed in both the Baltic and Ukrainian shields (Figs 1-3c (i, ii), 1-5c (i-iii)). Additional small Archean outcrops in the Don Valley belong to the largely buried Voronezh Anteclise. Elsewhere, substantial Archean-rich basement is buried beneath late Precambrian to Phanerozoic platform cover.

## 2.4.1 UKRAINIAN SHIELD

The Ukrainian Shield (Fig. 1-5c(ii)) includes five principal Archean-bearing structural blocks, of which the east–central Dneiprovian Block with its two bounding early Proterozoic synclinoria has been studied in most detail (Siroschtan et al 1978, Shcherbak et al 1984, Sivoronov et al 1984).

The *Dnieprovian Block*, 200 km by 180 km, is composed of predominant orthogneiss and massive granitoid plutons of the Dnieper Complex, with associated mafic–ultramafic xenoliths of the Auly 'Series', and greenstone belts of the Konka–Verkhovtsevo 'Series'.

Dome structures up to 60 km across (e.g. Saksagan Dome) are common in the Dnieper Complex. The domes vary in shape from round to oval, with a transition to arches. They are composed of older, structurally conformable tonalitic gneiss and younger, mainly cross-cutting granite, aplite and pegmatites. Associated fault zones contain numerous bodies of serpentinized gabbro–periodotite (Khain 1985).

*Konka–Verkhovtsevo* greenstone belts occupy a number of narrow structural interdomal keels in the Dnieper Complex. A representative stratigraphic section, 6000–7000 m thick (Shcherbak et al 1984, fig. 4), comprises a lower association (1500–2000 m) of amygdaloidal basaltic lava flows; a middle association (2000–3000 m) of basaltic flows and talc–carbonate schist with increasing upward intercalated intermediate–felsic tuff; and an upper association (about 1500 m) of felsic tuff, greywacke turbidites and banded magnetite quartzites (BIF). Amphibolite facies prevails with some greenschist facies removed from granitoid intrusions.

Konka–Verkhovtsevo metavolcanic rocks provide a zircon of ~3.2 Ga (Shcherbak et al 1984), regarded as a minimum for Konka volcanism (Bibikova 1984).

Dnieper Complex granitoids associated with Konka–Verkhovtsevo metavolcanic rocks have been variously interpreted as (1) older basement to the greenstones; and (2) younger intrusions. Tonalites and granodiorites of this possible basement have been locally dated at 3.0 Ga (Bibikova 1984). The data thus clearly indicate a post-volcanic age for these particular granitoid rocks without, however, eliminating the possibility of pre-volcanic basement elsewhere.

## 2.4.2 BALTIC SHIELD

### Introduction

Archean crust in the Baltic Shield is concentrated in the three easternmost Pre-Svecokarelide provinces: Kola Peninsula, Belomorian and Karelain (Figs 1-3c(i), 1-5c(i), 2-3) (Mikkola 1980, Simonen 1980, Bowes et al 1984, Gaál and Gorbatschev 1987, Turchenko 1992, Gorbatshev and Bogdanova 1993).

### Kola Peninsula Province

This province occupies nearly all of the Kola Peninsula. Common medium to high grade grey tonalitic–granitic gneisses include occasional charnockitic masses. Metasupracrustal (schist) belts, locally common in the gneisses, are represented by the Keivy assemblage, which comprises, from the base upward, metapelitic gneiss, quartzite and BIF; mafic volcanic rocks; and calc-alkalic volcanic rocks with tuffitic metaturbidites (Gorbunov et al 1985, Khain 1985).

Kola deformation is extremely complex, including many orders of steeply isoclinal folds, complicated by faults. Early granulite facies metamorphism was followed by later retrograde amphibolite facies. Migmatites, boudinage and blastomylonites are widespread (Lobach-Zhuchenko and Vrevsky 1984.

Granitoid plutons are common, as are intrusive bodies of gabbro, diorite, and occasional ultramafic rocks.

Numerous widespread U–Pb zircon dates and Sm–Nd model ages in various types of rocks, including both the so-called 'basement gneiss' of the Kola Group and the granitoid plutons, have yielded common ages of 2.9–2.7 Ga (Bibikova 1984, Daly et al 1993). Thus the available isotopic data do not support the existence of more ancient (pre-Lopian) crust in the Kola Peninsula Province.

### Belomorian Province

This comparatively narrow, NW-trending province, extending mainly along the southwestern coast of the White Sea constitutes a synformal belt, 700 km by 150 km in dimension, mainly composed of amphibolite-facies metapelites, amphibolites and granitoids. Together with the on-strike Lapland Granulite Belt (Fig. 2-3, inset), it separates the Kola Peninsula and Karelian provinces (Kratz and Mitrofanov 1980, Khain 1985, Gaál and Gorbatschev 1987). The Belomorian complex is subdivided into three groups: lower granitoid gneiss with subordinate amphibolites; median high grade metapelites and banded iron quartzites (BIF); and upper high-alumina garnet–kyanite- and mica–gneisses and schists. The basement to the Belomorian complex is not known. Belomorian rocks have been repeatedly deformed, the early phases characterized by recumbent folding with westward-verging nappes.

Zircon datings assign an age of 2.9–2.7 Ga to Belomorian deposition including protolith age (Gaál and Gorbatschev 1987) followed by intense magmatism–deformation (Rebolian) at 2.7–2.6 Ga, with widespread concurrent amphibolite-facies retrogression of former granulite facies rocks (Bogdanova and Bibikova 1993).

### Karelian Province

The Karelian granitoid–greenstone terrain extends southward for 1000 km from Lapland to Lake Onega, increasing in width southward from 300 km to 500 km (Fig. 2-3). Granitoid rocks predominate in the Karelian terrain (85%), the remainder taking the form of some 24 linear to oblong N–

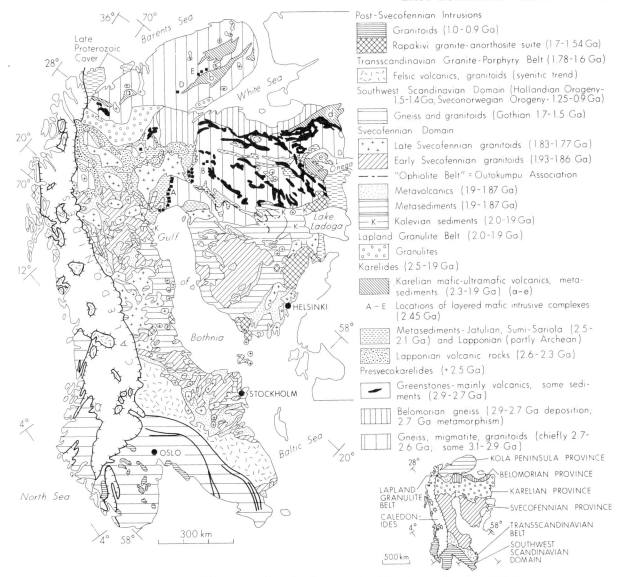

Fig. 2-3. Geologic map of Fennoscandia and adjacent parts of Russia showing the distribution of Precambrian lithologies by province, belt and domain in the Baltic Shield. A – Kemi, B – Koillismaa, C – Koitelainen, D – Monchegorsk, E – Fedorova and Pana fells; a – Pechenga, b – Imandra-Varzuga, c – Central Soviet Karelia, d – Vetienny Poyas, e – Suisaari. (From Gaál and Gorbatschev 1987, Figs. 1.2, and reproduced with permission of the authors).

NW-trending greenstone belts which range in length from 2–5 km up to 100–150 km and in width from 1 km to 15 km (Lobach-Zhuchenko et al 1986b). The metavolcanic–metasedimentary components of the greenstone belts belong to the Lopian Supergroup (Fig. 1-3c(i)). Unconformably overlying this Karelian basement are local patches of both early Proterozoic Sumi–Sariolan volcanic–sedimentary cover and, in the north, more extensive Jatulian platform cover.

Karelian greenstone belts of the Lopian Supergroup are assembled into four geographic zones: East Karelia, Central Karelia, West Karelia and East Finland (Vidal et al 1980, Auvray et al 1982, Bernard-Griffiths et al 1984, Jahn et al 1984, Martin et al 1984a, Saverikko et al 1985, Blois 1989). In general, the greenstone sequences are 2000–4000 m thick. Thick uniform basalt lava flow sequences predominate (40–70%) in the lower part of most belts. East and West Karelian zones are characterized by bimodal (mafic–felsic) volcanism, and Central Karelian and East Finland zones by multimodal (mafic–intermediate–felsic) volcanism, the latter including significant andesite. Basaltic and peridotitic komatiites are common locally. Andesite–dacite–rhyolite pyroclastic rocks are also common both in the lower and upper parts of some sequences. The greenstone belts are complexly deformed, with a predominance of vertical movements. The rocks have undergone mainly low to medium grade metamorphism. Contacts with adjoining granitoid rocks are typically intrusive (Lobach-Zhuchenko et al 1986b). The age range of greenstone accumulation is 3.0–2.7 Ga, coeval with Belomorian sedimentation (Gaál and Gorbatschev 1987). The nearby metamorphic complexes contain gneisses which yield ages to 3.14 Ga (Lobach-Zhuchenko et al 1993).

### Lofoten–Vesteralen Islands, Norway

This group of islands in the Caledonide Fold Belt of the Atlantic coast exposes an unusually deep section through the continental crust. Gravity and seismic surveys show NE-trending, ridge-like up-warping of the Moho to within 25 km of the surface (Griffin et al 1978). The oldest rocks are migmatitic gneiss of generally intermediate composition, probably largely of supracrustal origin, metamorphosed in granulite facies, and intruded by large monzonitic–charnockitic plutons.

Quartzofeldspathic migmatites have yielded a Pb–Pb age of 2685 Ma (Griffin et al 1978). This supersedes an earlier age on Vikan gneisses of 3460 Ma (Taylor 1975). The data are interpreted as evidence of an important crustal accretion event at ~ 2.7 Ga (Griffin et al 1978).

## 2.5 GREENLAND SHIELD (NORTH AMERICAN CRATON)

### 2.5.1 INTRODUCTION

The Archaean Block of southern Greenland is roughly triangular in shape, narrowing eastward as a result of the convergence of two adjoining early Proterozoic mobile belts, Nagssugtoqidian on the north and Ketilidian on the south (Figs 1-3(i), 1-5(i), 2-4).

More than 80% of the exposed Archaean Block is composed of quartzofeldspathic gneiss derived mainly from intrusive granitoids emplaced during at least two major episodes of plutonism at ~3.7 Ga and 3.1 Ga. Supracrustal rocks, mainly amphibolites of volcanic parentage, make up 15% of the Archean terrains. Layered maif igneous complexes, dominantly leucogabbro and anorthosite, make up the remaining 5%.

Isolated remnants of similar Archean basement, some of substantial size, occur within the adjoining early Proterozoic mobile belts. This suggests that the Archean terrain was once much larger and that a substantial part of it was reworked to varying degrees during post-Archean tectonic–metamorphic events (Kalsbeek and Taylor 1985, Gorbatschev and Bogdanova 1993, Kalsbeek et al 1993a) (see Fig. 3-2).

In common with many other medium to high grade gneissic terrains of the world, structural, geochemical and petrological evidence points to former depths of burial of the presently exposed rocks corresponding to at least 20–50 km, with considerable crustal thickening attributable in large part to horizontal tectonic movements.

### 2.5.2 SOUTHERN WEST GREENLAND

Geologic studies in the Godthåbsfjord–Isukasia region of southern West Greenland, involving unusually detailed isotopic investigations (Gulson and Krogh, 1972, Black et al 1973, Moorbath et al 1973, Pankhurst et al 1973a, Moorbath et al 1981, Hamilton et al 1983, Baadsgaard et al 1984, Kinny 1986, Collerson et al 1989, Nutman et al 1993), have established that the Archean gneiss complex is composed of six major early to mid-Archean lithochronostratigraphic units, ranging in age from at least 3870 Ma to 3000 Ma; and a similar number of late Archean events to 2600 Ma (Table 2-4). This locally defined sequence of events is, by extrapolation, used as a general framework for the broad Archean terrain of Greenland.

### Isua–Akilia supracrustal rocks

Amongst the oldest rocks so far identified in Greenland are the Isua assemblage of mafic to ultramafic schists, metasediments and quartzofeldspathic gneisses (Bridgwater et al 1976, Boak and Dymek 1982, Dymek 1984, Nutman et al 1984, Nutman 1986, Dymek and Klein 1988). They are exposed in a semicircular arc 10–20 km in diameter around a dome of gneiss near the western margin of the Inland Ice at Isukasia, 150 km northeast of Godthåb (Fig. 2-4). Contacts with the enclosing gneiss are sharp and near-vertical. The surrounding gneisses are interpreted as derived from younger granitoid plutons.

Widespread small enclaves of similar metasupra-

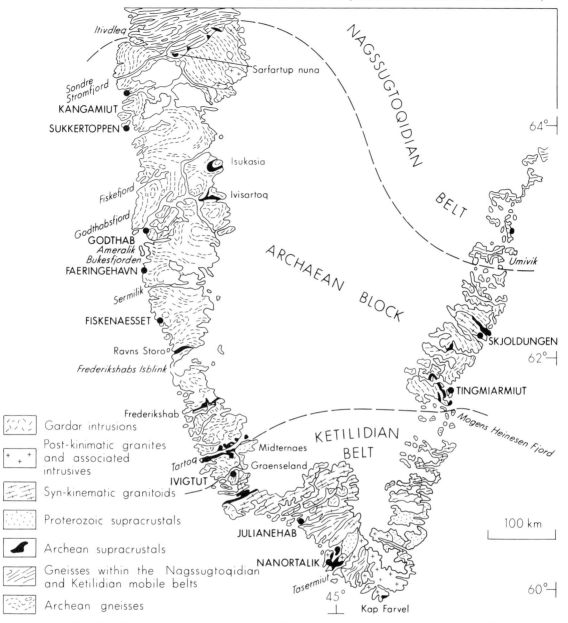

**Fig. 2-4.** Simplified geologic map of the Archaean Block and adjoining belts (in part) Greenland. (From Bridgwater et al 1976, Fig. 4. Published with the permission of the Geological Survey of Greenland).

crustal rocks scattered throughout Amîtsoq gneiss in the Godthåb–Bukesfjorden region are grouped together as the Akilia association (Chadwick and Coe 1983, Chadwick and Crewe 1986). Metamorphic grades range from middle greenschist–amphibolite for the Isua assemblage to amphibolite–granulite facies for the Akilia association at large. Estimated P–T conditions for Isua main-stage metamorphism are about 550°C and 5 kbar (Dymek 1984).

A major band of quartz–magnetite–amphibole–chlorite ironstone (BIF) in the easternmost Isua outcrop is estimated to contain at least $2 \times 10^9$ t of iron ore (Nutman et al 1984, Dymek and Klein 1988).

Planar structures and contacts between the different lithologic units dip steeply and follow the trend of the belt itself. The rocks have a strongly marked, steeply plunging, linear fabric defined by pencil-like rods and elongated conglomerate pebbles. The Isua Belt is cut off to the northwest by a major NNE-trending fault (Moorbath et al 1973, Michard-Vitrac et al 1977). However, Hamilton (1993) is unconvinced that these dates are not source ages rather than depositional ages, and that the depo-

*Table 2-4.* Simplified table of events for the Archean of southern West Greenland.

| Event | Age (Ma) |
|---|---|
| (1) Early crust providing source rocks for the Isua-Ikasia sediments | |
| (2) Deposition of the Isua-Ikasia supracrustals | >3820 |
| (3) Intrusion of syn- and late tectonic granites (parents of the Amîtsoq gneiss) | ~3820 |
| (4) Deformation and metamorphism of the Amîtsoq gneiss and Isua-Ikasia supracrustals | |
| (5) Intrusion of Ameralik mafic dykes | |
| (6) Deposition of Malene supracrustals (dominantly volcanic); intrusion of ultramafic-mafic bodies | >3040 (possibly 3820 or earlier) |
| (7) Emplacement of major stratiform anorthosites and gabbro-anorthosites, e.g. Fiskenaesset complex | >3040 |
| (8) Intrusion of major suites of syn- and late tectonic calc-alkalic rocks as subconcordant sheets (Nûk gneisses) | 3040 |
| (9) Intense deformation with the formation of major nappes, followed by less intense deformation which produced upright folds and widespread dome and basin interference patterns | 3040-2800 |
| (10) Emplacement of late granites | 3000-2800 |
| (11) Granulite facies metamorphism | 3000-2700 |
| (12) Emplacement of Qôrqut granite. Widespread post-tectonic pegmatite swarms | ~2600 |
| (13) Mafic dyke swarms | |

Adapted from Bridgwater et al (1976)

sitional age may be younger than 3.0 Ga. This suggests the need for still further field studies to more accurately define the protoliths and the field relationships of rocks in these assemblages.

### Amîtsoq gneiss

Amîsoq gneisses in the Godthåbsfjord region are the older of two major groups of quartzofeldpathic gneiss (Bridgwater et al 1976, Chadwick and Coe 1983, Nutman and Bridgwater 1986). They are differentiated in the field from younger Nuk gneisses by the presence in Amîtsoq gneisses of abundant bodies of amphibolite derived from mafic dykes called the Ameralik dykes (McGregor 1968, Moorbath et al 1972).

Amîtsoq gneisses have yielded very consistent dates of 3750–3650 Ma (Moorbath et al 1973, Hamilton et al 1983, Baadsgaard et al 1984). Ion-microprobe (SHRIMP) measurements from tonalitic gneiss are at least 3822 Ma old (Kinny 1986).

The typical rock is biotite–oligoclase gneiss, with variable, but generally low, microcline. The reworked gneisses are very inhomogeneous, with thin, well-developed pegmatite layering. All primary structures have been lost to recognition.

### Ameralik dykes

Ameralik dykes represent a suite of amphibolite dykes, discordant layers and lenses in Amîtsoq gneiss (McGregor 1968). As stated above, they are used to distinguish the older more restricted Amîtsoq gneisses, in which the dykes are abundant, from younger widespread Nûq gneisses, from which the dykes are considered to be absent.

Ameralik dykes vary from under a metre to 20–30 m across, typically occur every few metres or tens of metres across strike, and can be traced individually for 7 km or more along strike. Most are fine to medium grained metadiabase. They are concordant with the enclosing gneiss, grading from rectilinear bodies to trains of amphibolite fragments.

### Malene supracrustal rocks

Widespread conformable amphibolite units derived from mafic–ultramafic volcanic and intrusive rocks with metapelite–quartzite–marble associates were originally named Malene by McGregor (1973). They account for some 15% of the Archean gneiss complex of West Greenland. They are described by Chadwick and Coe (1983), Nutman and Bridgwater (1983) and Chadwick (1986, 1990), amongst others. Malene–Nûk relations in the field are extremely complex.

### Fiskenaesset Complex and associated Ultramafic–Anorthosite bodies

Metamorphosed calcic anorthosite and associated leucogabbro and gabbro occur as concordant layers and trains of inclusions throughout the gneiss complex of southern West Greenland (Kalsbeek and Myers 1973, Windley et al 1973, Myers 1981, 1984). The meta-anorthosites provide one of the best marker horizons for tracing out complex structures and, locally, provide way-up criteria. The presence of extensive chromite layering, of potential economic value, leads to their interpretation as gravity stratified, mafic, igneous bodies.

Fiskenaesset anorthosites are injected by veins of Nûk gneiss and reportedly contain xenoliths of Malene schists. On this basis, the anorthosites fall in the 3.2–2.8 Ga range (Chadwick and Coe 1983).

The Fiskenaesset Complex, located 150 km south of Godthåb, is one of the best studied units. It is composed of metamorphosed anorthosite, leucogabbro and gabbro, with minor ultramafic rocks and chromite. The rocks are typically disposed in complex recumbent fold patterns.

### Nûk gneiss

All quartzofeldspathic gneisses in the Godthåbsfjord area that do not contain Ameralik dykes and that intrude the older suites, including Malene supracrustal rocks, anorthosites and Amîtsoq gneiss, qualify for the informal lithostratigraphic designation of Nûk gneiss (McGregor 1973, 1979). The probable ages of formation of Nûk gneisses fall in the 3.0–2.8 Ga range (Baadsgaards 1976, Baadsgaard and McGregor 1981).

Nûk gneisses are very common, accounting for more than half of the rocks exposed in the Godthåbsfjord–Bukesfjorden region (Bridgwater et al 1976, Chadwick and Coe 1983, Chadwick 1986, Friend et al 1987). They occur as deformed sheets, up to several kilometres thick, and concordant to subconcordant bodies, some of which are clearly transgressive.

Nûk gneisses were subsequently intensely deformed with the development of major nappes, followed by upright folds and widespread dome-and-basin interference patterns. High grade metamorphism accompanied and outlasted the deformation, resulting in widespread granulite facies metaphormism. Granulite-facies events have been dated at both 2999 Ma and 2738 Ma (Friend and Nutman 1994).

Recent studies of the Nûk region indicate that it is a composite of four definable tectonostratigraphic terranes (Akia, Tasiusarsuaq, Tre Brødre and Faeringehavn) that evolved independently prior to their present tectonic juxtaposition in the late Archean (Friend et al 1988, Chadwick 1990).

### Late granites

Late to post-tectonic granitoid intrusions and pegmatite swarms postdate the granulite facies metamorphism. Granitoid intrusions, ranging from major bodies such as the Qôrqut granite to minor dioritic plugs, were emplaced during the period of at least 300 Ma between the end of granulite facies metamorphism, and final stabilization of the Archean craton.

The Qôrqut Granite Complex forms an elongate body of mainly homogenous granite, 50 km by 18 km, extending parallel to the regional grain and located 20 km east of Godthåbsfjord (Brown et al 1981, Moorbath et al 1981). It is the youngest of the major quartzfeldspathic complexes in southern West Greenland and is largely undeformed and unmetamorphosed. An associated pegmatite yielded an age of about 2600 Ma, which is interpreted to be the age of emplacement (Pankhurst et al 1973b).

## 2.5.3 SCOTTISH SHIELD FRAGMENT (NORTH ATLANTIC CRATON)

The Scottish Shield Fragment (North–West Kratogen, Hebridean Craton), extending along the western seaboard of Scotland, includes many of the offshore Hebridean islands and adjoining sea floor (Figs 1-3d(i), 5-3, 5-4) (Watson 1975a, Park and Tarney 1987). It is bounded on the east by the orogenic front of the Caledonides (North–West Caledonian Front), represented by the Moine thrust zone. The basement rocks of the shield fragment form a crystalline complex referred to as the Lewisian. On this rest (1) a cover of mid to late Precambrian and early Paleozoic rocks which are, at least in part, equivalent to the cover successions of the main Caledonian Fold Belt to the east; and (2) piles of Tertiary volcanics overlying a thin Mesozoic succession (Anderson 1978, fig. 2).

The geology of this small shield fragment has been studied for more than three-quarters of a century, starting with the pioneer investigations of members of the Geological Survey of Great Britain. They provided the first well-documented accounts of basement structures that were later found to be characteristic of many polycyclic gneiss terrains of the world (Craig 1991).

### Lewisian Complex

The Lewisian Complex represents a tectonic province which was finally stabilized in mid-Proterozoic times, but which contains partly modified Archean remnants. The main Lewisian components comprise an older Scourian gneiss complex, formed during a late Archean (2.9–2.6 Ga) cycle and involving Badcallian granulite-facies metamorphism at 2.7 Ga; this was later intruded by mafic–picritic dykes (2.4 Ga) (Scourier swarm) and modified during the Inverian (2.6–2.4 Ga), Laxfordian (1.9–1.7 Ga), and, in minor degree, post-Laxfordian (–Grenvillian) (1.1–1.0 Ga) orogenic episodes. Of these later events, the Laxfordian resulted in widespread crustal modification. Lewisian chronology is conveniently summarized in 10 steps (Table 2-5).

Scourian rocks are also preserved in a number of relict massifs enveloped in regenerated rocks which display varying Laxfordian structural and metamorphic patterns.

## 2.6 NORTH AMERICAN CRATON (LESS GREENLAND SHIELD)

Unreworked Archean crust is present within the Canadian Shield in Superior and Slave provinces, Eastern Nain Subprovince (Nutak segments), and locally in north–central Churchill Province; it occurs peripheral to the shield in the Minnesota River Valley Inlier and adjoining parts of Wisconsin and Minnesota, and in the Wyoming Uplift of western USA. Reworked Archean crust is widespread in adjoining Proterozoic belts, as considered in subsequent chapters (see Fig. 3-2).

### 2.6.1 SUPERIOR PROVINCE

#### General relations

This province represents an exceptionally large and broadly studied Archean craton (Fig. 2-5) (Card 1990, Card and King 1992). With an area of $1.57 \times 10^6$ km$^2$, it accounts for almost one quarter of Earth's exposed Archean crust. The broad lithologic range present in the province is conveniently divided into five field-based groups, as listed in Table 2-6. Calculated average proportions (Table 2.6 column 3) are, in round terms, granitoid gneiss:massive plutons:metasupracrustals = 6:3:1; combining paragneiss and metasupracrustals, this translates to igneous:metasupracrustal components = 4:1 (Goodwin 1978).

Proportions of metamorphic facies across the province (Table 2-7) are, in round terms, amphibolite:granulite:greenschist = 6:2:1. With the exception of the small Pikwitonei Subprovince at the northwestern border of Superior Province and the transecting Kapuskasing Structural Zone in the centre, granulites are effectively restricted to the large Ungava Craton of Labrador–New Quebec in the northeast with rare local exceptions (Pan et al 1994). Greenschist facies characterizes the main greenstone belts in the southern and western parts of the province. Amphibolite facies predominates elsewhere (Ayres 1978).

The most significant lithologic contrasts by region are between the medium–high grade gneisses in Ungava Craton (Domain) in the northeast (Table

*Table 2-5.* Outline of Lewisian chronology.

| Event | Age (Ma) |
|---|---|
| (1) Formation of early Scourian metasediments and mafic-ultramafic rocks and incorporation into developing tonalitic plutonic complex with associated strong horizontal thrusting and deformation | 2900 |
| (2) Main Badcallian high grade granulite-facies metamorphism affecting deeper part of Lewisian crust | 2700 |
| (3) Initiation of Inverian shear zones associated with uplift and segmentation of Archean blocks | 2600? |
| (4) Late Badcallian (post-Badcallian) biotite pegmatites | 2500 |
| (5) Emplacement of Scourie dolerites and norites, and later (?2200 Ma) olivine gabbros and picrites. Continuing retrogression of the granulites and intermittent movement on shear zones | 2400 |
| (6) Crustal extension with extrusion of voluminous lavas of the Loch Maree Group and associated sedimentation. Emplacement of late Scourie dikes? Formation of South Harris igneous complex? | 2000 |
| (7) Early Laxfordian deformation and high grade metamorphism | 1900? |
| (8) Early Laxfordian migmatization and emplacement of granites and muscovite pegmatites | 1800 |
| (9) Late Laxfordian deformation and retrogressive metamorphism | 1600–1400 |
| (10) Late or post-Laxfordian brittle folds and crush belts | 1400–1000? |

From Park and Tarney (1987), Table 1

# NORTH AMERICAN CRATON (LESS GREENLAND SHIELD)

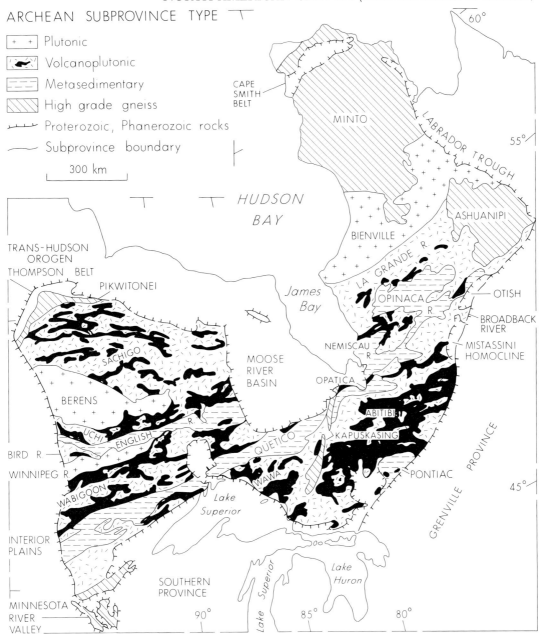

**Fig. 2-5.** Map of Superior Province showing volcanic-plutonic, metasedimentary-gneiss and plutonic subprovinces or superbelts and other subdivisions. (From Card, 1990. Published with permission of the author).

2-6, column 1) and the medium–low grade, alternating (1) granitoid–greenstone; and (2) metasedimentary–gneiss subprovinces in the southern–western parts (Western Superior Province) (Table 2-6, column 2; also Fig. 2-6, inset). These sharp lithologic–metamorphic contrasts by region across Superior Province reflect significantly greater tectonic uplift and erosion of the northeastern parts to expose deeper (catazonal), higher grade crustal levels therein (Goodwin 1978).

Summarily stated, Superior Province rocks range in established common age from ~3.1 Ga to ~2.6 Ga, with a single zircon date at 3.3 Ga. Major magmatic episodes at ~3.0–2.7 Ga, resulted in the accumulation, notably in central–western Superior Province, of (1) assorted mafic–felsic volcanic rocks, volcaniclastics and chemical sediments to form the existing greenstone-rich subprovinces, together with (2) the wide mainly turbidite-rich, now micaceous, intervening and alternating

68 ARCHEAN CRUST

*Table 2-6.* Lithologic proportions of Archean crust in Superior Province, Canadian Shield.

| Lithologic unit | | (1) Ungava Domain (%) | (2) Western Superior Province[1] (%) | (3) Average Superior Crust (%) |
|---|---|---|---|---|
| (1) | Banded gneiss, granitic gneiss, migmatite | 66 | 36 | 50 |
| (2) | Massive to slightly foliated granitoids | 24 | 38 | 31 |
| | Tonalite-granodiorite | 24 | 32 | 28 |
| | Granite-leucocratic | 0.4 | 6 | 3 |
| (3) | Paragneiss, veined gneiss, migmatite | 9 | 9 | 9 |
| (4) | Mafic to ultramafic intrusions | 0.1 | 1 | 1 |
| | Metasupracrustal rocks | 1 | 16 | 9 |
| | Volcanic rocks | — | 11 | 6 |
| (5) | Sediments | — | 5 | 3 |
| Size of area (km$^2$) | | 498 000 | ~231 000 | 1 572 000 |

[1]Weighted average of (a) Red Lake-Landsdowne, (b) Geotraverse and (c) Berens-Sachigo areas
Adapted from Goodwin (1978)

*Table 2-7.* Average proportions of metamorphic facies in Superior Province, Canadian Shield.

| Metamorphic facies | Percentage of total area (1 572 000 km$^2$) |
|---|---|
| Granulite | 22 |
| Amphibolite | 66 |
| Greenschist | 11 |
| Subgreenschist | 1 |
| | 100 |

Adapted from Goodwin (1985)

metasedimentary–gneiss subprovinces. Early magmatic episodes were associated with regional deformation and metamorphism. However, the effects of these early events were largely obliterated by the pervasive polyphase deformation, regional metamorphism and widespread plutonism of the last major orogenic event, the Kenoran Orogeny, at 2.73–2.65 Ga. This polyphase tectonism involved early ductile deformation under north to south, subhorizontal, regional compression and low pressure metamorphism, succeeded by increasingly brittle deformation, still under north to south compression, that culminated in transcurrent faulting, shearing and low grade metamorphism, with associated deposition of late alluvial–fluvial sediments and shoshonitic volcanic rocks (Timiskaming association) at ~2.71–2.70 Ga. This was followed by emplacement of post-kinematic granitic–syenitic intrusions ~2.70–2.67 Ga ago (Card 1990).

## Ungava Craton

Our knowledge of this large region (498 000 km$^2$), extending from Hudson Bay on the west to the Labrador Trough on the east and from Cape Smith Belt on the north to East Main River (James Bay) vicinity on the south (Fig. 2-5), is based on reconnaissance studies (Eade 1966, Stevenson 1968) supplemented by recent local studies (Skulski et al 1984, Avramtchev 1985, Percival and Girard 1988, Percival et al 1992, 1994, Stern et al 1994). Bedrock geology is dominated by granitoid rocks of at least five plutonic suites, both foliated and massive, with minor but widespread supracrustal remnants. Common ages range from 3.1 Ga to 2.7 Ga. Granulite facies terrains are widespread. Included are two large, granulite-rich subprovinces—Minto and Ashuanipi—and the intervening and adjoining plutonic Bienville Subprovince (Card 1990).

Calculated lithologic proportions in Ungava Craton (Table 2-6, column 1) are, in round terms, gneiss: massive plutons = 3:1. With allowance made for the proportion of gneisses of sedimentary origin (i.e. paragneiss), this translates to igneous:metasup-

Fig. 2-6. Distribution of major lithologies in Slave Province, Canadian Shield. The five major Archean supracrustal basins of accumulation are named in heavy print. The main supracrustal belts are named in light print, either once or repeated by segments as appropriate. (After Padgham 1985, Fig. 1). Inset map shows gross distribution of Archean metasupracrustal (greenstone) belts in Superior and Slave provicnes. (After Baragar and McGlynn 1976, Fig. 1). Both published with permission of the Geological Association of Canada.

# NORTH AMERICAN CRATON (LESS GREENLAND SHIELD)

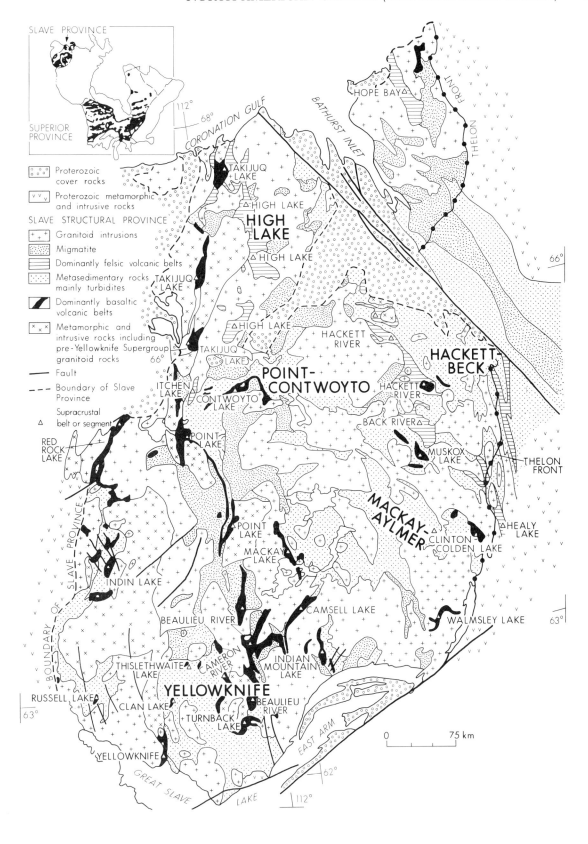

racrustal components = 9:1 thereby expressing dramatically deeper (catazonal) crustal levels in Ungava Craton.

The timing of the Kenoran Orogeny in the Ungava Craton is provided by recent studies in the Bienville and Ashuanipi domains which establish major deformation–metamorphism–plutonism at ~2730–2670 Ma, followed by plutonism to ~2650 Ma (Card 1990). U–Pb zircon and monazite geochronology indicates that the main anatectic intrusion and metamorphic event occurred at ~2667 Ma. Ashuanipi paragneiss provide ion-probe zircon ages of 3.35 and 2.6 Ga, tentatively interpreted as representing the presence of admixed detrital and metamorphic components respectively (Percival and Girard 1988).

## Southern–Western Superior Province

The principal granitoid–greenstone-bearing domain which lies in southern–western Superior Province is 1700 km (E–W) by 450–1300 km (N–S), or 926 000 km$^2$ in area (Fig. 2-5). It is subdivided across strike into a number of easterly trending subprovinces or superbelts, which are alternately granitoid-greenstone-rich (Abitibi–Wawa, Wagioon, Uchi and Sachigo) and metasedimentary–gneiss–pluton rich (Pontiac, Quetico–Opatica, English River and associates) with those in the west, partially enclosing the Berens plutonic subprovince. The NNE-trending early Proterozoic Kapuskasing Fault Zone, an obliquely eroded thrust-ramp that exposes progressively eastward down to uppermost lower crust, transects the east–central part of this large granitoid–greenstone-bearing domain. The granulite-rich Pikwitonei Subprovince forms the northwest boundary of Superior Province in contact with the post-Archean Thompson Belt of the Trans-Hudson Orogen (Churchill Province). Restricted counterparts of Superior Province occur as inliers to the south in Michigan and Minnesota, USA (Wilkin and Bornhorst 1992, fig. 1).

The boundaries between subprovinces (superbelts) commonly represent zones of structural and metamorphic transition of appreciable width, in which faulting and igneous activity have all but totally masked any primary lithologic transition.

Calculated lithologic proportions in a large (~231 000 km$^2$) representative part in western Ontario (Table 2-6, column 2) are, in round terms, gneiss:massive plutons: metavolcanics: metasupracrustal = 9:7:2; this translates into igneous:metasupracrustal components = 3:1; thereby expressing, in sharp contrast to Ungava Craton (9:1), higher metasupracrustal content due to shallower prevailing crustal levels (mesozonal–epizonal) at the present erosion surface in southern–western Superior Province.

### Granitoid–greenstone subprovinces

The greenstone-bearing subprovinces (or superbelts) contain at least 38 main greenstone belts, ranging in size from small schist units up to the unusually large volcanic-rich Abitibi Belt in the east, which measures 650 km × 225 km or 95 000 km$^2$ in area (Goodwin and Ridler 1970). The common cuspate-shaped greenstone belts reflect the presence of numerous intervening and adjoining round to ovoid, syn- to post-volcanic granitoid batholithic complexes. Deep seismic reflection profiling has revealed considerable information about the greenstone belts at depth (Jackson et al 1990, Ludden 1994).

The greenstone sequences comprise mainly mafic (–ultramafic) to felsic flows and pyroclastic rocks of mixed tholeiitic–komatiitic–calc–alkalic compositions, with intercalated and commonly overlying metasedimentary assemblages characterized by greywacke, mudstone, tuff, conglomerate and BIF (Goodwin 1977a, b, 1982a, Dimroth et al 1978, 1982, Jensen 1985, Sage 1990, Thurston and Chivers 1990, Barrett et al 1992, Chown et al 1992, Feng et al 1992, Turek et al 1992). Substantially older basement to the greenstones has not been positively identified in Superior Province. The greenstone belts feature common low grade greenschist to subgreenschist facies interiors and medium grade amphibolite facies margins in proximity to younger plutons. The syn- to post-volcanic granitoid intrusions include older, partly gneissic, plutons of TTG (Na-suite) composition and younger massive plutons, also mainly TTG in composition but including late K-granites (K-suite) (Table 2-6, column 2).

Archean sedimentation was dominated by the resedimented (turbidite) facies association of greywacke, mudstone–siltstone and conglomerate (Ojakangas 1985, Mueller and Donaldson 1992, Mortensen et al 1993, Mueller et al 1994). Deposition was commonly in submarine fans. An alluvial–fluvial facies association of cross-bedded sandstones and associated conglomerates, some very thick (1000 m or more) and traceable 50 km or more along strike, is widespread but markedly subordinate to the common turbidite association. An important pelagic–chemical association contains common BIF, including oxide-, carbonate- and sulphide-facies. Shelf-type lithologies locally stromatolitic (Hofmann and Masson 1994) are rare, indicating a paucity of broadshelf environments.

Across the granitoid–greenstone domains, composite granitoid batholithic complexes are, with rare exceptions, areally more abundant than the greenstone belts themselves (Table 2-6, column 2). The round to ovoid intravolcanic plutonic complexes, 50–100 km in diameter, typically comprise central gneissic batholiths, smaller gneissic domes and marginal crescentic granitoid plutons (Card 1982, Blackburn et al 1985). Some complexes contain up to 24 cross-cutting granitoid phases. The batholiths are mainly syn- to late-tectonic (Ayres and Thurston 1985), emplaced especially in the peiod 2750–2670 Ma, generally concurrent with, though slightly younger than, the main period of late Archean volcanism.

### Metasedimentary–gneiss–pluton subprovinces

The main subparallel subprovinces and associates of this type in southern–western Superior Province (Fig. 2-5) vary considerably in lithologic proportions (Beakhouse 1985). Berens Subprovince in the northwest is a predominantly batholithic domain; English River Belt and associates in the middle contain dominant paragneiss with subordinate yet locally significant (especially in the west) batholiths. Quetico (–Opatica–Nemiscau R.–Opinaca R.) Belt and Pontiac Belt, to the south and east, contain predominant paragneiss with substantial preserved metasedimentary schist. Amphibolite facies metamorphism prevails, with very rare granulite facies (Beakhouse 1985).

### Geochronology

Recent, high precision U–Pb zircon geochronology has provided a reliable general time framework for Archean volcanism, plutonism, and deformation in southern–western Superior Province (T. Krogh, F. Corfu and D.W. Davis in Ayres and Thurston 1985, Card, 1990, tables 1 and 2, fig. 4, Corfu 1993, Corfu and Stott 1993, Mortensen 1993a, b, Mortensen and Card 1993, Jackson et al 1994). The total dated time span is ~400 Ma (3040–2650 Ma), encompassing at least three episodes of volcanic activity: a particularly widespread episode at 2.76–2.70 Ga and two older apparently more local episodes at 3.0–2.9 and 2.85–2.80 Ga respectively. There was continuing igneous activity, culminating in the intrusion of late-tectonic plutons at 2710–2670 Ma, during the Kenoran Orogeny, involving province-wide an episode of deformation and metamorphism that continued with diminishing intensity to 2650 Ma. Thus the respective deeper level (catazonal) and shallower level (mesozol–epizonal) facies exposed across the entire province are essentially coeval, thereby expressing a province-wide constructional coherence, the net result being a > 1300 km-wide accretion of juvenile continental crust during ~500 Ma (~3.1–2.6 Ga).

### Petrogenesis and evolution

The lack of evidence for extensive older (pre ~3.0 Ga) sialic basement to the metavolcanic sequences, combined with the Na-rich character of the concurrent igneous activity, has led Davis et al (1987) and Card (1990), amongst others, to infer that the metavolcanic belts formed in an oceanic environment including active arcs. Age dates from supracrustal sequences marginal to the metavolcanic terrains are considered by them to indicate possible allochthonous–collisional relationships, and point to some form of plate interaction, such as small-scale arc-continent collision. However, others (e.g. Hamilton 1993) advocate a more primite voluminous magmatic–tectonic process (see further below).

Many fundamental petrogenetic problems of Archean volcanism remain to be resolved, including the relative roles of partial melting, crystallization differentiation, magma mixing, and sialic crustal assimilation and anatexis in derivation of the diverse mafic to felsic volcanic sequences in Superior Province (Ayres and Thurston 1985, Lafleche et al 1992, Barrie et al 1993, Sutcliffe et al 1993).

## Mineral deposits

Archean rocks in Superior Province contain exceptional mineral wealth. Numerous volcanogenic massive sulphide (Zn, Cu, Pb, Au) deposits and banded iron formations (Fe) are present in the volcano-sedimentary (greenstone) associations. Important lode Au–Ag deposits are found in quartz–carbonate veins, banded iron formations, schist-hosted sulphide deposits and Cu–Au sulphide veins. Ni, Cu, Cr, V, Pt, Pd deposits occur in layered mafic/ultramafic bodies. Occasional Cu, Mo, Au deposits are associated with porphyry granitoid masses. Rare-metal (Li, Cs, Be, Ta, etc.) pegmatite deposits are concentrated in certain regions. Most mineral deposits formed in the late Archean at ~2.7 Ga, corresponding to a period of construction and accretion of the granitoid–greenstone host rocks (Robert et al 1991, Poulsen et al 1992, Spooner and Barrie 1993).

## 2.6.2 SLAVE PROVINCE

The predominantly late Archean (3.1–2.6 Ga) Slave Province forms an irregular, N-trending ellipse about 800 km × 400 km in dimension or 225 000 km$^2$ in area (Fig. 2-6). Slave Province is bounded to the west by early Proterozoic rocks of Bear Province (Wopmay Orogen), to the southwest by Phanerozoic cover of the Interior Plains, to the south and southeast across the Great Slave Lake Shear Zone, by the East Arm Fold Belt of Bear (formerly Churchill) Province, to the east by metamorphic rocks of the Thelon Front, which marks the western limit of Churchill Province, and to the north and northeast Proterozoic strata of the Bathurst Plate Extension (Bear Province) and by the Phanerozoic cover of the Arctic Platform (Henderson 1981a, Easton 1985, Padgham 1985, Bevier and Gebert 1991, van Breemen et al 1992, Card and King 1992, Padgham and Fyson 1992, Isachsen and Bowring 1994).

### General geology

Slave Province is about equally underlain by gneissic to massive granitoid rocks including gneissic basement (3.1 Ga and 3.96 Ga) and younger granitic plutons (2.7–2.6 Ga) and by interspersed turbidite–volcanic (greenstone) assemblages of the Yellowknife Supergroup (2.72–2.66 Ga). This supergroup is composed of 80% metasediments and 20% metavolcanic rocks (greenstone belts) (Table 2-8, column 3). Most deformation and metamorphism occurred at 2.62–2.60 Ga, the latest post-tectonic granites at 2.59–2.56 Ga (Isachsen et al 1991, Van Breemen et al 1992).

Structural trends across the province are typically northerly in the western half of the province and northwesterly in the eastern half. Axial fold planes are vertical to subvertical, commonly with moderate plunges. Shear zones rich in chlorite–carbonate–sericite assemblages are widespread and faults of several ages and orientations are typically marked by mylonite and breccia (Fyson and Helmstaedt 1988).

The bulk of the metasupracrustal rocks are in the cordierite–amphibolite facies of metamorphism. The main areas of greenschist facies rocks lie in the larger metasedimentary belts, particularly where associated with pre-Yellowknife basement.

### Granitoid basement

Pre-Yellowknife basement gneisses, with ages to 3.15 Ga and even 3.96 Ga, have been documented locally, and several large areas of basement gneiss have been provisionally identified in the western part of the province (west of 110–112° lines of longitude) (Isachsen et al 1991). Basement gneiss at Point Lake has been dated at 3155 Ma (Krogh and Gibbons 1978, Nikic et al 1980). Acasta gneisses, exposed in the foreland and metamorphic zone of the Wopmay Orogen, ranging from massive to foliated granite to complexly interlayered tonalitic to granitic gneisses, provide SHRIMP zircon analyses of 3962 Ma, making them the oldest known intact terrestrial rocks (Bowring et al 1989b, Isachsen et al 1991). The apparent restriction of this old Slave Province crust to the western part is supported by Nd and Pb isotopic analyses (Thorpe et al 1992). The full extent of this old crust has yet to be established, but may prove to be quite sporadic.

Other than older basement, involving at least three distinct igneous suites (Davis et al 1994), the majority of plutonic rocks in Slave Province range from 2.58 to 2.70 Ga (van Breemen et al 1992).

### Greenstone belts

Slave Province contains at least 26 comparatively small volcanic-rich belts (Padgham 1985). Those in the western and southern parts of the province are composed of dominant mafic metavolcanic rocks with some felsic metavolcanic and metasedimentary associates (Yellowknife-type), while those in the northern and eastern parts are composed of dominant felsic metavolcanic rocks with assorted associates (Hackett River-type). The volcanic-rich belts characteristically lie along the margins of large, regional, greywacke–mudstone, turbidite-filled basins (domains). The majority of volcanic rocks were erupted in the interval 2.72–2.66 Ga with older components to 2.82 Ga and younger components to 2.60 Ga.

*Yellowknife-type* volcanic successions, each about 9000 m thick, are characterized by essentially bimodal (mafic–felsic) volcanic cycles. The stratigraphically thinner, *Hackett-River-type* volcanic belts contain 30–60%, or even higher, felsic vol-

Table 2-8. Generalized sequence of main Precambrian events in the North American Craton (excluding Greenland) by region.

| Time scale (Ga) | (1) Southern (Lake Superior) | (2) North-Central (Trans-Hudson) | (3) North-Western (Slave-Wopmay) | (4) Eastern (Nain-Grenville) | (5) South-Western (Wyoming-Central-Western USA) |
|---|---|---|---|---|---|
| 0.6 | | Ellesmere Group, ~0.57 Ga | | Avalonian Orogeny, 0.6 Ga | Windermere Sgp, 0.8–0.54 Ga |
| | Bayfield (Jacobsville) sediments, 1.0–0.6 Ga | Franklinian dykes-sills, 0.7 Ga | Windermere (Rapitan)—Ekwi—U. Tindir Sgps, 0.8–0.7 Ga. | Chilhowee Gp, 0.6–0.5 Ga | Bingham Sgp, 0.8–0.57 Ga |
| | | Kennedy Channel-Ella Bay Gps, 1.0–0.7 Ga | Natkusiak lavas, 0.7 Ga | Carolina Slate Belt, 0.6–0.5 Ga | Uinta Mountain Gp, 0.9–0.8 Ga |
| | | | Mackenzie Mts (L. Tindir) Sgps, 1.0–0.8 Ga | Ocoee Sgp, 0.8–0.6 Ga Crossnore Plutonic-Volcanic Complex, 0.9–0.8 Ga | |
| | | | | Harbour Main, Conception Gps George River, Green Head Gps | Pahrump Gp, 1.2–0.6 Ga |
| 1.0 | Grenvillian Orogeny, 1.1–1.0 Ga | Bylot Sgp, Thule Gp, 1.2–0.7 Ga | Shaler, Rae Gps, 1.2–0.7 Ga | Grenvillian Orogeny, 1.1–1.0 Ga | Grand Canyon Sgp, 1.25–0.8 Ga |
| | Oronto sediments, 1.1 Ga Keweenawan lavas, 1.11–1.10 Ga Duluth Complex, 1.1 Ga | Arctic Platform rifting, 1.2–1.0 Ga | Coppermine lavas, (Gp), 1.28 Ga Muskox Intrusion, 1.28 Ga | AMCG suite, mainly 1.5 Ga Grenville Sgp, 1.3–1.1 Ga Adirondack Massif, 1.2–1.0 Ga | |
| 1.4 | Midcontinent Rift system, 1.5–1.0 Ga | MacKenzie dikes, 1.28 Ga | Racklan (East Kootenay) Orogeny, 1.3–1.2 Ga Purcell-Wernecke Sgps, 1.5–1.25 Ga | Seal Lake Gp, alkaline magmatism, 1.2 Ga Elsonian Orogeny, 1.4 Ga | Belt Sgp, 1.5–1.2 Ga Apache-Troy Gps, 1.4–1.1 Ga |
| | Sibley arenites, 1.4–1.1 Ga Sioux-Baraboo quartzite, 1.76–1.63 Ga Killarney granites, 1.7–1.5 Ga Central Plains Orogeny, 1.8–1.63 Ga | Athabasca Group, 1.8–1.7 Ga Martin redbeds, 1.8–1.7 Ga Dubawnt-Thelon sequences, 1.85–1.72 Ga | Hornby Bay-Dismal Lakes Gps, 1.75–1.4 Ga | Anorogenic magmatism: 1.5–1.4 Ga Labradorian Orogeny, 1.7–1.6 Ga Bruce River Gp, 1.65 Ga | Anorogenic magmatism, mainly 1.5–1.4 Ga Uncompahgre Fm, 1.75 Ga Mazatzal Orogeny, 1.7–1.6 Ga Yavapai Orogeny, 1.74–1.68 Ga |
| | Whitewater Gp, 1.84 Ga Sudbury Irruptive, 1.84 Ga | | | Penrhyn, St. Mary Gps, 1.8–1.7 Ga | Central Belt—bimodal volcanics, turbidites, 1.8–1.6 Ga Harney Peak and Sierra Madre Granites, 1.7 Ga |
| | | | | Torngat orogen, 1.9–1.7 Ga | Central Plains Orogeny, 1.8–1.7 Ga |

Continued

canic units with correspondingly low mafic volcanic content (Lambert 1976, 1978).

### Clastic metasediments

Clastic metasediments, which form about 80% of the Yellowknife Supergroup, are dominated by uniform greywacke–mudstone sequences. Rare shallow-water siliciclastic sediments, carbonates and conglomerates are found in close association with the volcanic rocks.

The dominant greywacke–mudstone facies occupies the central parts of five major areas underlain by Yellowknife supracrustal rocks, each of which represents a basin of accumulation. Stratigraphic thickness estimates range from 3000 to 4500 m. The high proportion of felsic and intermediate vol-

*Table 2-8.* Continued.

| Time scale (Ga) | (1) Southern (Lake Superior) | (2) North-Central (Trans-Hudson) | (3) North-Western (Slave-Wopmay) | (4) Eastern (Nain-Grenville) | (5) South-Western (Wyoming-Central-Western USA) |
|---|---|---|---|---|---|
| 1.8 | Penokean Orogeny, 1.9–1.8 Ga<br>Marquette Range Sgp, Mille Lacs-Animikie Gps, 2.45–1.81 Ga<br>Wisconsin Magmatic Terranes 1.9–1.8 Ga<br>Kapuskasing Zone uplift, 1.9 Ga<br>Nipissing Diabase, 2.16 Ga<br>Creighton-Murray granites, 2.33 Ga<br>Huronian Sgp, 2.47–2.33 Ga | Trans-Hudson Orogeny, 1.9–1.8 Ga<br>Wathaman batholith, 1.85 Ga<br>Amisk-Lynn Lake volcanics, 1.9–1.85 Ga<br>Belcher Gp, 2.0–1.8 Ga<br><br>Hurwitz Gp, 2.45-2.11 Ga | Wopmay Orogeny, 1.9–1.8 Ga<br>Wopmay sequences (Coronation Sgp) 1.91–1.84 Ga<br><br>Nemo Gp, 2.4–2.1 Ga<br>Montgomery Lake Gp, 2.4 Ga<br>Union Island Gp, 2.4 Ga | Hudsonian Orogeny, 1.9–1.8 Ga<br>Cape Smith ophiolite, 1.86 Ga<br>Labrador Trough sequences, 2.2–1.9 Ga | Penokean Orogeny, 1.9–1.8 Ga<br><br>Black Hills sequence, 2.0–1.9 Ga<br>Mojave gneiss, 2.0–1.8 Ga<br>Snowy Pass Sgp, 2.5–1.8 Ga |
| 2.5 | | | **Kenoran Orogeny, ~2.65 Ga** | | |
| | MRV metamorphism, 2.6 Ga<br>Greenstones, granitoids, 2.8–2.6 Ga | Kaminak greenstone, 2.65 Ga<br>Pikwitonei granulites, 2.7–2.6 Ga<br>Prince Albert, etc. greenstones, 2.88 Ga<br>Melville Peninsula-Baker Lake gneiss, 3.0–2.6 Ga<br>Taltson-Queen Maud gneiss, 3.5–1.85 Ga | Yellowknife Sgp mainly 2.72–2.66 Ga<br>Slave Province granitoids, 2.7–2.6 Ga | Ungava granitoids, 3.3–2.6 Ga<br>Hopedalian event, >3.0 Ga | Stillwater Complex, 2.7 Ga<br>Wyoming greenstones, granitoids, 2.8–2.6 Ga<br>Metamorphism, granitoids, 3.0–2.8 Ga<br>Beartooth Orogeny, 2.9–2.7 Ga |
| 3.0 | MRV metamorphism, 3.1 Ga | Sugluk gneiss (Cape Smith), 3.2–3.0 Ga | Slave basement gneiss, 3.15 Ga | Major metamorphism, 3.1 Ga<br>Anorthosite-gabbros<br>Upernavik supracrustals, 3.2-2.6 Ga<br>Saglek dikes, Nulliak assemblage<br>Uivak gneiss: I, 3.8–3.6 Ga, II, 3.4 Ga | Cherry Creek metamorphites, 3.0–2.8 Ga<br>Beartooth gneiss, 3.56–3.25 Ga |
| 3.5 | Morton (-Michigan), TTG gneiss, 3.4 Ga<br><br>MRV etc. gneiss protoliths, 3.8–3.5 Ga | | Acasta gneiss 3.96 Ga | Uivak protoliths, 3.9 Ga | Wind River gneiss, 3.5–3.3 Ga<br><br>Wind River protoliths, 3.96–3.4 Ga |

MRV = Minnesota River Valley; Gp = Group; Sgp = Supergroup, FM = Formation; AMCG = anorthosite-mangerite-charnockite-rapakivi granite.

canic rock fragments and the abundance of quartz and feldspar indicate a mixed felsic volcanic and granitic provenance.

Most sediments were deposited concurrent with volcanism in the interval 2.72–2.66 Ga. Some younger sedimentary rocks yield ages of 2616 to 2590 Ma (Isachsen and Bowring 1994). Pyrrhotite-rich iron formation, locally auriferous, is locally abundantly associated with turbidites in a 200–300 km wide, ENE-trending zone crossing the heart of the province, including Contwoyto Lake (Padgham 1992).

## Mineral deposits

Volcanogenic massive sulphide deposits of varying sizes are present in nine of the greenstone belts. Of these, the High Lake deposit, the largest known, contains $5.2 \times 10^6$t of ore grading 3.5% Cu, 2.5% Zn, 16.7 gms/tonne of Ag, and minor Pb.

Other notable metallogenic components in Slave Province include gold in both quartz–carbonate–sericite–carbonate shear zones and quartz veins cutting mainly mafic volcanic rocks. Also, gold production recently commenced (Lupin mine) in amphibolite facies silicate–sulphide-facies iron formation in turbidites of the Point Lake–Contwoyto Lake Supracrustal Basin. Numerous pegmatites contain Li, Ba, Sn, W and rare metal (Li, Cs, Be, Ta, Sn, U, Th) deposits. Diamond occurrences of intriguing commercial interest are under active exploration.

## 2.6.3 NAIN PROVINCE

Archean rocks in Eastern Nain Subprovince, form a thin discontinuous sliver of deformed gneiss along the east coast of Labrador between 54°N and 59°N (Fig. 2-7). An original part of the pre-rift North Atlantic Craton, these rocks represent the faulted extension of southern West Greenland gneisses (Table 2-8, column 4). They lie in tectonic contact with the early Proterozoic Mahkovic subprovince to the southeast, the mid-Proterozoic Grenville Province (Labradorian high-grade terranes) to the south, and the Churchill Province to the west.

The Archean gneiss complex consists principally of quartzofeldspathic rocks of granodioritic composition. These are interlayered with subordinate metasedimentary and meta-igneous units intruded by mafic igneous rocks. These rocks, which were deformed and metamorphosed during the Kenoran Orogeny (2.6–2.5 Ga), form a steeply inclined, N-trending linear zone, cut by granitic veins and by younger diabase dykes. The western margin comprises an important cataclastic zone characterized by faults, thrusts, regional mylonites, blastomylonite and pseudotachylite (Morgan 1975, Knight and Morgan 1981).

In the northern Saglek Bay–Hebron Fjord area (Bridgwater et al 1975, Collerson et al 1976, Schiøtte et al 1990), where most of the gneiss complex is of amphibolite facies, the history of development is closely similar to that established in the Godthåb area of West Greenland (see above) (Schiøtte et al 1989a, b), with dated units ranging from 3.9 Ga to 1.8 Ga (Nutman et al 1989).

Recent SHRIMP dating of detrital zircons and overgrowths in three Upernavik units provide variable ages of 3235 to 2560 Ma (Schiøtte et al 1992). The results show that the Upernavik association is composite, and sediments in different units have widely different sources and metamorphic histories. This supports a tectonic collage for Nain Province according to which separate terranes were tectonically juxtaposed in the late Archean (Schiøtte et al 1993).

## 2.6.4 CHURCHILL PROVINCE

The main zones of unreworked Archean crust in Churchill Province are those of the Prince Albert–Woodburn/Ketyet groups of the Committee Bay and Armit Lake blocks, and the Kaminak Group of the Ennadai Block (Lewry et al 1985, Cavell et al 1992) (Figs 2-8, 2-9; Table 2-8, column 2). Other still larger areas of Archean crust, notably in Taltson and Queen Maud blocks to the west (Schau 1975, 1978, Ashton 1982, Frisch 1982, Cavell et al 1992), were extensively reworked in Hudsonian (early Proterozoic) time and so are considered in Chapter 3.

The dominantly metasedimentary Prince Albert, Woodburn and Ketyet groups are mainly composed of aluminous and orthoquartzites, calcsilicates, BIF, varied phyllitic pelites, chloritic schists, greywacke and arkosic arenites. Subordinate felsic to ultramafic metavolcanic rocks, the latter including spinifex-textured komatiites, occur locally. In the Melville Peninsula area to the northeast, volcanic rocks are more common, including felsic and mafic lava flows and pyroclastic units associated with greywacke and arkose. Metamorphic grade varies from greenschist facies in the southwest to upper amphibolite facies in the northeast.

Metamorphic grades in the Kaminak belt range from greenschist in the central parts to lower amphibolite facies at the margins. Kaminak felsic volcanic units are dated in the range 2700–2550 Ma and generally coeval granitoids at ~2700 Ma (Wanless and Eade 1975, Lewry et al 1985, Cavell et al 1992). The Kaminak Lake alkaline complex, with an intrusive age of 2659 Ma, is one of the oldest in the world (Cavell et al 1992).

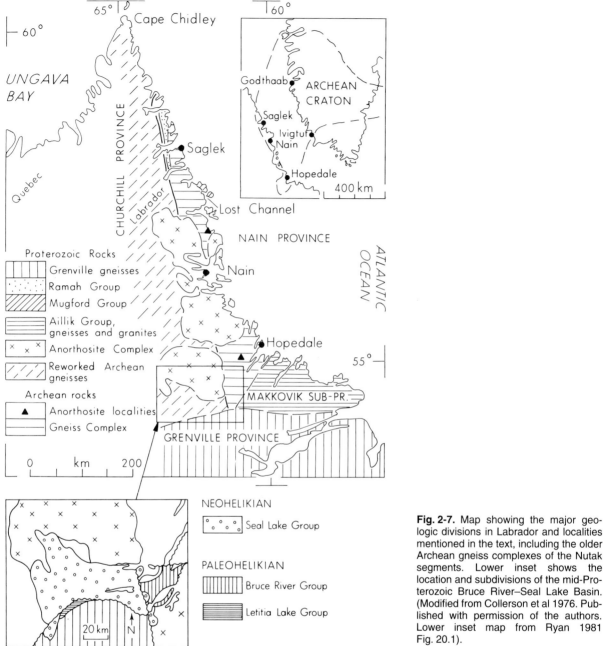

**Fig. 2-7.** Map showing the major geologic divisions in Labrador and localities mentioned in the text, including the older Archean gneiss complexes of the Nutak segments. Lower inset shows the location and subdivisions of the mid-Proterozoic Bruce River–Seal Lake Basin. (Modified from Collerson et al 1976. Published with permission of the authors. Lower inset map from Ryan 1981 Fig. 20.1).

### 2.6.5 MINNESOTA RIVER VALLEY INLIER

Precambrian rocks are exposed intermittently for about 150 km in the SE-trending valley of the Minnesota River (Fig. 2-10 Table 2-8, column 1). The Morton Gneiss, named for the outcrops in the vicinity of Morton, Minnesota, are typical. Similar mid–late Archean gneissic rocks, isotopically modified in the Penokean Orogeny (1.9–1.8 Ga), are exposed to the north and northeast in Michigan, Wisconsin and Minnesota (Sims 1980a, Sims and Peterman 1983, Sims et al 1993).

The Morton gneiss (Goldich et al 1980a, b), a hybrid rock characterized by highly contorted structure and varied textures and colours, comprises older tonalite–granodiorite gneisses, with amphibolite inclusions, and younger pink quartz monzonite–leucrogranite gneiss.

Discordant U–Pb ages on zircon from Morton

# NORTH AMERICAN CRATON (LESS GREENLAND SHIELD)

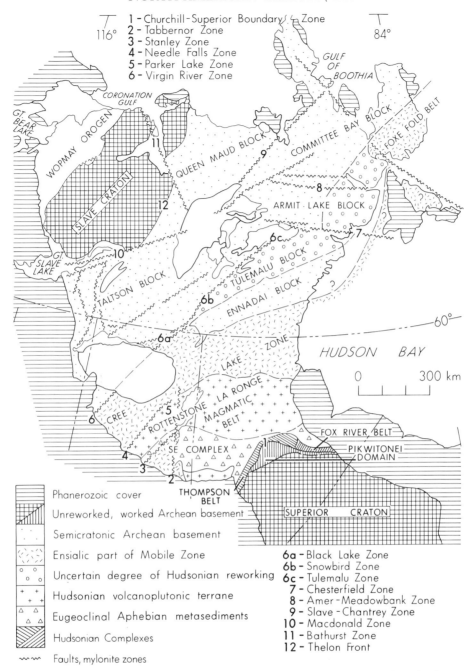

**Fig. 2-8.** Major crustal components of Western Churchill Province and adjacent terrains, showing general location of crustal blocks and other features discussed in the text. (After Lewry et al 1985, Fig. 4. Reproduced with permission of the Geological Association of Canada). See Fig. 3-2 for subsequent modifications.

gneiss give a minimum age of 3300 Ma with indications of an older protolith age (3500 Ma) (Goldich and Wooden 1980). Following early pervasive folding in the metamorphic complex, high grade metamorphism occurred at both 3050 Ma and 2600 Ma, and a thermal event accompanied by local igneous emplacement at 1800 Ma.

## 2.6.6 ARCHEAN GNEISS TERRAINS IN WISCONSIN AND MICHIGAN

Similar Archean gneisses are exposed 450 km east of the Minnesota River Valley in central and northern Wisconsin, Michigan and possibly Minnesota, all in the southern Lake Superior region (Fig. 2-10)

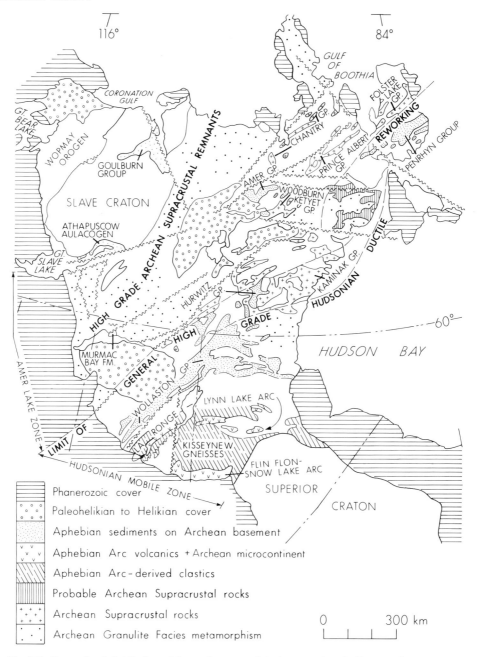

**Fig. 2-9.** Generalized distribution of the main areas of Archean and early Proterozoic supracrustal rocks in Western Churchill province, and of probable Archean granulite facies metamorphic parageneses recognizable through Hudsonian (early Proterozoic) overprint in the Taltson-Queen Maud Blocks to the northwest. (After Lewry et al 1985, Fig. 5. Reproduced with permission of the Geological Association of Canada).

(Sims 1980a, Sims and Peterman 1983, Boerboom and Zortman 1993, Boyd and Smithson 1993, Sims et al 1993). Collectively these form part of a large Archean gneiss segment, now mainly concealed by Proterozoic cover, which records a long complex mid–late Archean history. This mainly concealed gneiss segment is in contact across the Great Lakes Tectonic Zone (GLTZ) with the late Archean granitoid–greenstone terrain of southern–western Superior Province. The ENE-trending boundary (GLTZ) between these two lithologically contrasting Archean segments represents a fundamental crustal structure in the basement craton that has been repeatedly reactivated; it has been offset

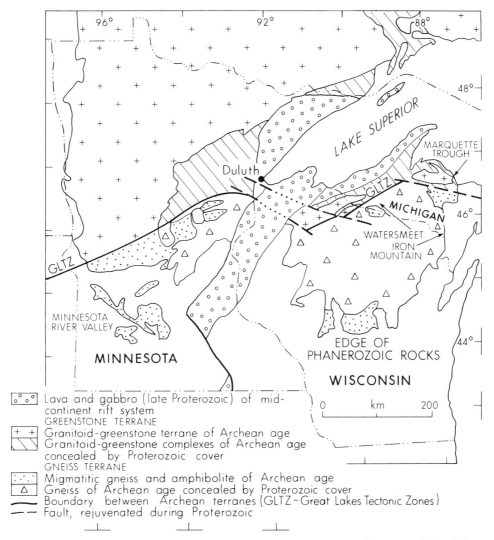

**Fig. 2-10.** Map showing known and inferred distribution of Archean crustal segments in the Lake Superior Region. Wolf River batholith omitted for clarity. (From Sims and Peterman 1983, Fig. 1. Reproduced with permission of the authors and of the Geological Society of America).

by two major NW-trending wrench fault systems, one passing through Duluth vicinity in the west and the other south of the Marquette Trough in the east.

Radiometric dating has established early-mid Archean ages and a complex subsequent history, including a strong Penokean (1.9–1.8 Ga) tectonothermal overprint. U–Th–Pb and Sm–Nd systematics provide a firm minimum age of 3410 Ma, with the possibility of a greater age to 3500 or 3800 Ma (Peterman et al 1980, McCulloch and Wasserburgh 1980). Also, an Archean mafic–ultramafic igneous body, dated at 2.9 Ga, lies to the west in northwestern Iowa, on the north flank of the Penokean orogen (Windom et al 1993).

## 2.6.7 WYOMING UPLIFT

Archean rocks are exposed in the cores of several young mountain ranges in Wyoming and environs that were uplifted during the Laramide Orogeny and which now collectively form the subcircular, ~700 km diameter Wyoming Province (Fig. 2-11, Table 2-8, column 5), (Condie 1976b, Peterman 1981, 1982, Hedge et al 1986, Frost 1993, Houston et al 1993, Mueller et al 1993).

The exposed Precambrian parts of the Wyoming Province are composed principally of gneiss–migmatite (60%) and granitoid batholiths (30%), with subordinate (10%) metasupracrustal (greenstone), amphibolite and ultramafic components, the latter

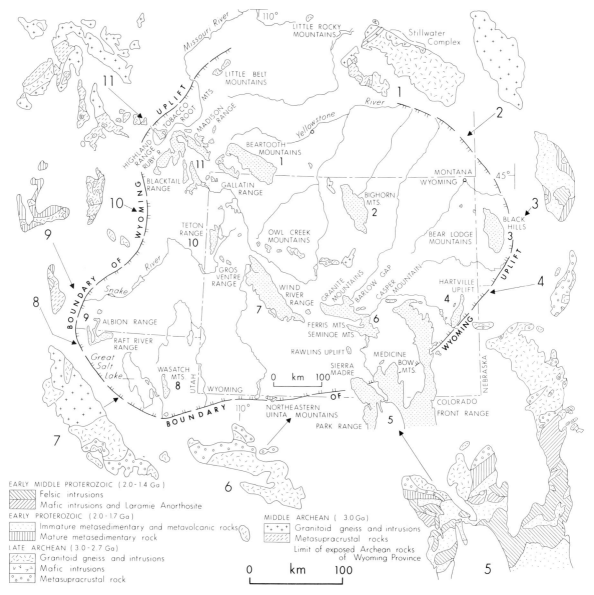

**Fig. 2-11.** Geologic map of Wyoming Uplift (Province) and environs showing the distribution of Precambrian inliers (Based on a recent map prepared by RS Houston and published with his permission).

including the renowned Stillwater Complex, a large, stratiform mafic–ultramafic intrusion (Lipin 1994), dated at 2701 Ma (De Paolo and Wasserberg 1979); the complex contains large reserves of high-iron chromite and nickel, copper and platinum sulphides in the ultramafic zone, and platinum values in the Merensky Reef-type (see below) sulphide layers in the upper part of the complex (Houston et al 1993). Regional structures are variable, with prevailing northeastward striking open folds, attributed to a single period of deformation (Condie 1976b). Amphibolite facies of metamorphism predominates. A complex ENE-trending mylonitized shear zone up to 10 km wide, the Cheyenne Belt, represents a fundamental discontinuity that separates the Wyoming Province on the north from early Proterozoic terrains to the south. Radiometric data including ion microprobe SHRIMP U–Pb zircon determinations reveal an extended Archean history from 3.96 to 2.69 Ga and beyond to 0.7 Ga (Peterman 1982, Mueller et al 1985, 1992). The major Archean event, including granitoid emplacements, occurred at 2.8–2.6 Ga, with lesser pulses at 2.64 and 2.59–2.57 Ga. Some distinctive Archean boundaries have been tenatively defined (Mogk et al 1992). A heating event at

1800–1600 Ma, corresponding to the Hudsonian (Penokean) Orogeny, reset mineral ages in the outer parts of the province. Several separate tectonic uplifts affected the province during the Laramide Orogeny.

## 2.7 SOUTH AMERICAN CRATON

Archean crust is exposed in all three component shields of the South American Craton (see Figs 1-3e, 1-5e, Table 2-9): (1) in the Guiana Shield, the (a) Pakairama and Xingu (northern segment) nuclei and nearby (b) Imataca Complex; (2) in the Central Brazil Shield, the (a) Xingu (southern segment) nucleus, (b) Serra dos Carajas inlier of the Maroni–Itacaiunas Belt and (c) Goias Massif of Tocantins Province; and (3) in the Atlantic Shield, the (a) Salvador–Juazeiro and (b) Belo Horizonte regions of the São Francisco Province, together with (c) several nearby small cratonic fragments.

### 2.7.1 GUIANA SHIELD

The *Pakairama and Xingu (northern segment)* nuclei (Cordani and de Brito Neves 1982) represent poorly defined Archean remnants, mainly surrounded by and transitional to enclosing Proterozoic mobile belt and cover (Fig. 2-12).

Both nuclei, as known, are composed of granitoid gneiss and migmatite with varied enclaves of amphibolite, quartzite and schist, together with younger, mainly granodioritic, plutons. Upper amphibolite–granulite facies of metamorphism prevails. However, the full extent of preserved Archean crust within the nuclei remains uncertain.

The *Imataca Complex* (Fig. 2-12) forms an ENE-trending, fault-bounded block bordering the Maroni–Itacaiunas Belt at the north–central margin of the Guiana Shield (Kalliokoski 1965, Dahlberg 1974, Hurley et al 1976, Gibbs and Barron 1983). To the south the complex is in faulted contact with Proterozoic rocks along the Guri Fault System, a zone of multiple faulting, shearing and mylonitization; it is elsewhere bordered mainly by Phanerozoic cover.

Quartzofeldspathic paragneiss is the predominant lithology in which the paramount structure is apparent bedding. The common granular to granoblastic textures are extensively overprinted by mortar, augen, flaser and mylonitic deformations. Imataca gneisses are in the almandine–amphibolite facies of regional metamorphism, except for local granulite facies relics.

Extensive BIF units, up to hundreds of metres wide, lie in a conformable sequence of quartzofeldspathic gneiss, migmatite and amphibolite. The hematite–magnetite-bearing BIF includes huge iron ore deposits, notably at Cerro Bolivar and El Pao.

Imataca rocks are complexly deformed. Folding is dominated by elongated or symmetrical domes, each one typically 3–5 km broad. The largest folds are broad sweeps defined by gneissic foliation and by continuous ridges of BIF. Small folds and wrinkle lineations are comparatively rare.

Transcurrent faults are important, notably the ENE-trending El Pao and Ciudata Pier–Guri faults. The Imataca Complex is, in all probability, entirely allochthonous.

Detailed geochronologic studies (Hurley et al 1976) demonstrate a protolith age in excess of 3.4 Ga and U–Pb ages in excess of 3.0 Ga (Montgomery 1979). The Imataca Complex is intruded by late Archean (2.8 Ga) and Transamazonian (~2.1 Ga) granitoid plutons.

### 2.7.2 CENTRAL BRAZIL SHIELD

The *Xingu Nucleus (southern segment)* (Figs 1-5e, 2-13) is lithologically similar to the nucleus of the same name in the Guiana Shield to the north. On the basis of a few isolated radiometric dates it is considered that the bulk of Xingu rocks were formed more than 2.5 Ga ago (Cordani and de Brito Neves 1982). However, Transamazonian (~2.1 Ga) overprinting is widespread, and the limtis of the nucleus are poorly defined.

The *Serra dos Carajás* area is located in the Carajás subprovince of the Tapajos province. It lies within the WNW-trending, early Proterozoic Maroni–Itacaiunas Belt near the NE contact of the Xingu Nucleus (Tassinari et al 1982, Cordani et al 1984, Hasui and Almeida 1985, Machado et al 1991) (Fig. 2-13, Table 2-9, column 1).

Archean (2.9–2.6 Ga) basement rocks at Carajás are dominantly tonalitic–granodioritic polymetamorphic gneiss, amphibolites and migmatites. Metamorphism is commonly upper amphibolite with local granulite facies (Tassinari et al 1982).

In the Carajás Ridge vicinity a number of WNW-trending metasupracrustal belts include those of the BIF-rich Graó–Para Group. The Graó–Para Group, forming a broad WNW-trending synclinorium, comprises both lower and upper mafic metavolcanic zones, each from 170 to 300 m thick, with the intervening BIF-rich Carajás Formation. The latter is a 100–300 m thick sequence of quartzite,

Table 2-9. Generalized sequence of main events in the Guiana, Central Brazil and Atlantic shields, South American Craton

| Time scale (Ga) | Guiana Shield | | Central Brazil Shield | Atlantic Shield | |
|---|---|---|---|---|---|
| | (1) | | (2) | (3) | (4) |
| 0 | Amazonian Craton | | Tocantins Province | São Francisco Craton | Barborema and Mantiquiero provinces |
| | | Guapore Craton | | | |
| 0.5 | | | Alto Paraguaia Gp | | Molasse deposits Granulites, migmatites, granites, 0.65–0.5 Ga |
| | | | **Brasíliano Orogeny, 0.7–0.5 Ga** | | |
| | Brasíliano fold belt (mainly buried) | | Brasilia and Paraguay–Araguaia belts sequences: Baixo–Araguaixan Sgp; Paranoa and Bambuí Gps | São Francisco cover: Bambuí Gp Macaubas Gp | Major transcurrent faulting. Repeated metamorphisms and deformations. Interior and marginal belt accumulations, e.g. Macaubas Gp |
| | | | **Uruaçuan Orogeny, 1.1 Ga** | | |
| 1.0 | Rondonian (Sunsas) Orogeny, 1.1–1.0 Ga | | | Espinhaço Orogeny | |
| | Rondonian (San Ignacio–Sunsas–Aguapei) Belt Complex | | | | |
| | | Sunsas and Vibosi Groups | | | |
| | | San Ignacio Orogeny, 1.3 Ga | | | |
| | | San Ignacio Schist Sgp | Paraguay–Araguaia sediments (?) | | |
| 1.5 | Jari–Balsino (Parguazan) Orogeny | | Niquelândia layered intrusion, ~1.56 Ga | Espinhaço–Chapada | Intermittent interior belt accumulation |
| | Parguaza rapakivi, 1.5 Ga | Canama syenite | | Diamantina sequences | |
| | Rio Negro– Juruena Belt, 1.7 Ga | | | | |
| | Roraima Gp, | Goritore Gp | Uruaçu Belt sequence | | |
| | Uatuma Volcanic–Plutonic Complex, 1.8–1.5 Ga | | | | |
| 2.0 | | | **Transamazonian Orogeny, 2.1–1.9 Ga** | | |
| | Granitoid plutonism | San Ignacio protoliths | | Serrinha, Capim greenstones. Contendas–Mirante, Jacobina Belts (Au, U, Mn) | Medium to high grade metamorphism. Interior belt accumulation, e.g. Paraiba, etc. Gps São Luis, Alves and Rio de la Plata fragments |
| | | | Ticunzal Group (U), ~2.0 Ga | | |
| | Maroni-Itacaiunas Belt | | | | |
| | Barama–Mazaruni and Pastora greenstones, ~2.2 Ga | Serra da Novio and Navio Gps (Mn) | | Minas Sgp (BIF), 2.4–2.0 Ga | |
| | | Rio Fresco Fm | Gabbro–anorthosites | Peridotites (Cr) | |
| | | | **Jequié Orogeny, 2.7–2.5 Ga** | | |
| 2.5 | Bartica gneiss, ~2.2 Ga | | | | Pre-Brasíliano basement, e.g. Seridó Belt, Caicó Complex |
| | Medium to high grade metamorphism | | Granulite metamorphism | Granitoid plutonism Rio das Velhas Sgp (greenstones), ~2.7 Ga, including Nova Lima Group (Mn, Au) | |
| | Xingu (N) and Pakairama granitoids and greenstone enclaves | Xingu (S) Complex Grao-Para, Alto Jaura, etc. Gps; major BIF; 2.8 Ga | Granitoid plutonism Pilar de Goias greenstones, ~2.9 Ga Goias mafic–ultramafic complexes | | |
| 3.0 | Granulite metamorphism, ~2.8 Ga | Carajas basement gneiss, 2.9–2.6 Ga | | | |
| | | | **Gurian Orogeny, 3.0 Ga** | Borrachudos Granite | |
| | Imataca Complex, major BIF | | Rio Porto gneiss–migmatite, 3.2 Ga | Brumado greenstone Jequié–Matuipe gneiss, ~3.2 Ga | |
| 3.5 | Imataca protoliths, >3.4 Ga | | | Boa Vista basement gneiss, ~3.4 Ga | |

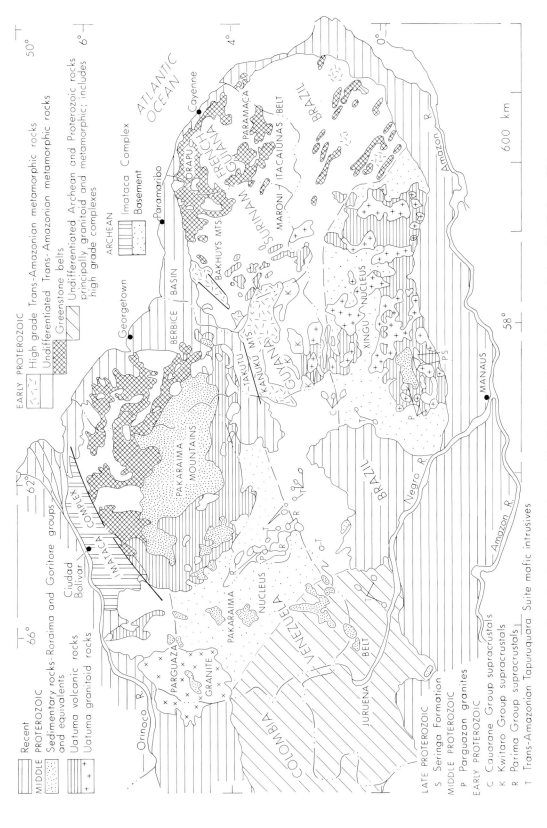

Fig. 2-12. General geologic map of the Guiana Shield, South American Craton. (Modified from Gibbs and Barron 1983, Fig. 1 and reproduced with permission of the authors).

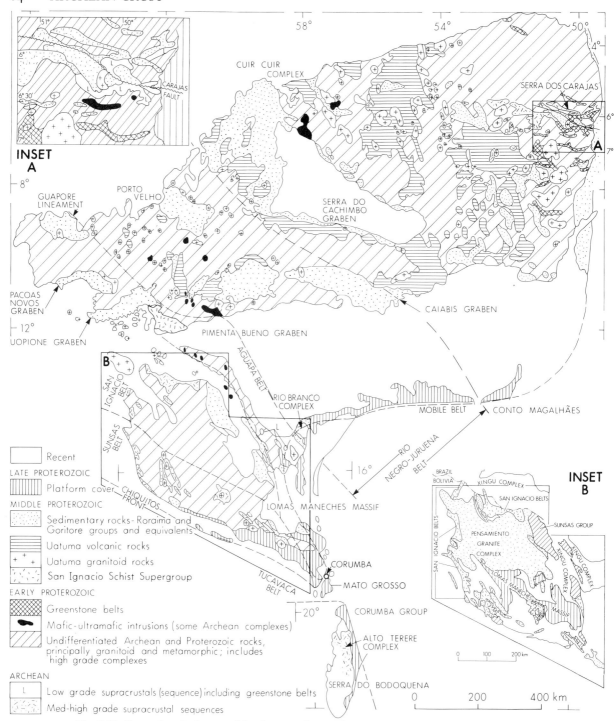

**Fig. 2-13.** General geologic map of the Guapore Craton, Central Brazil Shield, South American Craton. (Modified from Hasui and Almeida 1985, Fig. 3 and reproduced with permission of the authors). Insets illustrate, respectively, (A) Serra dos Carajas in the northeast and (B) Bolivian shield area in the southwest (from Litherland et al 1989, Fig. 4.).

conglomerate and phyllites with important zones of oxide facies BIF (jaspilites or itabirites). This BIF protore was intensively folded, metamorphosed and weathered to form the huge reserves of high grade iron ore of Serra dos Carajás estimated at $16 \times 10$ t and averaging 65% Fe. The iron ore was formed by deep Tertiary weathering and supergene enrichment which averages 100 m deep but extends to depths of 400 m (Beurlen and Cassedanne 1981, Teixeira et al 1994).

Zircons from rhyolites in the lower metavolcanic zone of the Graó–Para Group yield ages of 2758 Ma (Wirth et al 1985) and, more recently, 2759 Ma (Machado et al 1988). These zircon ages were the first indication that the Graó–Para Group is Archean in age, and not early Proterozoic as previously considered.

The *Goias Massif* (Fig. 2.14, Table 2-9 column 4), measuring 800 km by 100–300 km (Tassinari et al 1982), is aligned to the north–northeast for the most part along the centre of Tocantins Province with, however, a pronounced flexure near the southern end (Pirinopolis Megaflexure at 16°S). Medium grade migmatitic gneiss intruded by granitoid plutons predominate. However, the eastern part of the massif is dominated by a median series of tectonically disconnected mafic to ultramafic complexes, including Niquelândia intrusion, one of Earth's largest layered intrusions; the complexes are in places highly talcified and serpentinized and locally mineralized with copper and nickel sulphides (Nilson et al 1982, Danni et al 1982, Rivalenti et al 1982, Ferreira-Filho et al 1994). Granulitic terrains are closely associated with these mafic to ultramafic complexes. To the west is a parallel zone of low to medium grade granitoid–greenstone

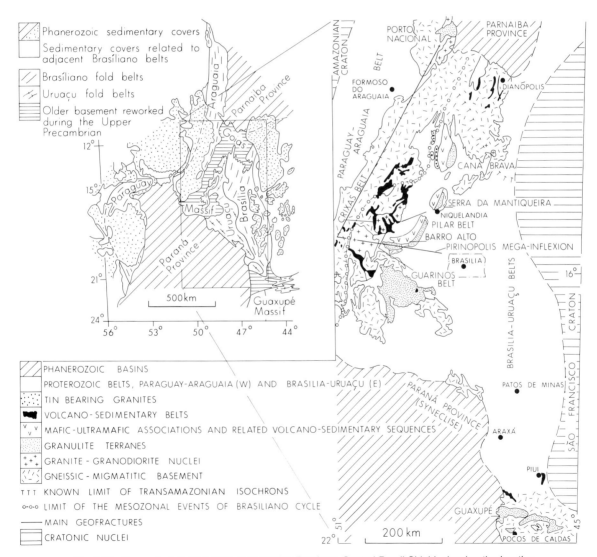

**Fig. 2-14.** General geological map of Tocantins Province, Central Brazil Shield, showing the location and composition of the Archean-rich Goias Massif, together with the adjoining mid-Proterozoic Uruaçu (Uruaçuanos) Belt and late Proterozoic Paraguay–Auaguais (to the west) and Brasilia (to the east) belts. (Modified from Almeida et al 1981, and reproduced with permission of the authors).

belts, extending along the massif for 700 km (Kuyumjian and Dardenne 1982, Montalvao et al 1982, Jost and Oliveira 1991).

The predominant migmatitic gneiss zones of the Goias Massif, representing Archean polymetamorphic complexes of general tonalite–granodiorite composition, were variously reworked during the Transamazonian (~2.1 Ga), Uruaçuan (~1.1 Ga) and Brasíliano (~600 Ma) events. Extensive shear belts, rich in cataclasites, are common. Granitic plutons have been dated at 2600–2900 Ma, and basement gneisses at 3200 Ma (Danni et al 1982). The Niquelândia intrusion provides an interpretive age of ~ 1565 Ma (Ferreira-Filho et al 1994). The age of granulite facies metamorphism in the massif is attributed by Danni et al (1982) to the late Archean Jequié event at ~2.7 Ga.

In the *Salvador–Juazeiro region* of the northeastern São Francisco Province (Figs 1-5e, 2-15; Table 2-9, column 3) both medium to high grade–granitic gneiss and medium to low grade granitoid–greenstone terrains occur in close juxtaposition (Mascarenhas and da Silva Sa 1982, Oliveira et al 1982, Bernasconi 1983, Cordani et al 1985). Some granitoids therein (e.g. Jequié–Matuipe) have yielded zircon dates up to 3403 Ma (Cordani et al 1985, Nutman and Cordani 1993).

Of several high-grade nuclei, the Jequié–Matuipe Complex, 480 km by 150 km, is composed of charnockitic gneiss with minor infolded khondalite, gondite, BIF, marble, calcsilicates, quartzite and amphibolite. Numerous dates in the range 2.7–2.4 Ga are attributed to Jequié metamorphism (Cordani et al 1985).

Several greenstone belts in the Brumado–Anaje area of south–central Bahia, are composed of ultramafic to felsic metavolcanic rocks and associated metasediments, all typically at greenschist–lower amphibolite facies of metamorphism. The age designation of the greenstone belts, whether Archean or early Proterozoic, has yet to be determined. However, nearby granitoid rocks have provided provisional dates of 3400 Ma, the oldest dates so far obtained in the São Francisco Craton (Cordani and de Brito Neves 1982).

The *Belo Horizonte region*, at the southern extremity of the São Francisco Craton (Fig. 2-15), includes a well-defined Archean granitoid–greenstone terrain which forms basement to the famous BIF-rich early Proterozoic Minas Supergroup of the Quadrilatero Ferrifero ('Iron Quadrangle'), a classical area of Precambrian geology in Brazil (Schorscher et al 1982, Teixeira 1982, 1985, Machado and Carneiro 1992).

Pre-Minas (i.e. Archean) granitoid rocks are divided into predominant gneiss–migmatite of tonalitic–trondhjemitic composition, and occasional granitoid dykes and small irregular to circular plutons dated at 2776 and 2721 Ma (Machado et al 1992).

The Rio das Velhas Supergroup, a typical pre-Minas greenstone sequence, is divided into three parts: (1) the lowermost Guebra Osso Group, less than 1 km thick, composed mainly of ultramafic lavas, including spinifex-textured periodotitic komatiites (Schorscher et al 1982, Jahn and Schrank 1993); (2) the median Nova Lima Group, a metavolcanic–greywacke–BIF–phyllite association more than 4 km thick also dated at 2776 and 2772 Ma (Ladeira and Noce 1990, Machado et al 1992), i.e. coeval with the above dated granitoids; and (3) the upper Maquine Group, about 2 km thick, of phyllite, quartzite and subgreywacke, with minor basal conglomerate.

The Lafaiete District in Minas Gerais State is one of the great manganese producers of the world. Mn-protores, correlated with the Nova Lima Group, include Mn-carbonate and -silicates in a metasediment–amphibolite sequence. The Lafaiete District, including San Joan de Rei and Saude areas, has produced over $11 \times 10^6$ t of high grade manganese oxide ore, obtained from over 20 individual deposits.

The Nova Lima Group is also the major gold-producing unit of the Quadrilatero Ferrifero. Gold is associated with sulphide minerals (pyrrhotite, pyrite, arsenophyrite and chalcopyrite) in carbonate-sulphide- and sulphide-facies BIF (Schorscher et al 1982).

Three small *cratonic fragments* distributed about the São Francisco Craton comprise Rio de la Plata, Luis Alves, and São Luis (Fig. 1-5e).

*Rio de la Plata* fragment, located mainly in Uruguay at 30°S and representing a part of the larger Rio Apa Massif, is composed of gneiss and migmatite, which are mainly products of the late Proterozoic Brasíliano Orogeny, but which may include some as yet undated Archean relics (Cordani and de Brito Neves 1982). The craton is bordered to the east by the Proterozoic Don Feliciano Belt.

The linear *Luis Alves* craton (24–27°S) is a typical domain of medium to high grade quartzofeldspathic gneiss, migmatite, ultramafic rocks, quartzite, BIF and calsilicates. Amphibolite–granulite facies prevail. The foliation trends NNE. Dates

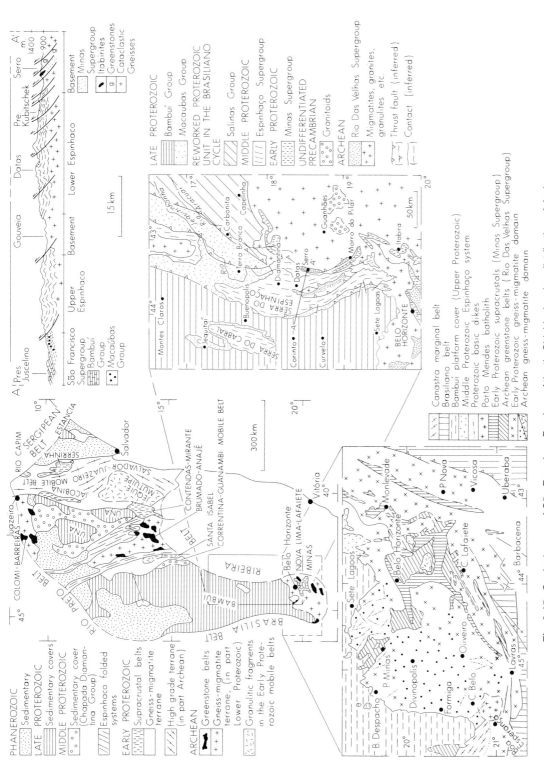

**Fig. 2-15.** Geologic map of São Francisco Province, Atlantic Shield, showing distribution of Archean granulite and granitoid-greenstone terranes, early Proterozoic fold belts (Espinhaço), and coeval sedimentary cover (Chapada Diamantino), and late Proterozoic–Paleozoic cover (São Francisco Supergroup). (After Almeida et al 1981, Fig. 1. Lower inset map of Minas Gerais area from Teixeira et al 1987, Fig. 1. Serra do Espinhaço map and structural cross-section after Uhlein et al 1986. Reproductions with permission of the authors).

of the Jequié cycle (2800–2500 Ma) are prevalent, together with those indicating extensive Transamazonian (2100–1900 Ma) rejuvenation (Cordani and de Brito Neves 1982). The NE-trending late Proterozoic Ribeira Belt adjoins to the north and northeast. The small elliptical *São Luis* fragment on the Atlantic coast is composed of gneiss, migmatite and schist of mainly early Protoerozoic age. No Archean dates have been reported. In the pre-Mesozoic Atlantic reconstruction, the São Juis craton represents the southern extension of the Birrimian domain of West Africa (see below) (Hurley and Rand 1969, Ledru et al 1994).

## 2.8 AFRICAN CRATON: SOUTHERN AFRICA

The Kalahari Craton (Domain) of southern Africa includes two Archean-rich cratons, Kaapvaal in the south and Zimbabwe in the north, separated by the E-trending, high grade Limpopo belt (Fig. 2-16, Table 2-10, together with overlying epicratonic basins (Table 2.11). The three basement terrains collectively contain one of Earth's best known and complete early Precambrian records, extending back to at least 3.64 Ga (Figs 1-3f(i), 1-5f(i,ii)).

### 2.8.1 KAAPVAAL CRATON

The Kaapvaal Craton is a rectangular area of 585 000 km² with, however, dominant (86%) Phanerozoic cover. The exposed basement (14%) is composed of predominant (91%) granitoid rocks and subordinate (9%) yet important greenstone belts (Fig. 2-16) (Hunter and Pretorius 1981, Tankard et al 1982, Anhaeusser and Maske 1986, Songe 1986, Hunter and Wilson 1988).

The early–middle Archean basement formed at different times over a span of ~600 Ma, starting before 3.6 Ga (Table 2-10). Although the geologic history of certain parts has been established in remarkable detail, the very limited rock exposure and uncertain tectonic relations of adjoining parts inhibit full reconstruction of its early crustal history. However, the Kaapvaal Craton was stabilized by 3.0 Ga; it represents the westernmost limit (present geography) of the hypothetical 3.0 Ga Ur Supercontinent (Rogers 1993).

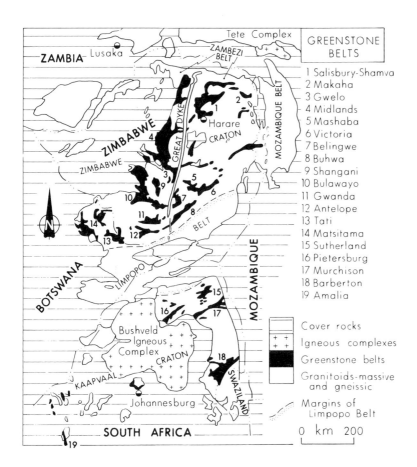

**Fig. 2-16.** Geologic sketch map of the Kaapvaal and Zimbabwe cratons and intervening Limpopo Mobile Belt, Southern Africa. Greenstone belts are numbered and named. (From Cahen et al 1984, Fig. 2.1, and reproduced with permission of Oxford University Press).

Table 2-10. Correlation chart of major Archean events in Southern Africa.

| Time Scale (Ma) | (1) Kaapvaal Craton | (2) Limpopo Belt | (3) Zimbabwe Craton |
|---|---|---|---|
| 2500 | 2514 Mpageni-type, Mhlosheni-type, Kwetta-type granite plutons, Pongola 'hood' to 2800 granite<br>2700 Ventersdorp lavas<br>2800 Witwatersrand Supergroup, including Dominion Reef lavas (3060 Ma)<br>2871 Usushwana Intrusive Suite<br>2940 Pongola Supergroup | 2514 Great Dyke satellites (Northern Marginal Zone) Major shear zones established (2650 Ma) Retrogression to amphibolite facies; deformation<br>2650 Vast volume granodiorite plutonism Rapid epeirogenic uplift and major decompression of crust<br>2700 High grade metamorphism and crustal thickening (Limpopo Orogeny)<br>2870 High grade metamorphism (North Marginal Zone) | 2514 Great Dyke<br>2595 Chilimanzi Granite Suite<br>2602 Sesombi tonalites, Shamvaian sediments<br>2660 Upper Bulawayan greenstones<br>2730 Gwenoro–Rhodesdale gneisses Lower Bulawayan greenstones<br>2817 Bangala and Chingez gneisses<br>2860 Mashaba Intrusive Suite |
| | *Ancient Gneiss Complex* | | |
| 3000 | ~3028 High level granites, e.g. Lochiel Batholith<br>~3200 TTG suite intrusions, strong deformation Mponono mafic–ultramafic intrusions<br>~3300 Granodiorite intrusions (Usutu suite)<br>~3350 Granulite metamorphism, ductile, deformation, thrusting, uplift<br>~3400 Deposition of shallow-water Mahamba and Mkhondo sequences<br>3420 Granulite metamorphism,<br>to 3460 thrusting, uplift, Tsawela Gneiss<br>~3500 Deposition of Dwalile, etc greenstones<br>~3550 Tonalite intrusion, metamorphism, deformation, intracrustal melting and K-granite emplacement<br>3644 Tonalitic intrusions | 3100 Mafic dikes<br>3150 High grade metamorphism (Central Zone)<br>~3200 Messina layered intrusion (Central Zone) and other mafic-ultramafic complexes Limpopo (Breitbridge) paragneisses (Central Zone) Sand River gneisses[1] (Central Zone)<br>3440 High grade TTG gneisses (marginal zones)<br>3450 High grade greenstone enclaves (marginal zones) | 3340 Mont d'Or Gneissic Complex<br>3369 Mushandike Granodiorite<br>3476 Tokwe–Shabani gneiss Sebakwian Group[2] |
| | *Barberton Greenstone Belt*<br>High level granites<br>Granodiorite suite, strong deformation<br>TTG diapiric plutons<br>(Kaapvalley and Stentor), deformation, 3350–3250 Ma<br>Deposition of Moodies Group, ~3200 Ma<br>Deposition of Fig Tree Group, deformation, ~3300 Ma<br>Deposition of Onverwacht Group, ~3440 Ma<br>TTG intrusion, deformation | | |
| 3500 | | | |

[1] Age of Sand River gneisses uncertain
[2] Relative ages of Tokwe–Shabani gneiss and Sebakwian greenstones commonly uncertain

*Table 2-11.* **Early Precambrian supracrustal cover and associated igneous rocks of the Kaapvaal Craton, South Africa.**

| | Supergroup | Lithochronostratigraphy | Approximate thickness (m) |
|---|---|---|---|
| Early Proterozoic | Waterberg | Waterberg, Matsap-Palapye-Umkondo basins: red arenites, local trachyte and rhyolite flows; 1790 ± 70 Ma. Soutpansberg Trough: basalt-trachyte lava flows; red arenites-pelites | Up to 5000 |
| | Transvaal | Bushveld Igneous Complex (2060–2050 Ma) and Rooiberg Felsic Suite (< 2250 Ma) | |
| | | Pretoria (Postmasburg) Group: quartzites, shale, carbonates; some volcanic rocks and mixtites (700 m max.); ~2224 Ma | |
| | | Chuniespoort (Campbell and Griquatown) Group: basal quartzite, carbonates, chert, major BIF; 3500 m max. | 5000–15000 |
| | | Wolkberg (Buffalo Springs) Groups: basalt flows, conglomerate, wacke, quartzite; (2552 Ma, 2432 Ma) | |
| | | ——— Major unconformity ——— | |
| Archean | Ventersdorp | Pniel Group: andesite lava flows, conglomerate, quartzite | |
| | | Platberg Group: quartzite, conglomerate, quartz porphyry lava and ash-flows; 2699 ± 16 Ma | 3000–4517 |
| | | Klipriviersberg Group: flood basalt-lava flows, agglomerate, tuff; 2708 Ma | |
| | | ——— Unconformity ——— | |
| | Witwatersrand | Central Rand Group: conglomerates, quartzites (auriferous fluvial fans), minor argillite and amygdaloidal flows | |
| | | West Rand Group: psammite-pelite alternations; amygdaloidal lava flows | 2500–7500 |
| | | Dominion Group: bimodal volcanic flows and pyroclasts; quartzites, conglomerates; 3060 ± 18 Ma | |
| | Pongola | Usushwana Intrusive Suite: mafic-ultramafic layered intrusions; 2871 ± 30 Ma | |
| | | Mozaan Group: quartz arenites, argillites, conglomerate | 3000–9000 |
| | | Nsuze Group: basalt, basaltic andesite, dacite, rhyolite; 2940 ± 22 Ma | |
| | | ——— Major unconformity ——— | |
| | | Granitoid basement 3000 Ma | |

### Granitoid complexes

Granitoid basement rocks have been mapped in detail in the Swaziland terrane, an area of 2500 km² located 100 km south of the Barberton Greenstone Belt. This terrane, underlain by various gneissic to massive granitoid rocks and associates and collectively known as the Ancient Gneiss Complex, provides an important key to unravelling the Archean history of southern Africa. (Key references: Hunter, in Tankard et al 1982, Hunter et al 1984, Kröner et al 1989, Kröner and Tegtmeyer 1994).

The *Ancient Gneiss Complex* contains a varied lithology with component ages as listed in Table 2-10 (Kroner et al 1987b, 1989, Compston and Kröner 1988, Hunter and Wilson 1988).

The dominant rock association (80%) of the Ancient Gneiss Complex is the *Biomodal Gneiss Suite* (renamed Ngwane gneiss), a tectonic assemblage of (1) leucocratic gneisses of trondhjemitic, tonalitic or granodioritic (TTG) composition that are complexly interlayered with (2) subordinate amphibolite (Hunter, in Tankard et al 1982, Kröner and Tegtmeyer 1994).

Recent geochronologic and isotopic dating reveal a complex history of development (Kröner et al 1987b, Compston and Kröner 1988, Kröner and Tegtmeyer 1994). The oldest dated component is preserved in zircon cores within tonalitic orthogneiss which provide a crystallization age of 3683 Ma. A succession of events (Table 2-10) led to the emplacement of high level granites at ~3000 Ma to complete the consolidation of the Kaapvaal Craton, followed by craton cover and granitoid intrusions to at least 2514 Ma.

Thus, the southeastern Kaapvaal province contains a fairly continuous record of crustal evolution from ~3.6 to ~2.6 Ga. It is worth emphasizing that a fundamental division in regional stratigraphy is provided by widespread intrusion of granites at ~3.0 Ga. This event, represented by the multiphase, tabular Lochiel Batholith, geographically separates the Ancient Gneiss Complex in the south from the well-known Barberton Greenstone Belt (see below) to the north. Emplacement of this batholith furthermore separates (1) middle Archean (3.5–3.0 Ga), pre-Lochiel assemblages of granitoids and metasupracrustal remnants from (2) late Archean (3.0–2.5 Ga) rocks including the moderately deformed Pongola cover sequence (see below) and numerous granite intrusions (Table 2-10). The thickening and strengthening of the crust required to achieve this crustal division was accomplished by successive intrusions of tabular granitoids that underplated the crust, culminating in final stabilization of the Kaapvaal Craton by 2.5 Ga (Hunter and Wilson 1988, Layer et al 1992).

*Granitoid domes*

Tonalitic gneisses of the *Johannesburg Dome*, a subcircular unit about 40 km in diameter (Fig. 2-19), comprise tonalite and trondhjemite gneiss (Hunter, in Tankard et al 1982). The gneisses locally show layering, which defines tight fold closures with axial-plane foliation.

The layered gneisses are dated at ~3.1 Ga (Allsop 1961, Anhaeusser and Burger 1982). They were later metamorphosed at 2.6 Ga (Allsopp 1961).

The almost circular, partially exposed *Vredefort Dome*, about 50 km across (Fig. 2-19), representsa massive uplift that has exposed a segment of earlier Archean basement centrally located in the late Archean Witwatersrand Basin (see below) (Bischoff 1988, Colliston and Reimold 1990, Nicolayson 1990, Nicolayson and Reimold 1990, Reimond and Levin 1991). The dome comprises a core of Archean granitoid gneiss and granulites (Old Granite) some 18 km in diameter, surrounded by a tilted and locally overturned girdle ('collar') of Witwatersrand and younger cover at least 40 km wide, and with a total stratigraphic thickness exceeding 15 km. The granitoid rocks originally acted as basement to these cover rocks but were later forced up to pierce through them though not as a simple igneous intrusion. Both the Old Granite and the 'collar' rocks are cut by an unusual and pervasive multiple-aged (2.2–1.1 Ga) breccia known as pseudotachylite.

Various models have been proposed to explain this enigmatic structure. Evidence of 'shock metamorphism' includes cataclasis, presence of coesite and stishovite, emplacement of mylonite, cataclasite and pseudotachylite veins, planar microdeformations in quartz, shatter cones and distinctive striated joint surfaces. On this basis, a meteorite impact origin has been proposed (Hamilton 1970, French and Nielson 1990, Martini 1991). However, this has been vigorously challenged by those supporting an endogenous origin, involving explosion detonation of rapidly expanding carbonate fluids from the mantle (Nicolaysen and Ferguson 1981), or unspecified unique post-Transvaal tectonism (Roering et al 1990, Reimold et al 1992).

The Vredefort cryptoexplosion event, whether its origin, is more or less contemporaneous with, though slightly postdating, the Bushveld activity (see below). The best recent estimates of its age, obtained for micas and zircons from the bronzite granophyre, fall in the time range 2050–1950 Ma (Walraven et al 1990).

## Greenstone belts

Relics of volcanosedimentary sequences metamorphosed at predominant greenschist facies are preserved in five main greenstone belts in the Kaapvaal Craton (Fig. 2-16). Of these, the Barberton Belt is the largest and best preserved and exposed. The supracrustal racks of all the belt are commonly correlated with the Swaziland Supergroup, for which Barberton is the type area.

*Barberton Greenstone Belt*

The Barberton Mountain Land comprises a rugged tract of country exposed in the low veld of the eastern Transvaal and Swaziland. The greenstone belt forms an irregular northeastward tapering triangular unit 120 km long and 65 km in maximum width (Fig. 2-17; Table 2-10, column 1). It contains a wide variety of volcanic, igneous and sedimentary rocks, collectively called the Swaziland Supergroup, which is subdivided up-section as follows: Onverwacht, Fig Tree and Moodies groups. (Key references: Viljoen and Vljoen 1970, Williams and Furnell 1979, Eriksson in Tankard et al 1982, Anhaeusser 1983, 1984, 1985, Barton et al 1983a, DeWit et al 1983, Lamb 1984, Lowe et al 1985, Kröner et al 1989, Lowe 1994.)

Bedding in rocks of the Swaziland Supergroup dips steeply to vertical. The principal large-scale

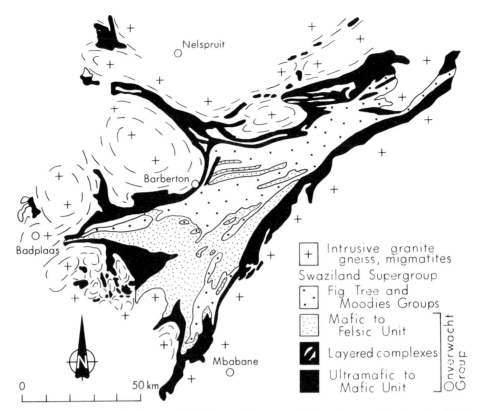

**Fig. 2-17.** Simplified geologic map of the Barberton Greenstone Belt and surrounding intrusive diapiric gneiss plutons. (From Anhaeusser 1981 Fig. 8-3, and reproduced with permission of the author and of Elsevier Science Publishers).

structures are a series of NE-trending folds with almost vertical axes. These folds are largely developed in Onverwacht and Fig Tree rocks. In the centre of the Mountain Land, rocks of the Moodies and Fig Tree groups form a series of large, tight, steeply plunging synclines separated by narrow, sheared, antiformal septa of Onverwacht rocks.

*Swaziland Stratigraphy*: The main subdivisions of the Swaziland Supergroup, with approximate preserved stratigraphic thicknesses in brackets are as follows:

(top)    (3)    Moodies Group (3500 m)
          (2)    Fig Tree Group (2000 m)
(base)  (1)    Onverwacht Group (~12 000 m)

The *Onverwacht Group* is distinguished by the abundance in the lower part of ultra-mafic and mafic rocks with a distinctive chemistry that distinguishes them from typical perioditite and picrite (Viljoen and Viljoen 1970). They are characterized by abundant pillowed and massive flows and sills of peridotite and basalt (komatiite, high-Mg basalt, tholeiite), together with subordinate interlayered felsic tuff, agglomerate and ironstone pods (de Ronde et al 1994).

The upper Onverwacht succession includes a variety of mafic–felsic calc-alkalic volcanic lava flows and pyroclastic rocks with thin persistent chert and carbonate sedimentary units (Middle Marker).

The original 'layer-cake' stratigraphic model of Viljoen and Viljoen (1970) has been replaced by one that allows for greater local complexity in the sequence of volcanic events in the Barberton Belt (see further below).

Metasedimentary rocks are widespread in the predominantly volcanic Onverwacht assemblage (Lowe and Knauth 1977, Lowe 1980, 1982, Eriksson, in Tankard et al 1982, Lowe et al 1985, Heubeck and Lowe, 1993), including terrigenous gravel, sand and silt altered to chert–sericite mosaic, coarse grained dolomite, now altered to homogenous chert beds, primary non-terrigenous silica.

Spherule layers in the Barberton Greenstone Belt previously ascribed to large asteroid or comet impacts on the early Earth (Lowe et al 1989), have been recently attributed to exclusively terrestrial processes (Koeberl et al 1993).

The *Fig Tree Group*, which gradationally overlies the Onverwacht in most areas, is some 2000 m thick. The 3.25–3.22 Ga group is composed of dominant greywacke and shale, with subordinate chert, BIF and pyroclastic rocks (Toulkeridis et al 1994). The group is interpreted as forming from progradation of a fan delta into a body of relatively quiet water (Nocita and Lowe 1990).

Fig Tree sediments generally coarsen upward into the *Moodies Group*, about 3500 m thick, of more mature arenaceous sediments (Eriksson, in Tankard et al 1982, Heubeck and Lowe 1994). Conglomerate and arkose are present at the stratigraphic base and dominate in the south (present geography). Subarkose and quartz arenite become more abundant northward, with shale and BIF becoming locally important. Moodies sedimentation (3.22–3.10 Ga) was dominated by interacting fluviatile, estuarine–deltaic, beach–offshore and tidal-flat depositional systems from a southeasterly source terrain such as the Ancient Gneiss Complex, with deposition on a shelf-rise continental margin.

Remapping of the Barberton Belt has led to some fundamental stratigraphic reinterpretation. According to some (De Wit and Stern 1980, De Wit 1982, De Wit et al 1983, 1987) the Onverwacht Group is much thinner (< 2 km) than earlier estimates due to folding, faulting, and facies thinning. The Onverwacht Supergroup has a preserved stratigraphic thickness of about 12 000 m. Thus, further work is required to establish the true preserved stratigraphic thicknesses and proper stratigraphic arrangements.

Detrital zircons from a presumed autochthonous metaquartzite underlying Onverwacht volcanic rocks yield a zircon age of 3456 Ma. A felsic volcanic flow from the lower mafic sequence provides a zircon age of 3438 Ma. These data effectively bracket Onverwacht mafic volcanism between 3438 and 3456 Ma (Kröner et al 1987b). Furthermore, recent argon ages (López-Martinez et al 1992) are consistent with the zircon ages and indicate that the original metamorphism of the ultramafic –mafic volcanic rocks occurred shortly after their eruption.

### Other greenstone belts

The nearby Murchison Belt (Anhaeusser in Hunter 1981, Anhaeusser and Wilson 1981) is characterized by important antimony and associated Cu mineralization. Two additional recently remapped belts both situated north of the Murchison Belt and near the north margin of the Kaapvaal Craton are the Pietersburg Belt (de Wit et al 1992, van Schalkwyk et al 1993) and Sutherland (Giyana) Belt (McCourt and van Reenen 1992). They essentially conform to the above established greenstone pattern. The Nondweni belt, 300 km south of the Barberton sequence, is a distinctive variant (Wilson and Versfeld 1994a, b).

### Pongola Supergroup

The Pongola Supergroup of southeastern Kaapvaal Craton is one of Earth's oldest known substantial epicratonic cover sequences. The rocks were deposited in an intracratonic basin upon Archean basement essentially stabilized at 3028 Ma (Table 2-11). They were intruded by the 2871 Ma old Usushwana Intrusive Suite. Pongola rocks and associates are exposed discontinuously over an area of 2500 km$^2$ (Fig. 2-18) (Matthews 1967, Watchorn 1980, Hegner et al 1984, Weilers 1990).

The Pongola Supergroup comprises the lower, thicker, predominantly volcanic Nsuze Group, up to 8 km thick, and the gradationally to unconformably overlying, thinner, dominantly sedimentary Mozaan Group up to 5 km thick. It has a common cumulative stratigraphic thickness of 3–9 km (Tankard et al 1982).

Mafic–ultramafic rocks of the Usushwana Intrusive Suite outcrop in three discrete areas, separated by younger granitoid plutons (Fig. 2-18).

The Pongola basin is attributed to a two stage thermo-tectonic sequence, the mainly volcanic Nsuze Group reflecting a first stage rift-forming event caused by uplift and crustal extension in response to lithospheric heating, and the sedimentary Mozaan Group a second stage broad crustal downwarping induced by thermal cooling and contraction (Weilers 1990).

### Witwatersrand Supergroup

The Witwatersrand Supergroup fills the second of the northward progressing and enlarging Kaapvaal epicratonic basins (Table 2-11, Fig. 2-18, inset). It is restricted to a basin located immediately south of Johannesburg and approximately centred on the Vredefort dome (Fig. 2-19). The three-part subdivision of Witwatersrand sedimentary and volcanic rocks underlies a NE-trending oval, measuring 350 km long by 200 km maximum width, in which the basal Dominion Group is restricted to the west-

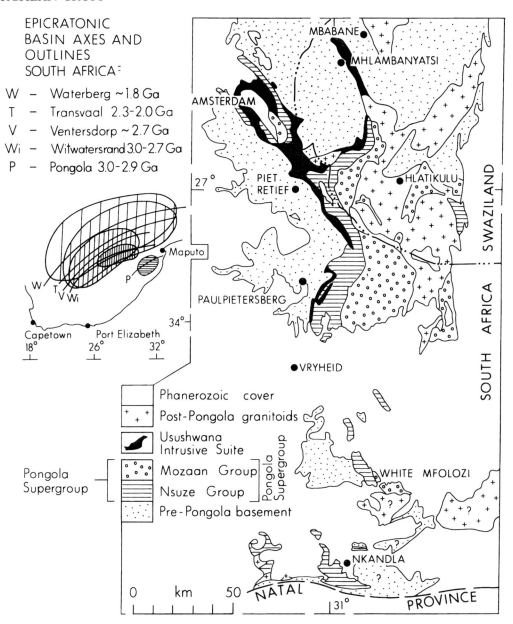

**Fig. 2-18.** Distribution of the Pongola Supergroup. Usushwana Intrusive Suite, and post-Pongola granitoid intrusions. (From Tankard et al 1982, Fig. 2-18, and reproduced with permission of authors and of Springer-Verlag Publishers). Inset map shows epicratonic basin axes and outlines upon the Kaapvaal Craton. (Adapted from Anhaeusser 1973a, Fig. 4, and reproduced with permission of the author).

ern part, but the two overlying groups are much more widespread (Jackson 1991). The original areal extent of the depository has been estimated at 80 000–100 000 km² (Pretorius, in Hunter 1981). The extent of outcrop is small and much subsurface data have been obtained during mining and exploring for gold and uranium deposits (Pretorius 1976, 1981a, Tankard et al 1982, Bowen et al 1986, Pretorius et al 1986, McCarthy 1990, McCarthy et al 1990a, Myers et al 1990a, Robb et al 1990, Stanistreet and McCarthy 1991, Moore et al 1993).

The composite thickness of Witwatersrand strata is estimated to be 11 000 m (Table 2-12). The maximum recorded thickness on the northwestern edge of the basin is 7500 m, and on the southeastern edge, 2500 m.

The Witwatersrand basin forms a NE-trending asymmetric synclinorium, with the depositional

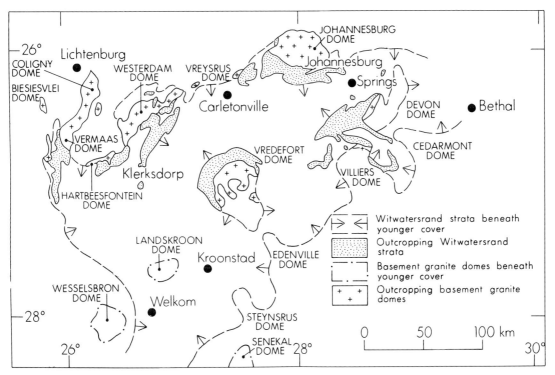

**Fig. 2-19.** Outcrop pattern of the Witwatersrand Basin showing Witwatersrand strata and basement granitoid domes. (From Pretorius 1981a, Fig. 9.4, and reproduced with permission of the author and of Elsevier Science Publishers).

axis being closer to the northwestern edge of the depository than to the southeastern rim. Dips of beds on both limbs decrease stratigraphically upwards, from very steep to vertical at the base, to less than 20° at the top.

Structural analysis indicates that during Witwatersrand time the cratonic basement became fragmented into at least 18 large blocks relative to each other by rotation and tilting (Myers et al 1990a), an important factor in interpreting the sedimentologic–metallogenetic history of the basin.

The three groups and seven subgroups of the Witwatersrand Supergroup (Table 2-11) represent a five-phase stratigraphic arrangement: (1) initial, very high energy, protobasinal volcanic phase (Dominion Group); (2) high energy, sedimentary phase (Hospital Hill and Government subgroups); (3) median, pivotal, low energy, sedimentary phase (Jeppestown Subgroup); (4) upper, high energy, sedimentary phase (Johannesburg and Turffontein subgroups); (5) terminal, very high energy, volcanic phase (Klipriviersbery Group of the overlying Ventersdorp Supergroup (Table 2-11)). Within this setting the reconstructive classification of the widespread yet variable Witwatersrand quartzites takes on added significance (Law et al 1990).

Mineral deposits

The Witwatersrand Basin contains Earth's greatest discovered gold deposits (fields), having produced close to 40% of all the gold mined in the history of mankind, together with important uranium production. Gold fields were developed within fluvial fans which formed at the interface of a river that flowed southeastward from a source-area on the northwestern side of the basin, and a shallow-water lake. By far the greatest concentration of gold and uranium is present in a number of horizons in the Johannesburg Subgroup of the uppermost Central Rand Group.

A placer theory of origin is heavily favoured for the gold deposits (Minter et al 1993). However, their exceptional size seems to call for a correspondingly exceptional gold-concentrating process, such as epithermal–volcanogenic or hydrothermal, to provide an adequate noble metal source. In this regard, recent dating of granitoid detritus in Witwatersrand sediments suggests that the detritus was not derived from the earlier Archean crystalline

Table 2-12. Sequence of principal early Precambrian events in the West African Craton.

| Time scale (Ga) | Man Shield (south) | Reguibat Shield (north) | | |
|---|---|---|---|---|
| | (1) Kenema–Man and Baoulé–Mossi domains | (2) Southwestern Province | (3) Eastern Province | (4) Composite Reguibat cycles |
| 1.5 | Supergroup I of the cover of the Taoudeni Basin (Base at ~1035 Ma—Atar Group) | | | |
| | | | Diabase dikes | (D) |
| | Post-tectonic granitoids, ~2.0 Ga Tarkwaian Molasse, ~2.0 Ga | | Post-tectonic granites, 1.8–1.6 Ga Guelb el Hadid Molasse Eglab volcanics, ~1.9 Ga Aftout ignimbrites | (C) Eglab Cycle |
| 2.0 | | Eburnean Orogeny, 2.1–1.9 Ga | | |
| | Birrimian Supergroup | Polyphase deformation | Yetti granitoids Folding and nappe thrusting Imourine Group, ~2.1 Ga Isoclinal recumbent folding Yetti Group and correlatives Chegga assemblage and correlatives | (B) Yetti Cycle |
| 2.5 | | Ghallman granites, 2.5 Ga *Amsaga Basement:* | | (A) Basement |
| | Liberian Orogeny, ~2.75 Ga Kambui Supergroup Liberian greenstones (uncertain pre-Liberian?) | Migmatitic Complex, 2.8 Ga | | |
| 3.0 | Leonean Orogeny, ~3.0 Ga Loko greenstones | Saouda granulites (W), 3.0 Ga Ghallaman gneisses (E), ~3.2 Ga | | |

basement, but rather from a younger more evolved hydrothermally altered, Au–U-enriched granitoid source with a commonality of ages ranging between 3200 and 3050 Ma (Robb et al 1989, 1990, 1992). Furthermore, the fact that the Central Rand Group, the overwhelming gold depository, is the only sequence containing zircons with ages younger than 3040 Ma, also suggests that the gold source was related to some event such as exhalative or shallow-level lode deposition, which coincided with granitoid emplacement in the ~3.0–2.9 Ga interval. Additionally, oxygen isotopic compositions of host rocks and pebbles suggest that uranium is derived from a granitic source, whereas gold has a 'mesothermal' greenstone lode-gold source (Vennemann et al 1992).

The basal Dominion Group of the Witwatersrand Supergroup has provided a maximum basin age of 3060 Ma, and the overlying Ventersdorp Supergroup a minimum basin age of 2709–2714 Ma, for a period not exceeding 360 Ma for accumulation of the full Witwatersrand basin (Armstrong et al 1991). Supporting age data are provided by Rundle and Snelling (1977), Minter et al (1988), Robb et al (1989), Barton et al (1989) and Walraven et al (1990).

Tectonic models

According to the classical taphrogenic-basin model (Pretorius 1981a), a pattern of interference folds was accentuated by subsidence of resultant basins and uplift of domes, the pattern and growth of the domes controlling the paleocurrent directions in Witwatersrand Basin. The environments of deposition relate to fan-delta distribution of braided, fluvial systems. The economically viable placer deposits are intimately related to unconformities within the asymmetric basin, in which the short side (northwest) displays high energy sediments, linked genetically to extensive, normal, strike faults. Reworking of sediments off domes is a major concentrating factor for placer mineralization.

*Table 2-13.* Sequence of principal events in the Tuareg Shield, Trans-Saharan Belt, Northwest Africa.

| Time scale (Ga) | (1) Western Hoggar (Pharusian Chain) | (2) Central Hoggar | (3) Eastern Hoggar | (4) Composite cycles and events |
|---|---|---|---|---|
| 0.5 | Nigritian and Purple Ahnet Groups molasse, 538 Ma | Granitoid intrusions to 515 Ma Late tectonic metamorphism, 590–580 Ma | Granitoid intrusions, 0.58 and 0.6 Ga | |
| | Granitoid intrusions at 572, 575 and 592 Ma Green Group and correlatives Deformation; mafic–ultramafic intrusions, 785 Ma | Syntectonic granitoids, 615–600 Ma | Deformation and metamorphism, 0.67 Ga Tiririne Formation | Pharusian Orogeny 615–580 Ma (plate collision) Pharusian II (plate extension) |
| 1.0 | Stromatolite Group, ~1.05 Ga (correlated with Atar Group of Taoudeni Basin) | | | Pharusian I |
| | Alkaline rhyolite at Adras Ougda, 1.1 Ga | Folding and metamorphism, 1.1 Ga Deposition of Aleksod Group | | (?) Kibaran events (?) ~1.1 Ga |
| | Ouallen granite, 1.8 Ga | Recumbent folding and granulite metamorphism, 2.0–1.8 Ga Oudenki mafic dikes, ~2.0 Ga Augen gneiss, 2.0–1.8 Ga | Issalane[1] gneiss-metasediments | |
| 2.0 | In Ouzzal metamorphism and anatexis, ~2.0 Ga Tassendjanet Group | Gour Oumelalen Group Arechchoum Group, 2.4 Ga | | Eburnean (Suggarian) Orogeny Suggarian Cycles |
| | In Ouzzal granulite metamorphism, 3.0 Ga | | | |
| 3.0 | | | | |
| | In Ouzzal granulite, 3.1–3.5 Ga | Red 'Series', ~3.5 Ga (?) | | Ouzzalian Event, ~3.0 Ga |

[1]Uncertain pre-Tiririne age

A proposed plate tectonic, extensional model (Bickle and Eriksson 1982) involves a downwarped, cratonic Witwatersrand Basin, without rifting, followed by a rifted Ventersdorp succession (see below). Burke et al (1985) ascribe Ventersdorp rifting to a collision between the Kaapvaal and Zimbabwe cratons, with the immediately preceding Witwatersrand site forming a foreland basin. Winter (1986) interprets the Witwatersrand deposition in terms of an Andean-style, subduction-controlled back-arc basin, as do McCarthy et al (1990a), with modifications.

## Ventersdorp Supergroup

The Ventersdorp Supergroup (Button 1981a, Tankard et al 1982, Burke et al 1985, Schweitzer and Kröner 1985, Visser and Grobler 1985, Bowen et al 1986, Myers et al 1990b) occupies a NE-trending belt, 750 km by 350 km or 26 000 km² in area (Figs 2-18 inset, 2-20; Tables 2-10, column 1, 2-11). Much of the Ventersdorp is covered by younger units (Transvaal and Karoo). Typically, Ventersdorp strata, some 3000–5000 m thick, are only gently deformed and even locally horizontal. However, they have the same structural pattern as the underlying Witwatersrand succession where deformed by folding and faulting, the latter especially important in developing Ventersdorp stratigraphy and sedimentation. The Ventersdorp Supergroup exhibits (1) rapid lateral facies variations; (2) a weakly developed bimodal tholeiitic basalt–dacite suite; (3) irregular basement topography; and (4) linear, faulted basin margins.

The Ventersdorp Supergroup is divided, from base to top, into the Klipriviersberg, Platberg and Pneil groups with relations and lithologies as illustrated (Table 2-11). Geochemical stratigraphy is an effective means of locating stratigraphic position

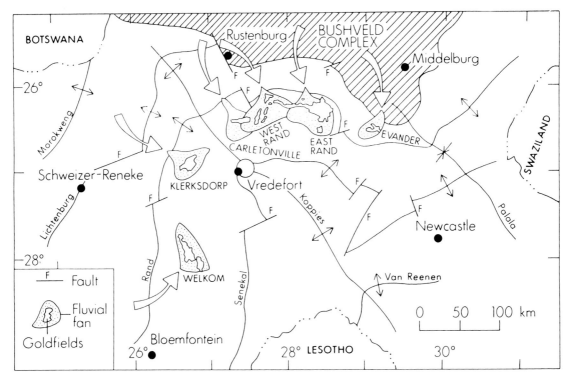

**Fig. 2-20.** Locality map for the Ventersdorp Basin (from Button 1981, Fig. 9.6, and reproduced with permission of the author and of Elsevier Science Publishers).

within this very homogeneous lava sequence, and of solving structural problems (Myers et al 1990b).

Ventersdorp rocks were deposited in a system of NE-trending grabens on the Kaapvaal Craton at ~2.7 Ga (Armstrong et al 1991). The presence of zircon xenocrysts in the flood basalts provides compelling evidence of crustal contamination (Nelson et al 1992). Sedimentation took place along the flanks of an actively eroding volcanic domal structure. The overall events are ascribed to an 'impactogenal' origin, resulting from the collisional impact of the converging Kaapvaal and Zimbabwe cratons across the Limpopo continental collisional zone (Burke et al 1985, Shackleton 1986). This interpretation follows on the earlier recognition that the Limpopo Belt is fundamental to any coherent evolution model of the Kaapvaal Craton (Hunter 1974a, b).

### Mineral deposits

Archean greenstone belts contain numerous Cr, Fe, Au and Ni and occasional Sb and W deposits. Emplacement of granitoid masses has resulted in the development of chrysotile asbestos and of enriched Au deposits. The epicratonic Witwatersrand basin contains uniquely valuable Au–U deposits (Songe 1986).

### 2.8.2 LIMPOPO MOBILE BELT

### Introduction

The intercratonic Limpopo Belt, 700 km × 320 km or 185 000 km² in exposed area, is an ENE-trending polycyclic strip of high grade metamorphic and igneous rocks variously called an orogenic, mobile or metamorphic belt, complex, domain or province (Figs 2-16, 2-21, Table 2-10, column 2). The Limpopo Belt is best viewed as an integral part of the larger Kaapvaal–Zimbabwe composite, which, as a result of its median position, underwent unusually complex polytectonic activities. Because the deformations within the belt extend uninterruptedly into the flanking cratons, the northern and southern limiting boundaries of the belt are taken to be orthopyroxene isograds marking the onset of granulite metamorphism. The Limpopo Belt displays a uniquely long history for an African mobile belt, dating back to at least 3.5 Ga and provides a unique view of deeper crustal levels. Limpopo

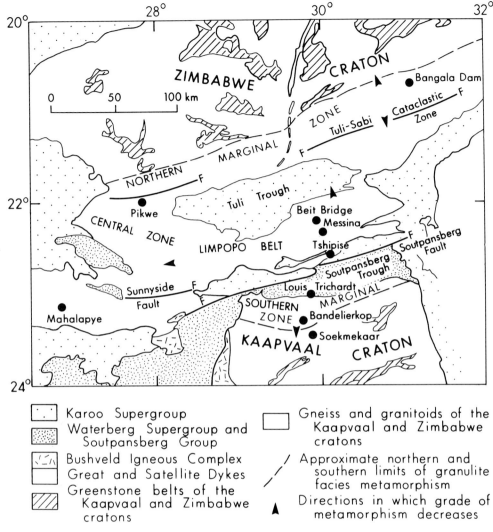

**Fig. 2-21.** The main geologic features of the Limpopo Belt. (Adapted from Cahen et al 1984, Fig. 2.5, and reproduced with permission of the authors and of Oxford University Press).

Orogeny refers to the common high-grade metamorphism and tectonic event during the interval ~2700 Ma to ~2650 Ma that affected the Archean rocks of the Central and Southern Zones (see below) of the Limpopo Belt and its influence on the adjacent portions of the Kaapvaal Craton (Key references: Barton 1981, Barton and Key 1981, Robertson and du Toit 1981, Tankard et al 1982, du Toit et al 1983, van Biljon and Legg 1983, Watkeys 1983, Watkeys et al 1983, Cahen et al 1984, Van Reenen et al 1987, Roering et al 1992a, b, Van Reenen et al 1992a, b, Barton and Van Reenen 1992, Barton et al 1994).

Zone construction

The Limpopo Belt is divided into (1) a central zone of distinctive lithostratigraphy and N-trending structures, separated by major shear belts or 'straight zones' from (2) northern and (3) southern marginal zones of dominantly ENE-trending structures (Fig. 2-21).

The Central Zone is characterized by complex relationships between apparently repeatedly metamorphosed 'basement' and intensely metamorphosed cover sequences (Limpopo Group). Major anorthosite complexes were intruded near the 'basement'-cover interface (Robertson and du Toit

1981, Tankard et al 1982, Watkeys et al 1983). The two marginal zones comprise reworked granitoid–greenstone basement raised to the granulite facies at 3.15 Ga in the southern marginal zone (Barton et al 1994) that are transitional peripherally across thrust zones to the lower grade granitoid–greenstone terrains of the adjoining cratons (Robertson and du Toit 1981, Watkey 1981, Ridley 1992, Tsunogae et al 1992).

The Central Zone underwent high grade metamorphism at ~3.1 Ga and again, at 2.0 Ga (Barton et al 1994). This was followed by rapid vertical uplift, metamorphic regression and vast granitoid plutonism (Steven and Ven Reemen 1992, Roering et al 1992a). The Central Zone was subsequently (~2.0 Ga) transported westward along respective sinistral (southern) and dextral (northern) wrench faults to its present relative position vis-à-vis the Kaapvaal and Zimbabwe cratons.

Geophysical studies prove that the terrane boundaries of the Limpopo Belt dip in towards the belt to give it the appearance of a 'pop-up' structure. The present boundary positions could be related to thin-skinned tectonics that affected the upper crust on the Kaapval Craton (de Beer and Stettler 1992, Durrheim et al 1992).

### Mineral deposits

Significant deposits of nickel–copper occur in Pikewe–Selibe troctolite sills in Botswana. Chromitites and magnetites are mined at Rhonda and Spinel mines in Zimbabwe. Sedimentary rocks contain small deposits of marble, dolomite and graphite.

Copper deposits occur in the Central Zone at Campbell, Messina and Harper mines in South Africa, all associated with the Messina Fault.

### Tectonic models

Most tectonic models of the Limpopo Belt have invoked continental collision—either south over north (e.g. Coward et al 1976) or east over west (e.g. McCourt and Vearncombe 1987). Van Reenen et al (1987) emphasize the overall symmetry of the Limpopo Belt and the absence of evidence for a resulting structural ramp, or crust on edge. Instead, these authors favour crustal thickening to at least 65 km in the high grade area of the Limpopo Belt at 2.7 Ga, as a direct result of two continents colliding in the manner of the Himalayas.

According to Watkeys (1983), two periods of transcurrent shearing were responsible for producing the present disposition of the Limpopo Belt. The earlier dextral movement predated the great Dyke (~2.5 Ga) and resulted in a juxtapositioning of the Zimbabwe Craton and the Central Zone. Associated with this is the over-thrusting of the Limpopo Belt on to the Zimbabwe Craton to the north. The later sinistral movement may have postdated the Bushvelt Complex (~2.0 Ga) and brought together the Central Zone (attached to the Zimbabwe Craton) and the Kaapvaal Craton.

Roering et al (1992b) emphasize that crustal thickening to at least 65 km during the Limpopo Orogeny (~2700–2650 Ma) the cause of the formation of the granulite terrane, probably resulted from the thrusting of the Kaapvaal Craton over the Zimbabwe Craton along the Tuli–Sabi Shear Zone. Peak metamorphism was superimposed on the thickened crust, followed by widespread thermal decompression during which rocks moved upwards and spread outwards on to the adjacent cratons thereby creating the regional 'pop-up'.

### 2.8.3 ZIMBABWE CRATON

The Zimbabwe Craton forms an irregular ellipse 750 km long, up to 400 km wide and 312 000 km$^2$ in area (Figs 2-16, 2-22; Table 2-10, column 3). Archean rocks, which underlie 63% of the craton, form a 'basement complex' comprising: (1) dominant tonalitic gneiss with some older (Sebakwian) metasupracrustal relics, collectively known as the older (~3.5 Ga) gneiss–greenstone association (83%); (2) younger (Bulawayan) greenstone belts with late granitoid plutons, collectively known as the younger (~2.9 Ga) greenstone and late granite association (17%). The greenstone (schist, 'gold' belts are generally arcuate or elongated synformal units about 200 km in length, forming intervening zones between subcircular batholithic complexes ('gregarious batholiths'). The Great Dyke, a major S-trending mafic–ultramafic association, represents the last major Archean event at 2514 Ma (Key references: Macgregor 1951, Nisbet et al 1977, Wilson et al 1978, Nisbet et al 1981, Wilson 1981, Stowe 1984, Hartnady et al 1985, Luais and Hawkesworth 1994, Nisbet and Martin 1994).

### Older gneiss–greenstone association

An older association of granitoid gneiss with infolded greenstone remnants has been identified mainly in the south–central part of the craton, in a small triangular area, about 75 km across, which

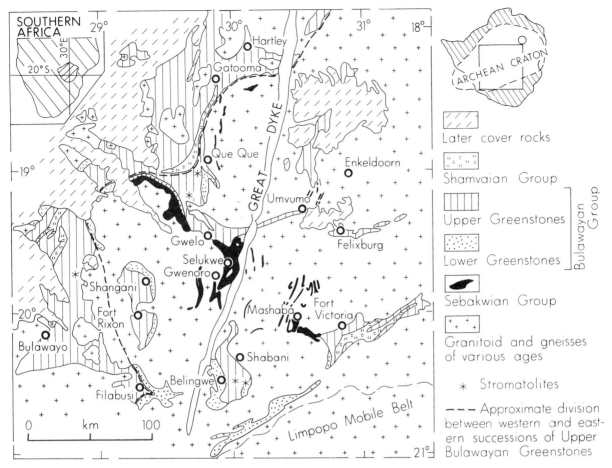

**Fig. 2-22.** Subdivisions of the greenstone belts in the central part of the Zimbabwe Craton. (After Wilson 1981, Fig. 8.8, and reproduced with permission of the author and of Elsevier Science Publishers).

has Selukwe, Fort Victoria and Shabani at the approximate corners (Fig. 2-22). This crustal segment consists predominantly of highly deformed ~3.5 Ga old tonalitic gneisses called Tokwe–Shabani. The gneisses contain infolded relicts, some of substantial size, of older Sebakwian greenstones, notably that of the structurally inverted Selukwe nappe (see below).

The tonalitic basement gneisses, or 'gregarious batholiths' of Macgregor (1951), comprise gneiss and migmatite varying in age between pre-Sebakwian and post-Bulawayan (Bliss and Stidolph 1969). Contacts with the schist (greenstone) belts are generally conformable, the two lithologically contrasting units having been deformed together. The gneiss complexes which contain numerous mainly ultramafic to mafic metavolcanic-rich, ~3.5 Ga-old Sebakwian Group remnants, range from the highly flattened banded, migmatitic variety of the Tokwe River exposures, west of Mashaba, to simpler homogeneous foliated types.

The best documented area of Sebakwian rocks is at Selukwe (renamed Shurugwi). Here the greenstone belt sequence is viewed as structurally inverted, forming the lower limb of a large allochthonous, recumbent fold structure known as the Selukwe nappe (Stowe 1984). This inverted terrain covers at least 1200 km², as now exposed, and horizontal movement may have exceeded 20 km. The igneous rocks range from peridotitic to basaltic and are largely metavolcanic; they include the chromite-bearing intrusion of the Selukwe ultramafic complex (Cotterill 1979, Prendergast 1987, Stowe 1994). However, reassessment of the structure indicates a more localized recumbent inversion with the major part of the stratigraphic succession essentially synclinal and readily correlatable with c. 2.7 Ga Upper Greenstones (see below)

(Tsomondo et al 1992). The chrome deposits are hosted in the large ultramafic–mafic sill-like complexes in simatic (Mg-rich) volcanic sequences (Stowe 1994) suggesting that early Archean oceanic crust was more ultramafic in composition than at present.

### Younger greenstone and late granite association

The *Bulawayan Group*, including Lower and Upper Greenstones, forms the most extensive greenstone belts in Zimbabwe (Wilson, Bickle and Nisbet 1993) (Fig. 2-22). At Belingwe in south–central Zimbabwe, situated near the south end of the Great Dyke, Upper Greenstones are folded into a major N-trending syncline, whereas Lower Greenstones flank this syncline on the western and, locally, southeastern sides (Kusky and Kidd 1992). Elsewhere more variable relations are observed.

*Lower Greenstones* comprise felsic flows and pyroclasts, mafic to ultramafic lavas, quartzite, conglomerate and BIF. Thick uppermost felsic volcanic units and derived schists are common. The unconformably overlying, predominating *Upper Greenstones* include, in the lower part, up to 7 km thick of both sedimentary and mafic–ultramafic volcanic rocks, locally with remarkable preservation of komatiite minerals and texturers (Renner et al 1994), conformably overlain in the western succession by up to 8 km thick of bimodal (basalt–dacite) volcanic rocks, andesites, and, in the eastern succession, by a thinner succession (2 km) of mafic–ultramafic lava flows).

Based on their study of greenstone data from a variety of Archean cratons, including Zimbabwe, Yilgarn, Pilbara, Superior, Slave and Wyoming, Bickle et al (1994) conclude that Archean greenstone belts do not contain ophiolites nor do they represent oceanic crust despite the possible presence of relict Archean oceanic crust preserved elsewhere in some Archean tectonic settings.

The type area of the *Shamvaian Group* is the Shamva Grits, northeast of Harare (Wilson 1981). Shamvaian sediments, dominated by arkose and subgreywacke, and products of a major transgression, were deposited unconformably on the Bulawayan successions and separated from them by major folding. Their distribution in the central cratonic area south of Harare is shown in Fig. 2-22.

*Younger potassic granites* postdate the main Bulawayan greenstone belts. Especially large masses occur east and northeast of Fort Victoria in the central cratonic area. Widespread ~2.6 Ga thermal disturbances over much of the southern part at least of the Zimbabwe Craton can probably be attributed to the widespread intrusion of these late granites.

Macgregor's (1951) 'gregarious' batholiths are now known to be but an idealistic representation of what is, in fact, a complicated tectonic–intrusion relation. The batholiths comprise both granitoid intrusions and gneiss of different ages, some postdating and some predating the Bulawayan greenstones. Diapiric intrusion undoubtedly played a deformational role but late granites commonly transgress most of the structures and postdate most of the tectonism. Nor are all the greenstone belts synclinal; in places they represent remnants of large structures disrupted by invading late granitoids. Furthermore, in the northern part of the craton, a NNW-trending fold interference pattern, superimposed on an earlier ENE-trend, explains the configuration of late granites. Thus configuration of the greenstone belts can no longer be explained merely in terms of multiple granitic intrusion (Snowden 1984).

### The Great Dyke

The unique Great Dyke (Fig. 2-16), a long narrow linear mass of mafic–ultramafic rocks, 530 km by 11 km in dimension, is composed of peridotites with chromite bands, pyroxenites and norites which are disposed in separate but continuous elongate, gently inward dipping masses emplaced in four main intrusive centres or complexes aligned along a NNE-trending graben structure. Uniformity of layers over many kilometres emphasizes the prevailing synclinal structures. The layers typically dip in from the margins of the dyke towards the central axis; the layers plunge gently south and north from the north and south limits respectively of each intrusive complex to produce canoe-like forms. (Worst 1960, Jackson 1970, Podmore 1970, Wilson 1981).

Schemes for the emplacement of the Great Dyke are reviewed by Wilson (1981). Earlier schemes involving single or multiple intrusion of successive pulses of ultramafic magma have been superseded by models involving crystallization from a large volume of liquid for each complex, with or without addition of original liquid to the crystallizing system (Wilson and Prendergast 1989, Wilson and Tredoux 1990).

The age of the Great Dyke emplacement including southern satellites in the Limpopo Belt has been documented at 2514 Ma (Robertson and Van Breemen 1970).

## Mineral deposits

Gold mining was the mainstay of Zimbabwe's economy for many decades and it was early established that gold favours the greenstone ('gold' or schist) belts. Gold deposits are present in both Sebakwian and Bulawayan greenstone assemblages (Anhaeusser and Maske 1986).

Chrysolite asbestos is mined in the southern part of the craton, with the bulk of the production coming from two major members of the Mashaba Ultramafic Suite, at Shabani and Mashaba respectively.

High grade stratiform chromitite deposits are mined at Selukwe and to a lesser extent at Mashaba. The deposits are hosted in large ultramafic-mafic sill-like complexes (Stowe 1994).

Important nickeliferous sulphide deposits, associated with gabbro–metagabbro masses, appear to be confined to the later (Bulawayan) greenstone belts.

Beryllium–lithium pegmatite mineralization is a feature of some of the Chilimanzi Suite of late granites.

## 2.9 AFRICAN CRATON: CENTRAL AFRICA

The ring of 'ancient' cratons surrounding the Congo Basin (Fig. 1-5f(i,ii)), broadly stable since the end of the Archean, includes Kasai–Angolan, Chaillu, Gabon, Bouca (Yadé), Bomu–Kibalian, Tanzania and Zambian (Bangweulu Block) Cratons. They are separated one from the other by intervening Proterozoic belts or cover.

The *Kasai–northeastern Angolan Craton*, about 1000 km by 800 km, comprises a highly metamorphosed Precambrian basement, widely buried beneath thin upper Paleozoic to Mesozoic–Cenozoic cover (Cahen et al 1976, 1984). The exposed Kasai Craton comprises (1) extensive tracts of NE–ENE-trending gneiss–migmatite (2.65 Ga) and calc-alkalic granites (2.59 Ga) of the Dibaya Complex; and (2) the Kasai–Lomami Complex of gabbro-norite and charnockite with ancient gneiss remnants (Luanyi) (3.4 Ga). Extensive granulite-grade metamorphism occurred at 2.82 Ga.

The southern continuation of the *Angolan Craton* extends for 900 km to Namibia with a width of some 1300 km, almost to the Atlantic coast. Much of the central–eastern part is obscured by Phanerozoic cover.

Angolan basement gneisses, with dates to ~3.4 Ga are overlain by early Proterozoic metasupracrustal rocks and associated migmatitic gneiss, themselves widely deformed and metamorphosed by Eburnea-age (~2.1 Ga) granitoid plutons. The vast Kunene gabbro–anorthosite complex in the southwest is about 2.1 Ga old. Post-tectonic magmatism in the region continued to 1.75 Ga. Anorogenic intrusions at 1.4 Ga old and a tectonic event at 1.3 Ga may correspond to the Kibaran event elsewhere.

To the north, the *Chaillu Massif* (Cahen et al 1984, Ledru et al 1989), a small (400 × 200 km) rectangular block, is composed predominantly of older Archean grey gneiss and late Archean (2.7 Ga) pink migmatitic granite masses. Some thin greenstone septa predate the granitoid rocks. The massif is unconformably overlain by flat-lying to gently folded early Proterozoic Francevillian–Ogooué beds on the east, and by late Proterozoic West Congo strata on the southwest.

The nearby *Gabon (Ntem) Massif*, a subcircular mass about 500 km across also belongs to the foreland of the N-trending West Congo Mobile Belt. The 3.0–2.6 Ga massif is dominated by mid to late Archean (2.9 Ga) granulite facies metamorphites (Toten et al 1994), commonly retrograded (2.1–2.0 Ga, i.e. Eburnean age) along zones of cataclasis, and intruded by post-tectonic granite plutons (Cahen et al 1984, Toten et al 1987, 1994, Caen-Vachette et al 1988, Nedelec et al 1990).

The Gabon massif is transitional northward by increasingly pervasive Pan-African tectono-metamorphic activities (620–590 Ma) (Toten et al 1994) to the E-trending regional Central African Mobile Belt.

To the east, the *Bomu–Kibalian Craton* (also known as West Nile Craton) forms the northern border of the Congo Basin (Legge 1974, Saggerson 1978, Lavreau 1982, Cahen et al 1984). The small intervening Yadé (Bouca) Massif includes the Bandas greenstone belts. Starting with the small Bomu extension on the west, the main Kibalian Craton extends eastward for 1000 km. The main craton is about 300 km wide extending from Lindian cover rocks on the south to the Central African Mobile Belt on the north, of which it forms the principal foreland. Dominant tonalitic to granodioritic gneiss–migmatites (~2.9 Ga) intruded by granite plutons (2.5 Ga) include at least two ages of greenstone belts: (1) older mafic volcanic-rich

Ganguan greenstones, which unconformably overlie ~3.3 Ga old Bomu mafic gneiss and are intruded by 2.9 Ga old plutons; and (2) younger Kibalian greenstones, which include both an older volcanic-rich facies, intruded by 2.9 Ga old tonalites, and a younger felsic volcanic–quartzite–BIF-rich facies intruded by 2.5 Ga old plutons.

The N-trending, oblong *Tanzania Craton*, 1000 km by 500 km, is entirely bounded by early Proterozoic and younger mobile belts (Cahen et al 1984, Gabert 1984). The southern–central parts of the craton contain dominant medium to high grade gneiss–migmatite with contained Dodoman schist belts, widely intruded and metamorphosed by 2.6 Ga old granitoid plutons. The northern part comprises older Nyanzian schist belts, intruded by ~2.8 Ga old granitoids and unconformably overlain by Kavirondian metapelite–arkose–conglomerate sequences, themselves intruded by 2.5 Ga old granitoids.

The triangular *Zambian Craton (Bangweulu Block)* to the southwest, about 500 km on the side, is composed of NE-trending, 1.1 Ga old orogenic belts that represent the foreland zones of the mid-Proterozoic Irumide Belt to the south. The fold-and-thrust and shear zones of the craton tectonically overlie an early Proterozoic crystalline basement that may include Archean components. The predominant northwestward-verging cover represents major crustal shortening in that direction above a décollement on the basement (see further below).

## 2.10 AFRICAN CRATON: WEST AFRICAN CRATON AND TRANS-SAHARAN MOBILE BELT

Archean crust forms the western parts of both the Man and Reguibat shields of the West African Craton, and occurs locally in the western branch and central part of the nearby Tuareg Shield (Fig. 2-23). Rare early Archean components (> 3.5 Ga) are present in local 3.0 Ga orthogneiss in northern Nigeria (see Chapter 3).

### 2.10.1 MAN SHIELD

Man (Ivory Coast) Shield is divided into the Archean-rich Kenema–Man (Liberian) domain to the west and the early Proterozoic-rich Baoulé–Mossi (Eburnean) Domain to the east (Fig. 1-5f(i,ii); Table 2-12, column 1), the latter with scattered Archean basement patches.

The *Kenema–Man Domain*, of which the Sierra Leone terrain is typical, is characterized by numerous (more than 60) small relict BIF-bearing greenstone belts distributed in granifoid gneiss and plutons (Rollinson 1978).

In the Kenema assemblage of northeast Sierra Leone, two separate greenstone suites are recognized on the basis of the intervening Leonean event (~3.0 Ga) (Macfarlane et al 1981) (Table 2-13, column 1): the older Loko Group composed of amphibolite, serpentinite, quartzite and BIF, and the younger Kambui supergroup, more than 6 km thick, composed of lower mafic–ultramafic volcanic rocks, and upper tuff, psammite, pelite and commercially important BIF. However, despite the evidence favouring this intervening event, Williams (1978a) and Williams and Culver (1988) interpret the evolution of the entire Kenema–Man Domain in terms of a single inhomogeneous Liberian event.

In Liberia to the east (White and Leo 1969, Hurley et al 1971, Hedge et al 1975), a crystalline basement (80%) of granitoid gneiss and plutons, paragneiss and schist is associated with NE-trending, BIF-rich Liberian greenstone belts. Large granitoid domes appear to be autochthonous. The basement complex varies in metamorphism from amphibolite to granulite facies. Liberian greenstone belts, including the 2 km thick succession at Nimba Mountains, comprise micaceous quartzite, amphibolite metapelite and important BIF, at prevailing greenschist to amphibolite facies with local granulite facies.

The westernmost part of Man Shield (Figs. 1-5f(i,ii), 2-23) comprises three narrow, N–NW-trending belts of the combined Rokel–Marampa–Kasila zone: (1) in the east, the Rokel River Group (Rokelides), ~7000 m thick, is composed of conglomerate, arkose, shales and mafic–felsic volcanic rocks; (2) in the centre, occur thrust-bound tectonic slivers or klippen of more deformed and metamorphosed, recumbently folded, Marampa metasediments and metavolcanic rocks, of uncertain late Archean–early Proterozoic age (Williams 1988); and (3) in the west, the 300 km long and up to 60 km wide Kasila belt is composed of dominantly high grade (granulites) Archean supracrustal rocks (Williams 1988).

The Kasila Group is part of the Kambui Supergroup, a term defined to include all Archean supracrustal rocks in Sierra Leone. The mafic granulites, metaleucogabbroic intrusions, metasedimentary

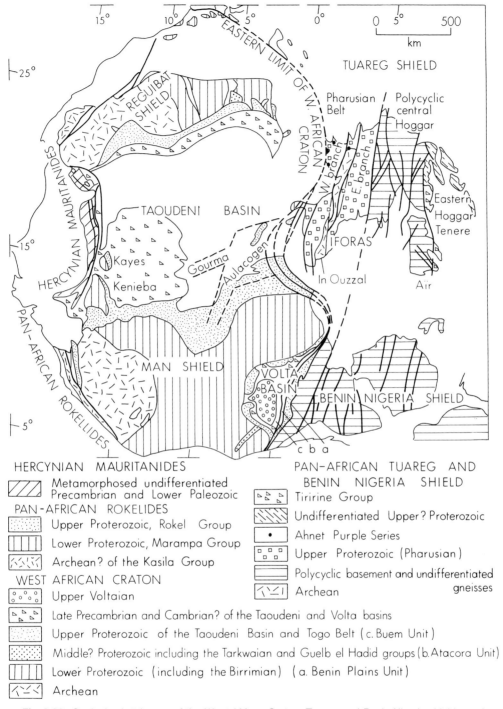

**Fig. 2-23.** Geologic sketch map of the West African Craton, Tuareg and Benin Nigeria shields, and late Proterozoic to Phanerozoic fold belts. (Adapted after R Black 1980, from Cahen et al 1984. Fig. 21.1, and published with permission of the author and of Oxford University Press).

granulites and migmatites, and meta-BIF of the Kasila Belt represent the deep-seated remains of a highly telescoped supracrustal succession. The eastern boundary of the Kasila Belt is a zone of highly deformed and mylonitized granulites at least 5 km wide which dips westwards at low angles. This zone has been plausibly interpreted as an Archean suture developed during tectonic collision of the Guiana Shield (South America) and the West African Craton (Williams and Culver 1988).

## 2.10.2 REGUIBAT SHIELD

To the north, the Reguibat Shield, some 1400 km long (east–northeast) and up to 40 km wide, also includes an Archean-rich Southwestern Province (15°W–12°/8°W) and an early Proterozoic-rich Eastern Province.

In the western region of the Southwestern Province (Cahen et al 1984), the Amsaga basement assemblage comprises two groups of rocks: an older high grade Saouda association of charnockites, pyroxene amphibolites, sillimanite gneiss, marble and widespread BIF (itabirite), and a younger migmatitic complex (Table 2-12, column 2).

In the eastern region of the Southwestern Province, the corresponding Ghallaman assemblage comprises N–NNW-trending patches of gneiss, migmatite, amphibolite, marble and quartzite. These are intruded successively by older diorite–granodiorite and younger adamellite–granodiorite plutons.

The Eastern province, lying astride the Mauritania–Algeria border, includes patches of Chegga (Table 2-12, column 3). Syn-tectonic granitoid plutons are common. Amphibolite facies prevails but with widespread greenschist retrogression and mylonitization. The assemblage is commonly equated with the Amsaga assemblage to the southwest.

## 2.10.3 TUAREG SHIELD

Archean rocks in the Tuareg Shield include a prominent submeridional granulitic slice, the In Ouzzal Domain, in the Western Hoggar region, as well as small exposures of Oumelalen gneisses in the Central Hoggar region (Fig. 2-23; Table 2-13, columns 1, 2).

### In Ouzzal Domain

The In Ouzzal Domain constitutes a long submeridional block thinning southward towards the Mali–Algerian frontier, then reappearing in a westerly displaced slice which widens to the south through the Adrar des Iforas (Bertrand and Lasserre 1976, Haddoum et al 1994). Granulites are common, including charnockites and leptynites (quartzofeldspathic gneiss), pelitic gneisses, marbles, BIF (itabirites) and calcsilicates, in addition to metaplutonites (charnockite, norite and lherzolite). Pan-African metamorphic retrogression, restricted to the southerly (Iforas) slice, is marked by amphibolite-facies assemblages.

Available age data, in brief, indicate granulite facies metamorphism at ~3.0 Ga on pre-existing quartzofeldspathic crust (Allègre et al 1972, Lancelot et al 1976, Ben Othman et al 1984, Carpena et al 1988).

The In Ouzzal basement is cut by 600–550 Ma old granites, representing the Pan-African Orogeny. However, the In Ouzzal-Iforas basement, along with the West African Craton to the west, underwent its last major tectono-thermal event at 2.1–1.9 Ga, and behaved thereafter as a comparatively inert block. Sedimentary basins overlie the In Ouzzal core in places and postdate a 600 Ma old regional uplift related to the Pan-African Orogeny (Caby 1972).

The Oumelalen-part of the Central Hoggar region includes a basement composed of homogeneous banded gneiss, called the Red 'Series' (Group), in which are distributed thin bands of marble, quartzites and metapelites (Table 2-13, column 2). The main metamorphic foliation is more or less gently dipping, often horizontal, and commonly associated with polyphase recumbent folds.

Granulites of the Red 'Series' have provided a single date of 3480 Ma (Latouche and Vidal 1974, Latouche 1978) with other results scattering in a 2500–3400 Ma age range. However, subsequent U–Pb zircon studies failed to confirm any pre-3.1 Ga date (Carpena et al 1988).

## 2.11 INDIAN CRATON

### 2.11.1 DHARWAR CRATON

The Dharwar Craton (Karnataka Block) of south-central India, which together with the South Indian Highlands (Granulite Domain) forms the Dravidian (South Indian) Shield, covers an area of 238 000 km² (Figs 1-3g, 1-5g, Table 2-14, column 1). The Dharwar Craton is underlain largely by (Peninsular) granitoid gneiss (80%), with numerous substantial NNW-trending schist (greenstone) belts and a plethora of small metasupracrustal enclaves (collectively 20%) (Naqvi 1981, 1982, Radhakrishna 1983, 1984, Naqvi and Rogers 1987, Sarkar 1988, Naha et al 1991).

Three divisions of the Dharwar Craton are recognized (1) a narrow, E-trending Gneiss–Granulite Transition Zone in the south; (2) Eastern Dharwar Domain; (3) Western Dharwar Domain; (4) Close-

*Table 2-14.* Generalized chronostratigraphic events in the Dharwar Craton, Granulite Domain, Eastern Ghats Belt, Singhbhum Craton and Aravalli Craton of India.

| Time scale (Ga) | (1) Dharwar Craton | (2) South Indian Highlands | (3) Eastern Ghats Belt | (4) Singhbhum Craton | (5) Aravalli Craton |
|---|---|---|---|---|---|
| 0.5 | | | | | |
| 1.0 | Kurnool Group, 1.0–0.5 Ga (Cuddapah Basin) | | Indian Ocean Orogeny, 0.7–0.45 Ga<br>High grade metamorphism (also in Sri Lanka), 1.1 Ga | | Vindhyan Supergroup 1.1–0.6 Ga |
| 1.5 | Nallamalai Group, 1.5–1.0 Ga (Cuddapah Basin) | | | Newer Dolerite dikes and sills, (1.6–0.9 Ga) | |
| 2.0 | Cuddapah Group, 2.0–1.5 Ga<br>Kaladgi–Godavari sediments | | Eastern Ghats orogeny, charnockites<br>Eastern Ghats Front, 1.6 Ga | Copperbelt Thrust, 1.6 Ga<br>Mayurbhanj Granite, 2.1 Ga | Delhi Supergroup 1.8–1.5 Ga |
| | | | | Kolhan Group, 2.2–2.1 Ga<br>Dhanjori Group, 2.3 Ga<br>Singhbhum Group, 2.3 Ga | Aravalli Supergroup, 2.5–2.0 Ga |
| 2.5 | Deformation, metamorphism, 2.6 Ga | Granulite facies metamorphism, 2.6 Ga | | | Metamorphism of Bundelkhand Igneous Complex and Bhilwara assemblage<br>Darwal granite |
| | Closepet granite, K-metasomatism<br>Granitoid plutonism<br>Dharwar–Kolar schist belts, 3.0–2.6 Ga | Mafic–ultramafic complexes<br>Platform sediments | Charnockites (2.6 Ga)<br>Khondite precursor deposition | | |
| 3.0 | Medium to high grade metamorphism, 3.1–3.0 Ga | | | | Metamorphism of BGC, 3.1–3.0 Ga |
| 3.5 | Sargur enclaves, 3.1 Ga<br>Kolar gneiss (W), 3.2 Ga<br>Peninsular gneiss, (3.3–3.0 Ga)<br>Gorur gneiss (3.3 Ga) | Tonalite intrusion(?) | | Singhbhum granite (Late Phase), 3.1 Ga<br>Iron Ore Group, 3.3–3.1 Ga<br>Older Metamorphic Group and Singhbhum granite (Early Phase), 3.4–3.3 Ga | BGC, 3.5–3.3 Ga |

BGC = Banded Gneissic Complex

pet Granite, separating the two Dharwar domains; and (5) pervasive Peninsular Gneiss.

(1) The *Gneiss–Granulite Transition Zone*, 30–60 km wide (N–S), is transitional from low grade granitoid–greenstone terrain in the north to high grade granulite terrain in the south (Fig. 1-5g) (Janardhan et al 1978, Naqvi and Rogers 1987). Metamorphic grade is characterized by upper amphibolite facies grading southward to granulite facies. The transition zone is underlain by ambient Peninsular gneiss–migmatite (see below) with numerous small, medium to high grade Sargur-type

Table 2-15. Sequence of principle early Precambrian events in the Singhbhum Craton, India.

| | | |
|---|---|---|
| | New Dolerite sills and dikes | |
| | Mayurbhanj granite 2.1 Ga | |
| | Gabbro–anorthosite intrusions | |
| | Ultramafic intrusions | |
| Kolhan Group, 2.2–2.1 Ga | Orthoquartzites, limestone, shale | |
| ——————————————————— Unconformity ——————————————————— | | |
| Dhanjori Group, ~2.3 Ga | Dhanjori-Simlipal lavas with quartzite, conglomerate | Jagannathpur Lavas |
| | | Malangtoli Lavas |
| ——————————————————— Unconformity ——————————————————— | | |
| Singhbhum Group, 2.4–2.3 Ga | Pelitic and arenaceous metasediments | Dhalbhum Formation: mica schists, phyllites. |
| | | Chaibasa Formation: mica schists, quartzites |
| ——————————————————— Unconformity ——————————————————— | | |
| | Singhbhum granite (Late Phase), 3.1 Ga, plus Nilgiri–Bonai granites | |
| | Deformation and metamorphism (Iron Ore Orogeny) | |
| | Diorite intrusions | |
| | Upper shales with volcanics | |
| | Banded hematite jasper with iron ore (BIF) | |
| Iron Ore Group, 3.2–3.1 Ga | Felsic volcanic tuffs, tuffaceous shales, mafic lavas with tuffs | |
| | Sandstone and conglomerate (local) | |
| ——————————————————— Unconformity ——————————————————— | | |
| | Singhbhum Granite (Early Phase), 3.3 Ga | |
| | Folding and metamorphism | |
| Older Metamorphic Group 3.3–3.5 Ga | Biotite-tonalite to granodiorite gneiss | |
| | Hornblende schists and amphibolites; metagabbros | |
| | Calc-magnesian metasediments; calcsilicates; hornblende schist; | |
| | muscovite–biotite schist; quartzite; quartz schist | |

schist units, the latter in the form of narrow bands, fold remnants, scattered enclaves and tectonic slices.

Sargur-type schist units (Fig. 2-24), up to 5 km long and 1–5 km wide, are characterized by (1) high metamorphic grade; (2) intense polycyclic deformation; (3) aluminous metasediments with BIF; and (4) chromite-bearing mafic–ultramafic complexes. No clear-cut evidence of basement-cover relations has been reported (Chadwick et al 1978, 1981, 1986, Janardhan, 1978, Srikantappa 1984, Naqvi and Rogers 1987).

Based on detrital zircon SHRIMP studies, deposition of Sargur protoliths and intrusion of the major gabbroic and peridotitic complexes occurred in the period 3130–2960 Ma (Nutman et al 1992). On the same basis, granitoid rocks in the age range 3580–3130 Ma were significant components of the Sargur sedimentary protoliths. Prior dating indicated two major episodes of deformation–metamorphism, at ~3.0 Ga and 2.6 Ga respectively.

(2) The *Eastern Dharwar Domain* (i.e. east of Closepet Granite) is characterized in the western part by small, sparse (about eight), locally auriferous Kolar-type schist belts, distributed in predominant Peninsular gneiss–migmatite terrains. Two main producing gold mines, Kolar and Hutti, lie in schist belts located in the southern and northwestern parts of the domain respectively (Fig. 2-24) (Radhakrishna 1983, Sarkar 1988, Krogstad et al 1991, Siddaiah et al 1994).

In general the schist belts are small, narrow, linear, complexly deformed units 10–50 km long by 1–5 km wide (Fig. 2-24). They are composed of

INDIAN CRATON 109

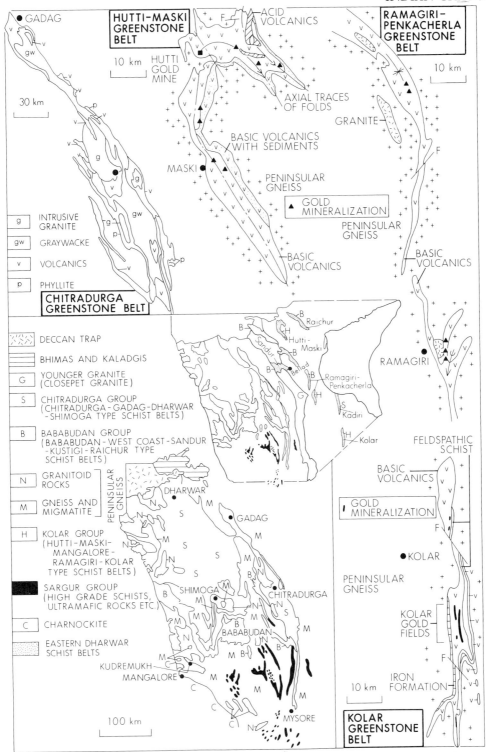

**Fig. 2-24.** Distribution of Dharwar, Sargur and Kolar greenstone (schist) belts in the Dharwar Craton of southern India. Individual maps show Western and Eastern Dharwar Domains, Chitradurga, Kolar and Hutti-Maski and Ramagiri-Penkacherla greenstone belts. (Based on Naqvi and Rogers 1987, Figs 2.1, 3.1, and Sarkar 1988, Fig. 11, and published with permission of the authors and of Oxford University Press).

predominant mafic–ultramafic volcanic assemblages with minor metasediments, including local conglomerate.

(3) The *Western Dharwar Domain* (i.e. west of Closepet Granite), the main metasupracrustal-bearing province in the Dharwar Craton, covers an area of 68 500 km$^2$ (Fig. 2-24). Metasupracrustal rocks therein form the type area for the Dharwar Supergroup (Chadwick et al 1981a, b, Naqvi 1981, Radhakrishna 1983, Chadwick et al 1985a, b, 1989, 1991, Naqvi and Rogers 1987, Chadwick et al 1988, Sarkar 1988, Manikyamba et al 1993, Khan et al 1993, Arora and Naqvi 1993, Manikyamba et al 1993).

Rocks of the Dharwar Supergroup overlie the +3.0 Ga old Peninsular gneiss with at least local regional unconformity and are, in turn, unconformably overlain to the north by flat-lying, unmetamorphosed 1.7 Ga old Kaladgi sediments. Eight main Dharwar schist belts, commonly BIF-bearing, are broadly distributed across the Province. They include the unusually large Dharwar–Shimoga belt in the north, Kudremukh–West Coast and Bababudan belts in the centre–west, Chikadurga–Gadag belt in the east, and Sandur belt still farther east: probable correlative belts (Kustigi and Raichur) lie near Hutti in the Eastern Dharwar Domain.

Dharwar schist belts are typically broad, open, curvilinear, elongate structures (Swami Nath and Ramakrishnan 1981). They are distributed in well-defined, oval-shaped depositional basins, typically with basement unconformities at the western and southwestern boundaries but with faulted, mylonitized and otherwise tectonically distorted contacts marking the eastern and northern boundaries. They feature, in general, shallow-water shelf facies sediments at the base (to the west) and thick volcanic–greywacke assemblages above (to the east). The composite Dharwar Supergroup, about 8 km thick, comprises two cycles, the lower Bababudan Group and the unconformably overlying Chitradurga Group. Facies relationships are complex and variable from place to place.

Mafic metavolcanic rocks in the Kudremukh–West Coast schist belt provide an age of ~3.0 Ga (Drury et al 1983), and nearby basement gneiss an older age limit of ~3.3 Ga (Beckinsale et al 1980). A granite intrusion (Chitradurga) dated at 2605 Ma (Taylor et al 1984) sets the younger age limit. Additional Dharwar dates are listed in Naqvi and Rogers (1987, table 2-2).

(4) The mainly N–NW-trending *Closepet Granite Complex*, the subdivider of the Dharwar Craton into two domains, comprises a series of polyphase granite bodies with variable remobilized gneiss, collectively forming a linear belt between 10 and 50 km wide and 500 km long. Closepet granite is intrusive into Peninsular gneiss. For most of its length the granite escaped any metamorphism. However, charnockite metamorphism of Closepet granite is adjoining the Gneiss–Granulite Transition Zone to the south (Radhakrishna 1984, Friend 1984, 1985, Allen et al 1986, Jayananda and Mahabaleswar 1991).

The age of formation of the Closepet Granite Complex is 2500–2600 Ma with a precise age of 2513 Ma interpreted as the time of crystallization of the granite (Friend and Nutman 1991).

(5) *Peninsular Gneiss Complex* is the name applied to assorted gneiss–migmatites that underlie broad areas throughout Peninsular India (Fig. 1-5g) (Pichamuthu 1967, Radhakrishna 1984, Ramakrishnan et al 1984). The gneiss complex represents a long period of time and affords evidence of several distinct episodes of intrusion, injection, granitization, metamorphism and tectonic deformation. It has proved extremely difficult to differentiate all the component types and to determine the sequence of their formation. Recent age determinations (see below) indicate that Peninsular gneiss formed mainly in a 3300–3000 Ma cycle of events prior to accumulation of the Dharwar Supergroup.

Peninsular gneisses are generally banded and often strongly contorted. Rock types vary from coarse-grained gneissic granite to strongly foliated migmatitic gneiss. Agmatites are common, as are pegmatites and quartz veins. The gneisses characteristically contain mafic inclusions of diverse size and composition. Especially in the Shimoga district, granodioritic and gneissic facies are pervasively retrogressed (Chadwick et al 1985a).

Peninsular gneiss underlying the regional unconformity at the base of both Bababudan and Chitradurga belts provide reliable dates of 3.2–3.0 Ga (Taylor et al 1984). U–Pb zircon data on Peninsular gneiss in Karnataka yield an age of 2963 Ma for the gneiss protolith (Friend and Nutman 1991).

## 2.11.2 GRANULITE DOMAIN (SOUTH INDIAN HIGHLANDS)

The late Archean high grade terrain of southern India extending southward for 350 km from the Gneiss–Granulite Transition Zone to the southern continental tip, is one of Earth's largest granulite provinces, with granulite facies rocks underlying some 70 000 km² (Buhl et al 1983, Condie and Allen 1984, Hansen et al 1984, Srikantappa et al 1985, Narain and Subrahmanyam 1986, Kumar and Raghavan 1992, Radhakrishna et al 1992) (Fig. 1-5g; Table 2-14, column 2).

Common rock types in the Granulite Domain include grey tonalitic to trondhjemitic gneiss, pink granitic gneiss, pegmatites, local migmatites and charnockites. Anorthositic complexes occur sporadically. Field relations strongly suggest that both granitoid gneiss and charnockite of the transition zone developed from tonalitic and metasupracrustal (see below) protoliths by metamorphic and metasomatic processes.

Great thicknesses of granulite-facies metasedimentary rocks occur in the Granulite Domain, particularly in the Khondalite Belt of southernmost India where layered granulites are exposed over an area 150 km by 80 km. This high grade supracrustal sequence is composed of graphitic biotite–garnet gneiss, khondalites (garnet–sillimanite metapelites) and lepynites (leucocratic gneiss with garnet) with minor quartzites, marbles and mafic granulites. Felsic orthogneisses in the Khondalite Belt are invariably charnockitic, except where retrogressed. It is reasonably postulated that the Khondalite Belt and adjoining high grade terrains, including much of the Eastern Ghats Belt to the northeast (see below), represent the lateral facies transition (southward–eastward) to continental shelf deposits of the main Dharwar greenstone (schist) belts (3.0–2.6 Ga) to the north. Although no direct age dates on the metasediments are available, the metasupracrustal sequence can be presumed to be older than 2600 Ma. Nd isotopes provide model ages from 3.4 Ga to 2.4 Ga with indications of significant crustal growth during the late Archean (3.0–2.5 Ga) (Harris et al 1994).

## 2.11.3 EASTERN GHATS BELT

The NE-trending, 100–200 km wide, high grade Eastern Ghats Belt extends for some 1200 km along the east coast of India (Figs 1-5g, 5-17; Table 2-14, column 3). At the northern end of the belt, the dominant NE trend changes abruptly to ESE, parallel to the Mahanadi Valley (Mahalik 1994). South of Bezwada (south–central part), one branch continues to the southwest, the other branch striking away from the land towards Sri Lanka and may re-appear there as the Highland Complex (charnockite–khondalite assemblage) (see below). A major thrust along the western margin of the belt, expressed in intense shearing and crushing, has been confirmed by deep seismic sounding (Naqvi and Rogers 1987). The Eastern Ghats Belt may have been affected by a late Archean metamorphism, as well as other younger metamorphic events.

The Eastern Ghats Belt is mainly composed of subparallel alternating layers of charnockite (pyroxene granulite), various manganiferous rocks (khondalite, kodurite, gondite), and associated massive to gneissic granitoids together with anorthosites/layered complexes and post-tectonic alkaline intrusions. All but the last are products of high grade, deep seated regional metamorphism under essential anhydrous conditions (Ray and Bose 1975, Roy 1981, Halden et al 1982, Park and Dash 1984, Naqvi and Rogers 1987).

In general, charnockites occupy the western marginal part of the Eastern Ghats Belt, whereas khondalites and other pyroxene-free rocks predominate to the east (Fig. 5-17). Within the belt, charnockites and mafic granulites are generally conformable and interlayered. At least two, and possibly more, periods of deformation have been demonstrated. Polyphase metamorphism is also likely (Naqvi and Rogers 1987).

## 2.11.4 SINGHBHUM CRATON OF SINGHBHUM–ORISSA

The ovoid Singhbhum Craton (Nucleus), 250 km by 170 km, forms the southern part of the combined Chotanagpur–Singhbhum Craton (Figs 1-5g, inset 3, 2-25; Table 2-14, column 4). The Singhbhum Craton and adjoining rocks developed in three main cycles (Table 2-15), two Archean and one early Proterozoic (Sarkar and Saha 1977, Sarkar et al 1981, Sarkar 1982, Sarkar and Saha 1983, Saha and Ray 1984, Saha et al 1984, Naqvi and Rogers 1987, Sengupta et al 1991).

*Cycle 1*: the centre–east part of the Singhbhum Craton is occupied by the Singhbhum Granite Complex, a multiphased, mid-Archean (3.3–3.1 Ga) intrusive suite which covers about 10 000 km². The oldest recognizable rock units in this complex are represented by enclaves of the Older Metamorphic Group (OMG) composed of mica schists, quartzite, calcsilicates and amphibolites, all metamorphosed under amphibolite facies (Sarkar and Saha 1983).

*Table 2-16.* Generalized sequence of main events by region in the Australian Craton.

| Time scale (Ga) | (1) West Australian Shield | (2) North Australian Province | (3) Central Australia | (4) South Australia |
|---|---|---|---|---|
| 0.6 | | Georgina Basin, 0.9–0.5 Ga | Officer Basin ⎫ 0.9–0.5 Ga<br>Amadeus Basin ⎭ | Delamerian Orogeny, 0.52 Ga<br>Adelaide Basin, 1.0–0.5 Ga |
| 1.0 | Bangemall Basin, 1.3–1.0 Ga | Victoria River Basin, 1.2–1.1 Ga<br>Argyle diamond pipe, 1.18 Ga<br><br>Dolerites, 1.76 and 1.69 Ga | Musgrave-Fraser Mobile Belts,<br>Polycyclic Orogenesis: 1.8–1.6, 1.5–1.3, 1.2–1.0 Ga<br>Musgrave-Arunta Volcanics, 1.6–1.5, 1.4–1.3 Ga | Gawler Craton stabilized, 1.5 Ga<br>Olympic Dam Formation, 1.6 Ga |
| 1.5 | Isothermal uplift, 1.7–1.5 Ga<br>Gascoyne Province<br>Earaheedy sub-basin, 1.8–1.7 Ga<br>Capricorn Orogen, 2.0–1.7 Ga<br>Nabberu Basin, 2.2–1.7 Ga | Northeast Orogens, 1.6 Ga<br>McArthur-Mount Isa Basins, 1.69–1.67 Ga<br>Kimberley Basin, 1.85–1.76 Ga<br>Barramundi Orogeny, 1.9–1.8 Ga | Aileron event, 1.7–1.65 Ga<br>Strangways event, 1.86 Ga<br>Albany-Fraser sediments, ~1.8 Ga<br><br>Arunta volcanics, 1.9–1.8 Ga | Gawler Range Volcanics, 1.6 Ga<br>Willyama Supergroup, 1.69 Ga<br>Kimban Orogeny, 1.8–1.7 Ga<br>Hutchison Group 2.0–1.8 Ga |
| 2.0 | Wyloo sediments, 2.0 Ga<br>Glengarry sub-basin, 2.2–1.8 Ga<br>Turee Creek Group, 2.4 Ga<br>Hamersley Group, 2.5 Ga | Pine Creek-type basins, 2.1–1.9 Ga | | Sleafordian Orogeny, 2.5–2.3 Ga |
| 2.5 | Yilgarn | Pilbara | Pine Creek Basement, e.g. Rum Jungle | Carnot gneisses, > 2.5 Ga |
| | Granitoids, 2.7–2.6 Ga | Hamersley Basin, 2.8–2.4 Ga<br>Fortescue Group, 2.8 Ga | | |
| | Greenstones, 3.0–2.7 Ga | | | |
| 3.0 | | Whim Creek Groups, 3.3–2.8 Ga<br>Gorge Creek | | |
| | Narryer Gneiss Complex, 3.7–3.35 Ga | Recurrent granitoids, 3.5–3.3, 3.0–2.8 Ga<br>Warrawoona Group, 3.5–3.3 Ga | | |
| 3.5 | | | | |
| | Manfred Complex, 3.73 Ga | | | |
| 4.3 | | | | |
| | Mount Narryer zircons, 4.28 Ga | | | |

Available geochronologic and supporting data indicate that OMG gneisses represent continental crust newly generated at ~3.4 Ga, practically coeval with the Singhbhum granite (Early Phase) (Moorbath and Taylor 1988).

*Cycle 2*: the Iron Ore Group, the oldest coherent supracrustal suite, occupies the remainder of the craton. In more detail, these structural relations are very complex and, in the absence of reliable dates, a variety of stratigraphic interpretations have been proposed, as reviewed by Naqvi and Rogers (1987). The Iron Ore Group proper is composed of shales, phyllites, banded hematite quartzites and jasper (BIF), mafic to felsic volcanic rocks, and mafic igneous bodies. Major iron and manganese ore deposits are present. The entire Iron Ore sequence was folded about NNE-trending, asymmetric to isoclinal axes and underwent low grade (greenschist) metamorphism.

This was followed by emplacement of the extensive Singhbhum granite (Late Phase) (Saha and Ray 1984, Saha et al 1984). Compositions range from adamellite to granite. The Singhbhum granite, in brief, is currently considered in terms of an Early Phase at ~3.35 Ga and a Late Phaes at ~3.1 Ga.

*Cycle 3*: subsequent Proterozoic events which mainly affected the northern parts of the Singhbhum Craton (i.e. Singhbhum–Dhalbhum Mobile Belt and Chotanagpur–Satpura Belt) are considered in Chapter 3.

## 2.11.5 RAJASTHAN AND BUNDELKHAND BLOCKS (ARAVALLI CRATON)

Older Precambrian rocks of Rajasthan (Aravalli Domain), located in north–central and northwestern India, are bounded on the west by desert sands and scattered outcrops of younger rocks, and on the northeast and south respectively by the Indo–Gangetic alluvium, the Vindhyan Basin and the northern fringes of the Deccan Traps (Figs 1-5g, inset 2, 2-25).

A significant part of the total area is underlain by a complex of granitoids–migmatites–metamorphites assigned to the Banded Gneissic Complex (BGC) (Naha 1983, Sharma 1983a, b, Basu 1986, Naqvi and Rogers 1987) (Table 2-14, column 5).

The BGC of Rajasthan is composed of granitoid gneiss, charnockite, migmatite, pegmatite, aplite and metabasic rocks, together with a large number of metasedimentary bands which remain as unabsorbed relics. Isotopic evidence favours an early

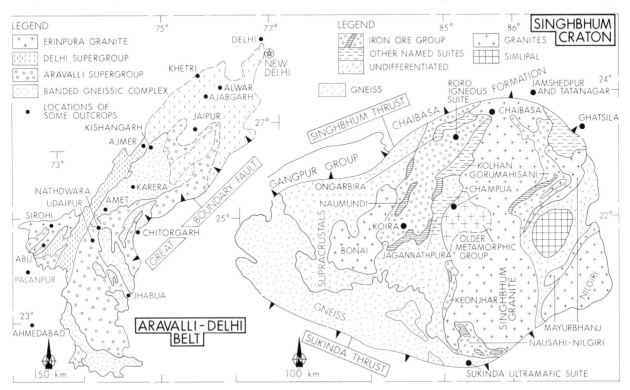

**Fig. 2-25.** Geologic maps of the Singhbhum Craton and Aravalli-Delhi Belt, northern India. (Based on Naqvi and Rogers 1987, Figs. 7.2 and 9.2, and published with permission of the authors and of Oxford University Press).

Archean presence to 3.3–3.5 Ga (Ahmad and Tarney 1994). Local BIF enclosed in BGC of Rajasthan have field-setting and petrochemical characteristics comparable to the BIF in Sargur Group rocks of South India (see above) (Sahoo and Mathur 1991).

The Bundelkhand region to the east, a little studied area of admixed gneissic and supracrustal rocks of Archean–earliest Proterozoic ages, contains a similar basement-cover assemblage (Sharma 1983b, Basu 1986, Naqvi and Rogers 1987).

## 2.11.6 BHANDARA CRATON

The southern Bastar region of the Bhandara Craton (Fig. 1-5g, inset 4) is dominated by granitoid rocks, tonalitic–granodioritic gneisses, small (schist) belts of metasupracrustal rocks, and late K-feldspar-rich leucocratic granitoid plutons. This assemblage has been intruded by Proterozoic mafic dykes and covered by late Proterozoic sedimentary sequences (Sarkar et al 1993).

A suite of trondhjemitic gneisses occurring in an enclave within a vast expanse of quartzofeldspathic gneisses provides zircon ages of 3509 Ma which are considered to be dates of primary crystallization (Sarkar et al 1993). Adjoining leucocratic granites yield an age of 2480 Ma.

The 3509 Ma age for the trondhjemite gneiss supports previous evidence that 3.5–3.3 Ga was an important period of crust formation (cratonization) in India. Similar dates have come from the Older Metamorphic Group (3.5–3.3 Ga) of the Singhbhum Craton and the Banded Gneissic Complex of Rajasthan (Table 2-14). The leucocratic granite age of 2480 Ma corresponds to widespread intrusions that concluded the Archean evolution of this craton.

It is noted that five middle Archean cratons in India with stabilization ages ≥ 3.0 Ga (Granulite Domain, West Dharwar, BGC of Rajasthan, Bastar and Singhbhum) constitute the central part of the proposed 3.0 Ga supercontinent of Ur (southern Africa–southern Madagascar–India–northern Antarctica–northwestern Australia) (Radhakrishna 1989, Meen et al 1992, Rogers 1993) (see below).

## 2.11.7 MINERAL DEPOSITS

Most gold occurrences in Peninsular India belong to Kolar-type schist belts in the East Dharwar Domain. Large, high grade iron ore deposits are associated with Singhbhum and Dharwar BIF.

Manganese ore deposits are located in the Sargur metasupracrustal rocks of Karnataka, in kodurite of Andhra Pradesh, in other metasedimentary manganiferous granulites of the Eastern Ghats, and in the Mn-bearing BIF of Singhbhum.

Chromite and nickel deposits are enclosed in peridotites in Karnataka and in Orissa. Titaniferous magnetites are locally vanadium bearing. The principal copper ore reserves are in the Chitradurga belt where massive and bedded Cu–Pb–As-sulphide ores of suggested volcanogenic origin (Naqvi and Rogers 1987) occur in volcaniclastic and chemical sediments.

## 2.12 AUSTRALIAN CRATON

Most Archean crust is concentrated in the Pilbara and Yilgarn blocks of West Australia (Figs 1-3h, 1-5h, Table 2-16). Archean basement may underlie mid-Proterozoic cover of the Kimberly Basin, north Australia Province. Local Archean basement to the early Proterozoic Pine Creek Inlier, Northwest Territories, is exposed in the Rum Jungle–Alligator River and Nanambu areas. Substantial patches of Archean gneiss occur in the Gawler Block and Cape Carnot area of southern Eyre Peninsula, South Australia.

## 2.12.1 PILBARA BLOCK

Pilbara Block, which occupies the northern margin of the combined Pilbara–Yilgarn Province of West Australia (Fig. 2-26; Table 2-16, column 1) constitutes an Archean low–medium grade terrain, in which dominant domal granitoid–gneiss batholiths up to 100 km across are separated by broadly coeval greenstone belts of the Pilbara Supergroup. Abundant exposure and simple structure have permitted a stratigraphic analysis of the entire supracrustal succession (Hickman 1983, Blake and McNaughton 1984, DiMarco and Lowe 1989, Horwitz 1990, Glover and Ho 1992, Tyler et al 1992, Krapez 1993, Myers 1993).

Granitoid–gneiss batholiths, which underlie 60% of the Pilbara Block, fall into three main lithologic categories, of generally decreasing ages: (1) migmatitic, gneissic and foliated granodiorite and adamellite with minor tonalite and trondhjemite; (2) foliated, porphyritic granodiorite and adamellite; and (3) unfoliated, post-tectonic granite and adamellite (commonly tin-bearing). Based on the more robust isotopic system (U–Pb and Sm–Nd), two distinct age clusters of batholith emplacement are recog-

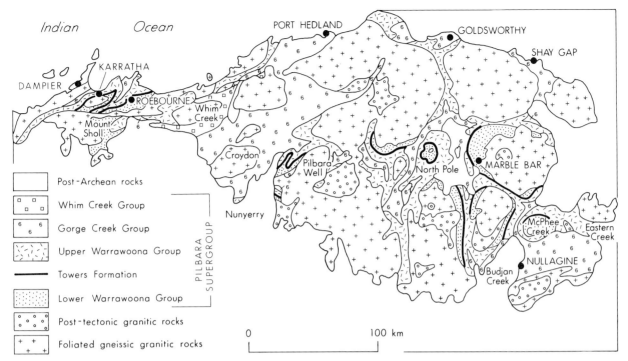

**Fig. 2-26.** Stratigraphic map of the Pilbara Block, West Australia. (From Hickman 1981, Fig. 1, and published with permission of the author and of the Geological Society of Australia).

nized: (1) ~3500–3300 Ma; and (2) ~3050–2850 Ma (de Laeter et al 1981, Collerson and McCullock 1983, Williams et al 1983, Blake and McNaughton 1984).

Pilbara Supergroup

In general, the Pilbara Supergroup shows remarkable stratigraphic continuity across the block. The supergroup comprises three groups, up-section: Warrawoona, Gorge Creek and Whim Creek (Table 2-16, column 1); the first two are distributed block-wide, with the third restricted to the west (Hickman, 1983, DiMarco and Lowe 1989, Horwitz 1990, Barley et al 1994). There is no unequivocal direct evidence (e.g. basement unconformity) for a sialic basement to the greenstone belts (Groves 1982) (however, see below).

Sedimentologic and paleontologic data indicate that much of the Pilbara succession was deposited in shallow, in part evaporitic water, with local chert–barite near the base (Lowe 1982, DiMarco and Lowe 1989, Dunlop and Buick 1981, Eriksson 1981, Barley 1993).

The calc-alkalic volcanism and depositional styles represented by this sequence are grossly similar to those associated with many Phanerozoic island arcs. However, the abundance of shallow-water facies, the non-linear and closely spaced distribution of felsic volcanic centres, sparse syn-volcanic deformation, and absence of associated granitoid detritus all suggest that the setting of this Archean volcanism was, in many significant details, unlike that of most modern arcs (DiMarco and Lowe 1989).

Available geochronologic evidence indicates that the Pilbara Supergroup accumulated chiefly between 3.5 and 2.8 Ga, with a diachronous westward growth of component successions, collectively coeval with the associated granitoids (Hickman 1983, Blake and McNaughton 1984, Williams and Collins 1990, Thorpe et al 1992, Barley 1993, Barley et al 1994) (Table 2-16). A Warrawoona xenocryst zircon age of > 3724 Ma, the oldest date so far obtained in Pilbara granitoid–greenstones, may reflect the existence of pre-Pilbara Supergroup sialic basement (Thorpe et al 1992).

A study of sequence stratigraphy of the Pilbara Supergroup has led to the interpretation of two almost identical megacycle sets (East Pilbara—3.49–3.11 Ga, and West Pilbara—3.11–2.77 Ga) which are similar in longevity, geotectonic components and predicted sea-level cycles (Krapez 1993).

### Granitoid–gneiss batholiths

The large domal granitoid complexes of the Pilbara Block are conveniently termed 'batholiths', but they owe their present geometry mainly to tectonic processes rather than direct magmatic intrusion (Hallberg and Glikson 1981, Hickman 1981, Bickle et al 1985, 1993). Few of the domes are simple structures; most contain internal synforms. An important metamorphic event occurred at 2950 Ma. This accompanied doming at ~3000–2950 Ma. A large part of the batholiths is composed of rocks 3500–2900 Ma old.

Granitoid intrusions have been divided into 12 types (Hickman 1981), ranging from early migmatitic complexes, through granodiorite, remobilized migmatitic rocks, granophyre and alkali-feldspar granite, to massive to poorly foliated granite and adamellite with sharp intrusive contacts against the older rocks.

In comparing the geochronologic data for the granitoid–gneiss batholiths and the layered succession of the Pilbara Supergroup, it is clear that the older granitoids are indistinguishable in age from the Warrawoona Group. However, the relationship between later granitoids and the younger layered succession is not yet firmly established.

### Mineral deposits

The Pilbara granitoid–greenstone system is the host for important gold, copper, iron, tin and barite deposits, and for some chrystile, antimony–gold, molybdenum–copper, copper–lead–zinc and tantalum–columbite deposits (Barley et al 1992). Minor occurrences of wolframite, beryl, lepidolite, spodumene and fluorite have been mined. Silver, copper, lead, zinc and cobalt are minor by-products of gold processing (Hallberg and Glikson 1981).

The younger, ~2.8 Ga old adamellites, commonly referred to as the 'Tin Granites', include pegmatites rich in cassiterite, tantalite–columbite and, locally, beryl, spodumene, lepidolite and gadolinite. Mafic–ultramafic intrusions locally contain titaniferous and vanadiferous magnetite.

### 2.12.2 HAMERSLEY BASIN

As considered above, Pilbara and Yilgarn blocks of the West Australian Shield are separated by partly deformed depositories, notably the BIF-rich Hamersley and Nabberu basins, respectively flanking the Pilbara and Yilgarn blocks (Fig. 1-5h; Table 2-16, column 1). The following descriptions and correlations of the Hamersley Basin draw heavily from the works of Trendall and Blockley (1968, 1970), Trendall (1975, 1976), Gee (1979), Trendall (1983), Blake and McNaughton (1984), Plumb (1985) and Blake and Groves (1987), Simonson (1992), Morris (1993) and Schmidt et al (1994).

*Lithostratigraphy*: The Mount Bruce Supergroup of the Hamersley Basin comprises thick clastic-chemical sediments and volcanic rocks that outcrop over an area of about 100 000 km$^2$ (Fig. 2-27). The Mount Bruce Supergroup comprises the lowermost Fortescue Group of dominantly mafic volcanic rocks with subordinate clastics, the middle Hamersley Group of BIF, shale, dolomite, dolerite intrusions and felsic volcanic rocks, and the upper Turee Creek Group composed of shales and litharenites including tilloids.

Of these, the economically important *Hamersley Group*, about 2500 m thick, is composed of alternating BIF, felsic volcanic rocks and dolerite sills, and shale with minor dolomite and tuff. It is notable for the remarkable stratigraphic continuity of component BIF, of which there are five major stratigraphic units. The Dales Gorge Member of the Brockman Iron Formation, the best-studied unit in the Hamersley Group, has been divided into 33 major alternating BIF and shale macrobands.

BIF units of the Hamersley Group tend to have a uniform 30% total iron content, despite wide variations in mineral composition.

*Structure and metamorphism*: The basal Fortescue Group dips gently off the southern western and eastern edges of the Pilbara Block (Fig. 2-27). The gentle southwesterly dip on the southern margin is maintained up to the axis of a major regional synclinorium—the Hamersley Range Synclinorium—south of which dips remain low but generally northeastward. Within the area of the synclinorium, local alterations of dip, mainly less than 10°, are common.

Low grade metamorphism prevails throughout the basin. A line drawn 10–20 km south of and parallel to the trace of the Hamersley Range Synclinorium separates prehnite–pumpellyite facies on the north from greenschist facies and even local amphibolite facies on the south, the southward increase reflecting proximity to the Ashburton Fold Belt.

The base of the Fortescue Group is about 2765 Ma old, thereby marking the initiation of the Hamersley Basin (Trendall 1983, Blake and

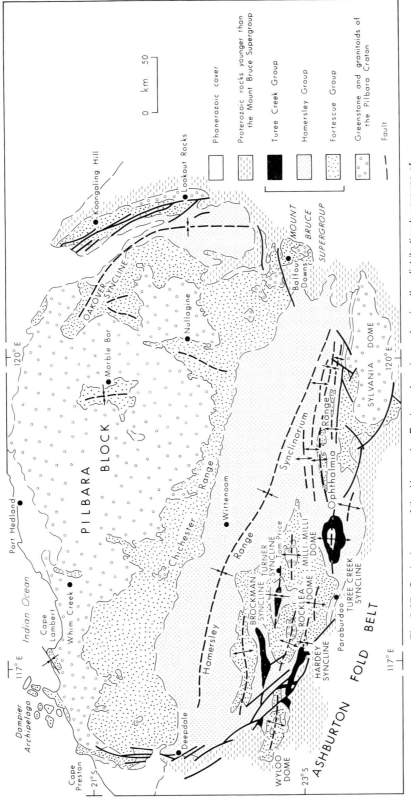

**Fig. 2-27.** Geologic map of the Hamersley Basin and environs showing the distribution by groups of Mount Bruce Supergroup and enclosing rocks. (From Trendall 1983, Fig. 3.1, and published with permission of the author and of Elsevier Science Publishers).

McNaughton 1984, Arndt et al 1991). The Dales Gorge Member of the Hamersley Group provides a reliable depositional age of 2490 Ma (Compston et al 1981). The Woongarra Volcanics of the Hamersley Group are dated at 2440 Na (Pidgeon and Horwitz 1991). The age of the Turee Group is poorly constrained but may be ~2400 Ma. Intrusive sills in the Hamersley Group are dated at 2300 Ma, which marks the approximate end of basin accumulation (Plumb and James 1986), for a total life span of the basin of about 500 Ma (i.e. 2765–2300 Ma).

### 2.12.3 YILGARN BLOCK

Yilgarn Block to the south forms a rectangle 1000 km by 700 km or 650 000 km² in area (Figs 1-5h, 2.28; Table 2-17, column 1). Following Gee et al (1981), the major subdivision is between predominant high grade gneiss and associated rocks of the Western Gneiss Terrain forming a westerly arc around three large low–medium grade granitoid–greenstone areas: Murchison, Southern Cross and Eastern Goldfields provinces (Fig. 2.28). Greenstone belts in the latter three provinces are essentially similar in lithology and age, and subdivision into separate provinces is based on changes in dominant belt size and structural trend. The Western Gneiss Terrain is also distinct in containing a unique, epicontinental to continental, metasedimentary sequence, in lacking significant volcanic components, and in including generally older isotopic dates than in the greenstone belts to the east (Hallberg and Glickson 1981).

A major tectono-thermal event at 2.7–2.6 Ga resulted in widespread granitoid emplacements with accompanying thermal and structural effects throughout the entire Yilgarn Block.

### Western Gneiss Terrain

The distinctive gneissic lithology in the Western Gneiss Terrain (Gee 1979, Gee et al 1981, Groves 1982, McCulloch et al 1983, Myers 1988a, b, Maas and McCulloch 1991) mainly represents metamorphosed quartzofeldspathic sediments, layered migmatite and porphyritic granitoids. The lithofacies in the Narryer Gneiss Complex in the north, in particular, represents a shallow-water sequence, including metamorphosed conglomerate, quartzite, pelite and calcareous sediments.

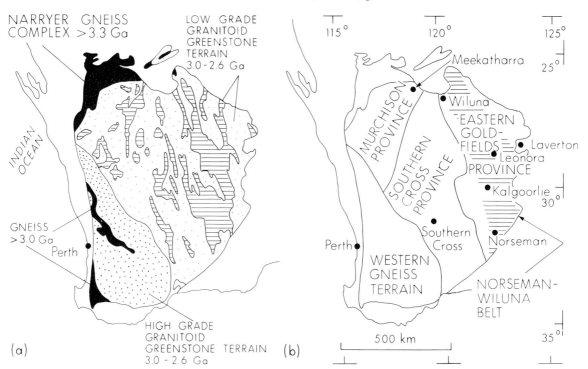

**Fig. 2-28.** General geology of the Yilgarn Block, Western Australia showing (a) regional subdivision into western high grade terrain (Western Gneiss Terrain) including gneiss complexes, and eastern low grade granitoid-greenstone terrain and (b) tectonic subdivision of the eastern low grade granitoid-greenstone terrain into component provinces including the Norseman–Wiluna Belt. (Based on Gee et al 1981 with modifications after Myers 1988, and published with permission of the authors and of the Geological Society of Australia).

Clastic zircons in Mount Narryer quartzites, which were probably deposited and metamorphosed about 3600–3350 Ma ago, provide some U–Pb ion-microprobe ages between 4100 and 4200 Ma (Froude et al 1983, Compston et al 1985, Myers and Williams 1985, Kinny et al 1988); one grain from Jack Hills registers the exceptionally old date of 4276 Ma, which may still be a minimum value for its original age (Compston and Pidgeon 1986). The age of these ancient minerals constrains the time of the earliest preservation of Earth's solid crust (Moorbath 1986).

Myers (1988a) divides the arcuate Western Gneiss Terrain into (1) the 3.73–3.0 Ga Narryer Gneiss Complex in the northwest; (2) to the south, two smaller lenses of > 3.0 Ga high grade gneiss that are contained in (3) a large southerly enclosing Archean (3.0–2.6 Ga) high grade granitoid–greenstone terrain that underwent granulite facies metamorphism in the narrow time range 2649–2640 Ma (Nemchin et al 1994) (Fig. 2-28a).

## Murchison, Southern Cross and Eastern Goldfields provinces

These three provinces are collectively underlain by about 70% granitoid rocks, both foliated and massive, and 30% metasupracrustal rocks. The latter are distributed in numerous greenstone belts, most of which are N–NNW-trending synformal remnants between granitoid domes. By far the largest greenstone belts lie in the Kalgoorlie Subprovince of the Eastern Goldfields Province (Fig. 2-28b) (Hallberg and Glikson 1981, Groves 1982, Hallberg 1986, Watkins and Hickman 1990, Passchier 1994, Hill et al 1992).

The greenstone belts in the three provinces are essentially similar in lithology and age. In general, mafic rocks, mainly low-K tholeiites, make up about 50% of the successions with ultramafic rocks common at lower levels; felsic volcanic rocks occur in the upper levels of the successions; thick sequences of clastic sedimentary rocks are present at the top. Dates of 2.98 Ga from the Murchison Province, 3.05 Ga from the Southern Cross Province, and 2.78–2.69 Ga from the Eastern Goldfields Province (Fletcher et al 1984, Swager et al 1992) indicate that the greenstone belts are younger than the Narryer Gneiss Complex of the Western Gneiss Terrain.

In the Kalgoorlie area several three-part polycyclic sequences involving mafic–ultramafic and felsic volcanic rocks with sedimentary associates are interpreted to be present. The repetitive cycles are especially characteristic of the Kalgoorlie subprovince, where the total stratigraphic thickness is estimated to be at least 15 km thick, and could be twice that amount (Hallberg and Glickson 1981). However, Kalgoorlie stratigraphy has been reinterpreted as a thinner but structurally duplicated, unicyclic stratigraphic sequence (Martyn 1987).

Gee et al (1981), while denying a uniform trans-Yilgarn stratigraphy, recognize, instead, a coherent stratigraphic sequence within each of the three provinces, which takes the form of three discrete basins of accumulation, each with distinctive proportions of komatiite, BIF and felsic volcanic components. Hallberg (1986) emphasizes the importance in the Eastern Goldfields Province of the Norseman–Wiluna Rift Zone (Belt) (Groves and Batt 1984, Cassidy et al 1991) (Fig. 2-28b), a graben superimposed on an earlier platform that developed rapidly under a tectonic regime initially involving high total extension and significant crustal thinning.

Mafic intrusions, probably co-magmatic, accompany the mafic volcanic rocks in the greenstone belts. Included are undifferentiated dolerite and gabbro sills, differentiated norite–granophyre sills, porphyritic gabbros and layered mafic sills. Stringers and pods of dunite/peridotite are common. Massive dunite to peridotite dykes are especially common in the northern part of the Kalgoorlie Subprovince and in the Southern Cross Province.

Metamorphic grades in the Murchison and Eastern Goldfields greenstone belts range from prehnite–pumpellyite facies to upper amphibolite facies with lower to middle greenschist facies predominant.

## Granitoid intrusions

Granitoid plutons account for about 70% of the exposed Yilgarn Block. In general, they comprise a three-part association of (1) early granodioritic to tonalitic, heterogeneous, strongly foliated gneisses and migmatites; (2) later composite, concordant to disconcordant intrusive batholiths of granodiorite to adamellite and granite; and (3) still younger massive discordant plutons of granite to syenite.

Granitoid ages across the Yilgarn Block fall in the 2.9–2.6 range, with 2.7–2.6 Ga dates most common (see Hallberg and Glickson 1981, table 2v). Major granitoid phases in Murchison Province provide ages of 2919–2602 Ma (Wiedenbeck and Watkins 1993). In the Norseman region of the Eastern Goldfields Province, large tonalite–granodiorite plutons were emplaced at 2690–2685 Ma, and smaller granite plutons at 2665–2660 Ma (Hill and Campbell 1993). In the Kalgoorlie region to the north the main regional deformation occurred at 2.68–2.61 Ga (Swager et al 1992).

Tectonics

Cassidy et al (1991) have concluded that the nature and distribution of granitoids in the Norseman–Wiluna Belt (Fig. 2-28b) are consistent with previous interpretations (Barley et al 1989) that this belt formed in a subduction-related convergent margin setting. Myers (1993), emphasizing the presence of numerous tectonostratigraphic terrane all indicating intense tectonic, volcanic, plutonic, and metamorphic activities between 2780 and 2630 Ma, interprets this as a major episode of plate activity which swept together and amalgamated diverse crustal fragments to form the Yilgarn Block (craton).

Passchier (1994), on the other hand, emphasizes the absence of any recognized relic oceanic crust, accretionary complexes and molasse basins. Also, the craton-wide extent of contemporaneous igneous–metamorphic–deformational activities and rapid mafic volcanic deposition with subsequent felsic volcanism are unlike that of any modern convergent plate-tectonic setting, despite geochemical similarities. Furthermore, the same author, while emphasizing the great number of actualistic macrotectonic models that have been proposed to explain Yilgarn relationships, concludes that no exact actualistic Phanerozoic equivalent exists in which the Yilgarn granitoid–greenstone terrains developed.

Mineral deposits

The West Australia 'nickel province' is largely confined to ultramafic rocks of the Eastern Goldfields Province (Barley et al 1992). Palladium, platinum, ruthenium, gold, copper, silver and cobalt are recovered from the nickel sulphide ores as by-products (Glover and Ho 1992). Auriferous quartz veins are found within all major rock types in the Yilgarn Block. Lode-hosted gold deposits at Wiluma are attributed to a fluid infiltration mechanism within a strike-slip fault system (Hagemann et al 1992). BIFs are the source of iron ore, lateritic ochre and manganese. Archean granitoid rocks and pegmatites have yielded small amounts of beryl, bismuth, lithium, molybdenum, niobium, tungsten and tin.

Laterization of Tertiary age over portions of the Yilgarn Block has produced significant deposits of magnesite, bauxite, tin and nickel laterite.

### 2.12.4 OTHERS

North Australian Province

Three small inliers of Archean massive to gneissic granitoids and metasediments, in the Rum Jungle, Litchfield and Nanambu complexes respectively, adjoin the early Proterozoic Pine Creek Inlier (Figs 1-5h; 3-21; Table 2-16, column 2). Similar rocks have been intersected by drilling beneath thin Mesozoic cover 65 km east of Darwin. In all cases Archean metasediments, schists, migmatites and gneisses were intruded by ~2.5 Ga old massive plutons (Plumb et al 1981, Needham et al 1988). Although the complexes unconformably underlie the Pine Creek succession, they were later reactivated and emplaced into the Pine Creek Inlier rocks as mantle gneiss domes during the ~1.8 Ga old Barramundi Orogeny (see below).

New evidence from zircon relicts in the superjacent early Proterozoic assemblages indicates that the late Archean sialic crust exemplified by these three small inliers may have been considerably more extensive (Page 1988).

South Australia

Several occurrences of Archean basement rocks are described in the Gawler Craton (Rutland et al 1981, Webb et al 1986, Fanning et al 1988, Parker et al 1988, Drexel et al 1993, Flöttman and Oliver 1994). Certain gneissic rocks of the Eyre Peninsula and northward extensions assigned to the Sleaford Complexes (2.7–2.3 Ga) and Mulgathing are considered to represent basement to the unconformably overlying early Proterozoic metasupracrustal rocks including the Hutchison Group (see below) Fig. 3-23, Table 2-16, column 4). Some granulite–upper amphibolite facies metasediments and associated augen gneiss at Cape Carnot are dated at 2.6 Ga (Cooper et al 1976, Fanning et al 1988).

An inferred Archean craton, the Nullarbor Block, may adjoin the Gawler Craton to the west (Groves 1982).

## 2.13 ANTARCTIC CRATON

The East Antarctic Metamorphic Shield contains a number of Archean cores within dominantly Proterozoic metamorphic areas: western Dronning Maud Land, Enderby Land, southern Prince Charles Mountains and Vestfold Hills, all located at or near the coast and within (or very close to) the 0–90°E longitudinal sector (James and Tingey 1983, Tingey 1991a, b, c). A late Archean–early Proterozoic remnant (~2.4 Ga) occurs at Commonwealth Bay (137°E) (Flöttman and Oliver 1994). In the Vestfold Hills, dominant felsic orthogneisses are attributed to activities in a 50 Ma period in the terminal Archean (2526–2477 Ma). Moreover, associated zircon cores and xenocrysts indicate a crustal source of at least 2800 Ma (Black et al 1991).

Of all the East Antarctic occurrences of Archean rocks, the Napier Complex, Enderby Land, is the largest and best known, ranking as one of Earth's better studied Archean domains.

### 2.13.1 NAPIER COMPLEX OF ENDERBY LAND

The first main mapping of Enderby Land was undertaken in the 1960s by Soviet geologists. They recognized the distinctive granulite-facies metamorphic character of the Napier Complex, subdividing it into a lower, presumably older 'orthogneissic' Raggatt 'series', and an upper, presumably younger mainly 'paragneissic' Tula 'series', the two series separated by a tectonic discontinuity (Ravich and Grikurov 1976). The great antiquity of the complex was first suggested by ~4000 Ma U–Th–Pb ages from the Fyfe Hills (Sobotovich et al 1976). Although this age was subsequently deemed isotopically meaningless (Black and James 1983), recent geochronologic investigations have, nonetheless, supported the great antiquity of these rocks.

As reviewed by Black et al (1986b), the Napier Complex consists of high temperature granulite facies assemblages, characterized by the common occurrence of calcic mesoperthite (indicative of water pressures less than 500 bar) and the presence in metapelites of the rare anhydrous associations sapphirine plus quartz, orthopyroxene plus sillimanite, and osumulite (Sheraton et al 1980). The dominant rock types are pyroxene–quartz–feldspar and garnet–quartz–feldspar gneisses, both representing deformed and metamorphosed rocks of originally igneous origin. Other rock types of igneous origin include mafic granulites and pyroxenite. Various siliceous, aluminous and ferruginous metasediments are widespread. Granitic intrusions and post-cratonic mafic dykes are also present.

A very well-developed horizontal layering and a parallel intense flattening and extension fabric are the prevailing characteristics of the Napier Complex in what is aptly termed a recumbent gneiss terrain. The large-scale layering is made up of thick units of paragneissic metasediments of largely epicontinental style (quartzite and garnet–quartz–feldspar gneiss of the Tula 'series'), interlayered with further finely laminated felsic, intermediate and mafic granulite of more indeterminate origin and also with homogeneous charnockitic orthogneiss (Raggatt 'series'). The tectono-thermal evolution of the various interlayered units reveals a complex association of superposed deformation events of varying intensity and style, including the production of overprinted tectono-thermal fabrics in a largely high grade environment. This culminated in the production of a dome-and-basin pattern which characterizes the craton. Tectonic evolution concluded with the development of retrogressive mylonitic shear zones, faults and the intrusions of mafic dykes.

Enderbites at Mount Sones indicate a well-defined protolith age of about 3800 Ma (Compston and Williams 1982), later determined to be $3927 \pm 20$ Ma (Black et al 1986a). This requires the Napier Complex to be a remnant of some of Earth's oldest preserved crust (Black et al 1986b).

The last major Archean event (2.5 Ga), which variably affected and stabilized the Napier Complex, is characterized by weak subhorizontal compression, which folded the high grade gneissic pile into large-scale, upright to steeply inclined, noncylindrical folds.

Isotopic studies in the basement rocks of Enderby Land, including the more robust U–Pb and Sm–Nd systems, have demonstrated that extensive isotopic resetting has occurred (Black 1988). Thus, anomalous results are present in all of Rb–Sr, U–Pb and Sm–Nd isochron systematics, results variously ascribed to isotopic resetting during transitional amphibolite to granulite fabric facies tectonism. These examples of isotopic resetting illustrate the folly of reliance on a single isotopic system and the dangers of isotopic dating without adequate geologic constraints.

# Chapter 3

# Early Proterozoic Crust

## 3.1 INTRODUCTION

### 3.1.1 DISTRIBUTION

The area of exposed plus buried (total) early Proterozoic crust is $20.6 \times 10^6$ km² (Table 3-1) or 19% of Earth's total Precambrian crust (Table 1-2). Of this amount, North America, including Greenland, accounts for 50%; South America and Africa each 11–12% ($\Sigma 22\%$); Asia, Europe and Australia each 8–9% ($\Sigma 25\%$); Antarctica 2%; and India < 1%. Of these, the four continents bordering the Atlantic (Europe, the Americas, Africa) collectively account for 80% of the total. In terms of pre-Mesozoic supercontinents, Laurasia (Asia, Europe and North America) accounts for 67% and Gondwanaland the remaining 33% (Fig. 3-1).

The area of exposed early Proterozoic crust is $6.5 \times 10^6$ km², or 22% of Earth's exposed Precambrian crust (Table 1-2). North America and Africa, about equal, together account for 63%, with the remaining six continents, each at < 1% to almost 13%, accounting for the remaining 37%. In this case, the four Atlantic continents account for 82%, Laurasia 44% and Gondwanaland 56%.

In brief, North America constitutes the dominant early Proterozoic craton, and the combined Atlantic cratons the dominant collectivity.

### 3.1.2 SALIENT CHARACTERISTICS (Table 3-2)

(1) Early Proterozoic crust underlies some 20 small to large, linear to irregularly ovoid terrains, typically adjoining and partly surrounding Archean nuclei. The largest such terrain by far is the reconstructed Churchill–Rinkian–Nagssugtoqidian Superprovince, together with buried extensions, of the North American Craton (Fig. 3-1). Elsewhere on Earth the distribution is uneven, with the remaining Atlantic cratons predominating, and India–Antarctica subordinate.

(2) Early Proterozoic crust is highly variable in composition and pattern. Certain lithologic elements closely resemble Archean precursors; others are practically unique to this era; still others appear for the first time to reappear consistently thereafter. The early Proterozoic era thereby represents a major watershed in Earth's continental evolution: on the one hand retaining certain Archean characteristics; on the other, introducing certain unique elements, products of exceptional endogenous–exogenous transitions; yet on the whole ushering in a new global scene that has persisted to the present.

(3) Early Proterozoic crust comprises 63% orthogneiss–migmatite, 12% massive granitoid plutons, and 25% metasupracrustal rocks, the latter dominantly metasediments but including significant metaigneous rocks. This estimated average compo-

*Table 3-1.* Distribution of preserved early Proterozoic crust by continent.

|  | Total crust (%) | Exposed crust (%) |
|---|---|---|
| Asia | 8.9 | 4.3 |
| Europe | 9.0 | 6.7 |
| North America | 48.8 | 33.4 |
| South America | 11.6 | 12.6 |
| Africa | 10.8 | 29.5 |
| India | 0.4 | 0.8 |
| Australia | 8.2 | 9.7 |
| Antarctica | 2.3 | 3.0 |
|  | 100.0 | 100.0 |

Total (exposed + buried) crust = 20 605 000 km²
Exposed crust = 6 518 000 km² (35% of total crust)

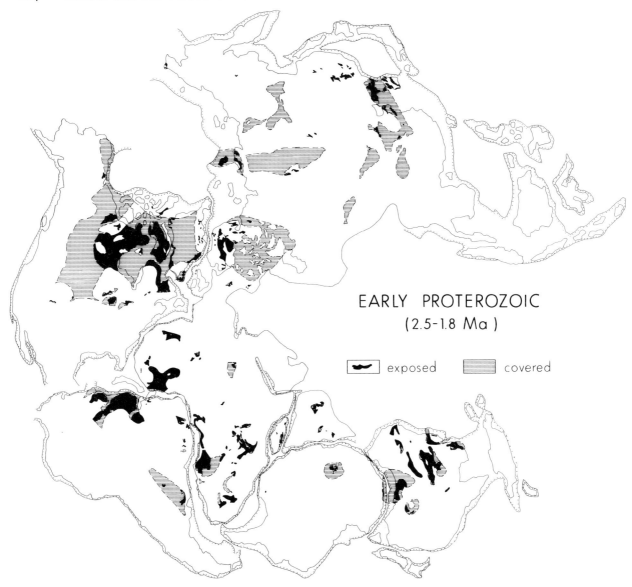

**Fig. 3-1.** Distribution of exposed and covered (i.e. sub-Phanerozoic) early Proterozoic crust in the Briden et al (1971) pre-drift Pangean reconstruction of the continents. Interior basement crust beneath late Proterozoic cover of the Siberian and East European platforms is included.

sition translates to gneiss:massive pluton: metasupracrustal = 5:1:2, or plutonites:metasupracrustals = 3.1 (cf. Archean ratio of 4:1). In terms of metamorphic grades, early Proterozoic crust averages approximately 55% amphibolite-, 28% granulite- and 17% greenschist– subgreenschist facies; this translates to, in round terms, 3:2:1. Thus early Proterozoic amphibolite facies orthogneiss continues to predominate, as in Archean terrains. However, early Proterozoic crust contains significantly higher ($\times 2.5$) supracrustal (excluding paragneiss) proportions which are present mainly in low to medium metamorphic grade foreland fold belts.

These fold belts include, for the first time, those with distinctive asymmetric platform–shelf–slope–rise facies, a manifestation of widespread stable shelves, themselves a product of extensive late Archean granitoid-induced cratonization, perhaps the single most important step in the growth of the continental crust.

(4) Early Proterozoic medium to high grade gneiss terrains resemble Archean counterparts in gross aspects. However, they commonly involve dominant remobilized Archean infrastructure with or without early Proterozoic cover. The Early Proterozoic gneiss terrains also exhibit highly variable

# INTRODUCTION

Table 3-2. Salient early Proterozoic characteristics.

| | |
|---|---|
| Distribution of preserved crust | Highest proportions in North America, Africa and South America; paucity in India and Antarctica; elsewhere comparatively even |
| Composition and metamorphism | Highly variable with both Archean counterparts and original elements, some unique to this era. Mainly amphibolite facies orthogneiss-migmatite gradational to subordinate (1) high grade granulite terrains and (2) low grade foreland fold belts. Gneiss terrains mainly represent remobilized Archean infrastructures with or without Proterozoic cover; some represent juvenile crustal accretions. Gneiss terrains accordingly display both low and moderately high initial Sr:Nd ratios. Lower La:Yb ratios imply cooler partial melt regimes |
| | Asymmetric foreland fold belts with distinctive platform-shelf-slope-rise facies in response to widespread stable continental shelves. Orthoquartzite-carbonate-BIF-pelite sequences common. Mega-BIF deposition. Clastics display uniform critical trace element abundance levels in response to the presence of newly acquired post-Archean K-granitoid-rich upper continental provenances |
| | Fold belts include local greenstone-turbidite assemblages representing juvenile magmatic arcs |
| | Oldest established ophiolites, glaciogenic mixtites and continental red-beds |
| | Mafic-ultramafic igneous complexes of exceptional size, complexity and economic value |
| Deformation | Fold belts, increasingly deformed towards the hinterland, display high degrees of foreland vergence. Gneiss terrains have highly variable deformation patterns marked by complex fold patterns, recumbency and napperian structures, attesting to widespread horizontal tectonic regimes, especially at deeper crustal levels |
| Mineralization | Placer Au-U; Fe-Mn; Ni-Cu-platinoids-Cr-Sn-Ti-V; asbestos; Cu-Pb-Zn |
| Paleoenvironment | Stable enhanced oxygenic environment, capable of supporting and sustaining aerobic microorganisms, resulting from enhanced biogenic photosynthesis. Dominant continental flux signature |
| Paleobiology | Diverse cyanobacteria and stromatolites. $O_2$-producing photosynthesizers spread rapidly. Greatly increased photosynthetic biomass. Concomitant burial of organic carbon by enhanced sedimentation. Environmental $Po_2$ under biologic control |
| Paleotectonics | Widespread horizontal-vertical motions of substantial lithospheric plates, including widespread stable shelves, ensialic fold belts and interplate collisional mobile belts. Mid-era mafic magmatic underplating. End-of-era global orogeny (1890–1850 Ma), including vertical accretion in intracontinental environments |

deformation patterns characterized by recumbency, large-scale low-angle thrusting, and high intensity of deformation. Complex interference structures and napperian folds attest to widespread horizontal tectonic regimes, especially operating at deeper crustal levels.

Geochemically, the Early Proterozoic gneiss terrains display both low and moderately high initial ratios (Sr and Nb), implying the presence, probably in about equal proportions, of both juvenile and recycled older crust, the former representing significant new crustal additions. Their significantly lower La:Yb ratios, however, imply a cooler partial melt regime than their Archean counterparts.

(5) Closely juxtaposed low to medium grade foreland fold belts are commonly asymmetric, with distinctive mio- and eugeoclinal facies disposed in increasingly complex deformational–metamorphic patterns towards the hinterland, the persistent craton-directed vergence leading in the extreme to allochthonous granulite-grade thrust-slices.

Other equally numerous ensialic fold belts occupy aulacogen-type structures in the Archean basement, evidence of broad regional tensile regimes. The structures are typically filled with transgressive–regressive continental–marine deposits expressed in terrigenous clastic–bimodal volcanic–pelite–carbonate sequences. Other closely related successions range from quartz-rich turbidites, through shallow-marine and evaporite–carbonate complexes, to fluviatile facies. The strata are invariably tightly to isoclinally folded, slaty cleavage or schistosity is ubiquitous, and superimposed folding is common. Most resulting fold belts lie within the greenschist facies of metamorphism, but some reach amphibolite and even low-granulite facies.

(6) In marked contrast to the Archean, early Proterozoic sediments include a high proportion of stable-shelf facies rich in orthoquartzite, carbonates and pelites. This reflects not only the presence of widespread stable shelves but also large-scale tectonic (epeirogenic) uplift and erosion of nearby craton provenances to produce vast quantities of resistate–chemical components, both consequences of widespread, terminal Archean cratonization.

Geochemically, early Proterozoic terrigenous clastics have very uniform abundance levels for several insoluble trace elements (e.g. rare earth elements (REE), Th, Sc). This reflects the composition of the newly developed post-Archean granitoid-rich upper continental crust, the main source of the detritus.

(7) Early Proterozoic greenstone belts, which are closely similar though not identical to Archean counterparts, are present in certain terrains, including those in North America (Saskatchewan–Manitoba–Wisconsin), South America (Guyana–Brazil) and West Africa (Birrimian). Some belts may have developed in a similar manner to their Archean counterparts and thereby represent a continuation of certain Archean-style tectonic processes. Others may have actualistic plate-tectonic implications.

(8) Well-documented ophiolites, some with recognizable accretionary wedges, fore- and back-arc basins, and magmatic arcs, make their appearance at least by 1.9 Ga. This may mark the introduction of some 'modern-style' plate-tectonic process to the Precambrian roster.

(9) Mafic–ultramafic layered igneous complexes of exceptional stature and complexity, including Bushveld (2.0 Ga) and Sudbury (1.9 Ga), both of proposed though dubious meteorite impact origin, together with the smaller terminal Archean Stillwater (2.7 Ga) and Kola Peninsula (2.5–2.4 Ga) complexes, collectively contain exceptionally rich nickel–copper–platinum–chromium–vanadium–titanium mineral deposits.

(10) Distinctive early Proterozoic sedimentary associations, unique in type and/or stature, include BIF, Au–U placers and tillites. Early Proterozoic BIF is developed on a truly heroic scale, in fact accounting for more than 90% of all known BIF on Earth. BIF deposition at this time is attributed in significant degree to the irreversible increase in oxidation potential (hydrosphere–atmosphere) resulting from greatly enhanced biogenic photosynthesis upon the newly developed continental shelves, the precipitated iron representing the last available great '$O_2$-sink'.

Placer Au–U deposition with quartzite–oligomictic conglomerate associates, which was initiated ~3.0 Ga on a substantial scale in the apparently one-of-a-kind Witwatersrand Basin upon the newly cratonized Kaapvaal crust, continued to accumulate elsewhere in local depressions and craton margins upon maturely eroded and peneplaned K–granite–greenstone-bearing foreland terrains following the widespread terminal Archean cratonization. Notable other examples include the Ventersdorp (South Africa), Huronian (North America), Jatulian (Fennoscandia), Jacobina (South America) and Nullagine (Australia), all but Ventersdorp (~2.7 Ga) representing 2.5–2.0 Ga age accumulates.

Glaciogenic mixtites, the oldest presently identified products of low temperature (glacial), high relief (mountainous) environments, are preserved in several continents, notably the Gowganda Formation (2.4 Ga) of North America, but including coeval deposits in Fennoscandia, Siberia, Southern Africa and Australia.

(11) The continuing saga of southern African epicratonic depositories, initiated by the Pongola Basin (3.0 Ga) and coeval Dominion Group of the Witwatersrand Basin, continued with development of the Ventersdorp (2.7 Ga), Transvaal–Griquatown (2.5–2.3 Ga) and Matsap–Waterberg–Soutpansberg (1.8 Ga) systems. Collectively, these sequentially enlarging, northward progressing basins may represent 'parageosynclines', whose depocentres migrated cratonward (northward) with progressive foreland (Zimbabwe Craton?) uplift and tectonism.

(12) Early Proterozoic mineral deposits feature iron, manganese, gold, uranium, nickel, copper, platinoids, cobalt, chrome, vanadium and titanium. Collectively, they represent the most productive Precambrian metallogenic era, once more attributable in large part to the effects, direct and indirect, of widespread late Archean cratonization.

(13) The tectonic climax of the early Proterozoic era, at 1900–1800 Ma, took the form of pervasive orogenesis involving widespread foreland overthrusting and imbrication, metamorphism, and granitoid intrusion–extrusion at least partly in response to plate collision activities, followed by unusually extensive post-orogenic and anorogenic magmatism. This collective tectonic–magmatic activity significantly expanded Earth's budget of tectonically stable continental crust to an estimated 80% of the present amount. The role of continent–continent collision is uncertain but was probably increasingly effective across the era. Processes of mantle-derived subcrustal additions (mafic magmatic underplating) were also probably important. This terminal era orogeny probably ranks only after the terminal Archean orogeny as a developer of juvenile continental crust. Together they provided the essential continental environment, including widespread cratonic basement, for younger Precambrian–Phanerozoic events of essentially on-going character.

## 3.2 CATHAYSIAN CRATON

Early Proterozoic rocks of the composite Cathaysian Craton are mainly concentrated in the component Sino–Korean Craton in close association with Archean basement (Figs 1-3a, 1-5a, 2-2). Initially, restricted volcano-sedimentary units accumulated in local basins, troughs and rift-depressions, which thereafter enlarged to form substantial sediment-dominated depositories. Early era (pre-2.2 Ga) pelitic–turbiditic sediments, included basal conglomerate–arkose, greywacke, pelites, Na-volcanic rocks and significant BIF. Later era components (2.2–1.8 Ga) are mainly pelite–stromatolitic dolostone rhythmic cycles, which accumulated in widespread marine basins and troughs upon comparatively stable shelves.

Prevailing greenschist–amphibolite facies metamorphism with local migmatization is attributable mainly to the Luliangian episode (1900–1740 Ma).

### 3.2.1. SINO–KOREAN CRATON

Representative stratigraphic sections are located in (1) Wuati–Taihang (central); (2) Yinshan (northern); and (3) Eastern Hebei (northeastern) districts (Cheng et al 1982, 1984, Yang et al 1986, Sun et al 1992).

**Wutai–Taihang District (Table 2-3, column 2)**

Early Proterozoic rocks are well exposed in the mountains of Shanxi Plateau, where both the Wutai and unconformably overlying Hutuo groups cover Archean basement rocks.

The Wutai Group, 5000–6000 m thick, is divided into three stratigraphic parts; a lower subgroup (~3000 m), of pebble-bearing quartzite, assorted schists, mafic volcanic rocks (amphibolite), biotite 'granulite' (amphibolite facies gneiss), serpentinite, carbonate–pyrite-bearing BIF, and tremolite-bearing marble; a middle subgroup (also ~3000 m) of

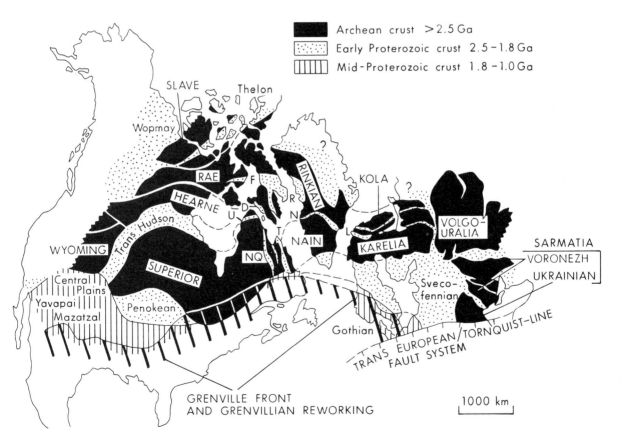

**Fig. 3-2.** North Atlantic Precambrian reconstruction showing pre-Grenvillian craton configurations as discussed in the text (after Gorbatschev and Bogdanova, 1993, Fig. 4). The letter symbols are: NQ – New Quebec; U – Ungava; D – Dorset; F – foxe; R – Rinkian; N – Nugssugtoqidian; T – Torngat; L – Lapland Granulite Belt.

metaconglomerate, schists and volcaniclastic units; and a comparatively thin, yet extensive upper subgroup (800 m), composed of graded phyllite, siltstone, quartzite and graphitic schist with important BIF. Wutai rocks are at greenschist to amphibolite facies metamorphism and are locally migmatized. The age limits extend from 2.5 to 2.3–2.2 Ga (Cheng et al 1984, Sun et al 1992). An island-arc setting has been proposed (Sun et al 1992).

The unconformably overlying Hutuo Group is estimated to exceed 10 km thick. It also comprises three subgroups, each 2000–5500 m thick and mutually separated by unconformities: a lower subgroup composed of basal conglomerate, quartzite and pelite with rare mafic to intermediate volcanic rocks, a middle subgroup (~5500 m) of dominant stromatolitic dolostones; and an upper subgroup of sandy, gritty conglomerate units. The presence of an intermontane 'redbed' formation at the base of the upper subgroup is interpreted to indicate the presence of a hot, dry climate and strong denudation adjacent to a rift-depression. Hutuo rocks are mainly in greenschist facies. Their age is ~2.4 Ga (Sun et al 1992).

### Yinshan District (Table 2-3, column 3)

The Sanheming Group (~3000 m), correlated with the Wutai Group of the Wutai–Taihang Region, is composed of amphibolite, schist, 'granulitite' and BIF. The unconformably overlying Erdaowa Group, also ~3000 m thick, is composed of conglomerate, pelite, carbonates, volcanic rocks of various composition and some BIF. These rocks are mainly at greenschist facies, locally mildly migmatized, and intruded by 1.9–1.7 Ga granitoids (Cheng et al 1984).

### Eastern Hebei District (Table 2-3, column 4)

The early Proterozoic Shuanshansi Group, ~3300 m thick, is in unconformable-faulted contract with Archean basement. The group is composed of amphibolite, 'granulitite', micaschists and, in the lower part, cummingtonite-bearing BIF. Low amphibolite to greenschist facies metamorphism prevails. A metamorphic age of 2.2 Ga has been reported (Cheng et al 1984). Some of the rocks from the lower part have been affected by K-rich migmatization.

Lying unconformably on Shuanshansi rocks is the Qinglonghe Group, a varied metasedimentary assemblage > 1350 m thick. A basal metaconglomerate with garnet–mica schist is overlain by 'granulitite', garnetiferous schist and intercalated cummingtonite–magnetite quartzites. Greenschist facies metamorphism prevails. The rocks yield a metamorphic age of 2.0 Ga (Cheng et al 1984).

## 3.3 SIBERIAN CRATON

Early Proterozoic activities in the Siberian Craton (Figs 1-3b, 1-5b) involved both (1) extensive regeneration and overprinting of Archean infrastructure, as in the Stanovoy Belt to the south, with substantial but as yet poorly defined lateral equivalents in the largely buried interior platform to the north; and (2) shelf- and slope-rise sedimentation in peripheral fold belts that partly frame the platform, including (a) in the south, the main Baikal Fold Belt; (b) in the southwest, the East Sayan Fold Belt; and (c) in the west, the Yenisei and Turukhansk fold belts. These peripheral belts typically comprise folded and metamorphosed schist, phyllite, quartzite and marble sequences up to 12 km in stratigraphic thickness. Additional early Proterozoic crust is found in the Taymyr Fold Belt and its northern extension, the North Zemlya Massif, both located beyond the northern boundary of the craton (Fig. 1-5b).

Significantly, early Proterozoic cover is absent from exposed parts of the craton, including the Anabar Shield in the north, the Aldan Shield and north slopes in the south, and along the northern slopes of the East Sayan Ridge in the southwest, wherein Sinian (mid to late Proterozoic) platform cover directly overlies deeply eroded Archean-rich basement; this implies that much of the present interior basement was topographically high and undergoing active erosion during early Proterozoic time.

### 3.3.1 CRUSTAL REGENERATION

At the southeastern margin of the Siberian Craton, major latitudinal faulting (Stanovoy Fault) occurred at the Aldan–Stanovoy boundary, resulting in structural separation of the Stanovoy Fold Belt from the Aldan Shield to the north. Abundant granitoid plutons were intruded in the Stanovoy Belt; very large gabbro–anorthosite plutons were emplaced along the intervening Stanovoy Fault. The Stanovoy Belt is replete with wide, linear folds with oval brachyform domes and bowls, early Proterozoic (1960–1870 Ma) 'granitization', and medium to high grade metamorphism (Bibikova et al 1984, Khain 1985, Nutman et al 1992).

Extensive tectonic reworking, metamorphism and granitoid intrusion of the pre-existing Archean crust resulted in a craton-wide network—as yet poorly defined—of early Proterozoic metamorphic belts including some juvenile early Proterozoic additions, locally at granulite facies, which adjoin variably modified Archean nuclei. By these means, the crust was extensively cratonized during the Stanovoy thermotectonic event (~1.9 Ga).

### 3.3.2 FOLD BELTS

Early Proterozoic 'fold belts' within the Aldan Shield take the form of narrow restricted troughs or paleoaulacogens (Khain 1985), each a few tens of kilometres wide by hundreds of kilometres long, that are filled with volcaniclastic–BIF sequences ranging from 2 to 7 km thick. These early rifts were subsequently 'regenerated' as wider, flatter troughs or basins, which were thereafter deformed into narrow linear greenstone-type belts that resemble the similar (Saksagan) structures of the East European Craton (see below). Other presently undisclosed early Proterozoic belts of this type may be widely distributed across the buried basement of the Siberian Craton.

#### Baikal Fold Belt

Fold belts of this type are well illustrated by the Udokan Group, which constitutes the shelf stratotype in the northern–western sectors of the Baikal Fold Belt (Salop 1983). The complementary and transitional slope-rise (eugeoclinal) facies of the Muya Group lies in the southern–eastern sectors of the same fold belt.

The *Udokan Group*, a complete shelf cycle up to 13 000 m thick, is situated in the Olekma–Vitim Highlands, where it forms the western frame to the Aldan Shield. Three equally thick, conformable subgroups are recognized composed respectively of metasandstone–amphibolite–marble–metasiltstone; sandstones–jaspilites–dolostones–partly cupriferous siltstones with crystal moulds after halite and gypsum; and sandstone–siltstone–local polymictic tilloids (Salop 1983).

Udokan rocks are deformed and metamorphosed increasingly northward towards the uplifted Aldan–Chara foreland. Udokan strata overlie Archean basement, dated at 2.8–2.6 Ga, and are intruded by late-stage granitoid plutons dated at ~1.9 Ga (Salop 1983). The succeeding mid-Proterozoic strata were deposited on the eroded surface of these early Proterozoic late-stage granitoid plutons.

The coeval, transitional *Muya Group* to the south, up to 6000 m thick, unconformably overlies Archean gneissic basement. The group is also subdivided in three subgroups, composed respectively of conglomerate–sandstone–volcanics–marble; mafic to felsic volcanic rocks (spilite–keratophyre association) and jaspilites; and quartzite–siltstone–marble–volcanic rocks.

Finally, the craton equivalents of the Muya (euogeoclinal) and Udokan (shelf) sequences are represented by local grabens, filled with metasandstone, phyllite and marble with common subaerial volcanic associates. Granitoid intrusions are dated at 1900 Ma, and the still younger alkaline Ulkan laccolith at 1660 Ma (Khain 1985).

#### Other fold belts

Similar peripheral fold belts including East Sayan, Yenisei and Taymyr unconformably overlie Archean basement, have a metamorphic age of 1.9 Ga and are typically transgressively overlain by mid-Proterozoic strata (Khain 1985).

## 3.4 EAST EUROPEAN CRATON

Early Proterozoic crust is an important component of the Baltic Shield in the northwest, Ukrainian Shield in the southwest, and Voronezh massif and adjoining buried basement in the east (Figs 1-3c(i), 1-5c(i), 2-3).

### 3.4.1 BALTIC SHIELD   (Fig. 2-3)

Early Proterozoic rocks, largely products of the Svecokarelian (Svecofennian) cycle, are widespread across the centre–east part of the Baltic Shield, in Kola Peninsula, Karelian and Svecofennian provinces (Fig. 2-3). Svecokarelian rocks are subdivided, from east to west, into two main facies: (1) Karelian epicontinental facies, itself subdivided into (a) platformal sub-facies remnants in Kola Peninsula and eastern Karelia provinces, transitional westward to (b) thick, continuous shelf sub-facies in the main Karelide (Karelian) Belt of eastern Finland and adjoining Russia; and (2) Svecofennian flysch facies of western Finland–Sweden (Svecofennian Province), including the easternmost Kalevian cover (see below). The Svecofennian–Kalevian association

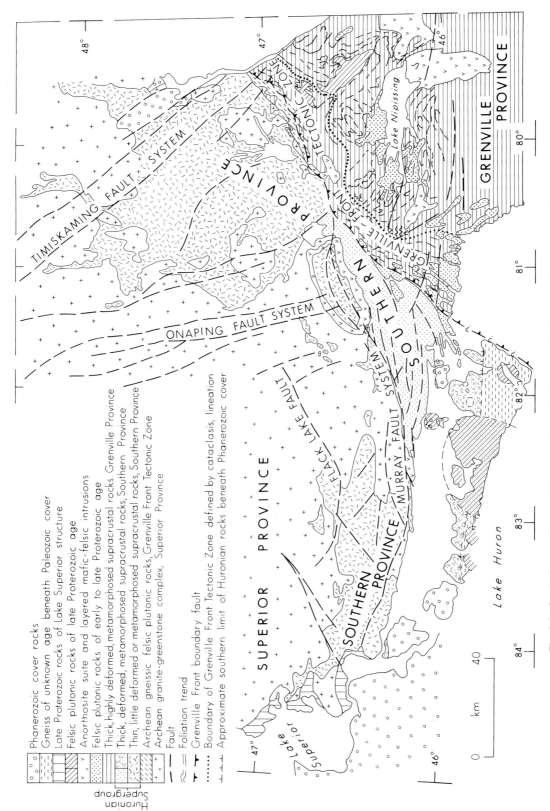

**Fig. 3-3.** Geologic-tectonic map of the Lake Huron Region including the centrally disposed, elliptical Sudbury Structure, North America. (From Sims et al 1981, Fig. 21.7, and reproduced with permission of the Geological Survey of Canada and the Minister of Supply and Services Canada).

is characterized by variably deformed and metamorphosed flysch (intermediate–felsic volcanic–greywacke–pelite) and abundant granitoid intrusions, which collectively represent juvenile accretions to the pre-Svecokarelian (Archean) craton to the east. (Key references: Lundquist 1979, Simonen 1980, Bowes et al 1984, Gorbunov et al 1985, Khain 1985, Park 1985, Gaál and Gorbatschev 1987, Gorbatschev and Bogdanova 1993.)

In round terms, early Proterozoic crust of the Svecofennian Province (Domain) is estimated to comprise 75% intermediate–felsic magmatic rocks, 20% metaturbidites and metapelites, and 5% minor basalt, gabbro and other types of sediments (Patchett and Kouvo 1986). Furthermore, it is noted that post- or anorogenic intermediate felsic plutonic intrusions, e.g. rapakivi granites, are unusually widespread in the Baltic Shield.

Svecokarelian rocks are now considered in terms of their respective (1) Karelian (epicontinental); and (2) Svecofennian (Kalevian) (flysch) facies.

### Karelian (epicontinental) facies

#### Karelian Supergroup

Karelian platform deposits are preserved in several NW-trending elongate troughs and basins, which unconformably overlie Archean basement in Kola Peninsula, and the White Sea–Lake Onega region of Karelia (Fig. 2.3). The stable platform accumulations, mainly comprising quartzite, conglomerate, arkose and bimodal volcanic rocks, are thin, discontinuous, only slightly metamorphosed, and replete with internal stratigraphic breaks. These rocks pass westward into thicker, more continuous Karelian shelf-deposits of the main NNW-trending Karelide Belt in Karelia–eastern Finland (Bowes et al 1984, Park 1985, Gaál and Gorbatschev 1987).

Karelian shelf deposits are here divided, up-section, as follows (Figs 1-3c(i), 2-3): Lapponian, Sumian, Sariolian (Sariolan), Jatulian and Suisarian groups, the last of very restricted distribution. The unconformably overlying Kalevian Group is variably interpreted as belonging to (1) the Karelian Supergroup (Park 1985); and (2) as is followed here, the Svecofennian association (Gaál and Gorbatschev 1987).

The oldest basement cover which is concentrated in the north is represented by mafic–ultramafic pyroclastics and lava flows, quartzites and pelites of the *Lapponian Group*, of estimated age ~2.6–2.45 Ga (Gaál and Gorbatschev 1987).

Farther south in central Finland and Karelia, the oldest Karelian sequence is represented by *Sumian* immature clastic sediments with mafic–felsic eruptive rocks, distributed in small, NW-trending, linear troughs. Sumian strata are commonly overlain by *Sariolian* conglomerates, quartzites and tuffitic sandstones, possibly an uppermost member of a single Sumian–Sariolian succession. Lacustrine–glacial and fluvioglacial tillites in the age range 2.4–2.3 Ga, are reported locally (Marmo and Ojankangas 1984, Strand and Laajoki 1993). The Pechenga and Imandra–Varzuga troughs, with thick terrigenous–volcanic fill, include Cu–Ni-bearing mafic–ultramafic intrusions of ages 2.5–2.4 Ga (Balashov et al 1993). The minimum age of Sariolian deposits is estimated to be 2.3 Ga (Gaál and Gorbatschev 1987). The succeeding *Jatulian Group* fills numerous northwesterly elongated Karelian depressions, ranging up to 150 km wide and several kilometres deep (Khain 1985). Lower to middle Jatulian sequences consist of locally uraniferous quartzites and conglomerates and tholeiitic basalts, whereas upper parts are dominated by dolomites, black slates and sandstones with local BIF, ferruginous manganese- and phosphate-bearing carbonates, and aplite-rich quartzites.

In western Finland, close to the junction with the Svecofennian Province, geologic relations are very complex, owing to tectonic imbrications, compounded by Svecofennian metamorphism (see below).

The local, conformably overlying *Suisaarian Group* (Gaál and Gorbatschev 1987), restricted to the western shore of Lake Onega, is composed of metabasaltic–picritic lava flows, breccia, tuffs and minor shales. These are overlain, in turn, by up to 2000 m of reddish terrigenous quartzites, conglomerate sandstone and siltstone of the *Vepsian (Subjotnian) Group*, which represents epi-Karelian redbeds (late Molasse). These, together with widespread rapakivi granites and diabase intrusions dated at 1700–1600 Ma, attest to the complete cratonization of the Svecokarelian terrains (Kratz and Mitrofanov 1980, Salop 1983, Khain 1985).

#### Karelian–Kalevian relations

In the Finnish Karelides, the older essentially autochthonous Karelian units (Sariolan and Jatulian) rest with modified unconformity on a late Archean (Presvecokarelian) gneiss-greenstone basement (Sarmatian). In sharp contrast, the overlying Kalevian units, rich in turbidites, tholeiites and 1.96 Ga old ophiolites (Kontinen 1987), are practically entirely allochthonous, having been thrust eastward into their present positions in accordance with a well-established sequence of events (Campbell et al 1979, Bowes 1980a, Koistinen 1981, Park et al 1984, Luukkonen 1985a, b, Park 1985). Pertinent geochronologic data show that the Karelian–Kalevian association is bracketed by 2.6–2.5 Ga maximum and 1.87–1.86 Ga minimum ages (Rickard 1979, Silvennoinen et al 1980, Lauerma 1982, Vivallo and Rickard 1984, Park 1985).

To summarize Karelian–Kalevian relations, the 100–150 km wide, NW-trending Karelidic 'Schist Belt' includes both older Jatulian and somewhat younger Kalevian strata, the latter composed mainly of metaturbidites and tholeiitic volcanites. The older Jatulian rocks belong to Karelian cover of the cratonic foreland to the east, the bulk of which unconformably underlies Kalevian metaturbidites. Accordingly, a major boundary is placed between the older Jatulian and the younger Kalevian groups (Gaál and Gorbatschev 1987).

In the Outokumpu District of central Finland, a tectonically displaced terrane of mica schists and gneisses, which is situated between the Kalevian belt to the east and the Ladoga–Bothnian Bay Zone on the west, marks the easternmost limit of the Svecofennian Province (see below). These rocks are interpreted as distal-type metaturbidites deposited in a separate coeval basin, originally sited west of the Kalevian Belt proper (Ward 1987).

## Svecofennian (flysch) facies

### Svecofennian Supergroup

The predominant flysch facies of the 1200 km-long by 1000 km-wide Svecofennian Province (Domain), predominates around the Gulf of Bothnia in both Sweden and western–northern Finland (Fig. 2.3). Stratigraphic relations are obscured by complex polyphase deformation, abundant migmatites and unusually widespread granitoid intrusions. As reviewed by Park (1985), the facies is characterized by (1) bimodal metavolcanic–volcaniclastic–turbidite belts at prevailing low amphibolite–greenschist facies of metamorphism, which separate (2) equivalent-sized infracrustal belts composed of migmatite-dominated trondhjemite–granodiorite–granite igneous rocks, with minor metasediments at high amphibolite–granulite facies (Simonen 1980). None of the belts shows significant physical or chemical signs of an Archean inheritance, giving evidence, rather, of juvenile crustal accretion (Patchett and Kouvo 1986). Of these belts, the northernmost (Skellefte province) is also the oldest, ~2.5 Ga (Wilson 1982); succeeding belts to the south (Tampere, Skaldo and Orebro), are progressively younger (1.90–1.84 Ga) (Wilson 1982).

In the Bergslagen Region of southern Sweden (Fig. 4.4), the major supracrustal rocks belong to the Leptite Formation, dominated by partly reworked felsic metavolcanic rocks. The best preserved aphanitic volcanic rocks, called halleflintas, contain primary porphyritic and glass-shard textures. Intercalated marbles, stromatolitic dolostones, and iron ore (apatite skarn types) and sulphide ore deposits occur sporadically, especially in the Bergslagen mining district (Sundblad et al 1993). Northward, in central Sweden–Tampere (western Finland), dominant flyschoid metasediments including reworked leptites and mafic metavolcanic rocks, all variably migmatized, underlie vast areas. The Skellefte field in north central Sweden contains at least 80 massive sulphide ore deposits, including the famous Boliden mine, now exhausted (Vivallo and Claesson 1987).

In northern Sweden (Norrbrotten, Fig. 4-4), including the well-known Kiruna area, dominant metavolcanic rocks are transitional northeastward to Karelian-type epicontinental facies. Kiruna polyglot felsic metavolcanic rocks (rhyolite–trachyte porphyries) generally overlie partly spilitic Kiruna greenstones. The latter include many stratiform copper (chalcopyrite–pyrrhotite–sphalerite) ore deposits. The 1900–1880 Ma apatite–magnetite ores at Kiruna (Cliff and Rickard 1990, 1992, Romer et al 1994) which collectively form particularly large iron ore reserves, occur as an inclined sheet lying between syenite porphyry footwall and quartz porphyry hanging wall.

Finally, in the Tampere Region, western Finland, dominant metagreywacke with mafic–felsic metavolcanic intercalations include copper-rich sulphide ore deposits in lower mafic volcanic rocks, and massive copper–lead–zinc ores in upper felsic volcanic rocks. Mg-alteration zones, with associated cordierite–anthophyllite, are characteristic (Colley and Westra 1987).

Svecofennian metavolcanic rocks have been dated at 2.1–1.8 Ga (Huhma 1984, Skiöld and Cliff 1984, Mearns and Krill 1985, Skiöld 1986, Perdahl and Frietsh 1993). This 3.00 Ma-long accumulation interval immediately preceded the ensuing Svecofennian Orogeny (1.9–1.8 Ga). Detrital zircons in Svecofennian metasediments (Sweden–Finland) provide dominant (65%) ages of 2.1–

1.9 Ga, common (30%) ages of 3.0–2.6 Ga, and rare ages of 3.4–3.3 Ga (Claesson et al 1993). This implies that major areas of 2.1–1.9 Ga crust were under erosion at 1.9 Ga. However, there is no evidence of the presence of any 2.6–2.1 Ga protoliths, an interesting 500 Ma-long terminal Archean–early Proterozoic hiatus. Furthermore, the age data do not support a pattern of westward accretion of juvenile island arcs, as had been previously proposed.

The concluding Svecofennian Orogeny is characterized by the rapid formation of great amounts of juvenile continental crust and extensive remobilized older crust. More than 50% of the exposed Svecokarelian Domain is underlain by granitoid plutons of granite, granodiorite and tonalite compositions belonging to different intrusive epochs. Amphibolite facies is the dominant metamorphic grade. Migmatites are extensively developed.

Synorogenic plutonic rocks, emplaced during the main deformation phase, are mainly quartz diorites, granodiorites and trondhjemites with minor associated mafic rocks. The late orogenic plutonic rocks are mostly K-rich granites forming migmatites with the older rocks (Simonen 1980).

Recent zircon dating constrains the main Svecofennian igneous activities between 1930 and 1870 Ma (Skiöld et al 1993), and even 1830–1800 Ma (Ehlers et al 1993, Romer and Smeds 1994). The data suggest that pre-Svecofennian rifting of the pre-existing Archean craton, with recently isotopically defined boundary in northern Sweden (Öhlander et al 1993) had created a passive continental margin and that the transition to an active margin with subsequent island-arc magmatism and presumed subduction beneath the Archean crust commenced prior to 1930 Ma.

Plutonic rocks of the Finnish Svecokarelian crust are considered by Patchett and Kouva (1986) to consist of ~90% juvenile mantle-derived material. This observation may apply to the great bulk of the Svecofennides, which would thereby collectively represent a massive accretion of juvenile early Proterozoic crust.

### Lapland Granulite Belt

The Lapland Granulite Belt in the northeastern Baltic Shield, up to 60 km wide and 500 km long (Fig. 2-3), has been overthrust southwestward along a sinuous, gently northeastward dipping fault. The dominant rocks of the belt are high grade metapelites–metapsammites with mafic metavolcanic intercalations, intruded by intermediate-ultramafic plutonic rocks (Barbey et al 1984, Gaál and Gorbatschev 1987). The metasediments form a khondalite suite of flysch affinity, whereas the igneous rocks of the charnockite complex have calc-alkalic and tholeiitic compositions.

The rocks of the Granulite Belt were metamorphosed at high temperatures and low to intermediate pressures. The internal structure is characterized by large isoclinal folds. The component supracrustal rocks accumulated at 2.4–2.0 Ga (Barbey et al 1984), with metamorphism and igneous activity to 2.0–1.9 Ga (Meriläinen 1976, Gaál and Grobatchev 1987).

Barbey et al (1984) attribute the Lapland Granulite Belt to the product of Svecofennian continent–continent collision involving the 'Inari–Kola' craton to the northeast and the 'South Lapland–Karelia' craton to the southwest.

### 3.4.2 UKRAINIAN SHIELD AND VORONEZH MASSIF

Early Proterozoic metasupracrustal rocks form a number of narrow to substantial N-trending interblock belts in the Ukrainian Shield (Fig. 1-5c(ii)). Similar belts have been located in the Voronezh Massif across the Dnieper–Donetz Aulacogen to the northeast (Fig. 1-5c(i)), especially in the Kursk region. These metasupracrustal belts are famous for their very large iron ore deposits, especially at Krivoy Rog and The Kursk Magnetic Anomaly (KMA).

### Krivoy Rog

Iron-bearing strata are concentrated in an area of 250 km by 300 km, outlined by the big bend of the Dnieper River and called the Bolshoy Krivoy Rog. At least five main N-trending zones or belts, each up to 150 km long and 10–20 km wide, have been identified in the region. Of these, the Krivoy Rog–Kremenchug Zone (Block) is renowned. Each zone is divided along strike into a number of iron ore regions. Other smaller zones are present in outlying parts and beneath younger cover.

The Krivoy Rog–Kremchug Zone is one of the biggest and oldest centres of iron ore industry in the world. It involves complicated N-trending fold structures in rocks of the Krivoy Rog Supergroup which are continuous along strike for approximately 70 km. According to the fivefold subdivision of Belevtsev et al (1983), the supergroup, some 8500 m thick, represents a complete transgressive–

regressive cycle. In brief, lower sandstones and metavolcanic rocks are overlain successively by: arkosic sandstones, phyllite, carbonates and talc–chlorite schists; by the main iron ore bearer (Saksagan Formation), 750–2000 m thick, dominated by alternating jaspilite and slate; and by unconformably overlying ferruginous sandstones, schists, dolomites and polymictic conglomerate.

Deformation resulted in development of the complicated N-trending Krivoy Rog synclinorium, later much affected by major faults, block uplifts and subsidence (see Belevtsev et al 1983, fig. 5-3). All the folded structures of the Krivoy Rog synclinorium plunge northward at 18–20°.

The age of metamorphism is 2000 Ma (Shcherbak et al 1984), associated with the formation of granitoids of the Kirovograd–Zhitomir Complex and analogues. This correlates with the final consolidation of much of the Ukrainian Shield.

Voronezh Massif

In the Voronezh Anteclise (Massif) to the northeast, the Kursk Group and overlying Oskol Group are close analogues to the Krivoy Rog Supergroup (Khain 1985).

The Voronezh Massif also includes very large reserves of iron ores, closely resembling those at Krivoy Rog (Salop 1983).

### 3.4.3 BURIED BASEMENT

The previously considered buried basement pattern involving major Archean massifs and subordinate intervening early Proterozoic fold belts with variable younger overprinting (Fig. 1-5c(iii)) has been reinterpreted on the basis of recent drilling–dating results from the western part of the buried craton, specifically west of a line joining Kiev–Moscow–St Petersburg. Representative basement samples from southern Estonia–Belorussia are dated at 2.18–1.80 Ga (Puura and Huhma 1993, Bogdanova et al 1994). In the resulting reconstruction (Fig. 3-2), major Archean crust is concentrated in the northwestern (Karelia), eastern (Vologo–Uralia) and southern (Ukrainian–Sarmatia–Voronezh) parts of the craton, with juvenile early Proterozoic crust underlying about one-third of the craton in the central–western part (Gorbatschev and Bogdanova 1993).

## 3.5 GREENLAND SHIELD (NORTH AMERICAN CRATON)

Early Proterozoic crust in Greenland is mainly represented by three E-trending mobile belts, each composed of variably remobilized Archean infrastructure with early Proterozoic cover: Nagssugtoqidian and Rinkian belts lying successively northward of the Archaean Block, and Ketilidian Belt to the south (Figs 1-3d(i), 1-5d(i), 3-2).

### 3.5.1 NAGSSUGTOQIDIAN MOBILE BELT

This ESE-trending, ~250 km wide tectonic zone consists mainly of variably reworked Archean gneisses, together with pristine Proterozoic crust in the form of gneiss and metasupracrustal patches (Escher et al 1976). The belt is characterized by a pronounced regional planar fabric characterized by a strong parallelism for all structural elements, which alternate across strike with areas of less deformed rock, in which older, more open fold structures are preserved. The rocks bear the isotopic imprint of the Nagssugtoqidian (Hudsonian) Orogeny, ~1.85 Ga (Kalsbeek and Taylor 1985, Kalsbeek et al 1987) superimposed upon an older deformation–metamorphism that occurred ~2.7 Ga (Escher and Watt 1976).

Post-supracrustal intrusions of granitoids, anorthosite and ultramafic rocks, all of limited size, are widespread throughout the belt. However, no large granitic plutons or migmatite assemblages have been reported.

Throughout the Nagssugtoqidian Mobile Belt, a weak deformation can be seen which locally refolds the typical Nagssugtoqidian ENE-trending structures by concentric type buckling around NW-trending axes. Brittle transcurrent or thrust movements postdating this deformation have produced mylonites and pseudotachylites (Bridgwater et al 1973a).

### 3.5.2 RINKIAN MOBILE BELT

The Rinkian belt generally comprises Archean crystalline infrastructure and early Proterozoic cover collectively deformed and metamorphosed during the Rinkian (Hudsonian) Orogeny ~1.85 Ga (Kasbeek et al 1988). The Rinkian–Nagssugtoqidian boundary is marked by an important sinistral transcurrent fault zone. Characteristic Rinkian patterns are as follows: in the south, very large gneiss domes with rim synclines of metasupracrustal rocks; in the centre, recumbently folded gneiss domes with, largely in the west, extensive early Proterozoic metasedimentary tracts; and in the north, granulite facies gneisses with large granite–charnockite plutons.

Early Proterozoic metasediments are dominated by the Karrat Group, an unusually thick (8000 m) cover sequence which is about equally divided into lower quartzite–staurolite schist–amphibolite–marble, and conformably overlying remarkably uniform pelites with intercalated calcsilicates (Escher and Pulvertaft 1976).

The Rinkian belt generally comprises Archean crystalline infrastructure and early Proterozoic cover collectively deformed and metamorphosed during the Rinkian (Hudsonian) Orogeny ~1.85 Ga (Kalsbeek et al 1988).

### 3.5.3 KETILIDIAN MOBILE BELT

The Ketilidian is a 300 km wide, E-trending mobile belt crossing the southern tip of Greenland, specifically occupying the region between Ivigtut in the west, Ikermit in the east, and Kap Farvel in the south (Fig. 1-5d(i)). It comprises gneiss, granitoid plutons and metasupracrustal rocks, and is characterized by the presence of numerous very large, late intrusive granitoid plutons (Windley 1991).

The mobile belt is divided into four zones of about equal width: (1) the Northern Border Zone features Ketilidian metasupracrustal rocks overlying crystalline basement with Archean metasupracrustal infolds; transitional southward to (2) the Central Granite Zone, a complex Ketilidian granite–diorite plutonic assemblage including many large granite bodies; succeeded southward by (3) the Folded Migmatite Zone, composed of intricately folded granite, gneiss and migmatized metasupracrustal rocks at amphibolite facies metamorphism; in turn transitional to (4) the Flat-Lying Migmatite Complex, forming the southern tip of Greenland (Allaart, 1976b, fig. 138).

Age dating reveals a range for Ketilidian granitoid rocks of ~1850 to ~1700 Ma (Kalsbeek and Taylor 1985). Granitoid plutons in the border zone of the Ketilidian Belt adjoining the Archaean Block to the north contain large proportions of Archean crustal materials. Granitoid rocks from the central part of the Ketilidian Belt show no evidence that they were contaminated with Pb and Sr derived from much older rocks at depth; this indicates that most of the Ketilidian Mobile Belt is not underlain by Archean crust. The same conclusion was reached earlier on the basis of Sr isotopic evidence (van Breemen et al 1974). On this basis, it is concluded that the Ketilidian Mobile Belt consists almost exclusively of pristine early Proterozoic crust and that the southern border of the main Archaean Block in early Proterozoic time was a continental plate margin (Kalsbeek and Taylor 1985).

In this connection, the islands of Colonsay, Islay and others situated southwest of Mull (Fig. 5-3), also include a basement of deformed alkalic igneous rocks (Rhinns complex), which represents juvenile mantle-derived additions to the crust at ~1.8 Ga. This provides a Greenland–Scandinavia link (Muir et al 1992).

### 3.5.4 SCOTTISH SHIELD FRAGMENT (NORTH ATLANTIC CRATON)

Following initiation of Inverian shear zones with selected uplift and segmentation of Archean blocks (~2.6 Ga), and late Badcallian pegmatite emplacements (2.5 Ga), the Laxfordian cycle of events affected the Scottish Shield Fragment situated to the west of the Moine Thrust (Fig. 5-3, Table 2-5).

The cycle was initiated by emplacement of Scourie dolerites (diabase) and norites at 2.4 Ga, followed by olivine gabbros and picrites at 2.2 Ga, along with continuing retrogression of Lewisian granulites and intermittent movement on shear zones. Loch Maree sediments in the southern Lewisian outcrop area, comprise greywacke, sandstone, pelite and some ironstones and limestone, now mainly in the form of mica schists, graphitic schists, quartz–magnetite schists and marble. Intercalated mafic volcanic rocks are metamorphosed to amphibolites, the metamorphism dated at 1975 Ma (Anderton et al 1979, Park 1994).

The Lewisian as presently exposed in the Scottish Shield Fragment thus represents the relics of the Laxfordian orogeny including thrust slices and other basement fragments that have been deformed and migmatized together with associated local patches of metasupracrustal cover. East of the Moine Thrust, Laxfordian evidence is all but lost due to younger events.

## 3.6 NORTH AMERICAN CRATON (LESS GREENLAND SHIELD)

### 3.6.1 INTRODUCTION

Early Proterozoic (Aphebian) gneiss–migmatite terrains and associated metasupracrustal fold belts predominate in three structural provinces of the Canadian Shield and their buried extensions (Figs 1-3d(ii), 1-5d(ii), Table 2-8); (1) the small yet complex Southern Province of the Great Lakes region (Sims et al 1981, 1993, Sims and Peterman 1983, Young 1983, Barovich et al 1989); (2) the very large and complex Churchill Province, rich in diverse granitoid rocks and foreland belts and basins (Henderson 1984, Stauffer 1984, Lewry et al 1985, Hoffman 1989, 1990); and (3) to the northwest, the tectonically important Wopmay Orogen including the Bear Province (Hoffman 1980, Hoffman and Bowring 1984). Additional Aphebian crust is exposed in median massifs of the Cordilleran Orogen (see Chapter 5).

Granitoid rocks, mainly gneissic but including significant massive components, predominate by far. Most gneisses represent recrystallized Archean basement; but others represent juvenile early Proterozoic accretions. Metasupracrustal rocks of early Proterozoic age are disposed both in distinctive BIF-bearing foreland fold belts marginal to the main Archean cratons such as Superior Province, and as tectonically discontinuous cover upon hinterland infrastructure. All these were variably affected by the Hudsonian (Penokean) Orogeny which peaked at 1.85 Ga.

Older Aphebian sequences (e.g. Huronian Supergroup) are characterized by thick, locally uraniferous sandstone–pelite–conglomerate sequences with minor basalt flows, typically disposed in off-craton thickening wedges. Placer uranium deposits are especially important in the Huronian Supergroup.

Younger Aphebian fold belts (2.1–1.8 Ga) of Churchill, Southern and Bear provinces lie grouped about Superior and Slave cratons respectively. The fold belts typically comprise asymmetric arrangements of thin platform-shelf (miogeoclinal) facies (orthoquartzite–conglomerate–BIF) and thicker off-shelf-slope rise (eugeoclinal) facies (turbidites–mafic–ultramafic volcanic rocks and intrusions).

Major iron ore deposits lie in the Animikie–Marquette and Labrador–Belcher sequences of the circum-Superior belts (Fig. 3-8). Significant Cu–Zn–Au deposits occur in some greenstone belts in Saskatchewan–Manitoba and Wisconsin–Michigan. Exceptional Ni–Cu–Pt–Cd ores occur in the Sudbury Lopolith (Card et al 1972, Krogh et al 1982, Pye et al 1984) of Southern Province. Important Ni–Cu concentrations occur in the Thompson and Cape Smith belts of Churchill Province.

In the Mesozoic pre-drift continental reconstruction, Churchill Province, including the Trans-Hudson Orogen (Hoffman 1989, Lewry and Stauffer 1990) which, in the pre-drift reconstruction, extends eastwards into Greenland, to incorporate the Nagssugtoqidian and Rinkian belts and southward into mid-continental USA, for a total strike length exceeding 5000 km (Figs 1-5d(i,ii), 3-2). Collectively, these form Earth's largest existing early Proterozoic terrain. The main tectonic–metamorphic imprints therein are related to thermo-tectonic events of the Hudsonian (–Rinkian) Orogeny, peaking at ~1.89 Ga.

### 3.6.2 CIRCUM-SUPERIOR FOLD BELTS

The Huronian Supergroup (Basin), which constitutes an older Aphebian sequence (2.47–2.33 Ga) together with the adjoining Sudbury Structure (1.85 Ga), are geographically associated with the Lake Superior segment of the younger Aphebian (2.2–1.8 Ga) Circum–Superior fold belts (Figs 3-3, 3-8; Table 2-8, column 1).

#### Huronian Supergroup (Basin)

The Huronian Supergroup, Lake Huron region, consists of clastic sedimentary rocks, subordinate chemical sediments and volcanic rocks (Frarey 1977, Sims et al 1981, Young 1983). The strata were deposited unconformably on Archean basement in the interval 2470–2330 Ma (Noble and Lightfoot 1992, Roscoe and Card 1992, 1993, Lightfoot et al 1993, Buchan et al 1994).

Huronian supracrustal rocks were deformed, metamorphosed, and intruded by mafic and felsic plutonic rocks during a protracted period from 2330 to 1700 Ma. The entire assemblage is overlain to the south by Paleozoic cover. Huronian rocks are transitional southward beneath this cover to gneisses of unknown primary age, subsequently affected to the south by the E-trending Penokean Orogeny (~1.8 Ga) and to the southeast by the SW-trending Grenvillian Orogeny (~1.0 Ga) (Fig. 3-3). The full original disposition of Huronian rocks, which may, in fact, reappear in Wyoming (Roscoe and Card 1993), is uncertain due to these superimposed events.

#### Lithostratigraphy

The Huronian Supergroup comprises a southeastward-thickening dominantly sedimentary wedge which on-laps the southern flank of Superior Province. This supracrustal wedge directly overlies an

ENE-trending tectonic zone, now marked by the Murray Fault System (Figs 3-3, 3-4), a probable eastward extension of the regionally significant Great Lakes Tectonic Zone (Figs 2-10, 3-6). The Huronian Supergroup is notable for: (1) paraconglomerate units including tillites; (2) abundant detrital minerals, including pyrite, uranium minerals, sparse gold and other heavy minerals which are concentrated in oligomict conglomerate beds; (3) exceptionally high ortho-quartzite content; and (4) cyclic sedimentary pattern, each cycle (group) comprising, up-section, para-conglomerate, shale, and sandstone units.

Near the Grenville Front Tectonic Zone, southeast of Sudbury (Fig. 3-3), the lower part of the Huronian Supergroup pinches out, suggesting the former presence of a pre-Huronian positive topographic elements in that direction.

### Paleoenvironment

The lower part (three groups) of the Huronian Supergroup is interpreted to represent a series of major regressive marine cycles controlled by syndepositional faulting and differential movements of basement blocks. The upper part (Cobalt Group) constitutes a comparatively thick clastic wedge comprising lower coarse clastics grading up to shallow-water sandstones, this group was probably deposited under control of glacial and glacial-related processes.

Sedimentologic data indicate that Huronian detritus was derived mainly from Archean basement to the north (Sutton and Maynard 1993), and was deposited in polymodal pattern by southward flowing currents. The remarkable concentration of both orthoquartzite and placer uranium minerals in the supergroup points to an exceptional degree of Archean foreland uplift, erosion and sedimentation during the preceding 2.6–2.4 Ga erosional interval (Eparchean Interval of Lawson (1934)) to produce such resistates on a uniquely grand scale.

The supergroup exhibits a change in the character of detritus suites from those containing abundant pyrite with uranium minerals and gold below, to those dominated by iron oxides and lacking uranium minerals above. These relationships are considered evidence for a change in atmosphere at ~2.3 Ga from one that lacked free oxygen to one that was beginning to accumulate free oxygen (Roscoe 1973, Sutton and Maynard 1993, Krupp et al 1994).

On a larger regional scale and based on positive matches of stratigraphic units, a plausible paleoreconstruction in Huronian time returns the Wyoming Uplift to a position contiguous to the Western Huronian terrain of Ontario. This implies that Huronian strata were deposited 2.4–2.2 Ga ago in an ensialic repository developed astride the boundary between the two now-separated structural provinces (i.e. Superior and Wyoming) (Roscoe and Card 1993).

### Sudbury Structure

The Sudbury Structure is located in the eastern part of Southern Province where it straddles the main Huronian Supergroup–Archean basement contact (Fig. 3-3) (Naldrett 1984, Dressler 1984). The Sudbury Igneous Complex (Irruptive), a major component of the structure, was emplaced at 1850 Ma (Krogh et al 1982, 1984b). The origin of this elliptical complex has provoked vigorous debate (Muir and Peredery 1984, Rousell 1984, Stevenson 1990). On the one hand, the site marks a long-lived focus of exceptional regional tectonic elements (Fig. 3-3), thereby pointing to endogenic processes. On the other hand, the presence of abundant shock metamorphic textures and structures has led to the inference of an important extraterrestrial (meteorite impact) role.

In brief, the elliptical cryptoexplosion Sudbury Structure (Fig. 3-5) measuring 60 km by 25 km in dimension, comprises, in inward order, (1) breccias in the Archean–early Proterozoic footwall rocks of the Sudbury Igneous Complex; (2) the Sudbury Igneous Complex, a three-part norite–gabbro–micropegmatite ring association emplaced at the base of (3) the enclosed Whitewater Group, a three-part up-section assemblage of (a) heterolithic breccia (Onaping Formation); (b) mudstones (Onwatin Formation); and (c) wackes (Chelmsford Formation). Of these, the lowermost Onaping breccias, a possible fall-back breccia, are directly related to the origin of the Sudbury Igneous Complex. The Cu–Ni sulphide ore deposits of the complex are found in the Sublayer and the Footwall including sublayer offsets—all located at or close to the lower contact of the Igneous Complex. The ore deposits are mined for Ni and Cu, as well as Co, Fe, platinum group metals, Au, Ag, Se and Te. Shock-metamorphic features including (1) shatter cones which are distributed in the footwall rocks for as much as 10 km from the complex; and (2) planar features and devitrification textures in the overlying Onaping Formation, demonstrate exceptionally high energy (50–500 kbar pressures) involvement.

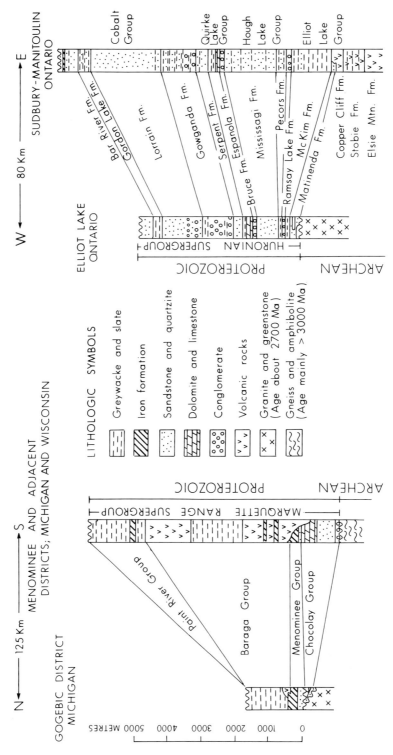

**Fig. 3-4.** Selected stratigraphic sections of early Proterozoic rocks of the Lake Superior and Lake Huron regions. (From Sims et al 1981, Fig. 21.6, and reproduced with permission of the Geological Survey of Canada and the Minister of Supply and Services Canada).

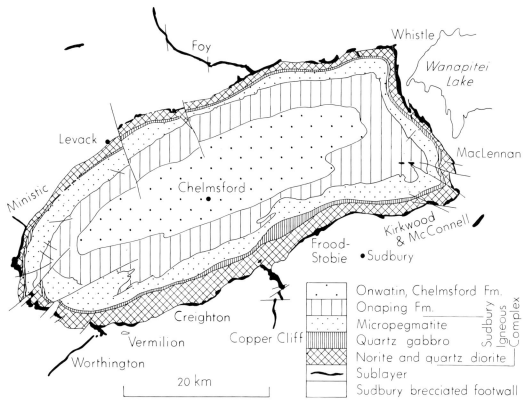

**Fig. 3-5.** Sketch map of the Sudbury Igneous Complex showing the distribution of the complex, sublayer, brecciated footwall, and enclosed Whitewater Group. (Adapted from Grand and Bite 1984, Fig. 12.1, and reproduced with permission of the Ontario Geological Survey).

Coincidentally, the Wanapitei Lake structure, 5 km east of Sudbury, is a well established ~37 Ma old meteorite impact site (Dressler 1984).

Recent seismic reflection data (Milkereit et al 1992) show the Sudbury structure to be markedly asymmetric at depth. Thus, whereas the North Range comprises the anticipated shallowly south dipping assemblage, the seismic image of the South Range is dominated by a series of south dipping reflections interpreted as thrust faults/shear zones on which severe telescoping and imbrication have occurred, resulting in considerable NW–SE shortening of the Sudbury structure. The original shape of the structure, whether circular (Hirt et al 1993) or otherwise, is thereby obscured. This apparent decollement rearrangement of the South Range units must be accommodated in any paleoreconstruction of the Sudbury structure.

Arguments in favour of meteorite impact stand upon the evidence that a shock wave, several hundreds of kilobars in magnitude, has affected the country rocks, both in the Footwall for distances up to 10 km from the complex, and as fragments in the overlying Onaping Formation (Muir and Peredery 1984, Rousell 1984). The Sudbury Breccia has close analogies in pseudotachylite veinlets described from the Ries and Manicougan structures, both of widely accepted impact origin. Also, the so-called Grey Member (Onaping Formation), with its preponderance of country rock fragments and devitrified glass, general lack of stratification, and heterogeneous composition of altered glass shards and flow-banded aphanitic-textured inclusions, closely resembles the suevite breccia of the Ries crater.

Points contrary to an impact origin and favouring an endogenous origin include, first and foremost, the location of Sudbury upon an elliptical gravity high (> 30 mgals) that forms part of an E-trending chain of gravity anomalies extending for more than 350 km along an arc from Elliot lake to Engelhart, Ontario. This enhances the unique regional tectonic setting of Sudbury, including its location at the junction of orthogonally disposed long-lived regional fault zones (Card et al 1984).

Since the non-impact origin of the pseudotachylite-bearing Vredefort Dome, South Africa, is now gaining favour, a similar endogenic interpretation may become increasingly viable at Sudbury. In this regard, the large amount of mafic–intermediate igneous rock present at Sudbury, whether mantle- or crustal-drived, is unmatched at all known terrestrial impact sites, an added complexity to the impact model.

Clearly, although the impact site theory remains an attractive model to explain several otherwise inexplicable relations at Sudbury, it is still best viewed as a stimulant to further research rather than as a fully satisfactory explanation of this unique cryptoexplosion structure.

### Lake Superior Region

The broad geologic framework of the Lake Superior Region, itself the southwestern segment of the Circum-Superior fold belt association (Fig. 3-8, Table 2-8, column 1), incorporates (1) an Archean basement unconformably overlain by (2) ENE-trending late Aphebian (2.5–1.8 Ga) BIF-bearing metasupracrustal rocks (Animikie Group, Mille Lacs Group, Marquette Range Supergroup) and to the south, the Wisconsin Magmatic Terranes (Volcanic Belt), themselves (3) intruded and metamorphosed during the Penokean (Algoman) Orogeny (1.9–1.8 Ga), and later transected and partly overlain by (4) mid-Proterozoic mafic magmatic rocks (Keweenawan) of the midcontinent rift system (~1.1 Ga) (Figs 3-4, 3-6).

#### Penokean Orogeny/Fold Belt

The Penokean Fold Belt is defined as the zone of early Proterozioc and Archean rocks, adjacent to and along the southern margin of Superior Province, which was deformed and metamorphosed during the Penokean Orogeny at mainly ~1850 Ma but extending to 1760 Ma (Sims and Peterman 1983, Sims et al 1993). The NE-trending fold belt, up to 250 km wide, is mainly confined to the older Archean gneiss terrain and associated early Proterozoic sequences lying south of the Great Lakes Tectonic Zone (Fig. 3-6), but does include a 30–40 km wide zone of deformed Archean and Proterozoic rocks to the north.

The fold belt is divided into a stable foreland area

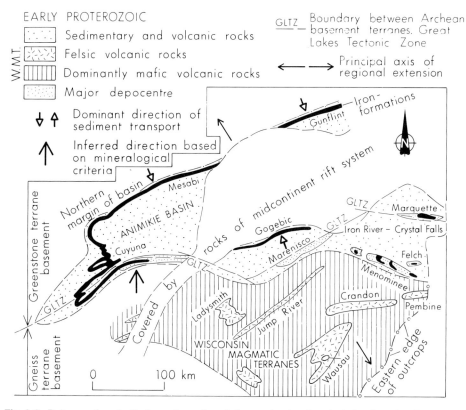

Fig. 3-6. Patterns of early Proterozoic sedimentation and volcanism in Lake Superior Region, as inferred prior to late mid-Proterozoic (Keweenawan) rifting. Major iron formations shown in black. Wausau shown for reference. W.M.T. = Wisconsin Magmatic Terranes. (From Sims and Peterman 1983, Fig. 4, and reproduced with permission of the authors and of the Geological Society of America).

on the north, and a main fold-thrust belt, involving compression and N-directed transport, on the south. The main fold-thrust part is itself subdivided into a northern subzone of deformed Animikie-Marquette epicratonic rocks and associated basement gneisses, and a main southern subzone of deformed volcanic and magmatic rocks (Wisconsin Magmatic Terranes). The boundary between the two subzones is a high-angle reverse fault (Niagara or Florence–Niagara fault zone) (not illustrated).

The Penokean Fold Belt is characterized for the most part by penetrative foliation, with local diversely oriented folds, mantled gneiss domes, and a nodular distribution pattern of metamorphism (Bayley and James 1973, Sims et al 1989, Holm et al 1993).

### Animikie Basin

The Lake Superior Region, a major depocentre of BIF and derived iron ore (Morey 1983a), has shipped $4.6 \times 10^9$ t since 1848, predominantly (96%) from early Proterozoic strata, and contains ore reserves of $271 \times 10^9$ t. The BIF-rich early Proterozoic rocks occupy the Animikie Basin, a broad ENE-trending and -plunging oval-shaped repository (Fig. 3-6) (Morey 1983b), in which the principal depocentres lie along the Great Lakes Tectonic Zone (GLTZ) (Fig. 2-10). The preserved parts of the basin are about 700 km by 400 km or 220 000 km² in area.

The Animikie Basin itself is tectonically bounded on the south by the coeval mafic–felsic volcanic-plutonic associations of the Wisconsin Magmatic Terrane (Fig. 3-6). However, these thick, yet poorly exposed volcanic sequences are nowhere interlayered, as known, with strata of the Animikie Basin proper (Sims and Peterman 1983, Barovich et al 1989).

As a result of Penokean deformation and Pleistocene cover, the BIF of the region is exposed in a number of separate 'iron ranges'. These include, in the northwest, the Gunflint, Mesabi and Cuyuna ranges of Minnesota and Ontario; and, in the southeast, the Gogebic, Marquette, Menominee and Iron River–Crystal Falls ranges of northern Wisconsin and Michigan (Fig. 3-6).

### Lithostratigraphy

The Midcontinent Rift System with its ~1.1 Ga Keweenawan volcanic-rich assemblage, effectively divides the east-plunging Animikie Basin, into northwestern (Gunflint–Mesabi–Cuyuna) and southeastern (Gogebic–Marquette–Iron River–Crystal Falls–Menominee) segments which are respectively composed of miogeoclinal (platform) facies and eugeoclinal facies. Strata in the northwestern platform segment form a 1.5–6 km-thick quartzite–BIF–greywacke–shale succession, variably assigned to the Animikie and Mille Lacs groups, whereas those in the southeastern eugeoclinal segment are assigned to the Marquette Range Supergroup, a comparatively thick (up to 9 km), lithologically diverse, internally interrupted, conglomerate–quartzite–BIF–carbonate–greywacke–volcanic association which is divided, up-section, into Chocolay, Menominee, Baraga and Paint River groups (Fig. 3-7) (Morey 1983b). Stratigraphic relations in the eastern part of the southeastern segment, e.g. Menominee Range, are far more complex, with the component groups much thicker and diverse, especially in the upper Baraga and Paint River groups which are dominated by thick turbidite–mafic volcanic sequences.

*The ages* of Animikie stratified rocks are poorly constrained in detail. Approximate depositional ages are as follows: Animikie Group: 2.12–1.81 Ga; Mille Lacs Group: 2.45–2.25 Ga; and Marquette Range Supergroup: 2.45–1.86 Ga (Sims et al. 1993). It is interesting to note that these dates would not necessarily preclude some overlap in deposition of the lowermost Chocolay Group and the uppermost Huronian Supergroup (> 2150 Ma) of Ontario (Fig. 3-4) (Young 1983). However, the data strongly point to the main Huronian deposition substantially preceding the main Animikie–Marquette deposition.

### Wisconsin Magmatic Terranes (Wisconsin Volcanic Belt)

Early Proterozoic assemblages southeast of Lake Superior are, as stated above, divided by the E-trending Niagara Fault Zone into the comparatively narrow (30–100 km) northern Penokean Domain underlain by the Marquette Range Supergroup, and, in the south, the broader (up to 200 km) Wisconsin Magmatic Terranes (Penokean Volcanic Belt) which is characterized by a volcanic-plutonic assemblage containing abundant calc-alkalic metavolcanic units (Basalt, andesite and rhyolite) and lesser shallower-water metasedimentary rocks, with abundant tonalite–granite intrusions. Regional greenschist facies prevails, with local amphibolite facies in the north (Morey 1983a, Greenberg and Brown 1983, Sims et al 1989, 1993). One of the component belts in north-central Wisconsin, termed the Ladysmith–Rhine-

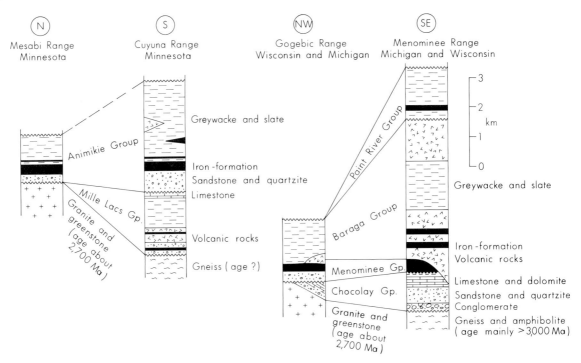

Fig. 3-7. Generalized correlations of lower Proterozoic strata in the northwestern segment (Animikie and Mille Lacs Groups) and in the southeastern segment (Marquette Range Supergroup) of the Animikie Basin of the Lake Superior Region. (From Morey 1983b, Fig. 3, and reproduced with permission of the author and of the Geological Society of America).

lander metavolcanic complex, is especially rich in felsic volcanic concentrations (centres) with some associated Cu–Zn–Pb-bearing massive sulphide deposits (e.g. Flambeau, Crandon, Lynne, Ladysmith) (De Matties 1994).

Volcanic–plutonic ages in the Wisconsin Magmatic Terranes fall mainly in the 1890–1830 Ma range including significant magmatic concentrations at 1889–1860 Ma and 1845–1835 Ma, tailing off to a younger less significant plutonic episode at 1760 Ma, the latter approximately coeval with the terminal phase of the Penokean orogeny (Sims et al 1989).

Tectonic interpretation

Early Proterozoic successions of the Great Lakes Region have been varyingly ascribed to: (1) intracratonic deposition in rifts (Sims et al 1981); (2) deposition on a passive continental margin followed by northward dipping subduction of oceanic crust (Van Schmus 1976); (3) deposition on a rifted continental margin followed by continent–continent collision involving a southward dipping subduction zone (Cambray 1978). More recently, Sims et al (1989, 1993) interpret the Wisconsin Magmatic Terranes in terms of arc-continent collision.

According to Barovich et al (1989), the Marquette Range Supergroup represents a predominantly miogeoclinal continental-margin sequence deposited before 1.85 Ga on Archean basement. Felsic metavolcanic rocks of the Wisconsin Magmatic Terranes to the south represent ~1.88 Ga evolved island-arc types that were intruded by 1.87–1.76 Ga granitoids. Barovich et al (1989) propose that the Niagara Fault Zone represents a Penokean collisional boundary marking convergence of the Wisconsin Magmatic Terranes in the south with a continental margin to the north (northern Penokean domain). Accordingly, the 1.9–1.7 Ga Penokean events would have involved major growth of new mantle-derived crust.

Labrador Trough

The Labrador Trough, the largest and continuous of the Circum-Ungava fold belts and an integral part of the eastern Churchill Province (Wardle et al 1990), extends along the entire eastern margin of Ungava Craton (Fig. 3-8, Table 2-8, column 4). Iron formations of the trough merge from component basin to basin to form a continuous stratigraphic unit more than 1200 km long, probably the

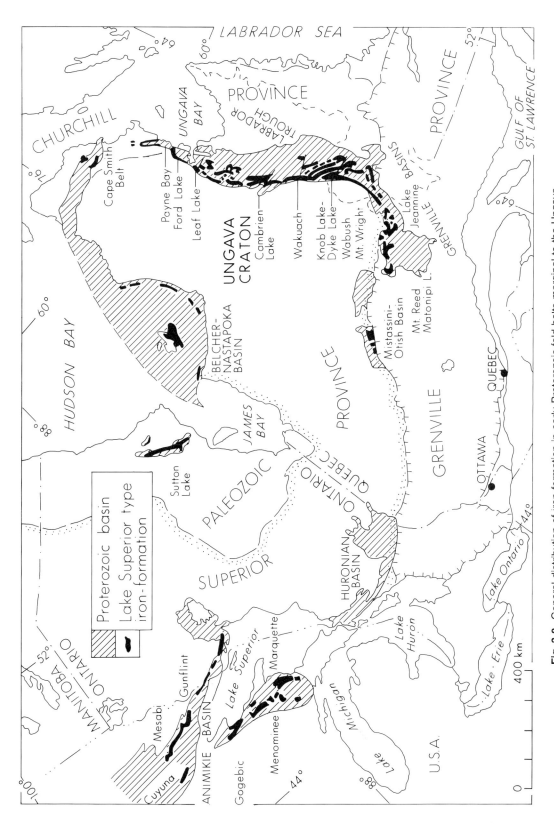

**Fig. 3-8.** General distribution of iron formation in early Proterozoic fold belts marginal to the Ungava Craton. (After Gross and Zajac 1983, Fig. 6-1, and reproduced with permission of the authors and of Elsevier Science Publishers).

most continuous such belt in the world (Gross 1968, Dimroth et al 1970, Wardle and Bailey 1981, Gross and Zajac 1983).

Structurally, the trough is characterized by low dipping homoclines of iron formation and associated shelf sediments unconformably overlying Archean basement. Locally, the shelf sediments tectonically overlie basement upon low-angle thrust planes. These marginal structures typically pass outward (eastward) from the craton to broad open folds that are deformed in their crestal parts by complex isoclinal folds and faults developed by thrusting and tectonic transport directed westward toward the craton. Thick accumulations of off-craton turbidites and mafic–intermediate–ultramafic–volcanic and intrusive rocks are interlayered with and overthrust upon the shelf sediments. Outer boundaries of the fold belts are commonly marked by overthrusts upon older basement rocks.

The southeast border of the Ungava Craton is truncated by the ~1.0 Ga old, NE–E-trending Grenville Front Zone, which transects the southwestern extension of the Labrador Trough (Figs 3-8, 3-9). However, the iron formation and association shelf sediments continue southwestward into the Grenville Province for more than 200 km, where they are complexly folded and more highly metamorphosed (amphibolite facies), and form economically important BIF-rich isolated structural segments (e.g. Wabush, Mount Reed) (Gross 1968).

Metamorphism of the iron formation and enclosing supracrustal rocks in the Labrador Trough and extensions varies from subgreenschist to greenschist facies, except, as mentioned, where amphibolite facies are found in the Grenville Province to the south.

Very large iron ore resources are of three main types: (1) hematite–geothite deposits, derived by secondary enrichment of iron-formation protore; (2) highly metamorphosed oxide facies iron formation (magnetite–hematite concentrates); and (3) fine grained, cherty iron formations amenable to processing and concentration of the iron minerals (taconite ore) (Gross 1968).

Lithostratigraphy

The Aphebian succession of the Labrador Trough comprises sedimentary and mafic volcanic rocks of the Kaniapiskaa Supergroup, which are intruded by gabbroic–ultramafic sills of the Montagnais Group (Fig. 3-9). The Kaniapiskau Supergroup, a complex, multiphased, in part structurally distorted, assemblage, is divided into a western, predominantly sedimentary shelf succession, the Knob Lake Group ($\geqslant 6500$ m), and an eastern predominantly mafic volcanic succession, the Doublet Group ($\geqslant 5000$ m). The Knob Lake Group, in turn, is subdivided into a western zone of mainly shallow-water BIF-rich sediments (Sokomon iron formation), and an eastern zone dominated by deeper-water sediments and abundant mafic volcanic rocks. The Doublet Group locally conformably overlies the Knob Lake Group on the eastern margin of the trough, but the two are generally in tectonic contact along the Walsh Lake Fault (Fig. 3-9). A thick succession of pelitic–sempelitic schist and amphibolite, termed the Laporte Group, is in overthrust contact with the eastern margin of the Doublet Group. Despite these structural complications, the Laporte Group is considered to represent the offshore basinal facies equivalent to the Knob Lake shelf facies, the Doublet Group forming a mainly younger sequence developed across the Knob Lake (west) to Laporte (east) transition. Detailed stratigraphic relations are provided by Wardle and Bailey (1981) and Le Gallais and Lavoie (1982).

Magmatic events (dykes, sills and lava flows) in the Kaniapiskau Supergroup have been dated at 2169 Ma (lower cycle) and 1900 (upper cycle) (Rohon et al 1993). The main Sokomon iron formation and coevals have provided dates of 1870–1880 Ma (Chevé and Machado 1988).

Tectonic setting

The Labrador Trough has been established as a shelf-basin slope-basin system, whose evolution was controlled by crustal rifting (Wardle and Bailey 1981). In conventional plate tectonic terms, the most obvious interpretation is that the trough represents an ancient continental shelf-slope-rise system developed on the passive margin of an oceanic rift system, as suggested by Wilson (1968) and Burke and Dewey (1973). However, in the absence of preserved remnants of oceanic crust to confirm the original presence of oceanic crust to the east, Wardle and Bailey (1981) speculate that the trough formed as a continental shelf-slope-rise system on the western edge of a proto-oceanic rift system. Whether this system evolved into a large ocean basin or remained as a narrow rift is unknown.

Others

Other tectonically similar BIF-bearing fold belts and basins of the Circum-Superior system (Fig. 3-

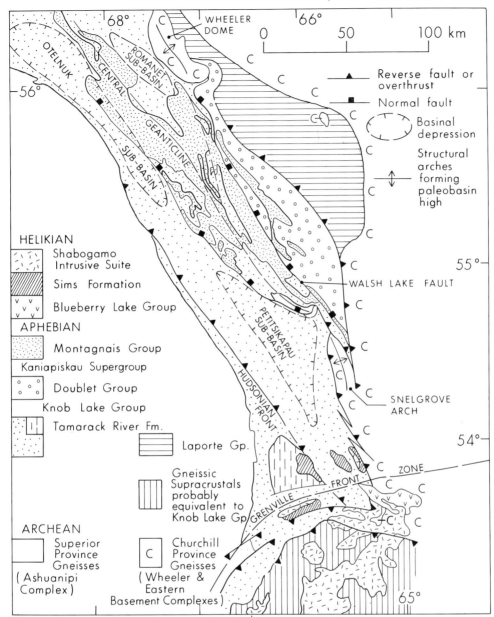

**Fig. 3-9.** General geology of the central Labrador Trough, showing the distribution and stratigraphic arrangement of Archean, Aphebian and Helikian components. (From Wardle and Bailey 1981, Fig. 19.2, and reproduced with permission of the Geological Survey of Canada and Minister of Supply and Services Canada).

8) include Mistassini–Otish Basin (Wynne Edwards 1972, Chown and Caty 1973); Cape Smith Belt (Baragar and Scoates 1981, 1987, Hynes and Francis 1982, Francis et al 1983, Hoffman 1985, Scott et al 1991, 1992, Lucas and St-Onge 1992, St-Onge et al 1992, Machado et al 1993); Belcher Belt (Ricketts and Donaldson 1981, Baragar and Scoates 1987, Legault et al 1994); Sutton Lake Inlier (Baragar and Scoates 1981); Fox River Belt (Scoates 1981, Baragar and Scoates 1981, 1987); and Thompson Belt (Coats et al 1972, Peredery et al 1982).

Of these, Cape Smith Belt is notable for the presence of well-documented 1.86 Ga ophiolites, metabasalts and komatiitic sills, which strengthens the case for a southward-directed imbrication of oceanic crust involving early rifting (1991 Ma), active arc magmatism (1874–1860 Ma), and later intrusions (1845 Ma). The Belcher–Nastapoka Basin contains a thick (7–9 km) but remarkably regular succession of distinctive sedimentary–volcanic units, including the Kipalu iron formation; the succession is divided into 14 units which are assembled into three major depositional cycles, all products of an evolving rift-arc magmatism environment.

Tectonic interpretation

Baragar and Scoates (1981), modifying Wilson (1968), Gibb and Walcott (1971), and Burke and Dewey (1973), proposed the following plate tectonic origin for the entire Circum-Superior belt system. Stretching of a sialic crust produced fracturing around the pre-existing stable craton of the ancestral Superior Province, followed by necking along fractures to produce annular sedimentary troughs. With continued separation, volcanism was initiated to on-lap on to the earlier sediments. Subsequently, the southern part (Lake Superior–Grenville) of the encircling rift system expanded at the expense of the northern part, with subduction occurring at the southern continental edge. The northern rift system (Labrador Trough, Cape Smith, Belchers, Fox River, Thomson) ceased activity and eventually closed with terminal deformation.

### 3.6.3 KAPUSKASING STRUCTURAL ZONE

The 500 km long, up to 50 km wide, NNE-trending, WNW-dipping, transecting, granulite-bearing, polycyclic Kapuskasing Structural Zone (Fig. 2-5) (Percival and Card 1983, 1985, Percival and McGrath 1986, Percival 1994, Percival and West 1994) provides an important structural link between Lake Superior to the south and James Bay off Hudson Bay to the north. The zone transects the 2.75–2.65 Ga, E-trending Wawa–Quetico (to the west) and Abitibi-Opatica (to the east) subprovinces (Fig. 2-5). The 120 km wide E–W transition from the low-grade Michipicoten greenstone belt on the west to the high grade granulite-grade Kapuskasing Structural Zone represents an oblique section through some 20–30 km thickness of Archean crust, uplifted at 1.9 Ga along the NNW-dipping Ivanhoe Lake thrust fault against the 2.7–2.6 Ga low-grade Swayze greenstone belt on the east, the thrusting involving ~37 km of crustal shortening. The Kapuskasing Zone is composed of 2.7–2.6 Ga paragneiss, mafic gneiss, tonalite and anorthosite, with 1.9 Ga and 1.1 Ga alkaline intrusions (carbonatites). The zone is marked by strong positive gravity and aeromagnetic anomalies.

The coincidence of Proterozoic events along the Kapuskasing structure with major coeval orogenic activity elsewhere in the Canadian Shield suggests that the structure represents an intracratonic basement uplift related to the ~1.85 Ga Hudsonian collision in Churchill Province to the north and northwest (Percival and McGrath 1986).

### 3.6.4 WESTERN CHURCHILL PROVINCE

The 1000–1500 km wide, NE–E-trending Western Churchill Province (Trans-Hudson Orogen of Lewry et al (1985)) (Fig. 2-8, Table 2-8, column 2) is divided into two composite zones: (1) in the southeast, the *Hudsonian Mobile Zone*, a 500–1000 km broad, two-part domain extending from the northwestern boundary of the Cree Lake Zone southeastward to Superior Craton; this zone is underlain, from northwest to southeast, by (a) thoroughly remobilized Archean basement crust (infrastructure) of the Cree Lake Zone (–Foxe Fold Belt) with variable Aphebian (early Proterozoic) miogeoclinal sedimentary cover (Fig. 3-10), all intensely deformed during the Hudsonian Orogeny (1.9–1.8 Ga); and (b) a major juvenile magmatic belt characterized by arc volcanism, turbiditic sedimentation and major calc-alkalic batholiths (Rottenstone–La Ronge Magmatic Belt and Southeastern Complex); and (2) in the northwest, the *Amer Lake Zone* incorporating Taltson, Tulemalu, Ennadai, Armit Lake, Committee Bay and Queen Maud blocks (Fig. 2-8). This zone constitutes an equally broad, poorly delimited cratonic foreland region composed of Archean basement, much at granulite facies metamorphism, and remnant Aphebian miogeoclinal and local foreland basin cover. It is characterized by generally low grade Hudsonian (1.9–1.8 Ga) metamorphic overprint, restricted brittle to plastic zonal reworking, huge shear zones, extensive Hudsonian granitoid plutonism, and pervasive isotopic resetting (Lewry et al 1985).

Subsequently the Trans-Hudson Orogon (Lewry et al 1985) was redefined to include only the major juvenile magmatic belt of the Hudsonian Mobile Zone (Rottenstone–La Ronge Magmatic Belt + Southeastern Complex), with additional extensive renaming of the overprinted units to the northwest, as follows: (1) Hearne Province = Cree Lake Zone + Tulemalu, Ennadai and most of Armit Lake blocks; (2) Rae Province = Taltson, Committee Bay

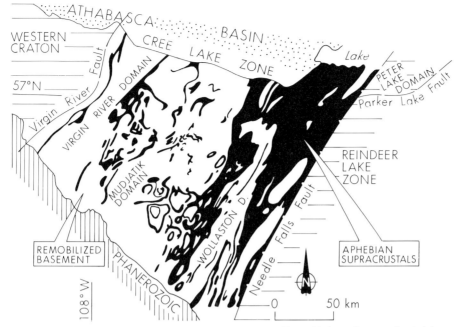

**Fig. 3-10.** The southwestern part of the Cree Lake Zone, Trans-Hudson Orogen, Saskatchewan. (From Stauffer 1984, Fig. 13, and reproduced with permission of the author).

and Queen Maud blocks, except for the comparatively narrow Thelon, Snowbird, Dorset and, to the east, Foxe and Dorset orogens (Fig. 3-2) (Hoffman 1989, 1990b, Lewry et al 1990, 1994, Lewry and Stauffer 1990, Meyer et al 1992, Crocker et al 1993).

Considered in general, available Trans-Hudson (i.e. Magmatic Belt + SE Complex) zircon dates fall in the period 1890–1835 Ma, with volcanic ages (1890–1875 Ma) clearly preceding plutonic ages (1870–1835 Ma), the latter moreover generally concentrated in the internal 1865–1850 (Lewry et al 1987, Van Schmus et al 1987a, Gordon et al 1990, Ansdell and Kyser 1991). On the other hand, gneisses in Hearne–Rae provinces to the northwest record a long and complex crustal history from possibly 3.5 Ga to 1.85 Ga including at least four major thermotectonic episodes in the interval 2.9–1.8 Ga with granulite facies metamorphism at ~2.3 and 2.0 Ga (Crocker et al 1993, Bickford et al 1994).

### Trans-Hudson Orogen (Southeastern Complex + Rottenstone–La Ronge Magmatic Belt)

Of the two component domains the Southeastern Complex comprises the massive-sulphide-bearing volcanoplutonic Flin Flon–Snow Lake Arc and, 150 km to the north, the similar Eastern La Ronge Domain, a possible telescoped fore-arc prism, together with the intervening Kisseynew Domain, a broad, inter-arc volcanogenic sedimentary basin; also included are two nearby small Archean microcontinents—Hanson Lake and Glennie Lake blocks (Fig. 2-9, Syme and Bailes 1993, Thomas and Heaman 1994).

The three-part Rottenstone–La Ronge Magmatic Belt on the northwest includes (1) the restricted La Ronge–Lynne Lake volcanoplutonic arc at the south margin; (2) the adjoining Rottenstone tonalite–migmatite complex; and (3) the uniquely large and predominating Wathaman Batholith, a compositionally asymmetric zoned tonalite–granodiorite–granite complex of age 1865–1850 Ma (Fig. 2-9, Lewry et al 1981, 1994, Fumerton et al 1984, Stauffer 1984, Symons 1991, Meyer et al 1992, White et al 1994b).

The Trans-Hudson Orogen, so defined, has been traced southward beneath Phanerozoic cover mainly by magnetic and seismic techniques augmented by basement samples from wells and boreholes. It extends between the buried westerly extension of Superior Province on the east and the eastern boundary of Wyoming Craton on the west to younger E-trending Central Plains Orogen (1.78–1.64 Ga) (Sims et al 1993). To the north and east the Trans-Hudson Orogen may connect with the Circum-Superior belts including Ungava and New Quebec orogens (Fig. 3-2). Available isotopic ages suggest that the Trans-Hudson, Wopmay (see below) and Penokean orogens all developed simultaneously (1.9–1.8 Ga) as a consequence of plate convergence (Sims et al 1993).

Nd isotopic studies in selected parts of the Trans-Hudson Orogen indicate that more than 90% of that particular terrain represents juvenile mantle-derived accretion (Chauvel et al 1987). However, in the larger Laurentian continental reconstruction, relevant data suggest that major early Proterozoic mobile belts such as the Hudsonian (North America)–Rinkian (Greenland) comprise about equal parts new mantle-derived crust and old recycled Archean crust (Patchett and Arndt 1986).

Tectonic interpretations

According to Lewry et al (1985), by the end of the Kenoran Orogeny (~2.5 Ga), a major sialic craton, 900–1500 km broad and at least 2100 km long (NE–SW), was established including Slave Province with westerly extensions in the adjoining Wopmay Orogen and Amer Lake and Cree Lake zones (renamed Hearne and Rae provinces) (Figs 2-8, 2-9, 3-2). The Slave–Churchill 'continental' plate thus defined (Lewry et al 1985) and the Superior craton to the southeast may have been independent late Archean to early Proterozoic crustal entities.

The various blocks making up Western Churchill Province show sharply contrasting crustal erosional levels and lithotectonic regimes, marked especially by contrasting metamorphic grades, and their present relationships may have been largely established by the end of the Archean eon. Boundaries between blocks are generally either major shear zones or loci of abrupt change in metamorphic grade and structural style (Fig. 2-8); they were active by at least the late Archean (~2.7 Ga), with major reactivation occurring during the Hudsonian Orogeny (~1.8 Ga) and also possibly earlier during the mid-Aphebian (~2.3 Ga).

According to Hoffman (1989, 1990), the Archean provinces were welded by early Proterozoic collisional orogenies characterized by deformed passive-margin and foredeep sedimentary prisms, and foreland thrust-fold belts. Their hinterlands bordered by magmatic arcs display basement reactivations, thrusting and transcurrent shearing accommodating collisional foreland indentations. Only the Trans-Hudson Orogen preserves a significant width of juvenile early Proterozoic crust including relicts of island arcs and obducted oceanic crust. The Thelon Orogen welds Slave and Rae provinces; the Trans-Hudson Orogen welds the Hearne and Superior provinces; and the New Quebec Orogen (see below) welds the Superior and Rae provinces.

### 3.6.5 EASTERN CHURCHILL PROVINCE

The eastern part of Churchill Province encompasses Baffin Island and those domains adjoining Ungava Craton (Superior Province), all contiguous to Greenland in the reconstructed Laurentia (Fig. 3-2). The region includes three cratonic elements (provinces) rich in overprinted Archean–early Proterozoic crust, and five intervening adjoining early Proterozoic orogenic belts (Fig. 3-2). Following Hoffman (1980) and Kranendonk et al (1993), the three cratonic elements are: (1) North Rae Province in the north; (2) Southeast Rae Province, a thin southerly trending sliver east of Ungava Craton; and (3) Nain Province (Fig. 2-7) at the Atlantic coast. The five intervening early Proterozoic belts (orogens) with encompassing cratons (provinces) in brackets are: (1) Ungava Orogen including Cape Smith Belt (Superior–Southeast Rae); (2) New Quebec Orogen including Labrador Trough (Superior–Southeast Rae); (3) Torngat Orogen (Southeast Rae–Nain); (4) Dorset and Foxe Orogens—Cumberland complex (Southeast Rae–North Rae); and (5) Makkovik Orogen (Nain–Grenville) (Fig. 2-7).

The three cratonic elements typically comprise basement gneisses and greenstones (2.9–2.7 Ga) intruded and metamorphosed at 2.7–2.6 Ga, and again at 1.9–1.8 Ga. Early Proterozoic orogenic belts, in turn, are characterized by large volumes of platformal–turbiditic metasupracrustal rocks (e.g. Lake Harbour, Penrhyn, Piling, Hoare Bay, St Mary groups), older gneisses (2.9–2.7 Ga), and younger metamorphic–plutonic associations (1.9–1.76 Ga). Of these, Torngat Orogen, Central Labrador, a product of Nain–SE Rae transpressional collision, records high-grade metamorphism at 1.86–1.82 Ga with subsequent events to 1.74 Ga (Bertrand et al 1993). Makkovik domain, Eastern Labrador, a small (800 km by up to 75 km) tectonically bound sliver of dominantly early Proterozoic crust is dominated by varied, mainly granitoid plutonic rocks and sparse metasupracrustal associates, including reworked basement gneiss (3.3–2.8 Ga), tonalitic orthogneiss (~2.0 Ga) metasupracrustal assemblages (pre-1.86 Ga), foliated granitoid rocks (1.8 Ga) and some post-tectonic granite plutons. These are associated with mid-Proterozoic bimodal volcanic rocks of the Bruce River Group (1.65 Ga) (Kerr et al 1992).

## 3.6.6 WOPMAY OROGEN

Three large correlative Aphebian tectonic units, all part of an early Proterozoic continental margin, lie at select sites of the Slave Province perimeter: Wopmay Orogen on the west, Kilohigok Basin of Bathurst Inlet region on the northeast and Athapuscow Aulacogen of Great Slave Lake (East Arm) on the south (Fig. 3-11). These tectonic units are remnants of an N-trending orogenic arc, convex to the west, which developed between 1.91 and 1.84 Ma, to the west of Slave craton (Hoffman 1973, 1980, Easton 1983a, Hoffman and Bowring 1984).

### Wopmay Orogen (Coronation Supergroup)

The four-zone Wopmay Orogen 600 km long by up to 220 km wide as exposed, unconformably overlies western Slave Province basement. It, in turn, is unconformably overlain by mid-Proterozoic supracrustal rocks of the Coppermine Homocline to the north, and Paleozoic strata of the Northern Interior Platform to the west.

The four Wopmay zones (1–4) are arranged at increasing distance from the Slave Foreland (Fig. 3-11; Table 2-8, column 3).

Zone 1 comprises thin (600 m) platform sediments. Zone 2, a thick (2–4 km) continental shelf

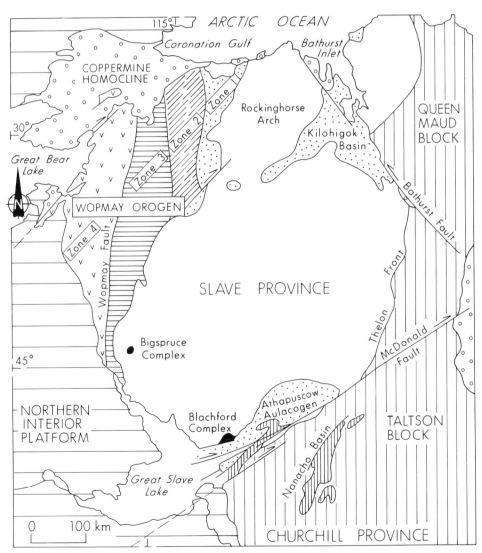

**Fig. 3-11.** Major tectonic elements in the northwest corner of the Canadian Shield. (From Hoffman 1980, Fig. 1, and reproduced with permission of the author and of the Geological Association of Canada).

facies, is allochthonous with many W-dipping, E-verging faults. Zone 3, comprising rift, rise and slope facies with associated batholiths, includes very thick (8–10 km) turbidite–volcanic assemblages overlain by shales with spectacular sedimentary breccias (olistostromes). Zone 4 (Great Bear Batholith) includes thick (+8 km) mainly ignimbritic volcanic rocks plus sediments, intruded by large epizonal granitoid plutons; the eastern boundary of this zone is a fundamental transcurrent discontinuity, the Wopmay Fault.

The Wopmay Orogen is interpreted in terms of a ~1.9 Ga continental margin that experienced a complete (open–close) Wilson cycle, with terminal collision that resulted in a wide swath of conjugate transurrent faults. The overall eastward tectonic transport of Wopmay Orogen is expressed in the many eastward nappes and folds (Hoffman 1980a, Hoffman and Bowring 1984).

### Kilohigok Basin

Kilohigok Basin, more than 7000 km$^2$ in area, preserves up to 7 km thickness of assorted strata of the Goulburn Group (Fig. 3-11) (Campbell and Cecile 1981). The sequence is rich in diverse sedimentary structures, notably stromatolites, olistostromes and paleosols. The complete basin comprises a SSE-trending axial zone, a main Western Platform and a subsidiary Eastern Platform, extending 300 km to the NNE. The basin, although located well within the Slave Craton, is connected to the continental margin on the north by a paleorift system including the SE-trending Bathurst Inlet Fault System. To the southeast, this tectonic system converges in the Baker Lake region with the related ENE-trending McDonald (Wilson) Fault of the Athapuscow Aulacogen system (Fig. 3-11). Following Hoffman (1973), the Kilohigok Basin is interpreted to be an intracratonic basin intimately related to the development of both the Wopmay Orogen and Athapuscow Aulacogen.

### Athapuscow Aulacogen

Athapuscow Aulacogen is a deformed ENE-trending basin 270 km long by up to 80 km wide, of little metamorphosed up to 5 km thick Aphebian sedimentary and magmatic rocks exposed in and around the East Arm of Great Slave Lake astride the southern Slave–Churchill mylonite-rich contact zone (Hoffman et al 1977, Hoffman 1981).

Athapuscow stratigraphy comprise two relatively conformable sequences: (1) the principal Great Slave Supergroup, a subgreenschist-grade arenite–carbonate–pelite sequence recording a complete transgressive–regressive marine cycle; and (2) the younger overlying virtually unmetamorphosed Etthen Group, comprising alluvial fanglomerate, local volcanic flows and pebbly fluvial sandstone (Hoffman et al 1977). A six-stage evolutionary model of the aulacogen has been proposed (Hoffman 1981).

The composite Athapuscow assemblage is characterized by many huge nappes, up to 70 km long, of recumbently folded strata. These were first moved at the end of Great Slave deposition northwestward into the East Arm from the region south of the McDonald Fault, and subsequently about 75 km of dextral strike-slip on that fault to their present site (Fig. 3-11) (Hoffman et al 1977).

## 3.6.7 WESTERN USA

Variably metamorphosed early Proterozoic sedimentary rocks are preserved in a string of synclinoria or basins along the southern and eastern margins of the Wyoming Uplift, notably the Medicine Bow Mountains and Sierra Madre of Wyoming and the Black Hills of South Dakota (Fig. 2-11; Table 2-8, column 5) (Hedge et al 1986, Houston et al 1993).

In the Medicine Bow Mountains–Sierra Madre region of Wyoming, the Snowy Pass Supergroup, at least 6 km thick, trends southward to the E-trending Cheyenne structural discontinuity. The supergroup is composed of quartzite, metaconglomerate, chlorite schist, mafic volcanic rocks, marble and diamictites. Loosely constrained ages range from 2492 to 2075 Ma in the lower part (Deep Lake and lower Libby Lake groups) and from 2075 to +1750 Ma in the unconformably overlying part (upper Libby Group). Deposition is cyclic, the cyclic units resembling those in the Huronian Supergroup of Ontario (Roscoe and Card 1993). The Snowy Pass pyritic quartz pebble conglomerates with their substantial uranium ore resources, resemble those in the Matinenda Formation, Huronian Supergroup.

The early Proterozoic (1.97–1.89 Ga) rocks of the Black Hills to the east, a broad Laramide anticline with Precambrian core, are more highly deformed and metamorphosed. The lower part of the section is rich in quartzite but also contains metaconglomerate, schist, marble and 'taconite' BIF. These rocks are overlain by abundant amphibolite (metabasalt) and metagreywacke, overlain in turn by dominant phyllites and quartzites. Uranium-bearing quartz pebble conglomerates occur in the lower shelf-type facies (Hedge et al 1986, Houston et al 1993, Roscoe and Card 1993).

Important mineralization is present in the early Proterozoic rocks of the Black Hills, notably iron ore deposits in BIF, uraniferous quartz pebble conglomerates and, most importantly, auriferous zones within carbonate- and oxide-facies iron formations in the Homestake, Montana Mine, and Rockford formations. More than 1150 t of gold have been produced from gold, arsenic and minor base metal-bearing strata at the Homestake Mine in the Lead district (Caddeg et al 1991).

Possible correlates of these Wyoming strata are present to the west in southern Idaho–northern Utah (Facer Formation) and east–central Idaho (Boehls Butte Formation (see Hedge et al 1986, Plate 1). Local early Proterozoic (~2.0–1.8 Ga) metasupracrustal rocks are reported in the Mojave province of Nevada–California (Reed et al 1993).

Finally, far to the northwest, the Kilbuck terrane in SW Alaska, a thin crustal sliver of amphibolite facies orthogneiss, provides dates of 2070–2040 Ma (Box et al 1990). Whether indigenous to North America or displaced from elsewhere is unknown.

## 3.7 SOUTH AMERICAN CRATON

Early Proterozoic rocks are concentrated in three main regions of the South American Craton: (1) the major NW–W-trending Maroni–Itacaiunas Mobile Belt of the Amazonian Craton (Guiana Shield plus Central Brazil Shield less Tocantins Province); (2) local metasedimentary sequences and mafic–ultramafic complexes in the western Goias region, Tocantins Province (Central Brazil Shield); (3) five small but important metasupracrustal terrains, including restricted fold belts, located respectively in southern and northeastern São Francisco Craton of the Atlantic Shield (Figs 1-3e, 1-5e; Table 2-9). Additional early Proterozoic crust occurs locally in the East Bolivian Shield area and as relict patches scattered throughout the younger Brasíliano fold belts of the platform. The closing Transamazonian Orogeny represents an important stage in the consolidation of the South American Craton (Fig. 1-3e), providing both widespread platformal environments for ensuing epicratonic–anorogenic activities and, in response to platform disruption, intracratonic geosynclinal-style settings, culminating in the late Proterozoic Brasíliano cycle of events (Almeida et al 1981, Cordani and de Brito Neves 1982, Danni et al 1982, Gibbs and Barron 1983, Litherland et al 1985, 1986, 1989, Brito Neves 1986).

### 3.7.1 MARONI–ITACAIUNAS MOBILE BELT, AMAZONIAN CRATON

The NW–W-trending bifurcating Maroni–Itacaiunas Mobile Belt underlies the centre–east Guiana Shield and, to the south, the northeastern corner of the Central Brazil Shield (Figs 1-5e, 2-12; Table 2-9, column 1). The preserved part of this composite Transamazonian belt exceeds 2500 km in length and 500 km in width, making it one of Earth's major early Proterozoic mobile terrains. It is terminated on the west successively by the NW-trending mid-Proterozoic Rio Negro–Juruena Belt and by the Imataca Complex, and on the east successively by the Atlantic coast and the N-trending late Proterozoic Paraguay–Araguaia Belt, Tocantins Province (Fig. 1-5e, insets 1A, 2). To the east, similar early Proterozoic assemblages are exposed in the tiny São Luis Craton of northeastern Brazil (Fig. 1-5e) and, in the pre-Mesozoic Atlantic reconstructions, still farther east in the Birrimian terrain of the West African Craton (see below).

*Granitoid terrains*

The Maroni–Itacaiunas Mobile Belt is characterized by extensive reworking of high grade Archean terrains, abundant granitoid gneiss–migmatites at mainly amphibolite facies, the presence of early Proterozoic greenstone and other linear metasupracrustal fold belts, and syn- and post-orogenic granitoid plutons.

A major cluster of NW-trending early Proterozoic greenstone belts extends for at least 1500 km along strike from eastern Brazil–French Guiana in the southeast to eastern Venezuela in the northwest (Fig. 2-12).

The Guyanese greenstone belts in the centre–east (Gibbs and Olszewski 1982, Gibbs and Barron 1983, Gruau et al 1985) display a common up-section volcanic progression from lower mafic, through intermediate–felsic, to upper tuffites with pelitic associates. Granitoid gneiss complexes, notably the Bartica gneiss, separate the greenstone belts.

Other major greenstone belts include, successively southward, the *Serra do Navio District* in

Amapa State (Guiana Shield), one of two major manganese producers in Brazil, and the *Serra dos Carajas Region* of the Central Brazil Shield (Fig. 2-13) wherein folded Archean basement rocks, including the major Graó Para BIF, are unconformably overlain. The greenstone-charged Maroni–Itacaiunas Belt, in its full pre-Mesozoic Guiana–West African reconstruction, apparently represents an asymmetric fold-and-thrust belt, developed about an older northern foreland, which included the Imataca Complex of Venezuela and the Liberian Craton of West Africa.

### 3.7.2. GOIAS MASSIF, TOCANTINS PROVINCE, CENTRAL BRAZIL SHIELD

East of Tocantins River, in the easternmost part of the State of Goias, the NNE-trending, at least 100 km-long, Ticunzal sequence, about 300 m thick, is composed of quartz–graphitic micaschist and muscovite–biotite gneiss (Table 2-9, column 2). The sequence unconformably overlies Archean basement and is unconformably overlain by mid-Proterozoic strata. High greenschist–amphibolite facies metamorphism prevails. The sequence contains important uranium mineralization.

In the nearby Serra de Santa Rita to the southwest, Archean greenstone belts are locally overlain by a metasedimentary sequence composed of placer gold–pyrite–uraninite-bearing basal conglomerates, schist, phyllite, metadolomite and minor BIF (Danni et al 1982).

Several mafic–ultramafic layered complexes near Goias, of uncertain early Proterozoic age, contain local Cu–Ni–Co sulphide deposits. Similar possible early Proterozoic gabbro–anorthosite intrusions tectonically overthrust the western borders of the Niquelandia and Barro Alto granulite complexes (Fig. 2-14).

### 3.7.3. SÃO FRANCISCO CRATON, ATLANTIC SHIELD

Early Proterozoic rocks have limited geographic distribution but major economic importance in four regions of the São Francisco Craton. Important itabirites (BIF), unusually rich in high grade iron ore deposits, are present in (1) the Minas Supergroup in the Quadrilatero Ferrifero at the southern tip of the craton. Comparatively small N-trending metasupracrustal units located in the northeastern part of the same craton include (2) Contendas–Mirante; (3) Serrinha and Capim; and (4) the mineral-rich Jacobina belts (Fig. 2-15; Table 2-9, column 3). A useful review is provided by Teixeira and Figueiredo (1991).

### Quadrilatero Ferrifero

The Quadrilatero Ferrifero (Iron Quadrangle) in Minas Gerais state, so named due to its 'squarish' geometric arrangement and impressive development of itabirite (BIF) in the early Proterozoic Minas Supergroup, is spectacularly visible on aerial photographs and satellite imagery (Dorr 1969, Schorscher et al 1982, Chauvet et al 1994, Chemale et al 1994).

The 'Iron Quadrangle' (Fig. 2-15, inset) covers an area of 7000 $km^2$ to the south and east of the city of Belo Horizonte and is well known especially for its large production of high grade iron ore. Extensions of Minas itabirite lie both to the southwest and especially to the northeast, where they include the important Monlevade and Itabira iron-ore districts.

Following Schorsher et al (1982), the Minas Supergroup, some 2000 m thick, is subdivided into four groups: Caraca (bottom), Itabira, Piracicaba and Itacolomi (top).

The *Caue Formation* of the Itabira Group is rich in itabirite (metamorphosed oxide-facies iron formation which is up to 350 m thick).

Minas Supergroup rocks are in exclusive tectonic contact with practically all the older rocks of the region. Typical tectonic contacts take the form of low angle thrust faults, characterized by intense mylonitization of the adjoining rocks. These first-order Minas structures are attributed to napperian transport and emplacement of deformed Minas rocks, the collective Quadrilatero Ferrifero representing an early Proterozoic allochthonous slice overthrust upon Archean basement from some undisclosed eastward source. The present structural framework is the result of three main tectonic events from 2000 Ma to 600 Ma (Chauvet et al 1994).

Detrital zircons from Minas greywacke provide an age of 2125 Ma. Post-Minas pegmatites reflecting the Transamazonian event were dated at 2060–2030 Ma (Ladeira and Noce 1990, Machado et al 1992).

### Contendas–Mirante Complex (Belt)

The Contendas–Mirante volcanosedimentary sequence constitutes a synclinorial belt, nearly

200 km long (north to south) and 65 km wide, situated on the western flank of the Jequié Craton (Fig. 2-15) (Sabate and Marinho 1982). The stratigraphic sequence, about 5000 m thick, is divided into equally thick lower volcanogenic and upper terrigenous sequences.

Rocks of the complex are intensely deformed, including two phases of folding accompanied by imbricate structures and severe shearing. Upper greenschist–amphibolite facies metamorphism prevails.

Serrinha and Capim greenstone belts

The small Serrinha or Rio Itapicuru greenstone belt is located 400 km north of Salvador, Bahia, within the Salvador–Juazeiro mobile belt (Fig. 2-15). Important stratabound gold deposits, including the large Fazenda Brasileiro deposit, are present in two principal isoclinally folded horizons of chlorite schists and quartz–feldspar breccia which include irregular quartz–carbonate segregations (Cordani and Brito Neves 1982, Davison et al 1988).

The still smaller Capim Belt, 150 km to the north, comprises a lower mafic volcanic part and an upper part of intermediate–felsic tuffs and breccias.

Jacobina Belt

To the west, the N-trending lenticular Jacobina Belt is about 450 km long and up to 50 km wide or 7120 km$^2$ in area (Beurlen and Cassedanne 1981, Mascarenhas and da Silva Sa 1982) (Fig. 2-15). The belt includes a very thick metamorphosed conglomerate arenite–pelite–volcanic sequence which has a gross monoclinal structure but is highly deformed about N-trending axes. Amphibolite facies metamorphism prevails. The lower Jacobina conglomerate–orthoquartzite association, at least 600 m thick, was deposited in an east-to-west fluvial paleosystem. Free gold, the principal mineralization, is distributed in the matrix of the conglomerate within at least four stratigraphic levels, usually accompanied by recrystallized pyrite and, in some levels, uranium minerals (uranite and torbenite).

In the *Serra do Jacobina Region*, important manganese deposits are contained in phyllites of the Jacobina volcanosedimentary complex. Nearby in the region of Carbaiba, a prominent contact metamorphic phlogopite schist belt is rich in emeralds and molybdenite. Barite and amethyst mineralization also occurs nearby, as do copper, asbestos, talc and marble deposits (Torquato et al 1978, Mascarenhas and da Silva Sa 1982).

In the *Campo Formoso District*, chrome ore deposits lie in serpentinized peridotites which are exposed on the west flank of the Jacobina range.

## 3.8 AFRICAN CRATON: SOUTHERN AFRICA

### 3.8.1 INTRODUCTION

On the Kaapvaal Craton two epicratonic basin complexes—(1) the Transvaal (~2.3–~2.0 Ga) with its enclosed Bushveld Igneous Complex (2.1 Ga); and (2) the Waterberg–Soutpansberg (2.0–1.75 Ga)—completed the northwestward migration of the successively enlarging depositories initiated by the Pongola and Dominion strata (Fig. 2-18 (inset); Table 2-11) upon the newly consolidated Kaapvaal Craton 3.0 Ga ago. The pericratonic Magondi and Kheis belts evolved concurrently (2.0–1.7 Ga), respectively occupying equivalent positions at the western margins of the Zimbabwe and Kaapvaal cratons (Fig. 1-5f(i,ii)).

### 3.8.2 TRANSVAAL SUPERGROUP BASIN COMPLEX

The Transvaal Supergroup occupies a large ENE-trending lobate basin complex extending for 1100 km from Prieska in the southwest to Lydenburg in the northeast. It is grossly divided into two separate structural units, the elliptical Bushveld (Transvaal) Sub-basin in the east and the Cape–Botswana (Griqualand West) sub-basin in the west (Fig. 3-12); these were recently renamed the Chuniespoort and Ghaap basins respectively (Barton et al 1994). The preserved area of the collective basin complex is 250 000 km$^2$, with the original extent probably exceeding 500 000 km$^2$. In the Bushveld Sub-basin, the Transvaal succession is up to 15 km thick, including an uppermost 3 km thick felsic volcanic pile (Rooiberg felsite) whereas in the Cape–Botswana Sub-basin, it is only 5–6 km thick (Fig. 3-13). The Transvaal Supergroup is noted for spectacular BIF development, which ranks it amongst the five mega-BIF depositories of the world. (Key references: Button 1981b, Tankard et al 1982, Beukes 1983, Klein and Beukes 1990, Truswell 1990, Barton et al 1994).

Structure

In the Bushveld Sub-basin in the east, the Transvaal sequence is characterized by gentle to moderate

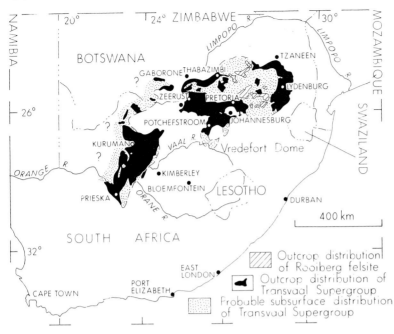

Fig. 3-12. Locality map for the Transvaal Basin. (From Button 1981, Fig. 9.8, and reproduced with permission of the author and of Elsevier Science Publishers).

dips inward towards the centre of the Bushveld Complex. Nearby Transvaal strata are deformed in a major discontinuous rim syncline around the Vredefort dome in the south.

In the main part of the Griqualand West Sub-basin, the strata dip 1–2° westward, the degree of deformation increasing gradually towards the NNW-trending lineament near Prieska, where Transvaal rocks are highly folded along N–NNW-trending axes.

The two sub-basins are separated by a major NNW-trending arch (Klerksdor–Vryburg) where pre-Transvaal rocks (mainly Ventersdorp lavas) outcrop.

### Lithostratigraphy

The Transvaal Supergroup as distributed across the two sub-basins is divided into (1) Lower Mafic Volcanic and Clastic Assemblage; (2) a median Chemical Sedimentary Assemblage; and (3) Upper Clastic Assemblage (Button 1981). In the eastern or Transvaal sequence these are named respectively (1) Wolkberg; (2) Chuniespoort; and (3) Pretoria groups, whereas in the Griqualand West sequence the corresponding names are (1) Buffalo Springs (Schmidtsdrif) (not developed everywhere); (2) Ghaap (Campbell and Koegas); and (3) Postmasburg groups, respectively. Respective lithologies and thicknesses are shown in Fig. 3-13.

In the *Lower Mafic Volcanic and Clastic Assemblage* (Wolkberg and Buffalo Springs (Schmidtsdrif) groups), stratigraphic units are strongly influenced by paleotopography, wedging out against basement-highs (usually granitic terrains) and thickest in ancient valleys (generally over greenstone belts).

The *Chemical Sedimentary Assemblage* (Chuniespoort and Ghaap groups) is developed throughout the complete Transvaal Basin, commencing with basal quartzite, transitional up to carbonates and chert, and terminating with major BIF known respectively as *Penge Formation* in the Transvaal (eastern) sequence and Asbesheuwels (–Koegas) Subgroup or Kuruman Formation in the Griqualand West sequence. Maximum BIF thicknesses are 600 m and 2000 m in the eastern and western sub-basins respectively. The BIF is composed of quartz, magnetite, hematite, stilpnomelane, riebeckite, minnesotaite, grunerite and carbonates (siderite, ankerite, ferro-dolomite, dolomite and calcite), either in mixed or monomineralic layers, which are disposed in macro-, meso- and microbanded sediments with very persistent layering. Riebeckite and grunerite are mined in their fibrous forms as crocidolite and amosite asbestos, respectively. Some of the stilpnomelane-rich layers contain altered volcanic shards.

The overall manganese content of the Transvaal BIF increases upwards (Beukes 1983, Roy 1992). Pyroclastic material is associated with most of the BIF and volcanism may have contributed to BIF deposition. In this regard, Beukes (1983) considers

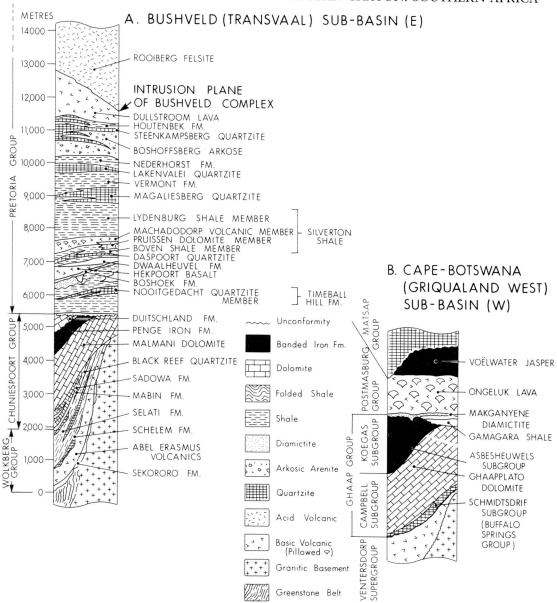

**Fig. 3-13.** Stratigraphic columns for the Transvaal Basin. (Modified after Button 1981, Fig. 9.10, and reproduced with permission of the author and of Elsevier Science Publishers).

that most of the iron and silica in the BIF were derived from the early Proterozoic oceans. However, Klein and Beukes (1990) attribute the ultimate source of iron and silica to a very dilute hydrothermal input in the deep ocean waters.

The *Upper Clastic Assemblage* (Pretoria and Postmasburg groups) is up to 7000 m thick in the Transvaal sequence but only 2000 m in the Griqualand West sequence.

The Pretoria Group comprises a marginal-marine cycle composed of quartzite and shale, with three carbonate and three volcanic units, and, in the north, a number of wedges of fluvial arkose.

Glaciomarine and fluvioglacial sediments (diamictites) are present in the lower Timeball Hill Formation. The thick (3000 m) uppermost Rooiberg felsites, which were extruded during the final stages of Transvaal sedimentation, grade up to a redbed sedimentary succession, representing the last pre-Bushveld unit in the Transvaal Basin.

The major post-Transvaal tectonic event affecting the Transvaal (Chuniespoort) Basin was the emplacement of the Bushveld Complex. The major phase of this complex was intruded in a number of arcuate lobes, which largely determine the present outcrop shape of the basin.

## Mineral deposits

An exceptionally wide variety of ore minerals are present in the Transvaal Basin. Iron, fluorite, asbestos, limestone and dolomite deposits are widespread. The northern Cape contains some of the largest manganese and high grade hematite deposits in the world. Very large tonnages of andalusite are present in Pretoria Group hornfels.

## Age

Based on earlier studies (Beukes 1983, Hartnady et al 1985, Minter et al 1988, Walraven et al 1990) Transvaal deposition was assigned to the ~300 Ma interval from 2350 Ma to ~2050 Ma. However, a recent zircon age of 2552 Ma from the upper Campbell banded tuff, Griqualand West sequence (Barton et al 1994) in conjunction with an earlier Pb–Pb age of 2557 Ma (Jahn et al 1990) from nearby Schmidtsdrif stromatolitic limestones and a zircon age of 2432 Ma (Trendall et al 1990) from a tuff layer at the base of the Kuruman Iron Formation, indicates that Transvaal deposition was underway by at least 2550 Ma, in this respect, closely corresponding to the Hamersley Iron Formation, West Australia, the two iron formations representing possible correlates in a super continental paleoreconstruction.

## 3.8.3. THE BUSHVELD COMPLEX

The Bushveld Complex (Hunter and Hamilton 1978, Tankard et al 1982, Vermaak and Von Gruenewaldt 1986, Naldrett 1989) lies within the Kaapvaal Craton, central to the Transvaal Sub-basin (Chuniespoort Basin). With a total area of 67 340 km², about half of which is exposed, it is by far the largest known layered mafic igneous complex in the world (Fig. 3-14; Table 2-11). Its petrology, chemistry and cyclic repetition of lithologic units are, however, comparable with other layered mafic intrusions. The complex is a major source of platinum group metals, chrome and vanadiferous iron ore. Production of tin and fluorite are also regionally important and its metamorphic aureole contains mineable deposits of numerous industrial minerals.

## Setting and form

The Bushveld Complex is mainly emplaced in Transvaal sediments accompanied by local Archean basement in the north. Formerly thought to be a simple lopolith, the complex is now regarded as comprising four separate intrusions (lobes) which collectively have a cruciform outline arranged symmetrically about two equidimensional axes, each 350 km long, aligned easterly and northerly respectively (Fig. 3-14). The earlier mafic layered rocks of the Complex lie on the outer peripheries but do not fully enclose the inner later acid core composed of Bushveld granite and granophyre.

The dip of the layered sequence in each of the four mafic lobes is centripetally directed, usually at low angles (10–25°) but with steeper local dips.

## Bushveld mafic phase

The layered mafic sequence consists of a 7–8 km thick pile of dunite, pyroxenite, harzburgite, norite, anorthosite, gabbro, magnetite gabbro and diorite. It is divided up-section into Lower, Transition, Critical Main and Upper zones each with subdivisions (Hunter and Hamilton 1978, Vermaak and Gruenewaldt 1986). The Critical Zone, in particular, displays a magnificent cyclicity of cumulus chromite, orthopyroxene and plagioclase, exemplified by the Merensky reef. Stratiform chrome deposits are hosted in the Critical Zone of the Bushveld mafic phase (Stowe 1994, Scoon and Teigler 1994).

Various studies suggest that the mafic magma of the complex was emplaced as a succession of magmatic influxes. Concepts such as density stratification, double-diffusive mixing, bottom growth, and lateral accretion of layers have been used in genetic modelling (Irvine et al 1983).

## Bushveld acid phase

The acid rocks of the complex comprising granophyres and granites are spatially associated with the older Rooiberg felsite succession. The dominant acid phase is the stratiform 2.8 km thick, K-feldspar-rich Bushveld granite, which variably intruded either the felsic roof, the Transvaal sediments, or the mafic sequence. Stocks of later red, medium to coarse grained granite cut across the Bushveld granite. Economic deposits of cassiterite occur as disseminations in the younger granite or in zoned pipes in the upper parts of the younger granite stock, the Bushveld granite and granophyre.

The spatially associated, up to 4000 m-thick, Rooiberg felsites comprise a comparatively homogeneous sequence of felsite flows that are exposed

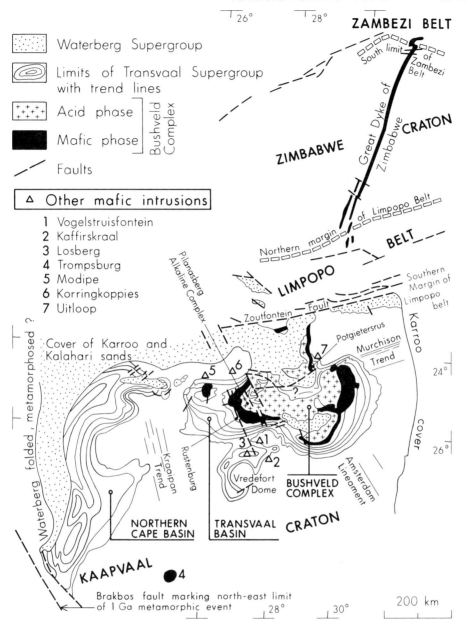

**Fig. 3-14.** The Bushveld Complex and locations of various coeval mafic intrusions in their broad regional setting. The Transvaal and Northern Cape basins are also called Bushveld and Cape-Botswana sub-basins respectively as specified in the text. (Modified after Hunter and Hamilton 1978, Fig. 1, and reproduced with permission of the authors).

mainly in the central and southern parts of the Bushveld Complex. The sequence has been regarded either as the terminal volcanic phase of the preceding Transvaal Supergroup, as generally related to the Bushveld acid phase, or as unrelated to either of these two entities. Whatever their genetic relationship may be, there is no question that the Rooiberg Group predates both the Rustenburg Layered Suite (layered mafic phase) and the Nebo (Red) granite (acid phase) of the Bushveld Complex.

### Petrogenesis

Theories to account for the layered stratigraphy can be broadly divided into two groups: (1) differentiation of a magma at depth with subsequent intrusion of the separate fractions; and (2) differentiation of magma *in situ* with or without additions of magma to the differentiating body. The critical relationships and factors have been carefully reviewed (Tankard et al 1982, Eales et al 1990).

Hamilton (1970), following Dietz (1961), speculated that the structural interpretation could be integrated with a meteorite impact model for the origin of the Bushveld complex. However, recent careful, detailed field studies emphasize the absence of any recognizable shock–metamorphic effects in the Bushveld Complex (French 1990) which would seem to negate this hypothesis.

### Age

Despite a wide range of published ages (Hamilton 1977, Coertze et al 1978), the Bushveld event, including emplacement of the complex, is now placed confidently in the 2060–2050 Ma time slot (Walraven et al 1990).

### Mineral deposits

The vast layered mafic complexes of the Great Dyke and Bushveld Complex contain Earth's largest resources of Cr, Pt and V, as well as Ni, Co, Fe and Ti resources. The later Lebowo Granite Suite contains Sn, W and F deposits. The nearby coeval Palabora Complex (south flank of Murchison range) contains Cu and P deposits (Songe 1986).

## 3.8.4. THE WATERBERG–SOUTPANSBERG–MATSAP–UMKONDO BASINS

### Introduction

Younger epicratonic cover on the Kaapvaal Craton is represented by the ~1.8 Ga old Waterberg family of basins (Fig. 3-15; Table 2-11) (Jansen 1981, Tankard et al 1982). Of these, the Matsap (Cape), Waterberg (Transvaal) and Palapye (Botswana) intracratonic basins are characterized by predominant arenaceous fill including flavial and shallow-water arenites with rare volcanic components. In contrast, the rift-induced Soutpansberg trough (northeast Transvaal), a possible aulacogen, contains roughly equal parts of arenites and basalt–andesite–trachyte lava flows (Jansen 1981). In addition, the Umkondo Basin, located at the eastern margin of the Zimbabwe Craton, is of Waterberg type. All basins are roughly coeval and characterized by the ubiquitous presence of the earliest

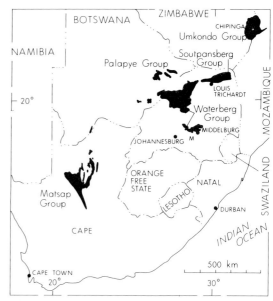

**Fig. 3-15.** Distribution of the Waterberg, Soutpansberg, Umkondo, Matsap and Palapye groups in southern Africa. (From Tankard et al 1982, Fig. 7.1, and reproduced with permission of the authors and of Springer-Verlag Publishers).

substantial redbeds on Earth (Tankard et al 1982). The basins owe their existence to downwarping and locally intense contemporaneous block-faulting along the basin rims.

## 3.8.5 MAGONDI MOBILE BELT

The comparatively small, northerly-trending Magondi (Lomagundian) Belt, bordering the Zimbabwe Craton on the west, is dominated by redbeds, volcanics, arenite–dolomite–angillite and greywacke with local stratabound Cu deposits of the Magondi Supergroup (Figs 1-3f(i), 1-5f(i,ii), 4-12). These early Proterozoic basinal accumulates, up to 5500 m thick, were deformed and metamorphosed during the ~1.8 Ga old Magondian Orogeny. The maximum exposed strike length of the belt is 250 km but deep borehole data northwest of Bulawayo to the south indicate the presence there of buried Magondi rocks which would at least double the real strike length (Stagman, in Hunter 1981, Treloar 1988). Magondi strata unconformably overlie either Archean granitoid–greenstone basement or intervening early Proterozoic pre-Magondi silicic paragneisses.

Available age data suggest an age of 2100–1800 Ma for this belt (Treloar 1988, Treloar and Kramers 1989).

### 3.8.6. KHEIS BELT

The Kheis Belt (Domain), also called the Eastern Marginal Zone of the Namaqua Province, extends northward along the western margin of the Kaapvaal Craton for some 400 km into southern Botswana (Figs 1-3f(ii), 1-5f(i,ii)). It represents an eastward vergent 'thin-skinned' fold-and-thrust belt comparable in most respects to the Magondi Mobile Belt. The sequence (~3000 m thick) is composed of metaquartzite, phyllite and amphibolite arranged in five formations. Much of the western and eastern belt boundaries are tectonic. The Kheis Group is in part at least stratigraphically equivalent to the Matsap (Waterberg) Group.

According to Hartnedy et al (1985), the belt evolved between 2000 and 1700 Ma. The small enigmatic Okwa Inlier (Block) to the north (Fig. 4.12) includes a basement complex of Eburnean-age (2.0 Ga) crust that was accreted to a nearby Archean craton at ~1.8 Ga (Aldiss and Corney 1992).

## 3.9 AFRICAN CRATON: CENTRAL (EQUATORIAL) AFRICA

Early Proterozoic rocks are present in four mobile belts, one basinal sequence and two reworked craton-cover assemblages (Figs 1-3f(i), 1-5f): in the east, (1) the NW–NNW-trending Ubendian (Ruzizian) Belt; (2) the nearby E–NE-trending Usagaran Belt; and (3) the nearby (to the north) E-trending Ruwenzori Belt; far to the west (4) the NNW-trending Kimezian Assemblage, part of the internal zone of the younger (late Proterozoic) West Congo Belt; (5) the nearby Francevillian–Ogooué basinal sequence; and, in the south, deformed early Proterozoic cover upon both (6) the Angolan Craton in the southwest; and (7) the Zambian Craton (Bangweulu Block) in the southeast. The small greenstone belt of Bogoin-Boali, located 100 km NNW of Banqui in western Central African Republic, has been recently dated as early Proterozoic (2.5–2.15 Ga) (Poidevin, 1994).

### 3.9.1 UBENDIAN (RUZIZIAN) BELT

The NW–NNW-trending Ubendian Belt, located in southeastern Equatorial Africa (Figs 3-16, 4-13), extends for at least 1000 km in exposed length with a maximum width of 250 km. The southern part in Tanzania is called Ubendian while the northern part, mainly in Zaire, is called Ruzizian (Mendelsohn 1981, Cahen et al 1984, Daly 1986a).

This broad fold belt lies between the Tanzania Craton on the northeast and the Zambian (Bangweulu) Block on the southwest. To the northwest, in Zaire, the belt is largely obscured by the NE-trending mid-Proterozoic Kibaran Belt, though relict Ruzizian patches emerge locally from beneath Kibaran cover. To the east, the Ubendian Belt together with adjoining coeval Usagaran strata lose their identity in the younger (Pan-African) N-trending Mozambique Belt.

The belt assemblage, according to Mendelsohn (1981), comprises quartzite, arkose, phyllites, schists and amphibolites. Metamorphic grade varies from greenschist facies in the south (Zambia–Malawi) through lower to middle amphibolite facies in the median part, to middle and upper amphibolite in the north (Tanzania–Burundi–Rwanda). Deformation occurred in two episodes, resulting in regional foliation-folds or isoclinal large-scale folds with SE-trending axial planar schistosity.

The Ubendian cycle of events started with post 2.6 Ga accumulation of stable shelf-type supracrustal rocks and peaked with the Ubendian Orogeny at ~2050 Ma (Cahen et al 1984), with subsequent reactivation including prominent strike-slips in Irumide time (~1.1 Ga) (Nonyaro et al, 1983, Daly 1986a).

### 3.9.2 USAGARAN BELT

The E–NE-trending Usagaran Belt of folded metasedimentary rocks, 500 km long and 50 km broad, represents the eastward extension of the southernmost Ubendian Belt (Fig. 3.16). The Usagaran Belt (Southern Highland) lies at the southeast border of the Tanzania Craton foreland to the belt, which in turn forms a local foreland to the younger Mozambique Belt to the east.

Psammitic–pelitic metasediments of the Usagaran Supergroup, ~4000 m thick, transgressively overlie amphibolite granulite facies basement gneisses of the Tanzania Craton. Usagaran strata, in turn, are unconformably overlain by up to 3500 m thick of mainly andesitic–rhyolitic volcanic rocks of the Ndembera Group.

Usgaran–Ndembera rocks strike essentially east–northeast to north–northeast and are intensely folded. The metamophic grade commonly reaches the staurolite–almandine subfacies but with scattered occurrences of granulite facies rocks. The

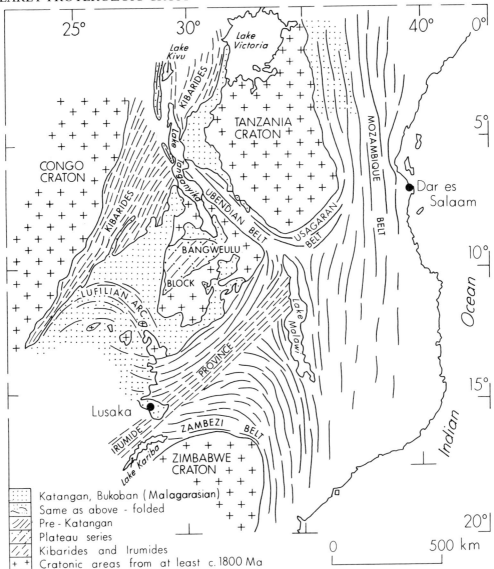

**Fig. 3-16.** Geologic sketch map of southeast Africa showing the distribution of the Tanzania, Congo (Kasai) and Zimbabwe cratons, and intervening Ubendian Belt (early Proterozoic), Kibaran and Irumide Belts (mid-Proterozoic), and adjoining Mozambique Belt (mid-late Proterozoic). (From Cahen et al 1984, Fig. 6.1, and reproduced with permission of the authors and of Oxford University Press).

junction between the SE-trending Ubendian and the ENE-trending Usagaran rocks occurs at the northeast end of Lake Malawi. However, this vergation may reflect the influence of the younger Mozambiquian Orogeny (~600 Ma), superimposed on the older rocks. In any event, still farther to the east, the Usagaran becomes progressively implicated in the younger N-trending Mozambique Belt.

Gemstones are found in the multiply deformed and polymetamorphic Usagaran terrains. The common gem types include varieties of garnet, almandine, rhodolite, pyrope and hessonite, ruby, sapphire, amethyst, kyanite, beryl, emerald, aquamarine, alexandrite, opalite and topaz. Gemstone mineralization in these metamorphic terrains formed either by metamorphic–hydrothermal processes or in connection with pegmatites and silicic vein rocks along fault zones (Malisa and Muhongo 1990).

### 3.9.3 RUWENZORI FOLD BELT

The name Ruwenzori was given by Tanner (1973) to an approximately E-trending belt, about 600 km long and 150 km wide, previously named the Buganda–Toro (–Kibalian) belt. The Ruwenzori Belt (Fig. 1-5f(i,ii)) extends westwards along the north shore of Lake Victoria to the Western Rift (Lake Edward), near which it is prominently exposed in the Ruwenzori Mountains. To the west of Lake Victoria, limited southward by the Ruwenzori Belt, is unconformably overlying mid-Proterozoic Karagwe–Ankolean (Kibaran) strata. Still farther west, Ruwenzori rocks can be followed discontinuously across the Western Rift into Zaire. To the east, Ruwenzori strata unconformably overlie the Tanzania Craton. To the north, they unconformably overlie the ancient basement of Uganda.

Buganda strata comprise basal conglomerate, quartzite, phyllite and assorted volcanic rocks. The Toro Group of western Uganda includes a schist and gneiss association (Igara schists) and the Toro quartzites. Greenschist facies metamorphism prevails. Structures are complex.

Buganda rocks are loosely constrained in the 2500–1800 Ma age range (Cahen et al 1984).

## 3.9.4. KIMEZIAN ASSEMBLAGE

The Kimezian zone, part of the internal (i.e. western) zone of the West Congo (Pan-African) Orogen, extends for 1300 km parallel and close to the Atlantic coast. The zone, which is characterized by pronounced east–northeast vergence, is up to 150 km wide at the south end but thins gradually to the north. It includes highly deformed and metamorphosed rocks of the Kimezian Supergroup (Cahen et al 1984, Porada 1989) (Fig. 5-14).

In the western part of the zone, a migmatite–gneiss complex is complexly folded along NE, NNE and N–S trends. Eastward towards the foreland original Kimezian sediments are still recognizable in the form of moderately deformed metapelitic schist, quartzite and limestone at prevailing greenschist facies metamorphism.

Kimezian dates of 2126–1952 Ma (Cahen et al 1979a) establish the general age of the Tadilian Orogeny, corresponding elsewhere to the Eburnean (Ubendian) Orogeny.

## 3.9.5. FRANCEVILLIAN SUPERGROUP

Francevillian sedimentary rocks, a dominantly conglomerate–arkosic sandstone assemblage with chert and local volcanic rocks, occupy an intercratonic (Chailu–Gabon cratons) basin complex situated on the easterly slope of the foreland of the West Congo Belt (Fig. 1-5f(i)). The Ogooué schists, which are presumed Francevillian equivalents, occupy an adjoining basin to the west with the two separated by the N-trending Okandu fracture system. The Francevillian Supergroup has been studied since 1954, especially due to the exceptional uranium concentrations around Mounana and Okla, which are remarkable for the unique natural nuclear event, as well as for large manganese deposits at Moanda (Bonhomme et al 1982, Gauthier-Lafaye and Weber 1989, Gauthier-Lafaye et al 1989, Ledru et al 1989, Bros et al 1992, Bertrand-Safarti and Potin 1994).

The Francevillian Supergroup, 1000–4000 in thickness, occupies several local basins and sub-basins formed by intervening N- and NW-trending basement ridges. Francevillian strata unconformably overlie Archean basement. The supergroup, subdivided into five lithostratigraphic units, comprises a complex arenite–pelite–carbonate–volcanic assemblage, including local stromotolitic carbonates and bitumen.

The uranium ore deposits consisting of black silicified sandstone with 0.1–10% U, are located in the upper levels of the basin, where they are capped by impermeable black shales (Gauthier-Lafaye and Weber 1989). Sixteen natural fission reactors are known in three different ore deposits, of which the most famous and first discovered are those at Oklo (Gauthier-Lafaye et al 1989).

The succession of events in the Francevillian sequence including deposition and lithification of sediments at ~2099–2036 Ma followed by nuclear reaction at 1970 Ma are summarized as follows:

| (1) | Stabilization of the basement | 2700 Ma |
| (2) | Deposition of Francevillian A (before N'Goutou intrusion) | Before 2140 Ma |
| (3) | Deposition of the rest of Francevillian | Before 2050 Ma |
| (4) | Nuclear reaction | 1970 Ma |
| (5) | End of diagenesis | 1870 Ma |
| (6) | End of cooling | 1700 Ma |
| (7) | Intrusion of dolerites (diabase) | 970 Ma |

The Ogooué Basin, roughly equivalent in size to that of the Francevillian, extends westward for 200 km from the Francevillian to join with the Kimezian internal zone of the West Congo Belt, running parallel to the Atlantic Coast. Ogooué rocks have been variably considered to represent (1) a metamorphic facies of the Francevillian Supergroup; (2) separated from the Francevillian by a major unconformity; and (3) divided into lower easterly metamorphosed and upper westerly less metamorphosed portions, with varying NE to N–S trends (Cahen et al. 1984).

## 3.9.6 ANGOLAN CRATON

South of the Malange Transcurrent Fault, at 9°S (Figs 1-5f(i,ii), 5-14), and within the northernmost Angolan Craton, numerous Archean basement inliers of granulite facies gneiss and migmatite exist within a vast expanse of southward increasing early Proterozoic metasediments, migmatites, gneiss and plutons, many of which bear the imprint of the Eburnean Orogeny (~2.0 Ga). This early Proterozoic cover upon the craton represents the southward extension of the Kimezian Assemblage (~2.0 Ga), which, as stated above, forms part of the internal zone of the West Congo Belt. The preserved Angolan Craton, which is 900 km long (N–S) and up to 600 km wide, is transitional southward across several early Proterozoic inliers in Namibia to the main E-trending Damaran (Pan-African) Belt (Carvalho 1983, Cahen et al 1984, Tegtmeyer and Kröner 1985).

Early Proterozoic cover on the Angolan Craton is composed of metamorphosed conglomerate, arenites, greywackes, siltstone, and volcanic and hypabyssal rocks of the Oendolongo Supergroup. Oendolongo deposition took place upon Archean basement in an early Proterozoic basin named the 'Eburnian geosyncline' (Torquato et al 1979).

In southwest Angola, between 15°S and the frontier with Namibia at 17.30°S, and west of 14°E, the E- to ENE-trending, so-called 'Metamorphic Series of SW Angola' (Carvalho 1972) comprising schists, metaquartzite, metagreywacke with marble lenses, amphibolite, talc schist, chlorite schist and associated migmatitic gneiss, is considered to be Oendolongo in age (Carvalho 1983). The rocks bear the imprint of the Eburnean Orogeny. Several small patches of flat-lying, barely metamorphosed, 500 m thick Chela strata, composed of conglomerate, sandstone, pelites and limestones, rest unconformably on a granitoid basement. Kröner and Correia (1980) correlate the Chela Group with the Nosib Group of the Damara Belt (Pan-African) in Namibia to the south (see below).

## 3.9.7 KUNENE ANORTHOSITE COMPLEX

The major portion (88%) of this huge gabbro–anorthosite complex (Vermaak 1981) is situated in southernmost Angola, where it straddles the Kunene River and extends into Namibia. The complex, which underlies an area of 17 378 km² of which at least 13% is covered, is estimated to exceed 14 km in thickness. The adjoining country rocks consist of gneiss, granulite and 'leptite', which enclose scattered rafts and remnants of quartzite, marble to calcsilicate rocks, jaspilite, phyllitic schists and metavolcanic rocks.

The complex consists of preponderant melano- to leucocratic anorthosites (72%), together with a more ultramafic marginal facies (3%) and granitic rocks (25%).

The main complex is composed of pale massive anorthosite and dark cyclic troctolite. The interrelationship of the two main rock types is controversial (Simpson, 1970). The prevailing view favours the anorthosite complex being an intrusion, with crystal fractionation the cause of the lithologic diversity (Silva 1972). Its age is ~2100 Ma.

## 3.9.8 BANGWEULU BLOCK (ZAMBIAN CRATON)

The triangular Bangweulu Block or Zambian Craton (Fig. 3-16) includes two exposed lithologic associations: (1) northerly and southerly terrains of early Proterozoic granitoid gneiss and volcanic rocks of present concern; and (2) a central strip of unconformably overlying mid-Proterozoic Muva (formerly Plateau 'Series') sediments, which belong to the foreland zone of the neighbouring mid-Proterozoic Irumide Belt (see below). (Key references: Anderson and Unrug 1984, Daly 1986a, b).

In the early Proterozoic gneiss–volcanic terrains, the Luapula felsic metavolcanic rocks with associated granitoid intrusions crop out along the Luapula River at the foot of the Plateau 'Series' escarpments. On the opposite (eastern) side of Lake Tanganyika, the same volcanic rocks are called the Kate porphyry.

Luapula volcanic rocks are mainly rhyolite–dacite–andesite lava flows and pyroclastics, with minor quartzite and schist intercalations. These volcanic and sedimentary rocks were deformed prior to deposition of the mid-Proterozoic Muva (Plateau) sediments (Thieme 1975). The prevailing grade of metamorphism is greenschist facies with local amphibolite facies. The age of the volcanic rocks is indicated to be 1900–1800 Ma (Brewer et al 1979a).

# 3.10 AFRICAN CRATON: NORTHWEST AFRICA

## 3.10.1 WEST AFRICAN CRATON

Early Proterozoic rocks are preserved in the Man and Reguibat shields of the West African Craton, as well as in the adjoining Tuareg Shield (Figs 1-3f(ii), 1-5f(i,ii)).

### Man Shield: Eburnean Domain

The Eburnean (Baoulé–Mossi) Domain, an irregular lobate region that underlies the central–eastern part of Man Shield (Fig. 2-23) (Bessoles 1977, Bessoles and Trompette 1980, Black 1980, Cahen et al 1984) is divided into (1) a smaller southeastern Birrimian (Birimian) subdomain, dominated by early Proterozoic metasupracrustal rocks of the Birrimian Supergroup (2.2–2.1 Ga) which are sediment-rich in the west and volcanic-rich in the east; and (2) the larger partly enclosing (western–northern–northeastern) 'basin and mole' subdomain, in which numerous discrete Birrimian-filled troughs and basins, including, especially in the northeast, volcanic-rich greenstone belts, are separated by granitoid domes collectively composed of both older gneissic basement and younger Eburnean (2.1–1.9 Ga) plutons (Table 2-13, column 1) (Cahen et al 1984, Hirdes et al 1992, Feybesse and Milési 1994). The erosional products of near-contemporaneous uplift were deposited as Tarkwaian molasse (~2.0 Ga) in long, narrow intermontane grabens formed by rifting in certain Birrimian belts (Ledru et al 1991, Taylor et al 1992).

In the Birrimian Subdomain, broadly distributed NNE-trending Birrimian belts are of two types: (1) a dominant volcanic–flysch type composed of mafic–felsic volcanic rocks with quartzite, phyllite and persistent gondite (Mn) horizons typically distributed in deep, elongated troughs (Type I) with Fe, Mn, Au and local diamond deposits; these are transitional to (2) arenaceous–rudaceous–felsic volcanic facies distributed in broad, shallow basins (Type II) (Cahen et al 1984, fig. 17.3, Leube et al 1990, Hirdes et al 1992, Taylor et al 1992).

The numerous Birrimian-correlated N–NE-trending troughs and basins in the larger adjoining 'basin and mole' subdomain range individually up to 90 000 km$^2$ in area. Clastic–flysch facies prevailing in the west and northwest are transitional northeastward to volcaniclastic–greenstone facies (Leube et al 1990).

Unconformably overlying Tarkwaian molasse occurs in several shallow basins scattered about the Baoulé–Mossi Domain. Undeformed to locally highly deformed, low metamorphic grade, mainly coarse grained arenaceous, locally auriferous sediments predominate, collectively representing alluvial fans with braided stream channel patterns. At Tarkwa goldfields, Ghana—the type area—the 1500 m thick sequence comprises up-section, basal conglomerate, quartzite, grits and conglomerate, argillites and sandstone.

The two isolated Kenieba and Kayes inliers to the northwest of Man Shield (Fig. 2-23) are composed of Birrimian volcanodetritic–flysch assemblages intruded by Eburnean granitoid plutons (Black 1980).

Structurally, Birrimians rocks are characterized by N–NE-striking, steeply dipping foliation. In many areas, foliation corresponds to bedding planes and the folds can be shown to be isoclinal, with subvertical to vertical axial planes (Vidal and Alric 1994). Metamorphic grade varies from prehnite–pumpellyite facies to almandine–amphibolite subfacies with greenschist facies prevailing. The overlying Tarkwaian sediments are less strongly deformed and commonly barely metamorphosed.

On a regional scale, Au and Mn occurrences, though geographically separated, are preferentially aligned along the flanks of Birrimian volcanic-rich belts. Most Au occurrences are concentrated in narrow corridors, 10–15 m wide and up to several hundred kilometres long, typically located at the lithofacies transition between volcanic belts and sedimentary basins where commonly associated with regionally extensive shear zones (Leube et al 1990).

Birrimian and Tarkwaian metasupracrustal rocks provide accurate zircon dates of 2195–2166 Ma (Taylor et al 1988). The younger granitoids and metamorphites are concentrated in the range 2170–2000 Ma which is attributed to the Eburnean Orogeny (Taylor et al 1988, Liégois et al 1991, Hirdes et al 1992, Davis et al 1994). Collectively these represent a major West African crust-forming episode from ~2.3 Ga to 2.0 Ga, attributable to differentiation from a depleted mantle source (Taylor et al 1992).

### Reguibat Shield: Eastern Province

The Eastern Province of Reguibat Shield to the north (Fig. 2-23) is about 750 km long and 250 km wide. This province is dominated by rocks and events of the Eburnean (Yetti) cycle (Table 2-12, columns 3, 4) (Bessoles 1977, Black 1980, Cahen et al 1984).

In brief, detrital, volcaniclastic and volcanic rock (Yetti) of common greenschist facies are intruded by high level syn-orogenic Eburnean granitoids with closely associated ignimbrites (Aftout). Unconformably overlying, postorogenic molasse (Guelb el Hadid) is distributed in up to 4 km thick units. Widespread Eburnean granitoid intrusions are dated (Rb–Sr) at 2.1–1.9 Ga (Clauer et al 1982).

Occasional Archean basement inliers also lie in the eastern part of the province. They are composed of gneiss–migmatite assemblages, partly at amphibolite facies metamorphism, though commonly retrogressed to greenschist facies, and frequently mylonitized. These rocks are generally correlated with Amsaga assemblages of the Southwestern Province (Bessoles 1977) (see above).

## 3.10.2 TUAREG SHIELD

Early Proterozoic rocks of the Tuareg Shield are recognized in three separate areas (Fig. 2-23, Table 2-13): (1) in western Hoggar, the Tassendjanet Group, an epi- to meso-zonal metasupracrustal assemblage with various granitoid intrusions (Bertrand and Caby 1978); (2) in central Hoggar, the Arechchoum Group, comprising migmatitic gneiss with marbles, quartzites and amphibolites (Bertrand et al 1984); (3) adjoining to the east, the Gour Oumelalan Assemblage, a varied assemblage of gneisses, leptynites, charnockites, marbles and quartzites (Latouche 1972). These early Proterozoic rocks at least locally unconformably overlie Archean basement, including Red Group ('Series') rocks. Together with assorted granitoid intrusions, the early Proterozoic rocks represent the Suggarian (Eburnean) cycle of events, culminating in deformation and granulite facies metamorphism at 1975 Ma (Fig. 2-23; Table 2-14).

Practically throughout the Hoggar, a Tassendjanet–Arechchoum-type basement dated at ~2000 Ma, has been tentatively identified, leading Bertrand and Lasserre (1976) to infer that the whole of the Hoggar was underlain at one time by Suggarian (early Proterozoic) or older basement.

## 3.10.3 BENIN NIGERIA SHIELD

The geological evolution of this large southeastern area is still poorly known (Fig. 2-23). The rock types include a variety of migmatites and gneisses (predominantly quartzofeldspathic), granitoids, a range of metasedimentary schist and gneiss, various amphibolites, pyroxenites and metagabbros, charnockite and the distinctive rock-type 'bauchite' (Grant 1978, Affaton et al 1980). The metasedimentary rocks occur in some cases intercalated with gneiss, and in others as meridionally elongated, epizonal, schist–quartzite units, locally including marble and conglomerate. Tectonic contacts between units predominate. In general, granulite facies metamorphism prevails in the west (southeast Ghana, Togo and Benin) and amphibolite facies farther east in Nigeria. Grant (1973) attributes this contrast in metamorphism to the presence of a major eastward inclined meridional thrust, along which the late Proterozoic 'Dahomeyan' rocks have been thrust up and over the early Proterozoic Birrimian rocks to the west.

Two older orogenic events in the Benin Nigeria Shield were earlier reported, namely, emplacement of granitic gneiss of 2100 Ma (Eburnean event), and a locally preserved Liberian event at 2750 Ma (Grant 1978). However, recent zircon dating on granodiorite orthogneiss from northern Nigeria provides ages of 3040 Ma, and a local zircon core age of > 3500 Ma, the oldest age for crust ever found in this area. Furthermore, no evidence for either Eburnean (2000 Ma) or Kibaran (1100 Ma) orogenies was detected, despite widespread Pan-African (~618 Ma) dates (Bruguier et al 1994).

## 3.10.4 ANTI-ATLAS DOMAIN

The Anti-Atlas Domain, with widespread folded Paleozoic rocks, grades southward into the Tindouf Basin, itself located on the northern edge of the West African Craton (Fig. 1-5f(i,ii)). The domain in question continues southeastward to link with the Hoggar. The Anti-Atlas Domain, only very slightly affected by Alpine movements, includes numerous Precambrian inliers, cropping out from under folded or subtabular Paleozoic (and latest Proterozoic) cover. About 10 separate older Precambrian inliers have been identified.

Available dates on varied lithologies in the inliers range from 2300 to 1700 Ma. These ages are grouped into an early Proterozoic metasupracrustal suite intruded during both an older event at ~1797 Ma (Charlot 1978, Hassenforder 1978).

## 3.11 AFRICAN CRATON: NORTHEAST AFRICA AND ARABIA

Much of the northern fringe of Africa and the eastern part of Arabia is mantled by thick sequences of Paleozoic and Cenozoic sediments which form a generally undeformed cover to a deep crystalline basement. The basement rocks are exposed (1) in the Arabian–Nubian Shield; (2) in isolated inliers farther west in the interior of the north African continent; and (3) to the south as merging terranes with the metamorphosed basement areas of Central Africa (Figs 1-5f(i), 3-17).

### 3.11.1 UWEINAT INLIER

At the Egypt–Libya–Sudan frontier the crescentic Uweinat (–Jebel) basement inlier of Archean–early Proterozoic age, some 250 km long (north to south) and 125 km wide, is unconformably overlain by Devonian strata and Cretaceous to Permian Nubian sandstones. The basement rocks of the inlier have

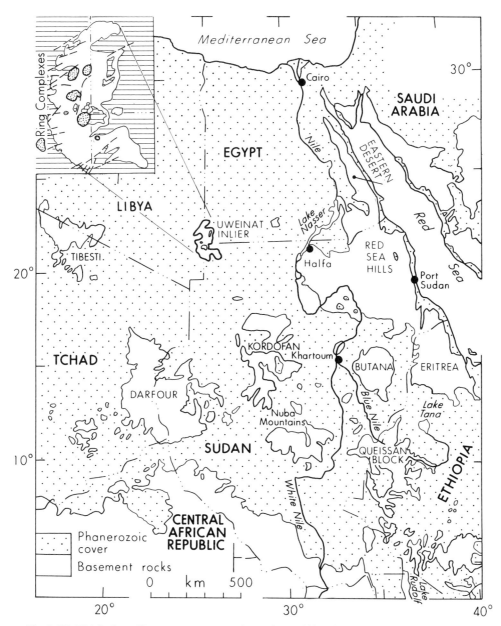

**Fig. 3-17.** Distribution of basement exposures in northeast Africa. (After Cahen et al 1984, Fig. 14.1, and reproduced with permission of the authors and of Oxford University Press).

been divided into a number of groups on the basis of metamorphism and structure (Klerkx 1971, Schürmann 1974): biotite and amphibolite gneiss, migmatitic gneiss and granitoid plutons; partly retrogressed pyroxene granulites with charnockitic, noritic gneisses and metaquartzites; and migmatites associated with a prominent NE-trending mylonite zone.

In brief, available data clearly indicate the presence of Archean basement rocks in this region, together with early Proterozoic (~1.8 Ga) migmatites and cross-cutting granitoid plutons. Cahen et al (1984) suggest that basement granulite facies metamorphism may have occurred at ~2900 Ma, coincident with high grade metamorphism elsewhere in Africa.

### 3.11.2 ARABIAN–NUBIAN SHIELD

It is now generally agreed that the greater part of the Arabian–Nubian Shield developed from oceanic crustal material, in a sequence of island arcs, during the late Proterozoic interval 900–600 Ma (Pallister et al 1988) (see Chapter 5). Nevertheless, there is convincing evidence of the involvement in this late Proterozoic shield growth of older Precambrian crust, including early Proterozoic and even Archean components. These probably represent fragments of older continental crust entrained in the accreting arc complexes.

In example, a granodiorite from the eastern margin of the Arabian Shield (Jabal Khida quadrangle) provides a zircon date consistent with an original crystallization age of 1600–1700 Ma (Stacey and Hedge 1984). Furthermore, single grain dating of detrital zircons from two localities in the Eastern Desert of Egypt yield two age clusters respectively at 2.65–2.4 Ga and 2.39–1.46 Ga, clearly revealing the input of old continental crust during Pan-African sedimentation (Wust et al 1987).

### 3.12 INDIAN CRATON

Early Proterozoic events in Peninsular India are mainly restricted to the northern and central parts (Figs 1-3g, 1-5g, Tables 2-14, 2-15). In the northeast (1) the Singhbhum (Copperbelt) Thrust Zone, which bisects the Chotanagpur–Singhbhum Craton, is associated with Singhbhum, Dhanjori and Kolhan sedimentary–volcanic groups (2.3 Ga). To the west, (2) the Aravalli–Delhi Belt contains thick sedimentary sequences of the Aravalli and Delhi supergroups (2.5–2.0 Ga). In the centre–east, (3) the Bhandara Craton includes the Sausar, Sakoli and Dongargarh supracrustal sequences (2.2 Ga). Also, the nearby Cuddapah Basin includes lowermost Cuddapah strata (~2.0 Ga) marking early-stage platform rifting, as considered below (Chapter 4).

### 3.12.1 COPPERBELT THRUST ZONE

The Chotanagpur Block (Craton), which forms the northern part of the combined Chotanagpur–Singhbhum Craton (Rogers 1986) (Fig. 1-5g, inset 3), is composed largely of poor known and undated gneissic and granitic rocks. The Singhbhum Craton (nucleus), to the south, and the Chotanagpur Block (Terrane) to the north, are separated by the northward dipping Copperbelt (Singhbhum) Thrust Zone (Fig. 1-5g, inset 1, Table 2-15) which involves a belt of compressed supracrustal rocks (Singhbhum mobile belt) representing the eastern extension of the Satpura Belt. The Copperbelt Thrust Zone, 170 km long and 20–30 km wide, consists of a sole thrust together with smaller thrusts and entrapped chaotic terranes. All transport has brought northern suites southward over southern ones. Movement occurred about 1600 Ma ago (Table 2-15, column 4).

To the north of the Copperbelt (Singhbhum) Thrust a melange of metasupracrustal rocks includes two main groups both of age 2.4–2.3 Ga: (1) Singhbhum Group including Dhalbhum phyllites, schists and quartzites, and Chaibasa mica schists, quartzites and metavolcanic rocks; and (2) Dalma Group composed of mafic–ultramafic tuffs, phyllites, ironstones (BIF) and limestone. These are locally intruded by granitic plutons (2.2 Ga).

To the south of the Copperbelt Thrust (Fig. 2-25) coeval (2.4–2.3 Ga) metasupracrustal assemblages include: (1) Dhanjori Group including the basalt-rich Ongarbira volcanic–sedimentary belt (Blackburn and Srivastava 1994) and Dhanjori conglomerates and quartzites; and (2) Simlipal–Jagannathpur (Dhanjori) mafic–ultramafic volcanic rocks (Alvi and Raza 1991). The unconformably overlying Kolhan Group (2.2–2.1 Ga) is composed of conglomerate, sandstone, limestone and shale. These groups are intruded by granite–gabbro bodies including the substantial Mayurbhanj Granite (2.1 Ga) in the northeast.

### 3.12.2 ARAVALLI–DELHI BELT

The NNE-trending combined Aravalli–Delhi Belt of Rajasthan, NW India, extends for 750 km from

Ahmedabad vicinity in the south to Delhi in the north, with an average width of 200 km (Fig. 2-25) (Roy 1988, Naqvi and Rogers 1987, Deb and Sarkar 1990, Banerjee and Bhattacharya 1994). Three major tectonic domains are recognized in the belt: the Archean Banded Gneissic Complex (BGC), the early Proterozoic Aravalli Complex (Supergroup), and the mid (–late?) Proterozoic Delhi Complex (Supergroup) (Table 2-14, column 5). The present distribution pattern of the three domains within the combined belt is a function of complex multiple deformations.

The Banded Gneissic Complex, a highly deformed and irregularly segmented central domain at amphibolite–granulite facies of metamorphism, forms remnant basement to the two metasupracrustal sequences, themselves tectonically disposed with the Aravalli Complex lying in the centre–east and the Delhi Complex in the west and northeast (Fig. 2-25).

The main Aravalli Complex in the east, a composite belt measuring 350 km by 150 km, comprises three parallel tectonically disposed zones mutually interspersed with BGC basement segments (see Deb and Sarkar 1990, fig. 1). Of these, the thin easterly zone (Bhilwara belt, near Chitorgarh, is itself divided into several deformed subzones separated by tracts of migmatized BGC. The other two adjoining (central–western) zones (combined Aravalli–Jharol belt) form a 250 km-long, N- to NE-trending arc reaching Nathdwara in the north; the two zones represent a single depositional basin, with shared deformations hence similar tectonic trends.

The Aravalli Supergroup is a highly variably thick sequence of mainly greywacke, arenite and argillite with intercalated stomatolitic carbonates, phosphorites and mafic volcanic rocks (Rajamani and Ahmad 1991, Banerjee and Bhattacharya 1994). The rocks have undergone very low grade metamorphism and display complex structures resulting from two major and several minor episodes of deformation. The stratigraphic sequence in the eastern (Bhilwara) belt is dominantly arenite-pelite–carbonate, whereas in the adjoining centre-east (Aravalli–Jharol) belt, an eastern carbonate-phosphorite-rich marginal facies is transitional westward to deep water pelites (Deb and Sarkar 1990). This pattern of facies transition, particularly that involving dolomite with stromatolitic rock phosphate, was controlled by paleo-sea-floor topography (Roy and Paliwal 1981, Banerjee et al 1992). The Aravalli Supergroup is interpreted as a product of a passive to active continental margin in a plate-tectonic setting (Banerjee and Bhattacharya 1994). The Aravalli–Delhi contact is interpreted as representing the closure of a Proterozoic ocean basin (Tobisch et al 1994).

The Delhi Complex, adjoining to the west and northeast, is divided into the comparatively thin linear, complexly deformed South Delhi belt (300 km by 50 km extending from Palaupur vicinity northward to Ajmer, and the broad irregular (400 km by 200 km), tight-to-open-folded, N–NW-plunging North Delhi belt of the Aravalli mountains (Jaipur–Delhi region) in the northeast. Delhi Supergroup strata unconformably overlie BGC basement and are in structural discordance with Aravalli strata. The main rock types are arenites, phyllites, slates and carbonates with some mafic volcanic rocks, dolomites and rock phosphorites. The Delhi Basin in the northeast, which is characterized by several fossil grabens, horsts and arches, comprises at least three sub-basins with about 10 km thickness of volcano sedimentary infill. Depositional environments featured an association of braided streams, tidal flats and barrier islands. Greenschist facies metamorphism prevails. Certain mafic volcanic units and tectonized mafic–ultramafic rocks in the South Delhi belt are assigned to the Phulad Ophiolite (Sinha Roy and Mohanty 1988, Volpe and Macdougall 1990). Nearby patches of metagreywacke contain relict blueschist mineral assemblages including glaucophane and lawsonite. These ophiolite–blueschist-bearing rocks are interpreted as an ophiolite emplaced in a melange zone, products of plate-tectonic processes (Sinha Roy and Mohanty 1988).

Aravalli-Delhi assemblages have not yet been adequately dated and correlated. The Aravalli Supergroup is commonly assigned an early Proterozoic age (2.5–2.0 Ga) with some 1.8–1.7 Ga model ages. The Delhi Supergroup, commonly assigned a mid-Proterozoic age (1.8–1.0 Ga), is loosely bracketed by the Aravalli Orogeny (to 2.0 Ga) and the Delhi Orogeny (to ~1.5 Ga) with post-Delhi tectonism extending to 850–750 Ma (Tobisch et al 1994).

The Aravalli–Delhi Belt accounts for the largest Pb–Zn ore reserves in India; it also contains significant Cu deposits, Mn concentrations, and some U and Sn–W reserves.

### 3.12.3 BHANDARA CRATON

Three major early Proterozoic supracrustal assemblages occur in the Bhandara Craton–Dongargarh,

Sakoli and Sausur (Fig. 1-5g, inset 4). The Dongargarh belt, 90 km by 150 km comprising equal parts metasupracrustals and granitoids, extends north between the Sakoli synclinorium on the west and the late Proterozoic Chattisgarh basin on the east. The Dongargarh Supergroup is divided into three conformable groups: lower schists, quartzites, gneiss and amphibolites; middle bimodal (tholeiite–rhyolite) volcanics; and upper shales, sandstones and mafic volcanic rocks. The entire assemblage has been affected by at least three phases of tight, complex folding. Metamorphism varies from greenschist to amphibolite facies. Rhyolites in the middle group are dated at 2.2 Ga (Sarkar et al 1981).

The Sakoli Group (~2.2 Ga) occupies a large synclinorium near the west–centre margin of the craton. The major rock types are metapelites, amphibolites and banded hematite quartzites (BIF), with sparse mafic–ultramafic volcanic associates. BIF includes both wavy-banded hematite–chert and evenly laminated magnetite–martite chert varieties that contain very large iron ore deposits.

The Sausar Group forms an E-trending complexly folded, arcuate belt, 30 km by 210 km, in the northern part of the craton. The group is composed mainly of metamorphosed sandy, shaly and calcareous sediments; it includes widespread Mn ore deposits making it the major Mn producing area in India (Roy 1992, Dasgupta et al 1992). A complex and controversial stratigraphy is characterized by basal quartzofeldspathic schists and quartzites overlain, successively, by calcsilicates, marbles, manganiferous–garnetiferous micaschists and phyllites, quartzites and mica schists, and uppermost marbles and calcsilicates.

Structures in the southern part of the Sausur belt feature isoclinal folds and nappes which are recumbent or overturned towards the north in the direction of complexly deformed gneisses of the central core zone. Large K-granite bodies were emplaced along the refolded structures. Greenschist–amphibolite facies metamorphism prevails with local granulite facies. The post-Satpura orogeny is tentatively dated at 1525 Ma (Naqvi and Rogers 1987).

## 3.12.4 OTHERS

The Hogenakal carbonatites form a series of discontinuous bodies within pyroxenite dykes in the northern margin (12°N latitude) of the Granulite Domain, southern India. Three main types of carbonatite are present. Sr ratios imply a subcontinental mantle source. With an indicated age of 1.99 Ga, the carbonatite is the oldest yet known in India but also one of the few oldest carbonatites known in the world (Natarajan et al 1994).

## 3.13 AUSTRALIAN CRATON

Early Proterozoic assemblages lie in (1) the West Australian Shield; (2) North Australian Craton; and (3) Gawler Province, South Australia. Additional limited supracrustal accumulation followed by orogenic events occurred in Central Australia, prelude to major mid-Proterozoic activities (Figs 1-3h, 1-5h, Table 2-17).

### 3.13.1 WEST AUSTRALIAN SHIELD (CRATON)

The Capricorn Orogen, a major 400–500 km-wide E-trending composite suture which effectively amalgamated the Pilbara and Yilgarn blocks (Myers 1993) is characterized by geosynclinal sedimentation, metamorphism, basement reworking and granitoid plutonism. It includes (1) the Gascoyne Province (Block) in the west; (2) the Ashburton Trough (Fold Belt) infilled with ~2.0 Ga Wyloo Group sediments in the north; (3) the transitional early to mid-Proterozoic Nabberu Basin in the south; and (4) the buried basement beneath the centrally disposed Bangemall Basin which is the product of a younger (~1.0 Ga) sedimentary cycle that blankets much of the Capricorn Orogen (Fig. 3-18, inset, Table 2-16, column 1).

Also, both the Yilgarn and Pilbara blocks contain extensive suites of largely undeformed mafic–ultramafic layered dykes of early Proterozoic age that are grossly similar to the Great Dyke of Zimbabwe, and dated at ~2420 Ma. Yilgarn representatives include the Binneringie and Jimberlana intrusions, each 150–300 km long and 2–3 km wide; Pilbara Block contains the Black Range dolerites. Campbell (1987) has investigated textural and petrogenetic aspects.

### Gascoyne Province

The Gascoyne Province (Block) represents the deformed and high grade metamorphic core of the Capricorn Orogen. It forms an irregular northwestward tapering triangle, some 500 km long (northwest to southeast) and 70–300 km wide, composed of voluminous granitoid intrusions, mantled gneiss domes, metamorphosed and partly

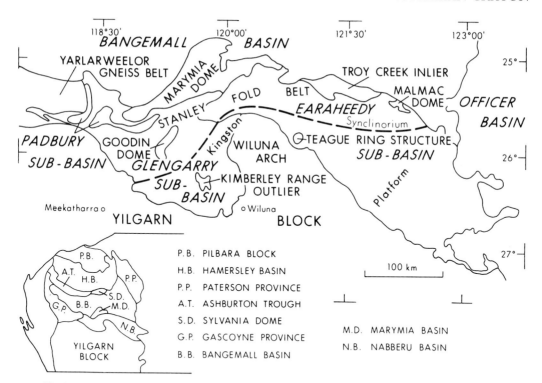

**Fig. 3-18.** Regional setting and tectonic subdivisions of the Nabberu Basin. Inset shows subdivisions of the Capricorn Orogen situated between the Pilbara and Yilgarn blocks and their associated basins. (From Bunting 1986, Fig. 2, and reproduced with permission of the author).

migmatized sediments, and remobilized Archean basement gneisses. It is tectonically divided along its length into three zones, each 100–200 km long which are composed, respectively northward, of (1) predominant Archean granulite-grade quartzofeldspathic gneisses; (2) similar Archean basement with early Proterozoic sedimentary cover; and (3) metasediments (Morrissey Metamorphic Suite) extensively intruded by Proterozoic granitoids. This last zone passes northward through a very steep metamorphic gradient into the Ashburton Trough, a deep, craton margin basin filled with 12–14 km thick of Wyloo metasediments (Gee 1979, Thorne and Seymour 1986, Williams 1986).

According to Muhling (1988), the Capricorn Orogen developed by collision of the Pilbara and Yilgarn blocks, rather than in an ensialic mobile zone as claimed by Gee (1979) and Libby et al (1986).

### Ashburton Trough

The Ashburton Trough is a 500 km long, SE–E-trending arcuate basin in abrupt tectonic–metamorphic contact with the Gascoyne Province to the south. The 12–14 km thick Wyloo Group, which fills the trough and extends beyond it as thinner unconformable cover on the older units of the Pilbara Block, comprises a mixed suite of sedimentary rocks with minor mafic and felsic vocanic components.

The Ashburton Fold Belt, coincident with the trough, is characterized by tight folding and regionally pervasive southward dipping slaty cleavage (Gee 1979). The belt is unconformably overlain to the south and east by younger Bangemall sediments.

Wyloo Group is correlated with the Glengarry Group of the Nabberu Basin, with an estimated depositional age of 2.0–1.8 Ga (Table 2-16, column 1).

### Nabberu Basin

The Nabberu Basin extends for at least 700 km along the northern margin of the Yilgarn Block, encompassing an area of approximately 60 000 km² (Fig. 3-19) (Gee 1979, Goode 1981, Goode et al 1983, Bunting 1986). The basin, which includes thick, extensive BIF of Lake Superior type (i.e. granular–oolitic or shallow-water), is divided into two main sub-basins—eastern Earaheedy and

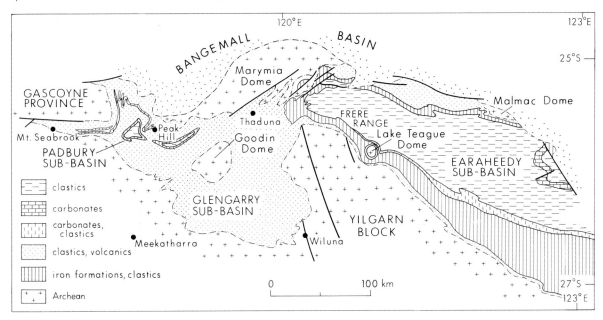

**Fig. 3-19.** Geologic map of the Nabberu Basin, West Australia. (From Goode 1981, Fig. 3.10, and reproduced with permission of the author and of Elsevier Science Publishers).

western Glengarry, plus a small westernmost Padbury subsidiary sub-basin. To the south, Nabberu strata on the Kingston Platform of the Yilgarn Block are virtually undeformed. A complex NE–E-trending synclinorial axis that crosses the length of the Nabberu Basin marks the buried northern limit of the Yilgarn Craton, to the north of which the strata are substantially deformed and metamorphosed to form the Stanley Fold Belt.

Lithostratigraphy

*Earaheedy Sub-basin* forms a large synclinorium opening to the southeast. The Earaheedy Group comprises about 6000 m thick of shallow-water sediments disposed in two cycles, named Tooloo and Miningarra subgroups respectively (Fig. 3-20). Both subgroups are composed of orthoquartzite, siltstone and stromatolitic carbonates. The lower

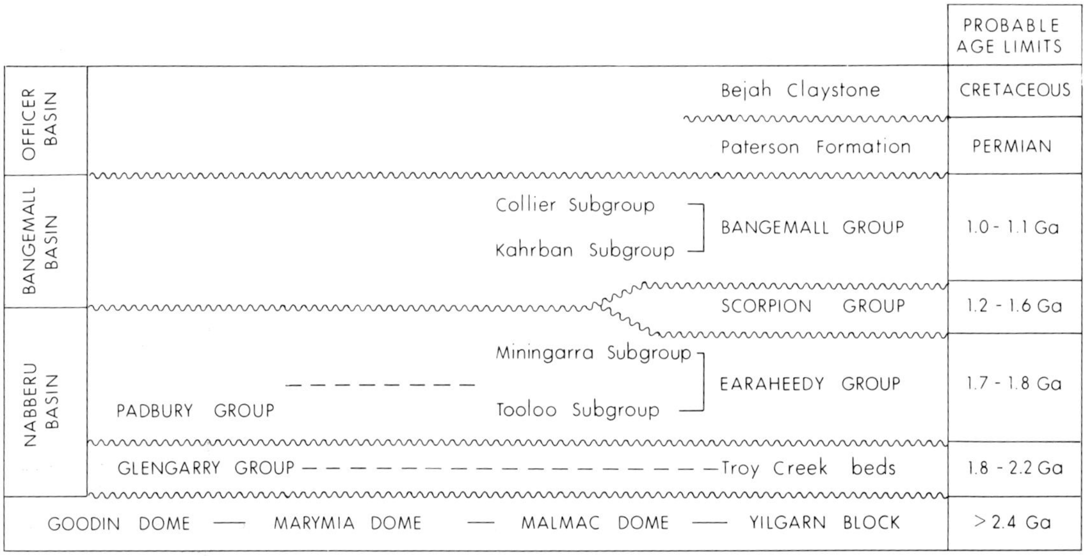

**Fig. 3-20.** Summary of main stratigraphic units within and around the Nabberu Basin. (From Bunting 1986, Fig. 3, and reproduced with permission of the author).

Tooloo subgroup also inclues thick (~1300 m) granular–oolitic BIF (Frere Formation).

The most common BIF in the Frere Formation is distinctly intraclastal (pelletal), comprising rounded intraclasts of ferruginous chert (hematite–quartz with or without magnetite) and/or chert up to 2 mm in diameter, in a chert or chalcedonic matrix. The intraclastic BIF are commonly internally unbanded and lenticular. Locally, abundant algal filaments and microfossils occur interbedded with pelletal and oolitic ferruginous cherts. Carbonate units, commonly dolomitic and stromalitic, occur near both the base and the top of the iron formations. Gypsum pseudomorphs, indicating an evaporitic environment (sabkha), are also present locally in basal dolomites.

*Glengarry Sub-basin*: most of the Glengarry Sub-basin in the west is occupied by the Glengarry Group, which consists of thin (~1000 m) shelf facies in the south, and a thicker (up to 6000 m) trough facies in the north. Unconformably overlying this, and infolded with it, is the Padbury Group, a 2 km thick sequence of shale, granular BIF, minor carbonate, sandstone, and conglomerate (Fig. 3-20) (Bunting 1986). The Padbury Group is correlated with the Earaheedy Group to the east.

Nabberu BIF falls in the age range 1.8–1.7 Ga (Bunting 1986) (Fig. 3-20). It is clearly younger, by about 700 Ma, than Hamersley BIF (~2.5 Ga) (Plumb and James 1986). Furthermore, as stated above, the Glengarry Group at age ~1.9 Ga (Plumb 1985, Bunting 1986) is broadly coeval with the Wyloo Group to the north and may represent a facies transition.

## 3.13.2 NORTH AUSTRALIAN CRATON

Early Proterozoic orogenic domains comprise a mosaic of linear geosynclinal belts which were deformed, metamorphosed, and finally cratonized during the tightly constrained, terminal-era Barramundi Orogeny (1890–1850 Ma). The domains are of similar age, rock type, deformation and sedimentation–orogenesis (Figs 1-5h, 3-21, 3-22; Table 2-16, column 2) (Page et al 1984, Etheridge et al 1987, Wyborn 1988). Component depositories are commonly interpreted as essentially ensialic structures. The assemblages display a common three-part development: (1) lower epiclastic–mafic volcanics interpreted to represent a rift phase of crustal extension; (2) middle finer-grained clastics with carbonates and BIF, representing the sag phase of basin subsidence; and (3) upper turbidite facies marking the syn-orogenic (Barramundi) phase. Greenschist facies metamorphism prevails with local amphibolite and even low-granulite facies. Both mafic and felsic syn- to post-orogenic intrusions are common.

The *Halls Creek Inlier* is exposed over about 45 000 km$^2$ to the east and southeast of the Kimberley Basin (Hancock and Rutland 1984, Plumb et al 1985, Page and Hancock 1988, Ogasawara 1988). In brief, a thick early Proterozoic supracrustal succession (Halls Creek Group) was later intruded by a variety of mafic–ultramafic and felsic igneous rocks and converted to high grade metamorphic rocks, collectively known as the Lamboo Complex. These rocks are all unconformably overlain by platform cover of the mid-Proterozoic Kimberley and younger basins. No older basement has yet been identified, although the presence of such basement is postulated from Nd isotopic data (Page and Hancock 1988).

Relationships in the Halls Creek Inlier are based on a fourfold tectonic zonation on the basis of structure and metamorphic grade, with the N-trending boundaries commonly marked by major faults (Hancock and Rutland 1984).

Recent precise U–Pb zircon dating (Page and Hancock 1988) indicates that the full cycle of events, from initial sedimentation to final deformation, metamorphism and intrusion, occurred in the interval 1856 to 1845 Ma, this marking the timing of the Barramundi Orogeny in this area.

The sub-elliptical *Pine Creek Inlier*, about 66 000 km$^2$ in exposed area, abuts the late Archean Litchfield–Rum Jungle and Nanambu complexes. The inlier represents part of a much larger area of early Proterozoic accumulation, which extended to the south and east, now concealed by younger rocks. A depositional sedimentary–volcanic domain (Pine Creek Geosynclinal Sequence) about 10 km thick and of age 2200–1880 Ma, rests on a ~2.5 Ga granitic basement. This deposition was succeeded by tectonic–intrusive episode, with the main activity at 1870–1850 Ma. Subsequent rift-related, late orogenic, felsic volcanic and volcaniclastic rocks are themselves unconformably overlain by platform cover sandstones of age ~1650 Ma. Pine Creek depositional environments ranged form fluviatile to intertidal to neritic, but flysch sediments in the upper part indicate that subsidence rates increased towards the end of sedimentation (Plumb et al 1981, Needham et al 1988).

The Pine Creek Inlier is of special economic importance due to the presence of large, high grade

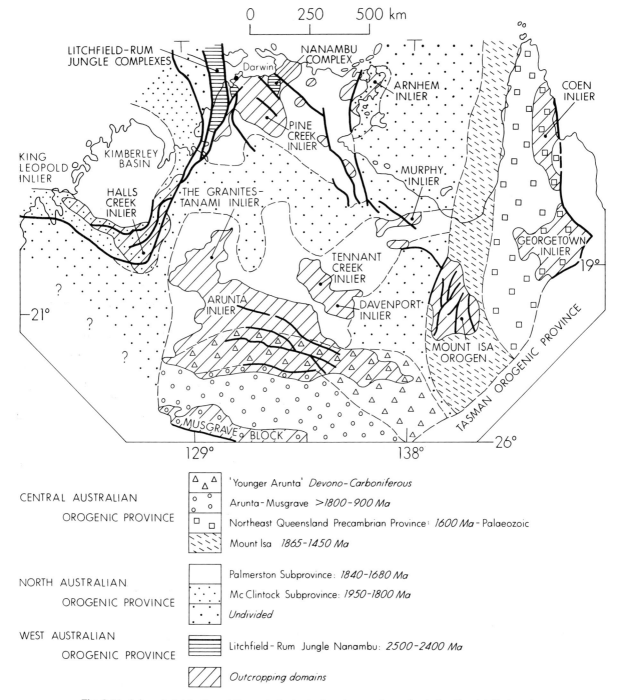

Fig. 3-21. Inferred distribution of the main tectonic domains, northern Australia. For detailed explanation of symbols see original publication. (From Plumb et al 1981, Fig. 4.4, and published with permission of the authors and of Elsevier Science Publishers).

uranium deposits. About 2500 t of $U_3O_8$ have been produced from the Rum Jungle and South Alligator fields, while measured reserves total about 250 000 t. The U, U–Au and smaller Au–Pt–Pd deposits are stratabound in brecciated carbonaceous pelitic rocks, associated with carbonate rocks after massive evaporites. They are situated immediately above the basement domes of the Rum Jungle and Nanambu complexes, and are attributed to complex epigenetic concentrations following primary sedimentation of the metals (Mernagh et al 1994).

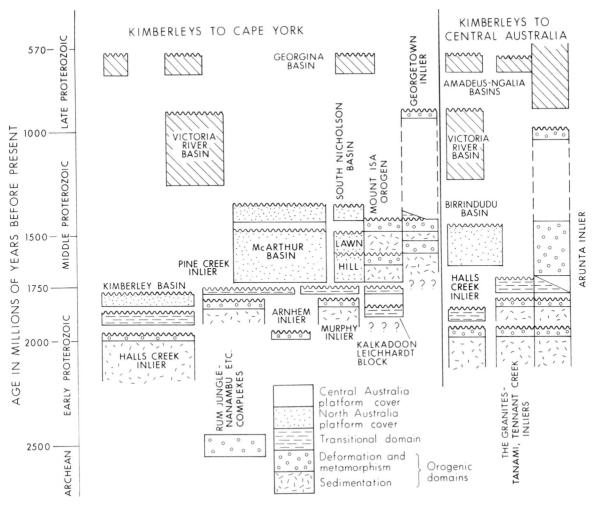

**Fig. 3-22.** Generalized correlations and critical isotopic age constraints of the principal Precambrian platform covers of Australia. (After Plumb 1985, Fig. 3, and reproduced with permission of the authors and of Elsevier Science Publishers).

Small base-metal deposits (Ag–Pb–Zn, Cu, Au) and large supergene Fe deposits also occur at the same stratigraphic level as the basement-associated uranium deposits.

*Other inliers* with similar lithostratigraphic and tectonic relationships include Murphy Tennant Greek–Davenport, Arunta, Georgetown inliers and Mount Isa Orogen (Figs 3-21, 3-22).

### 3.13.3 SOUTH AUSTRALIA

Early Proterozoic sedimentary rocks of South Australia, mainly concentrated in the Gawler and Willyama domains, are notable for their relatively shallow-water, mature and uniform character over wide areas (Fig. 3-23; Table 2-16, column 4). In the lower parts of the sequences, highly complex superposed deformations and high grade metamorphism have obscured regional correlations (Rutland et al 1981, Hobbs et al 1984, Fanning et al 1988).

### Gawler Domain

The Gawler Domain (Craton) is a NNW-trending elliptical mass, about 700 km by 250 km in dimension. To the east, a series of isoclinally folded early Proterozoic metasediments (Hutchison Group), together with granitoid gneiss (Lincoln Complex), form a N–NE-trending orogenic belt, up to 125 km wide, that is characterized by prominent mylonite zones.

The *Hutchison Group*, a mixed clastic and chemical metasedimentary association, is notable for its relatively shallow-water, mature and uniform character over wide areas (Cooper et al 1976, Rutland et al 1981, Fanning et al 1988, Parker et

**Fig. 3-23.** Main subdivisions of the Precambrian geology of South Australia. (From Rutland et al 1981, Fig. 5.1, and reproduced with permission of the authors and of Elsevier Science Publishers).

al 1988). Basal quartzites and pelitic schists are transitional up to a sequence of mixed chemical and semi-pelitic metasediments, ranging from massive dolomite and associated BIF and chert to quartzo-feldspathic gneiss and schist. This sequence is well developed in the *Middleback Ranges*, where iron ore has been mined for over half a century.

Deformation and igneous intrusion (1.8–1.6 Ga) resulting in amphibolite–granulite facies metamorphism, has greatly modified the apparent thickness of the Hutchison succession but it is estimated to be about 1500–2000 m. The age of Hutchison basin development, including deposition of the Middleback BIF, is estimated to be 1960–1847 Ma (Fanning et al 1988). The subsequent Kimban Orogeny, characterized by plutonism (gabbro–norite and I-type granitoids), deformation and upper amphibolite–granulite facies metamorphism, occurred at ~1840 Ma (Fanning et al 1988). Intrusion of post-Kimban Orogeny granitoids and volcanics completed the cratonization of the Gawler Craton by ~1.45 Ga (Parker et al 1988).

## Willyama Domain

In the Willyama Domain (Block), 400 km to the east (Fig. 3-23, Table 2-16, column 4) a grossly similar succession is present. In the Broken Hill mines area (block) at the centre, it has been shown that the lower part of the succession was inverted so that the major fold structures are downward facing. The Broken Hill orebody and enclosing host rocks have been displaced during nappe formation from their original site of deposition (Key references: Rutland et al 1981, Hobbs et al 1984, Wright et al 1987, Stevens et al 1988, Barnes 1988, Plimer 1994.)

The *Broken Hill Block*, comprising amphibolite to granulite facies metamorphic rocks of supracrustal origin, outcrops over 3000 km². It contains the well-known Broken Hill massive Pb–Zn–Ag deposit which, with over $120 \times 10^6$ t produced since 1883, represents one of the world's largest known base-metal sulphide ore bodies. In addition occur numerous small stratiform, strata bound and vein mineral deposits (Barnes 1988, Parr 1994).

The Broken Hill Block is considered to be composed originally of psammites and pelites of turbidite facies, with lesser felsic and mafic volcanic rocks, and minor zones of chemically unusual rocks which are rich in one or more of the following: Fe, Mn, Ba, Na and Zn. The Willyama Supergroup with its eight groups (suites) has an estimated total thickness of 7–9 km, with neither top nor basement exposed. The preserved sequence is estimated to be ~5200 m thick. No unconformities are recognized within it. Within the immediate mines area, the Willyama Supergroup contains three main stratigraphic suites (3–5 in ascending order as shown in Fig. 3-24.

Within Suite 4, the main ore-bearing 'Lode horizon' and its associated, 'Potosi' and 'Parnell' gneisses can be traced throughout the Broken Hill Block. Tourmaline-rich rocks are common in the Broken Hill Group including those that host the main Pb–Zn–Ag ores. These gneisses have been traditionally considered to represent metavolcanic rocks (especially metadacites) closely associated with BIF and with the Pb–Zn mineralization (Page and Laing 1992); however, they have also been interpreted as meta-arkoses and as rift sediments, volcanics and hot-spring precipitates (Plimer 1994, Plimer et al 1994), with significant meta-evaporite components (Cook and Ashley 1992, Slack et al 1993). The stratigraphically overlying sillimanite gneisses are generally regarded as mature non-volcanogenic metasediments.

*Ore genesis*: the Broken Hill Pb–Zn ore body is commonly ascribed to a volcanic exhalative origin in a deep water environment. The distinctive tourmalinites are ascribed to the same submarine hydrothermal processes as the main Pb–Zn–Ag lodes, with the boron derived from underlying evaporitic borates (Slack et al 1993). However, Wright et al (1987) interpret the succession as a series of progradational wedges interfingering with marine transgressive cycles deposited in a gradually deepening basin. The Pb–Zn mineralization, which is hosted by shallow-marine sands, was accordingly generated by compactive expulsion of metal-bearing brines during accumulation of the sedimentary pile. These conflicting interpretations demand further studies (Barnes and Willis 1989).

Recent ion microprobe (SHRIMP) studies focused on amphibolite-grade volcanic rocks of lower strain away from Broken Hill that correlate with the higher grade equivalents (Potosi gneiss) that contain the main Pb–Zn–Ag lodes. This has defined a major magmatic population of 1690 Ma, considered to be the age of zircon crystallization and eruption of felsic volcanics in this part of the Broken Hill Group. Additionally, zircons in the highest grade ore-bearing granulite facies rocks provide a pooled age of 1600 Ma, which determines the timing of this granulite facies event. Isotopic resetting at ~480 Ma reflects the Delamerian (early Paleozoic) orogenesis (Page and Laing 1992). The depositional age at Broken Hill of 1690 Ma provides for the first time, a firm chronologic tie between the Broken Hill Group on the one hand, and the sequences enclosing the major stratiform Pb–Zn–Age ore bodies at Mount Isa (1670 Ma) and McArthur River (1690 Ma) on the other hand. The Broken Hill ore body would, accordingly, be reclassified as mid-Proterozoic in age.

### 3.13.4. BARRAMUNDI OROGENY

The early Proterozoic terrains of Australia were affected in large part by the Barramundi Orogeny, widespread isochronous orogenic event at 1880–1840 Ma (Etheridge et al 1987, Wyborn 1988). It is particularly well defined in the Mount Isa and Pine Creek orogenic domains. The Barramundi Orogeny was immediately preceded by a widespread basin-forming episode, which, as described

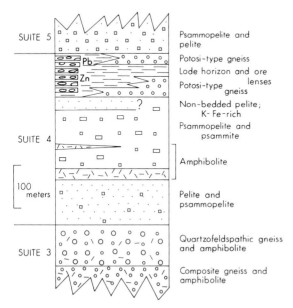

Fig. 3-24. Stratigraphic sequence in the main part of the Broken Hill mines area. The bar scale is very approximately 100 m. (From WP Laing in unpublished report GS 1979/062, Geological Survey of New South Wales, from Rutland et al 1981, Fig. 5.5, and reproduced with the permission of the authors, of WP Laing, and of Elsevier Science Publishers).

above, apparently resulted from local extension of Archean basement. The orogeny itself is characterized by a distinctive terminal igneous event of large magnitude involving the emplacement of rocks of remarkably consistent chemistry.

The orogeny was followed by further sedimentation in most provinces, but the style and timing of the younger episodes of basin formation varied between provinces (Etheridge et al 1987). Associated with the orogeny, or immediately postdating it, was a major magmatic episode represented by extensive batholith intrusions and co-magmatic, largely ignimbritic extrusives. The extrusives unconformably overlie rocks of the preceding sedimentary cycle. These late stage volcanic rocks and associated sediments appear to be largely related to rifts formed during the terminal phase of the orogeny (Needham et al 1988).

## 3.14 ANTARCTIC CRATON

Early Proterozoic (2400–1540 Ma) events are comparatively uncommon in the Antarctic Craton (Figs 1-3i, 1-5i). Granulite facies metamorphism has been recorded in the gneisses at Lutzow–Holm Bay, Enderby Land, Prince Charles Mountains and the Rayner Complex. In Dronning Maud Land, high grade metavolcanic and sedimentary sequences, including BIF, are cut by mafic intrusions dated at 1700 Ma (Bredell 1982). Whether or not these represent a number of intracratonic geosynclines, as present in other Gondwanaland platforms (e.g. Australian), remains to be established (James and Tingey 1983, Black et al 1987, Black 1988).

An extensive early Proterozoic mobile belt, locally known as the Rayner Complex, occurs in East Antarctica. Much of this belt is the product of early Proterozoic (~2.0–1.8 Ga) juvenile crust formation. Melting of this crust at ~1500 Ma produced the felsic magmas from which the dominant orthogneisses of this terrain were subsequently derived. Deformation and transitional granulite–amphibolite facies conditions produced syn-tectonic granitoids at ~960 Ma. Subsequent felsic magmatism occurred at ~770 Ma (Black et al 1987, Black 1988). Also, complex Proterozoic deformation and magmatism are defined in the Colbeck Archipelago of the Rayner Complex (White and Clarke 1993).

Two iron metallogenic subprovinces are defined in East Antarctica based on scattered exposures and glacial erratics—'iron-formation subprovince' and 'iron oxide vein subprovince'. They may be early Proterozoic or Archean in age but are summarized here for convenience (Ronley et al 1983).

The 'iron-formation subprovince' extends from Western Wilkes to Western Enderby Land and contains scattered exposures and erratics of BIF (jaspilite). Thick exposures of BIF occur in Prince Charles Mountains. The best exposures occur at Mt Ruker where abundant greenschist grade jaspilite beds as much as 70 m thick alternate with slate, siltstone, ferruginous quartzite, schist and metagabbro/volcanic rocks. The main sequence contains laminated-type iron-formation up to 400 m thick. Enclosing sequences contain lesser jaspilite. Concealed anomalies extend 120 and 180 km west from Mt Ruker under the ice, suggesting that the BIF deposits correspond in size with large-scale equivalents found elsewhere in the world. Minor jaspilites are found in Vestfold Hills. Bedrock exposures in the same area contain unspecified volumes of BIF of apparent Archean age. Small masses of magnetite occur in Larsemann Hills; and magnetite-bearing rocks, including BIF, are reported in Endersby Land, western Wilkes Land and Bunger Hills.

The smaller 'iron oxide vein subprovince' is centred on western–central Queen Maud Land where occur abundant iron and other metals (e.g. copper) disseminated in garnet and quartz–magnetite veins, garnet and pyroxene–magnetite veins and magnetite-bearing stockworks.

# Chapter 4

# Mid-Proterozoic Crust

## 4.1 INTRODUCTION

### 4.1.1 DISTRIBUTION

The area of exposed and buried (total) mid-Proterozoic crust is $24.6 \times 10^6$ km$^2$ (Table 4-1) or 23% of Earth's total Precambrian crust (Table 1-2). Of this total, North America accounts for 26%, Australia, South America and Antarctica each 15–18% ($\Sigma$49%), Asia and Africa each 8–11% ($\Sigma$19%), and Europe and India each 2–4% ($\Sigma$6%). The four Atlantic continents (Europe, two Americas, Africa) account for 54%, the three Laurasian continents (Asia, Europe, North America) 40%, and those of the reconstructed Gondwanaland 60% (Fig. 4-1).

The calculated area of exposed only mid-Proterozoic crust is $7.1 \times 10^6$ km$^2$ or 23% (coincidentally the same as above) of Earth's exposed Precambrian crust. Of this, South America accounts for 27%, North America 19%, Asia, Australia and Africa each 12–16% ($\Sigma$41%), and Antarctica, Europe and India each 3–6% ($\Sigma$13%). The four Atlantic continents account for 61% of the total, the three Laurasian continents 39%, and the five Gondwanaland continents 61%.

In brief, mid-Proterozoic crust is broadly though unevenly distributed across the continents, with high proportions in the two Americas, Australia and Antarctica, and low proportions in Europe, Asia, India and Africa.

### 4.1.2 SALIENT CHARACTERISTICS (Table 4-2)

(1) Transcontinental anorogenic belts, products of taphrogeneis (i.e. formation of rift phenomena) of large stable crustal blocks, and replete with the anorogenic 'trinity'—(a) anorthosites and layered gabbros; (b) mangerites (i.e. pyroxene monzonite), including charnockitic diorite to syenite suites; and (c) rapakivi granites with bimodal basalt–rhyolite associates—are a salient characteristic of mid-Proterozoic crust. The largest continuous belt, located in North America, extends from California northeastward for 3000 km to Labrador on the Atlantic coast and is 600–1000 km broad; somewhat older, now rifted extensions occur in southern Greenland and the Baltic Shield. In the southern continents a corresponding, though less well-defined, anorogenic belt extends southeastward across South America with possible rifted extensions in southern Africa and eastward. Of these anorogenic belts, the 3000 km long North American segment is the best documented and most impressive. The activity there occurred in the interval 1.5–1.0 Ga, but over 70% is tightly constrained at 1.5–1.4 Ga. The scale of emplacement of these anorogenic magmatic bodies as well as their lithologic–geochemical consistency are astonishing and, as known, occurred on this scale only once in the history of the continental crust.

*Table 4-1.* Distribution of preserved mid-Proteozoic crust by continent.

|  | Total crust (%) | Exposed crust (%) |
|---|---|---|
| Asia | 10.6 | 15.8 |
| Europe | 3.8 | 3.7 |
| North America | 26.1 | 19.2 |
| South America | 15.8 | 26.9 |
| Africa | 7.9 | 11.6 |
| India | 2.5 | 3.6 |
| Australia | 18.1 | 13.6 |
| Antarctica | 15.2 | 5.6 |
|  | 100.0 | 100.0 |

Total (exposed + buried) crust = 24 599 000 km$^2$
Exposed crust = 7 071 000 km$^2$ (31% of total)

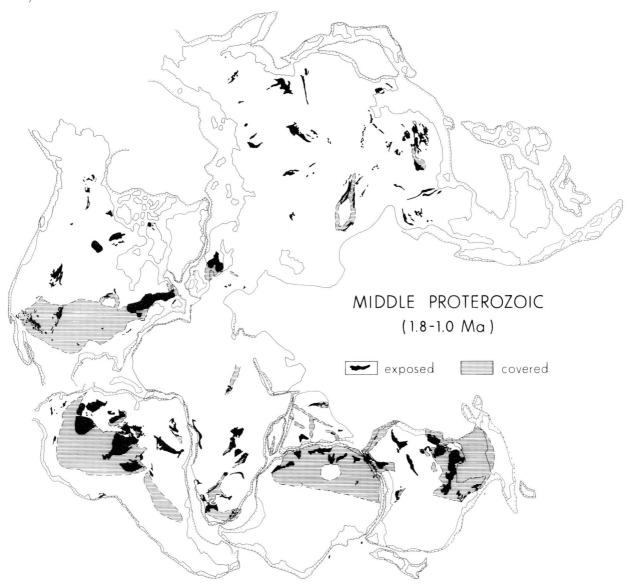

Fig. 4-1. Distribution of exposed and covered (i.e. sub-Phanerozoic) mid-Proterozoic crust in the Briden et al (1971) pre-drift Pangean reconstruction of the continents.

Rapakivi granites and extrusive associates predominate by far in the anorogenic belts. They represent mainly primary melts derived from the fusion of contemporary lower crust. Closely associated are the hundreds of individual round to ovoid anorthosite bodies, composed of predominant Ca-rich plagioclase, especially concentrated in the Grenville Province of the Canadian Shield. Representing juvenile mantle melts, the anorthosites, as known, although apparently emplaced as shallow crustal sheets, now bear the metamorphic imprint in the form of granulites of a deep crustal setting, marking a complex tectonic pattern involving repeated crustal-scale vertical motions.

The generation of such crust-derived granitoid rocks, notably rapakivi granites, under anorogenic conditions on this scale is attributed to linear thermal doming in the mantle, in some cases the prelude to continental rifting. The linearly associated, mantle-derived anorthositic and mangeritic magmas may have played an active role in generating the necessary thermal environment for fusion of the superjacent lower continental crust to produce the coeval felsic magmas. Furthermore, the presence of such large and numerous anorthosites required exceptionally stable mantle-crust environments to permit feldspar concentration on such a grand scale. This anorogenic 'trinity', in brief,

*Table 4-2.* Salient mid-Proterozoic characteristics.

| | |
|---|---|
| Distribution | Highest proportions in North America; moderately high in South America, Australia and Antarctica; uniform elsewhere. |
| Composition | Pericratonic fold belts variably composed of (1) low to medium grade, outward younging, juvenile plutonite-metasupracrustal, accretions, and (2) high to medium grade reworked basement with complex infrastructures. Late-stage magmatic (intrusive–extrusive) complexes with major Sn mineralization.<br>Craton-scale taphrogenic–epeirogenic rifting with horst–graben movements and selective platform subsidence, typically with basal redbed-bimodal volcanic infill transitional up to terrigenous blankets and/or thick marine pelite–carbonate sequences. Late-era mid-continent rift systems, the sites of immense flood basalt accumulations with gabbroic and carbonatite intrusions.<br>Mid-era (1.5–1.4 Ga) transcontinental anorogenic belts characterized by (1) dominant crust-derived rapakivi granites and basalt-rhyolites, (2) mangerites with charnockites and (3) mantle-derived anorthosites and layered gabbros.<br>Late stage (1.5–1.0 Ga) epicratonic rifting and continental fragmentation with development of sandstone–siltstone–carbonate shelf sequences. |
| Deformation | Fold belts with prominent foreland imbricates marked by complex stacking of allochthonous slices and inclined basement ductile shear zones. Basement domains repeatedly and intensely deformed as marked by mylonites, granulites and contorted patterns of superposed folds. Major imprint of end-of-era orogeny (~1.0 Ga). |
| Mineralization | Stratiform Pb–Zn sulphides; shale-hosted Cu; unconformity-related U (–Cu–Ni); Ti. |
| Paleoenvironment | Stable oxygenic environment (~4% PAL) capable of sustaining aerobic micro-organisms. Dominant continental flux signature. |
| Paleobiology | Cyanobacteria and late development of primitive eucaryota. |
| Paleotectonics | Repeated crustal extensions with unique mid-era anorogenic development of thick stable crust. Widespread platform taphrogenesis–epeirogenesis, including both craton elevations and basin subsidence reflecting thermal plume activities. Occasional plate collisions with pericratonic accretions. |

PAL = present atmospheric level

would relate well to, without proving the existence of, the proposed long-sustained, comparatively stable Pangean-type supercontinent in mid-Proterozoic time.

(2) Aulacogen-style taphrogenic–epeirogenic rifting with consequent block movements leading to platform subsidence also occurred on a grand continental scale. In this way, orthogonal or parallel rifting of the continental crust affected entire composite cratons with widespread horst-and-graben movements. The resulting rifts are typically infilled with arenites and bimodal volcanic associates overlain by thick, pluton-intruded terrigenous blankets or marine pelite–carbonate sequences, locally mineralized with economically significant Pb–Zn–Ag and/or Cu–U–V–Cd sulphides. Epicratonic redbed basins include important U–(Cu–Ni) ore deposits. In some cases, rifts developed into major continental depressions (syneclise), sites of continental–marine transgressions. Still others drifted apart to form discrete continents (plates).

Closely associated but generally younger continental rift patterns are characterized by alkaline ring structures, including syenites and nepheline syenites, with associated radioactive elements and even Al-rich cryolite concentrations, as in Greenland.

Still other spectacular mid-continental rifts (e.g. North American) became the sites of immense flood basalt accumulations with locally mineralized (Cu) arenite–pelite intercalations.

(3) Major pericratonic mobile belts, up to 3000 km long and 1300 km wide characterize some cratons, notably North and South America. Included are outward younging, mainly juvenile early-era (1.8–1.5 Ga) continental accretions, typically composed of calcalkalic plutonites with bimodal volcanic–clastic turbidite associates. Other younger gently to sharply intersecting continental mobile belts of equivalent or even greater stature (e.g. Grenville), products of terminal era (1.0 Ga) continent–continent collision, incorporate substantial reworked older crust. Such pericratonic mobile belts bear witness to on-going plate-tectonic movements in mid-Proterozoic time.

(4) Mid-Proterozoic terrigenous deposits include abundant, widespread redbeds, signalling the presence of a uniform oxidizing atmosphere. The oldest substantial redbeds at 1.8–1.7 Ga, found in south-

ern Africa (Matsap), heralded the arrival of thereafter-common oxidized continental accumulates.

(5) Typical mid-Proterozoic mineral deposits are shale-hosted Pb–Zn–Ag, stratabound Cu, unconformity related U–(Cu–Ni) deposits, and Sn–W-radioactive-element-bearing alkaline intrusions.

## 4.2 CATHAYSIAN CRATON

### 4.2.1 INTRODUCTION

Pre-Sinian (i.e. mid to late Proterozoic) (1.85–0.85 Ga) strata are broadly distributed across the Cathaysian Craton both as cover sequences on deformed basement and as inliers in adjoining fold belts. The three relevant systems—Changcheng, Jixian and Qingbaikou (Fig. 1-3a)—are not readily separable hence they are treated as a pre-Sinian mid-Proterozoic composite (Fig. 4-2). Pre-Sinian strata extend (1) discontinuously along the northern flanks of the Sino–Korean and Tarim cratons, and sparsely southward therefrom to the southern flanks; and (2) broadly interspersed with Sinian assemblages across the Yangtze Craton (southern China), with particular concentrations along the southeastern and western flanks. Pre-Sinian Proterozoic inliers are also common in the adjoining Paleozoic fold belts.

### 4.2.2 LITHOSTRATIGRAPHY

Three prominent pre-Sinian strata-types are present. *Type 1*: carbonate formations of stable cratonic regions. This, the dominant facies, is widespread in the two northern cratons where platform dolostones of littoral to shallow-marine facies predominate, with sparse clastic and argillaceous associates. *Type 2*: volcanosedimentary flysch

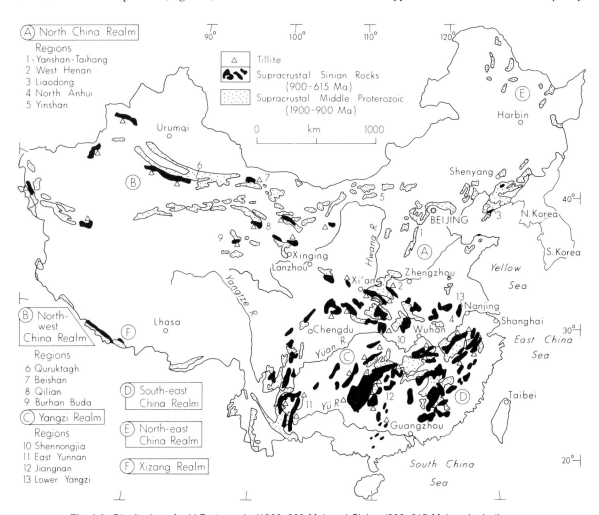

**Fig. 4-2.** Distribution of mid-Proterozoic (1900–900 Ma) and Sinian (900–615 Ma) rocks in the pre-Sinian Cathaysian Craton. (Modified from Chen et al 1981, Fig. 1, and reproduced with permission of the authors).

associations of mobile zones, as notably developed along the southeastern and western flanks of the Yangtze Craton. Products of rapid deposition, these include plentiful spilite–keratophyre lava flows and tuffs, with occasional ophiolite-bearing mafic–ultramafic assemblages. Prominent angular unconformities are common products of tectonically active source areas. Regional low grade metamorphism prevails. *Type 3*: limited intervening sandy–argillaceous and carbonate formations of the shelf–flysch transitional zones.

### Sino–Korean and Tarim Cratons

In *Yanshan Mountains*, North China (A1 in Fig. 4-2), the most complete section of pre-Sinian Proterozoic strata is found near Beijing on the southern slope of the Tanshan Mountains. Some 9700 m of clastic–carbonate strata are exposed in a continuous and regular succession in an E-trending subsiding basin, unconformably overlying Archean basement. The section is divided into three parts (Fig. 1-3a): (1) lower Changchengian (4266 m); (2) middle Jixianian (4936 m); and (3) upper Qingbaikou (500 m). Stromatolite assemblages and acritarch remains are common throughout.

The Baiyun Obo REE–Fe–Nb deposits, located in Inner Mongolia (41.42°N, 109.52°E) at the northern margin of the Sino–Korean Craton, contain Earth's largest known REE concentrations. The ore deposits occur in large tabular bodies and lenses in steeply dipping cherty dolomite, part of the 2000 m-thick quartzite–carbonate–shale Baiyan Obo Group. The deposits extend 18 km from east to west with a width of 2–3 km, and are mainly composed of magnetite, hematite and rare earth minerals, such as bastnaesite, monazite and aechynite. REE emplacement probably took place at 1400–1500 Ma and not later than 1260 Ma (Nakai et al 1989, Drew et al 1990, Conrad and McKee 1992). Jixinian strata at Wafangzi, 400 km east of Beijing, include large manganese deposits hosted in both red shale–silty limestone (oxide facies) and black shale (carbonate facies) (Delian et al 1992).

To the south in western Henan region (A2 in Fig. 4-2), thick sequences (+6 km) of terrestrial volcaniclastics including prominent ultra-potassic (up to 10% $K_2O$) trachytes, overlain by shales, glauconitic sandstones and sandy dolostones are transitional southward to marine carbonates, this composite assemblage representing an aulacogen-induced shelf opening southward to the Qinling paleomarine basin.

A peculiar feature of the Eastern Border sequence in Liadong Peninsula to the northeast (A3 in Fig. 4.2) is that only the uppermost Qingbaikouan strata are present, here unconformably overlying Archean to early Proterozoic basement. The Qingbaikouan section in Liaodong Peninsula, 2600 m thick, is about equally divided between lower rapidly deposited conglomerates–sandstones, and transgressively overlying littoral to shallow-marine glauconitic sandstones and variegated shales.

Far to the west, the well known Quruktagh section (B6 in Fig. 4.2) is generally correlated with the uppermost Changchengian, Jixianian and Qingbaikouan groups on the basis of stromatolites. The section (~6000 m) is subdivided into three groups: lower schists, phyllites and quartzites; middle dolostone with schist and quartzite: upper dolostone and crystalline limestone with thin intercalations of schist, phyllite and quartzite.

Similar lithostratigraphic sections with occasional migmatite–gneiss associates occur in the Kunlun and Qilian–Qinling mountains on the southern flanks of the Tarim and Sino–Korean cratons.

### Yangtze Craton

Three stratotypes are represented: a paracover type in the interior and northern parts of the craton; a miogeoclinal type in the southwestern part; and a eugeoclinal or island-arc type in the southeastern part along the Jiangnan Oldland.

The centrally disposed *Western Hubei region* (C10 in Fig. 4-2) contains an unmetamorphosed transitional carbonate-bearing paracover sequence about 6600 m thick, subdivided into three disconformable parts: lower sandstones–conglomerates–phyllites–dolostones; middle sandstones–slate–dolostone with locally abundant spilitic tuffs and basaltic flows; upper slates and cherts with common spilites and tuffs and uppermost dolostones. Minimum age is provided by diabase dykes dated at 1332 Ma and 950 Ma (Yang et al 1986).

In the *Kham–Yunnan (Dam) Oldlands* in the southwest (C11 in Fig. 4-2), the miogeoclinal-type section is also subdivided into three parts: lower (~2500 m) slates and siliceous dolostone with marly interbeds; middle slates, siltstones, rhythmic sandstones, andesitic–basaltic tuffs and stromatolitic carbonates; and upper unconformably overlying conglomerate–sandstones, carbonaceous slates and siltstones, and intermediate–felsic volcanic rocks. A number of dates cluster around 1050 and 850 Ma, which provide the group boundaries (Yang et al 1986).

To the southeast, the eugeoclinal lithotype in the NE-trending *Jiangnan Oldlands region* (C12 in Fig. 4-2), extending for some 1500 km along the southeastern margin of the craton, constitutes a complex, volcanic-charged mobile belt. The lower group (10 000 m thick) comprises flyschoid types with spilite–keratophyre extrusive and subvolcanic associates. The unconformably overlying group, a similar flyschoid succession, also at least 10 000 m thick, is variably composed of red, coarse-grained clastic rocks, green flysch, and spilite–keratophyre volcanic rocks. To the southeast the coeval lithologically similar so-called 'green facies' is over 10 000 m thick.

Andesites in the middle part of the overlying group are dated at 950 Ma. The subjacent older group is intruded by granite which is dated at 1065 and 1109 Ma (Yang et al 1986).

Facies equivalents to the southeast (south of Guangzhou) beyond the margin of the Yangtze Craton were metamorphosed during Caledonian folding to form complex gneissic associations now dispersed in massifs (inliers).

Finally, in the Himalaya Fold Belt to the southwest, coeval migmatized schist and gneiss with cherty carbonates are broadly correlated with the pre-Sinian platform cover.

## 4.3 SIBERIAN CRATON

### 4.3.1 INTRODUCTION

The 1.8–1.0 Ga rock association considered here encompasses the post-Stanovoyan association of Akitkan, Bourzianian and Yurmatinian systems (Fig. 1-3b). As such it is at variance with the official Russian classification with its 1.6 Ga cut-off, which thereby relegates the Akitkan, a post-Stanovoyan molasse, to the early Proterozoic era.

Akitkan geology has been reviewed by Salop (1977, 1983) and Khain (1985), and the Riphean–Vendian assemblages by Chumakov and Semikhatov (1981).

Mid-Proterozoic (i.e. 1.8–1.0 Ga) geology of both the Siberian and East European cratons is dominated by early taphrogenic structures, including aulacogens and pericratonic troughs, both characterized by bimodal volcanic–coarse clastic fill, typically transitional up to more widespread platform cover in response to subsequent platform subsidence (Fig. 4-3). The taphrogenic structures are typically located in tectonic depressions, including grabens, rift, aulacogens and troughs. Superimposed deformations resulted in highly complex structures; in particular, original steep to vertical faults were transformed into thrusts, and gentle folds variably deformed with local changes in orientation and vergence.

### 4.3.2. AKITKAN AND RELATED GROUPS

The felsic volcanic-rich Akitkan Group is developed mainly in an asymmetric trough that follows the northerly-trending boundary of the Baikal Fold belt (Western Pribaikalie) in contact to the west with the interior Siberian platform (Figs 4-3). The Akitkan trough extends north of Lake Baikal for more than 500 km, with a width of up to 20 km (Salop 1977, 1983). Taken together with the laterally equivalent sedimentary groups, these sedimentary–volcanic sequences, up to 60 km wide, can be traced northward for more than 1200 km along strike. In the Lake Baikal area, lower stratigraphic parts are bounded by the trough proper, whereas upper parts spill out and are bounded on the west by a series of high cataclastic late Riphean to early Paleozoic marginal faults and fractures. Akitkan volcanosedimentary rocks, which are of dominant continental affinity, everywhere overlie the basement with angular unconformity.

The Akitkan Group is subdivided into three mainly conformable parts: lower (600–1500 m) polymictic conglomerate, arkose, sandstone, tuff and siltstone, with quartz porphyry and amygdaloidal basalt intercalations; middle (up to 4500 m) grey to red quartz porphyry and trachyandesitic tuffs and ignimbrites, which alternate with sandstones up to 300 m thick, the entire sequence characterized by rapid facies changes, including volcanic wedge-outs; and upper grits themselves divided into (1) lower (2000–6000 m) grey to pink arkose, conglomerate, gritstone, siltstone and tuff with sheets of porphyry; and (2) upper (100–1300 m) grey to pink sandstone, gritstone and conglomerate with rare stromatolitic limestone.

Similar taphrogenic deposits occur in Eastern Siberia within the Ulkan Trough of the Uchur Depression located 200 km west of the Maya depression, itself located at the eastern margin of the Aldan Shield (Figs 1-5b, 4-3) (Salop 1983).

The three-part up-section fill, which unconformably overlies Archean basement, comprises basal conglomerate–sandstone upon Archean basement (200–450 m), median subaerial trachyandes-

Table 4-3. General stratigraphic scheme of the Precambrian of Russia.

| Eon | Chronostratigraphic units (boundaries in Ma) | Stratotypes |
|---|---|---|
| Phanerozoic | Lower Cambrian ($\mathcal{E}_1$) | |
| | ——— 570±20 ——— | |
| Proterozoic (PR) | Vendian | Vil'chitsy (Vil'chany), Volyn' and Valday Groups of western part of Moscow syneclise |
| | ——— (650–680)±20 ——— | |
| | Kudashian $R_4$ | Uk Formation, Krivaya Luka Group |
| | Karatavian (Upper Riphean) $R^3$ | Karatan Group |
| | ——— 1000±50 ——— | |
| Upper (PR$_2$) | Yurmatinian (Mid-Riphean) $R_2$ | Yurmatan Group |
| | ——— 1350±50 ——— | |
| | Burzyanian (Lower Riphean) $R_1$ | Burzyan Group |
| | ——— 1650±50 ——— | |
| Lower (PR$_1$)' | | Karelian Supergroup |
| | ——— 2600±100 ——— | |
| Archean (A) | | Lopian (Gimola Group) White Sea Complex |

After Chumakov and Semikhatov (1981)
'Pr$_1$ is divisible into older (2600-1950 Ma) Oudokan, and younger (1950-1650 Ma) Akitkan divisions.

ite–trachybasalt (250 m), and upper subaerial quartz porphyries with intercalated tuffs, ignimbrites and terrigenous clastics.

Other similar taphrogenic structures in Siberia occur in east Sayan and central Kazakhstan (Salop 1983).

## 4.3.3 BURZYANIAN GROUP (EARLY RIPHEAN)

The accumulation of craton cover, begun in Akitkan time, continued on a wider, yet still discontinuous scale in early to mid-Riphean times, to be succeeded only in Vendian time by continuous platform cover.

By Burzyanian ($R_1$) time (~1650 Ma), full cratonic conditions prevailed over practically the entire Siberian Craton, except the Turukhansk–Noril'sk zone in the northwest (Fig. 4-3). The principal craton boundaries were finally established at this time, with continued development in the western and southern parts respectively of pericratonic trough systems, including (northwest to southeast) the Turukhansk–Noril'sk (closed by late Riphean time), Yenisei, Eastern Sayan, Baikal–Patom Upland and Monogo–Amur systems, and (to the east) the Yudoma–Maya meridional trough, which separates the Aldan Shield from the Okhotsk Massif. In the north, the craton extended to the southern boundary of the Taymyr Fold Belt. The prevailing tectonic situation remains unclear in the northeastern and eastern boundaries due to superimposed tectonic events (Chumakov and Semikhatov 1981).

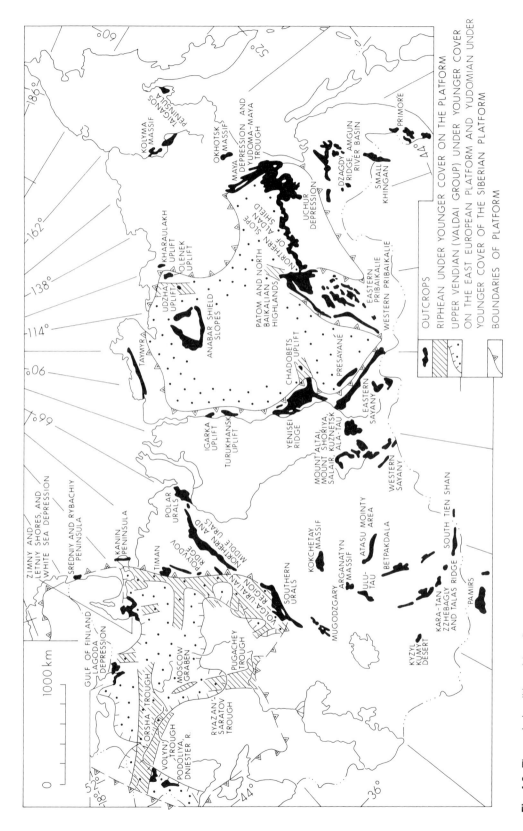

**Fig. 4-3.** The main localities of late Proterozoic (upper Vendian–Riphean) sequence in Russia and neighbouring territories (From Chumakov and Semikhatov 1981, Fig. 1, and reproduced with permission of the authors).

The structural pattern of the craton interior remained comparatively stable during the entire Riphean, most of it evolving by gentle and moderate subsidence, with the sedimentary response in the form of widespread, broadly repeated psammite–carbonate–siltstone cycles. The main positive element in the emerging craton was the Central Siberian Anteclise (Fig. 1-5b) (Khain 1985) which lacks any Riphean cover.

The lower boundary of the Riphean (i.e. Burzyanian Group) is drawn along the top of a characteristic late molasse of the Akitkan (Svecofennian) tectonomagmatic cycle.

### 4.3.4 YURMATINIAN GROUP (MID-RIPHEAN)

Yurmatinian ($R_2$) strata are much more extensive across Siberia than are underlying Burzyanian ($R_1$) strata. In the Uchur–Maya region of the eastern Aldan Shield (Fig. 4-3), Yurmatinian (Yurmata) strata include the Aimchan and Kerpyl groups, each corresponding to a large transgressive sequence composed of lower sandstone–siltstone and upper stromatolitic carbonates.

Lithologically similar Yurmatinian strata occur in the Patom structure north of Lake Baikal in the Yenisei Ridge to the west, and the Olenek Uplift to the north.

## 4.4. EAST EUROPEAN CRATON

Mid-Proterozoic plutonic rocks with sparse sediments characterize the centre–west Baltic Shield, notably the Transscandinavian Belt and Southwest Scandinavian Domain. Buried southward extensions reappear in the Ukrainian Shield and adjoining western margin of the buried parent craton. Elsewhere, buried mid-Proterozoic rocks occupy basement aulacogens of the interior craton, while pericratonic extensions crop out as median massifs in the adjoining fold belts, notably the Urals to the east (Figs 1-5c(i,ii), 2-3, 4-3, 4-4, 4-5).

### 4.4.1 BALTIC SHIELD

The pervasive Svecofennian Orogeny (1.9–1.8 Ga) (Fig. 1-3c(i)) was followed by a long period of erosion and cratonization, during which occurred: (1) extensive bimodal intracratonic rapakivi granite–gabbro/anorthosite magmatism (1.7–1.5 Ga), especially in the southern–central Svecofennian Province; and (2) concurrent emplacement of Gothian (sub-Jotnian) granite plutons and volcanic–clastic associates, especially in the Swedish part of the Svecofennian Province and adjoining Southwest Scandinavian Province (Domain).

To the west of the Svecofennian Province the 1600 km long Transscandinavian (Värmland–Småland) Granite Porphyry Belt, developed mainly by emplacement of granitoid intrusions of pronounced syenitic trend in the interval 1.75–1.60 Ga. This belt is separated by the Protogine Zone of major shearing and faulting from the Southwest Scandinavian Domain Province, formed mainly of Gothian granitoid plutons emplaced 1.7–1.5 Ga ago with the main crust-forming event of 1.7–1.6 Ga, followed by anotectic magmatism at 1.5–1.4 Ga. There was extensive reworking during the Hallandian (1.5–1.4 Ga) and Sveconorwegian (1.25–0.9 Ga) orogenies including crustal thickening, high-grade metamorphism and tectonic uplift in SW Sweden.

Finally, red Jotnian sediments were deposited ~1.4–1.3 Ga in downfaults and grabens upon the deeply eroded Svecofennian and older basement (Lundquist 1979, Gaál and Gorbatschev 1987, Johansson et al 1993, Johansson and Kullerud 1993).

### Rapakivi granites

Rapakivi massifs are distributed about the central Baltic Shield, mainly in the Svecofennian Belt of Finland and Sweden. Twelve main masses and numerous smaller units lie in a square about 800 km on the side extending westward from Lake Ladoga to Småland–Värmland, and northward from the Gulf of Finland to the Caledonides. The largest exposed rapakivi massif, about 120 km in diameter, lies east of Helsinki on the north shore of the Gulf of Finland (Fig. 4-4).

In southern Finland, rapakivi granite plutons cross-cut all the structures of the basement rocks but are themselves undeformed and non-metamorphosed. In addition to the classical porphyritic varieties, there are even-grained, granite porphyritic, quartz porphyritic, porphyry aplitic, aplitic and pegmatitic types of rapakivi. The oldest phases of rapakivi granites are dated at 1700–1640 Ma, and the youngest phases at 1530–1540 Ma (Simonen 1980, Rämö 1991); included are two major granite events at 1640 and 1630 Ma respectively. Globally, the rapakivi events represent one of Earth's largest pulses, if not the largest, of intracratonic magmatism, and may be a consequence of the

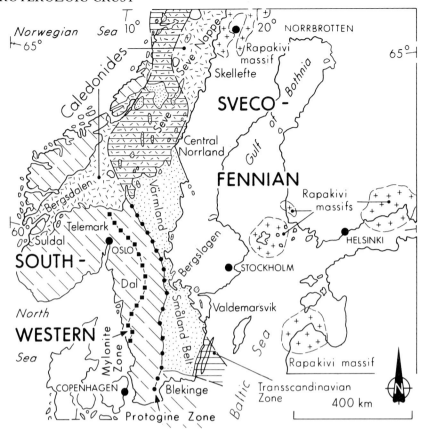

**Fig. 4-4.** The fundamental subdivisions of Precambrian crust in southern and central Scandinavia. The northern extension of the Varmland–Smaland Belt beneath the Caledonides is shown by stippled-hatched patterns. The Jotnian sandstone-dolerite cover has been omitted. Heavy dotting shows segregated granite massifs formed between 1700 and 1500 Ma. Additional rapakivi massifs may occur in the northern Baltic Sea. The Mylonite Zone and the Protogine Zone are the two largest fault zones in Southern Sweden. (From Gorbatschev 1980, Fig. 1, and reproduced with permission of the author).

preceding rapid growth of early Proterozoic continental crust (Vaasjoki et al 1991).

### Gothian (sub-Jotnian) complexes

Sub-Jotnian complexes include a variety of assorted volcanic, sedimentary and intrusive rocks, the latter including the Gothian granites. The Svecofennian Belt of Sweden is particularly rich in these complexes, locally difficult to distinguish from older Svecofennian components. Most fall in the age range 1650–1415 Ma (Lundquist 1979, Gorbatschev 1980, Gaál and Gorbatschev 1987).

### Transscandinavian (Småland–Värmland) Belt

The 1.75–1.6 Ga old Transscandinavian Granite Porphyry Belt, located at the western margin of the Svecofennian Province (Fig. 4-4), comprises both extensive supracrustal complexes and granitic–porphyritic intrusions that are characteristically unaffected by the penetrative deformation so common in Svecofennian rocks to the east (Gorbatschev 1980; Gaál and Gorbatschev 1987, Wilkström 1991).

The western boundary of the Småland–Värmland Belt is marked by a suture-like belt of tectonization called the Protogine Zone (also variously called Svecokarelian Front, Gothian Front and Småland suture). The Protogine Zone has been reactivated repeatedly at least from 1570 to 1000 Ma (Lundquist 1979).

### Southwest Scandinavian (Southwestern) Province

To the west of the Protogine Zone occurs a 500 km wide (east to west) gneissic domain, characterized by a long (~1.9–0.9 Ga) and complex Proterozoic history followed by Caledonian events (0.6–0.4 Ga) (Fig. 4-4). This domain is mainly formed of Gothian granitoid plutons, emplaced 1.75–1.50 Ga ago,

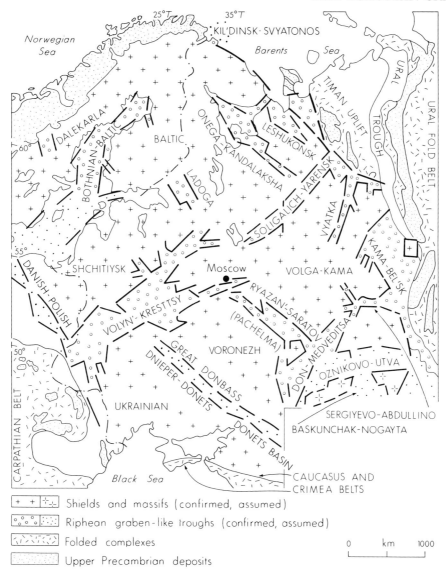

**Fig. 4-5.** Riphean structures of the East European Craton illustrating the distribution of the graben-like troughs (aulacogens) and intervening shields and massifs. (From Aksenov et al 1978, Fig. 1, and reproduced with permission of the authors).

that were successively reworked during the Hallandian (1.5–1.4 Ga) and Sveconorwegian (1.25–0.9 Ga) orogenies, together with local metasupracrustal sequences (Gorbatschev 1980, Gaál and Gorbatschev 1987, Eliasson and Schöberg 1991, Johansson et al 1993).

*Mylonite Zone*, a broad zone of strongly sheared and in part mylonitized rock, follows a N-trending sinuous course for about 450 km between the western Swedish coast and the Telemark region of Norway (Fig. 4-4) (Lundquist 1979). The Mylonite Zone is a composite feature, similar in that regard to the Protogine Zone. In the north, it takes the form of a steeply dipping, prominent lithotectonic boundary, whereas in the south it is replaced by a westward dipping thrust, dissipating eventually into several smaller faults, none of which represents drastic lithologic breaks.

### Jotnian sandstone

Molasse-type sequences of sandstone, arkose, conglomerate and shale of non-marine facies accumulated locally in several regions on the eroded surface of the Svecofennian crystalline complexes and their contained rapakivi granites (Simonen 1980). Jotnian (Dala) sandstones are preserved in fault-bounded basins of no great extent, often located not far from rapakivi centres. They are almost undisturbed and retain primary features—current-bedding, ripple-marks, rain prints and a prevailing red or brown colouring—which suggests accumulation in piedmont or flood-plain environments.

### 4.4.2 UKRAINIAN SHIELD

In the westernmost Volyn Block of the Ukrainian Shield occurs the much differentiated rapakivi granite–gabbroid Korosten pluton (Fig. 1-5c(ii)) (Semenenko 1972). The nearby somewhat older Osnitsk Complex (not illustrated) is a volcanoplutonic association composed of quartz porphyries, felsic volcanic tuffs (leptites), migmatites and granitoids.

To the north, a sequence of quartzite, sandstone, shales, porphyries and diabases fills the E-trending Ovruch synclinal trough. The Ovruch sequence is correlated with the Jotnian group of the Baltic Shield (i.e. 1.4–1.3 Ga).

Younger Riphean and Vendian deposits are absent on the Ukrainian Shield proper but transgressively overlie its slopes and fill the adjoining Dniester and Black Sea Downwarps, part of the Pripyat Trough and the Dnieper–Donents Aulacogen. On the Voronezh Massif, to the northeast of the Ukrainian Shield, the youngest basement rocks comprise mafic to felsic metavolcanic rocks, gabbroids and granitoids, the latter including granite- and syenite-porphyries, microcline granite and migmatites (Khain 1985). These rocks are 1450–1200 Ma old, thereby corresponding to the Gothian (sub-Jotnian) or Jotnian events of the Baltic Shield.

### 4.4.3 BURIED EAST EUROPEAN CRATON

Widespread drilling to basement across the East European platform has firmly established the general Riphean structure and stratigraphy (Fig. 4-5) (Aksenov et al 1978). The structural pattern is dominated by a remarkable orthogonal system of about equally NE- and NW-trending (present geography), graben-like troughs or aulacogens, which bisect this huge Precambrian craton into a number of intervening basement massifs. Of these, the E-NE-trending Volyn–Kresttsy and Soligalich–Yarensk troughs, parents to the late Proterozoic Moscow syneclise (see below), are the dominant mid-Proterozoic interior extensional elements in the craton. The ENE-trending Volyn–Soligalich (Central Russia) and NW-trending Pachelma diagonal aulacogens together divide the craton into (1) Fennoscandia including the Baltic Shield (NW); (2) Sarmatia (SW); and (3) Volgo–Uralia (Fig. 3-2) (Gorbatschev and Bogdanova 1993). Sarmatia, in turn, was subdivided by the NW-trending Dnieper–Donets aulacogen into the Ukrainian Shield and Voronezh Massif.

The western part of the buried craton exhibits substantial mid-Proterozoic reworking and magmatism, the latter typified by volcanoplutonic associations including rapakivi granites (Khain 1985) (Fig. 1-5c(iii)). Collectively this represents a southerly extension of western Baltic Shield reworking. Concurrently, major pericratonic troughs (geosynclines) developed about the craton, notably Galician–Caucasian (W) and Timan–Uralian (E).

Red clastic sediments, mainly continental and psammitic in the lower parts, accumulated in the evolving early Riphean aulacogens (Postnikova 1976). By mid-Riphean time (1.4 Ga) these red fluviatile gritstone–sandstone accumulates had been replaced by grey-black glauconite-bearing siltstones and clays. The appearance of glauconite suggests a transition from continental to littoral-marine environments. The early-mid-Riphean strata in the interior aulacogens are in the order of 2–3 km thick (e.g. Moscow, Pachelma) whereas in the coeval pericratonic depressions (e.g. Kama) they commonly reach 4–5 km in thickness. The interior assemblages suggest a single, concurrent early Riphean (to 1.4 Ga) trans-craton aulacogen development. The succeeding mid-Riphean (1.4–1.0 Ga) strata are also widely developed in the interior troughs of the central–northern buried platform, with a predominance of coarse-grained terrigenous and volcanogenic deposits.

### 4.4.4 SOUTHERN URALS

The full Riphean–Vendian complex of the Bashkir anticlinorium in the southern Urals represents the Riphean stratotype (Chumakov and Semikhator 1981, Salop 1983). This thick (12 km) composite succession is divided into three main Riphean groups—Burzyan, Yurmatan and Karatan—each 2–3 km thick and representing a complete arenite–pelite–carbonate marine transgression with minor volcanic associates, together with overlying Kudash and Vendian (Asha) groups (Fig. 1-3c(ii)). All divisions are characterized by distinctive stromatolites, oncolites and phyrolites (porphyries), and have been sufficiently dated to provide a reliable geochronologic framework (Chumakov and Semikhatov

1981, fig. 3). The correlative volcanogenic–terrigenous deposits of the interior basement troughs (e.g. Krestetsk and Salmi suites) consist of red, coarse-grained sandstones with dolerite dykes, basalt lava flows and tuffs.

## 4.5 GREENLAND SHIELD (NORTH AMERICAN CRATON)

Mid-Proterozoic events in Greenland are represented by (1) depositional, intrusive and structural activities in the Gardar Province of southern Greenland; and (2) the early development stages of the East Greenland Fold Belt (Figs 1-3d(i), 1-5d(i)).

### 4.5.1 GARDAR PROVINCE

Inland from Julianehab for 150 km in southern Greenland (Fig. 2-4) (60°45′–61°N), faulted outliers of sandstone and volcanic rocks of the Eriksfjord Formation (1600–1310 Ma) unconformably overlie a metamorphic–migmatite–granitoid basement in the Ilimaussaq area. Both basement and cover rocks are intruded by a remarkable series of largely alkalic dykes and central complexes (1330–1125 Ma), to form the distinctive 150 km by 70 km Gardar province (Emeleus and Upton 1976, fig. 170, Upton and Emeleus 1987).

### Eriksfjord Formation

The thickest preserved Eriksfjord succession (on the Ilimaussaq Peninsula) comprises some 3000 m of sandstones and lava-dominated volcanic rocks in approximately equal amounts.

Red to grey arkosic sandstones predominate. Local conglomeratic facies contain granitoid boulders up to 3 m in diameter. Cross-bedding, ripplemarks, sun-cracks and rain-pits are common. Quartz pseudomorphs after gypsum are reported. These fluviatile, lacustrine and aeolian deposits formed under intermittently arid conditions, within fault-bounded troughs.

Volcanic rocks make their appearance generally a few tens of metres above the basal unconformities. Most are olivine basalts, although a few monchiquitic, carbonatitic and trachytic flows are present (Escher and Wätt 1976). A sequence of rhyolitic rocks near Narssaq appears to be largely ignimbritic. Most lavas were erupted quietly on to a lowland terrain under subaerial conditions. Bedded ash and agglomerates are locally common.

Small diatremes are fairly common in the Tugtutoq–Narssarssuaq Zone, cutting both basement and nepheline syenite, and probably related to carbonatite or $CO_2$-rich ultramafic magmas.

### Central complexes

Ten major intrusive complexes lie in the designated Gardar Province, within an area 200 km by 70 km. Each is predominantly composed of alkaline rocks and the ensemble constitutes one of Earth's most remarkable alkaline igneous provinces (Emeleus and Upton 1976). The complexes are circular to elliptical in plan view and range in size from the diminutive Ivigtut intrusion, only 500 m in diameter, to the large Nunarssuit Complex, exceeding 1000 km$^2$ in outcrop area.

The Gardar complexes fall into two main categories: (1) oversaturated rocks, such as quartz syenite and granite; and (2) undersaturated foyaitic rocks. In general, the various units of the complexes are steep sided. Outcrops are commonly arcuate in plan. Generally, the annular outcrop form has resulted from younger subcylindrical stocks being emplaced with slight offset from earlier subcylindrical stocks. Cone sheets are entirely absent and true ring dykes relatively rare. However, partial ring dykes are not uncommon. Permissive emplacement by stoping and engulfment of the country rocks seems to have been the rule. Collapse or downsagging of the country rocks close to intrusive contacts is common at some complexes. Xenoliths and large rafts of earlier rocks are widespread.

The celebrated *Ivigtut peralkaline granite stock*, only 270 m in diameter and surrounded by intrusion breccia, contained a cryolite ore body that was explosively introduced, shattering the surrounding granitoid rocks. The ore body, prior to removal by mining, was a zoned pegmatite with extremely complex mineralogy. The major components were fluorides, carbonates and sulphides, with predominant siderite–cryolite rock.

The *Ilimaussaq intrusion*, by far the best known, is ovoid in plan, measuring 16 km by 7 km. The intrusion is faulted in such a way that successively lower erosion levels are exposed towards the southeast. Augite syenite, the earliest unit, forms a sheath around the western and northern margins of the complex. A subsequent intrusive event involved a more highly differentiated magma, which includes one of the most remarkable suites of peralkaline rocks known.

Kumarapeli and Saul (1966) discussed a branching zig-zag patterned rift system crossing the eastern part of North America. This St Lawrence rift system, which was active during the Mesozoic, is thought to have a Precambrian origin and to be related with the Lake Superior Keweenawan (late Proterozoic) rift system as well as the 'Gardar-age' alkaline complexes of Ontario. It now seems reasonable that a branch of this Precambrian rift system extended across South Greenland and that the Gardar activity represents a localized tectono–volcano province within this extended system, as suggested by Emeleus and Upton (1976).

### 4.5.2 EAST GREENLAND FOLD BELT

A considerable proportion of the East Greenland (Caledonian) Fold Belt comprises crystalline infracrustal gneisses, granitoids and metasupracrustal rocks. Included are remnants of the Carolinidian orogenic complex (~1.0 Ga), which contains the remnants of a mid-Proterozoic fold belt. Exposures are typically found at the western margin of the main fold belt, in tectonic windows through westward directed thrust sheets, which represent autochthonous or para-autochthonous foreland to the main Caledonian fold belt (Henriksen and Higgins 1976, fig. 193) (Figs 1-5d(i), 4-6).

In the southern part of the fold belt (70–72°N), rock units outcrop in a tectonic window below a major thrust with a westward displacement of at least 40 km. They comprise infracrustal basement of banded gneiss, augen gneiss and amphibolite, conformably overlain by a quartzite–schist–marble sequence, at least 2000 m thick (Steck 1971). Low greenschist facies prevailing in the south is transitional to high amphibolite facies in the north.

In the central metamorphic complex (72–76°N) (Henriksen and Higgins 1976), thick (1.5–2.5 km), rusty brown pelitic–psammitic metasediments of the Krummedal supracrustal sequence overlie and are tectonically interzoned with the older gneissic complex (Fig. 4-6). More or less migmatites occur at high structural level in proximity to extensive thrust sheets. A date of 1162 Ma (Hansen et al 1974) is attributed to post-depositional metamorphism.

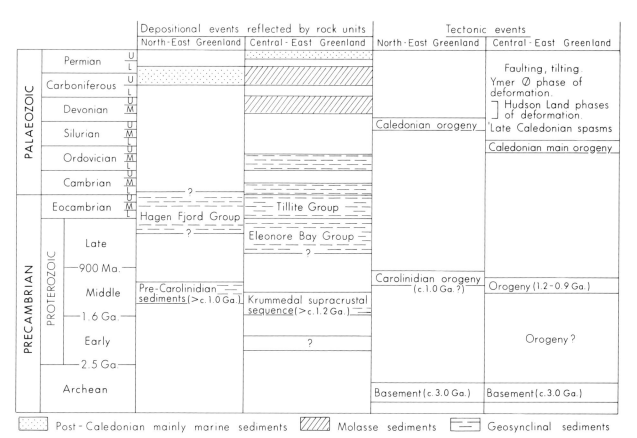

**Fig. 4-6.** Provisional chronologic scheme of deposition and orogeny represented in the East Greenland Fold Belt. (From Henriksen and Higgins 1976, Fig. 174, and reproduced with permission of the Geological Survey of Greenland).

The regions lying north of 76°N preserve traces of at least three orogenic episodes, namely Archean, Carolinidian and Caledonian (Fig. 4-6). Although the age of the Carolinidian Orogeny is not well established, it is considered to have occurred ~1000 Ma ago (Escher and Wätt 1976).

## 4.6 NORTH AMERICAN CRATON (LESS GREENLAND SHIELD)

Mid-Proterozoic (Helikian) activities in the North American Craton, characteristically taphrogenic–epeirogenic but including major pericratonic accretions, mark a significant stage in craton evolution. The era is conveniently punctuated by the Elsonian Orogeny (1.4 Ga) into Paleohilikian (1.8–1.4 Ga) and Neohelikian (1.4–1.0 Ga) suberas. Paleohilikian time is characterized by (1) the accretion of a major midcontinent mobile belt on the southern (present geography) flank of the craton; closely succeeded by (2) transcontinental emplacement of granitoid–mangerite–anorthosite anorogenic complexes; with (3) concurrent initiation elsewhere of intra- and pericratonic rifting with bimodal volcanic-redbed infill. The intervening Elsonian Orogeny is mainly confined to the east. Neohelikian time, in turn, is characterized by (4) the accretion of the 5000 km long, intersecting Grenville Mobile Belt on the eastern and southeastern flanks, to effectively complete the craton (Laurentia) and by (5) recurrent epi- and pericratonic rifting with extensive shelf sedimentation, including components in the present (a) Arctic; and (b) Cordilleran regions; together with (c) major midcontinental rifting with mega-flood basalt accumulations, notably in the Lake Superior Region, but also including the Bruce River and Seal Lake structures in the northeast and Coppermine Region in the extreme north. The net result of these varied and widespread activities was to complete North American Craton (Laurentia) in its present gross form and composition.

The several mid-Proterozoic elements of the craton are now considered consecutively in terms of Paleohelikian (1.8–1.4 Ga) and Neohelikian (1.4–1.0 Ga) suberas and intervening Elsonian Orogeny (1.4 Ga).

### 4.6.1 PALEOHELIKIAN SUBERA

#### Central Orogenic Belt

*Introduction*

The Central (Midcontinent) Belt (renamed Transcontinental Proterozoic provinces) constitutes a group of ENE-trending, southward younging mid-Proterozoic orogenic provinces, which occupy the south–central part of the North American Craton (Fig. 1-5d(ii)). They extend continuously along strike for at least 2600 km (east to west) between the Cordilleran and Appalachian (–Grenville) orogens, with additional deformed and disconnected northeastward extensions within the younger transecting Grenville Province and beyond, and for ~1200 km (north to south) in width from the E-trending Wyoming–Penokean (Lake Superior) 'join' to southern Texas where transected by the younger SW-W-trending Grenville Belt of Neohelikian age.

The Central Belt (Fig. 4-7) extends from California–Nevada on the west to Ohio–Tennessee on the east. It is mainly exposed in the southwest (California, Nevada, Arizona, Colorado and New Mexico); many drill holes penetrate basement in the western midcontinent region (Nebraska, Iowa, Kansas, Missouri and Oklahoma); fewer drill holes and exposures are present east of the Mississippi River. The belt is divided into two parallel provinces, named northern and southern, the latter itself locally subdivided, which become progressively younger to the south in the age range 1.8–1.5 Ga. As known, the Central Belt is composed of predominant (~90%) juvenile crust, thereby representing a major pristine mid-Proterozoic accretion. The northern and southern provinces respectively bear the tectonic imprint of two partly overlapping mid-Proterozoic orogenies—Central Plains Orogeny (1.78–1.69 Ga) in the north, and Mazatzal (1.66–1.60 Ga) Orogeny and associated Yavapai Orogeny (1.74–1.68 Ga) in the south (Condie 1982a, Anderson 1983, Silver 1984, Bickford et al 1986, Sims and Peterman 1986, Martin and Walker 1992, Bauer and Williams 1994). Other more detailed belt configurations have been developed in response to ongoing studies, including an overall division into Inner Accretionary (1.8–1.7 Ga) and Outer Tectonic (1.7–1.63 Ga) belts, and a westernmost subdivision into Mojave, Yavapai and Mazatzal provinces (Condie 1986a, 1987, Karlstrom and Bowring 1993, Van Schmus et al 1993). However, a simple twofold northern and southern subdivision is used here (Fig. 4-7).

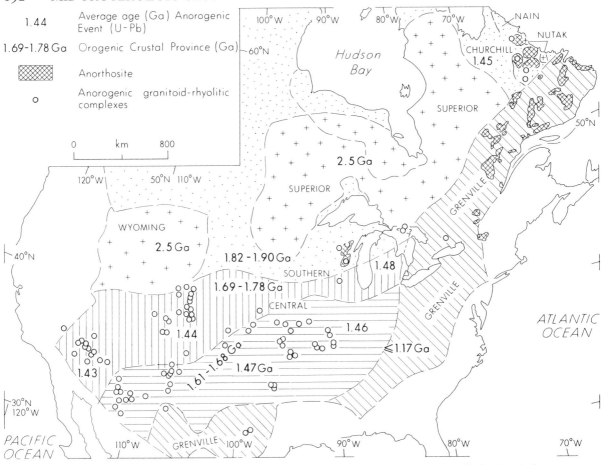

**Fig. 4-7.** Distribution of mid-Proterozoic orogenic provinces (Central Belts and Grenville Belt) and of anorogenic intrusions, (anorthosites and granitoid-rhyolitic complexes) in relation to older Precambrian subdivisions of North America. (Adapted from Anderson 1983, Fig. 1, and reproduced with permission of the author and of the Geological Society of America).

The characteristic lithologic assemblage in both provinces comprises (1) quartz–feldspar gneisses with amphibolites; and (2) interlayered biotite gneiss with migmatite (Reed et al 1987, Bickford 1988). Better preserved sequences reveal that the protoliths are bimodal volcanic rocks, intercalated with and overlain by thick sequences of volcanogenic, commonly turbiditic, metasedimentary rocks, subsequently metamorphosed in greenschist and amphibolite facies. Granulite facies gneisses are locally present. Migmatites are very common. Well-preserved metasupracrustal sequences in Arizona (Yavapai and Tonto Basin supergroups, ~1.7 Ga) include pillow basalt, dacite–rhyolite ash flow tuffs, arenites, greywacke and siltstone, and a reported ophiolite sequence (Dann 1991). These all form part of the pre-Belt (see below) basement. Some volcanogenic polymetallic (Cu–Pb–Zn–Ag–Au) deposits are present. Pegmatites locally contain mica, Be, Li, Ta, Nb mineralization.

These metasupracrustal rocks have been intruded by plutons, commonly tonalitic to monzogranitic, and by sills and stocks of gabbro. Discordant plutons of weakly foliated monzogranite and tonalite are locally common. Small plutons of gabbro, quartz diorite and trondhjemite are enclosed in amphibolite facies gneisses.

Foliation in most amphibolite facies supracrustal rocks parallels compositional layering, even around the hinges of isoclinal folds. Foliation dips are variably gentle ($30°$) or steep ($> 60°$) (Reed et al 1987).

The rocks are cut by a complicated array of faults and shear zones. Of these, large-scale thrusting carried rocks of the Central Belt northward against the Wyoming Uplift along the Cheyenne Belt.

Indicated age ranges in the northern and southern provinces are 1.78–1.69 Ga and 1.68–1.61 Ga respectively (Boardman and Bickford 1982, Bickford 1984, 1988, Reed et al 1987, Bowring and Karlstrom 1990). No pre-1.8 Ga basement has been recognized. Furthermore, isotope systematics suggest that no significant Archean crustal component was involved in the formation of most Central Province supracrustal or plutonic rocks (Nelson and De Paolo 1985, Bennett and De Paolo 1987). However, the data do reveal the presence of older crust in Mojave Province, either Archean, earliest Proterozoic, or both.

Available data suggest to Condie (1982a, 1987) that the supracrustal rocks developed along a continental margin, products of arc magmatism and related sedimentation in oceanic–continental margin settings.

A plate-tectonic model is also favoured by Bowring and Karlstrom (1990) and Van Schmus et al (1993), involving successive marginal basin closures and Andean-type orogenies associated with southward migrating arcs to incorporate numerous tectonostratigraphic terraines. The net result was to accrete 1300 km width of mainly juvenile crust in the 300 Ma interval, 1.8–1.5 Ga. Incorporating the on-coming, widespread anorogenic activity (see below) increases the accretion interval to 400 Ma (1.8–1.5 Ga). Furthermore, incorporating the southeasterly adjoining Grenville Belt increases the overall accretion width to 1600 km over an 800 Ma interval (1.8–1.0 Ga).

Although derivation of the Central Belt through arc magmatism and cannibalistic sedimentation is consistent with many aspects, it leaves the following important problems unanswered: (1) the paucity of andesites, so common in recent magmatic arcs; (2) the paucity–absence of recognized ophiolites and subduction-related melanges; (3) the lack of identified sutures, with the possible exception of the Cheyenne Belt (Reed et al 1987).

## Anorogenic complexes

The period from 1.49 to 1.03 Ga represents a unique stage in craton development, characterized by abundant anorogenic magmatic activity (anorthosites, layered gabbros, charnockitic diorite to syenite suites, rapakivi granites, bimodal basalt–rhyolite suites, and tholeiitic dykes and sills). Anorogenic complexes fall into three magmatic episodes: the main 1.49–1.41 Ga episode and lesser 1.41–1.34 Ga and 1.08–1.03 Ga episodes (Anderson 1983), restricted to a 600–1000 km-wide, 5000 km-long, N–NE-trending California–Labrador belt (Fig. 4-7). The majority (> 70%) of anorogenic activity falls in the main 1.49–1.41 Ga anorogenic episode (Bickford et al 1986, Van Schmus 1987b,

1993, Ryan 1991, Ashwal et al 1992, Karlstrom and Bowring 1993, Yu and Morse 1993). This particular belt appears to be unique to North America; it is temporally separated by a 200–270 Ma magmatic gap from similar but older complexes in south Greenland and Scandinavia, including the type Wiborg Rapakivi Massif in Finland, of age 1.70 Ga.

The most obvious regional lithologic variation is the quantitative restriction of mantle-derived anorthosites and gabbros to eastern North America, notably the 1600 km long Labrador–Adirondack tract. Crustal-derived anorogenic granitic intrusions, on the other hand, are widely scattered though locally concentrated throughout the entire 5000 km long Labrador–California belt.

### Granitoid–rhyolitic complexes

A large part of the southern midcontinent, extending from Ohio to the Texas Panhandle, is underlain by assemblages of rhyolitic–dacitic volcanic rocks and cogenetic epizonal plutons including large batholiths (Anderson et al 1980, Anderson 1983, Van Schmus et al 1987b, Anderson and Bender 1989, Rämö 1991, Lowell 1991, Sims et al 1993). These Precambrian granitoid–volcanic complexes are mainly buried under Phanerozoic cover except for local exposures such as at St Francois Mountains, SE Missouri. Most of the silicic volcanic rocks are probably pyroclastic accumulates, whereas associated subvolcanic plutons are characteristically micrographic or granophyric textured, both indicative of shallow crustal environments. The granites are relatively silicic, potassic, and ferriginous, are mostly metaluminous but locally peraluminous or peralkalic. Igneous rocks of intermediate to mafic composition as well as sedimentary rocks are subordinate to rare. Metamorphism is consistently low greenschist facies, and penetrative deformation is confined to local occurrences.

The anorogenic suties, also called anorthosite–mangerite–charnockite–rapakivi granite (AMCG) magmatic suites, commonly intruded associated orogenic belts within 100–600 Ma after cesstaion of the typically calc-alkalic formative crustal generation processes. However, despite this temporal proximity, the geochemistry and bimodal (i.e. anorthosite–granitoid) character of AMCG magmatism is markedly dissimilar to that of calc-alkalic associations (Emslie 1991).

Regarding the bulk of the silicic magmas, the involvement of a significant crustal component in their source seems mandatory, the evidence pointing to derivation of the especially large granite batholiths by fusion of the lower crust. A small amount of crustal fusion (10–30%) would account for their characteristic potassic, iron-rich, minimum melt composition. These comparatively dry anorogenic granite magmas were able to intrude into upper crustal levels with the shallower plutons yielding concurrent volcanic activity (e.g. St Francois and Montello batholiths). Furthermore, the granitoid suites are considered to have developed by way of extended crystal–liquid fractionation from voluminous parent melts (Emslie 1991).

However, the absence of any significant gravity-magnetic indicators in the crust makes it unlikely that complete rifting of continental crust was a factor in any associated scheme of crustal extension (Hamilton 1989). Also, Nyman et al (1994) propose abandonment of the term anorogenic with its implications to supercontinent breakup and regional extension (Windley 1993), ascribing the intrusion in question to compression or transpressional plate-margin tectonism in the interval 1.5–1.3 Ga along southern Laurentia.

*Anorthosite massifs*

The abundance of large anorthosite massifs is a unique mid-Proterozoic feature in Earth's overall crustal evolution. Major massifs fall in the age range 1.7–0.9 Ga. Older examples are, for the most part, not well documented; substantial Archean examples are unknown although small Archean units are reported in Africa (Moore et al 1993). Younger Paleozoic massifs are very rare (Herz 1969, Emslie 1985).

*Setting*: anorthosite massifs typically occur in Proterozoic terranes that have undergone relatively high grade metamorphism and deformation. The massifs are typically associated with sparse ferrodioritic rocks and common K-rich monzonite–granite anorogenic plutons. A notable feature of this compositional 'trinity' is the absence of significant volumes of mafic rocks. Pristine massifs of central Labrador (Nain Province) demonstrate intrusion into non-orogenic environments. On the other hand, the common metamorphosed massifs with variable Grenvillian metamorphic–tectonic overprint have ambiguous tectonic settings and uncertain absolute age.

The principal anorthosite suite forms about 15 large massifs and numerous small intrusive bodies in a broad NE-trending belt in eastern Canada and the USA, the exposed part extending for 1700 km from the Adirondack Massif in New York State on the southwest along the Grenville Province to central Labrador on the Northeast. Although it was previously considered that anorthosites crystallized from orogenic magmas intruded into deep crustal sites of granulite metamorphism, recent work on massif anorthosites has resulted in the picture of shallow emplacement of at least some anorthosites in a shallow continental rift environment, with the igneous ages predating the granulite metamorphic ages by 50–300 Ma (Emslie 1978, Newton 1987, Yu and Morse 1993). The bodies in the Nain Province to the north are unmetamorphosed but those in Grenville Province are metamorphosed and deformed in varying degrees. Wynne-Edwards (1972) estimates that this particular belt contains 75% of Earth's known anorthosite, a not-undisputed claim, however (Hamilton 1989).

Available data generally provide considerable support for a primary subcontinental mantle peridotite source for anorthosite parent magmas. However, subordinate contribution from mafic granulites of the lower crust seems likely (Emslie 1978, Morse 1982).

### Rifted redbed basins

Redbed basins of Paleohilikian age developed by aulacogenic (epi- to pericratonic) rifting of the post-Hudsonian craton within a broad NNW-trending transcontinental (Great Lakes–Arctic Ocean) swath. The basins include (1) those in north–central USA (e.g. Sioux Quartzites) and, successively northwestward across the Canadian Shield; (2) Athabasca; (3) Dubawnt (Thelon–Baker Lake); (4) Great Bear; and (5) Bathurst Inlet basins (Fig. 4-8). The fault-bounded basins are characterized by redbed–quartzite–bimodal volcanic assemblages with or without uppermost marine carbonates. Factors in common are the presence of (1) clay-rich quartz arenites and (2) pervasive hematite content, in (3) moderately deep fault-induced basin structures, within (4) an intracratonic setting. Paleocurrent directions in the basins are southward in the north–central USA but northwestward to westward in those to the north, broadly coincidental with present North American continental drainage patterns.

*Sioux–Baraboo–Barron quartzites*

Red quartzite sedimentary rocks occur in three principal fault-banded basins of north–central USA

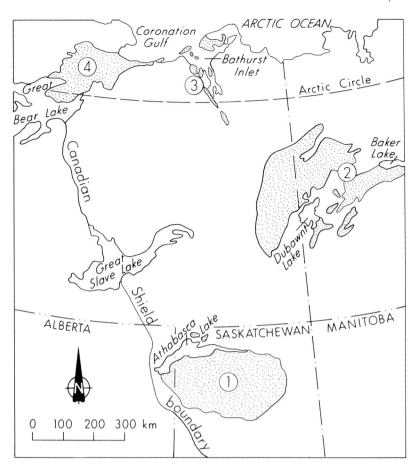

**Fig. 4-8.** Distribution of Helikian basins in the northwestern Canadian Shield: 1 – Athabasca; 2 – Thelon-Baker Lake; 3 – Bathurst; 4 – Coppermine River-Great Bear Lake. (From Fraser et al 1970, Fig. 1, and reproduced with permission of the Geological Survey of Canada and of the Minister of Supply and Services Canada).

(Table 2-8, column 1). The Baraboo, Sioux and Barron quartzites are so similar that they tend to be correlated. All postdate the Penokean Orogeny (~1850 Ma) and fall in the depositional interval 1760–1630 Ma. They represent redbed sequences deposited by braided streams flowing southward over deeply weathered land surfaces of moderate relief (Ojakangas and Morey 1982a, b, Dott 1983, Sims and Peterman 1986, Southwick et al 1986, Sims et al 1993).

Other younger, local redbed quartzites around Lake Superior include the Puckwunge of Minnesota, the Bessemer of Michigan–Wisconsin, and the Sibley of Ontario, all generally regarded to be 1340–1100 Ma old (Ojakangas and Morey 1982a, b).

Athabasca Basin

Associated with the Athabasca Basin, the main Helikian depocentre in the region, but situated on the north side of Lake Athabasca are a number of small, fault-bounded, redbed units (Fig. 4-8; Table 2-8, column 2). Typical of these is the Martin Formation, a redbed–volcanic association occupying two small basins, each about 80 km² in area. The sequence consists of some 5000 m thick of breccias, conglomerate, arkose, siltstone and mafic volcanics with associated gabbroic dykes and sills (Fraser et al 1970). The inadequately constrained age of deposition of the Martin Formation is 1730–1830 Ma (Cumming et al 1987, Miller et al 1989). Nearby pitchblende vein deposits are dated at ~1750 Ma (Koeppel 1968).

The Athabasca Basin is an E-trending elliptical unit, 450 km by 200 km, filled with 1000–1500 m thick of little deformed, fluvial to lacustrine–marine sediments, mainly first-cycle orthoquartzites of the Athabasca Group. The dips of beds exposed at surfaces are very gentle (Ramaekers 1981, Armstrong and Ramaekers 1985).

The basement underlying the Athabasca deposits was intensively weathered. The paleoweathering profile, up to 50 m in preserved thickness, has characteristics typical of present day lower level laterite (Ramaekers and Dunn 1977).

Substantial unconformity-related U–(Cu, Ni)

deposits occur near the base of the Athabasca succession (Sibbald et al 1990, Carl et al 1992). Most but not all deposits are confined to a narrow region near the present eastern erosional edge of the Athabasca Basin (Cumming and Krstic 1992, fig. 1). The Key Lake ore deposit, one of several such large high grade deposits located at the eastern margin of the basin, contains two main types of bedrock-hosted ore: massive U–Ni aggregates; and disseminated U–Ni impregnations of both Athabasca Group and subjacent basement rocks. The primary ore-forming phase is considered to have occurred in the 1380–1330 Ma age range followed by successively younger generations of U mineralization and periods of remobilization down to at least 1.0 Ga (Cumming and Krstic 1992, Phillippe et al 1993, Carl et al 1992). However additional older ages of mineralization to 1514 Ma and even 1800 Ma in the basin have been reported elsewhere (Williams-Jones and Sawiuk 1985). The Cigar Lake deposit is exceptional in terms of both large tonnage and high grade (Bruneton 1993).

### Baker Lake and Thelon basins

In the Baker Lake region, the 1.8–1.7 Ga older Dubawnt Group (Fraser et al 1970, Blake 1980a, Heywood and Schau 1981, LeCheminant et al 1987) occupies two fault-bounded basins, measuring 450 km by 150 km (Dubawnt Lake Basin) and 300 km by 50 km (Baker Lake Basin) (Fig. 4-8; Table 2-8, column 2).

Dubawnt strata form a mainly flat-lying to gently dipping cratonic cover upon an irregular, locally regolithic Archean–Hudsonian basement. The sequence comprises lower redbeds with alkalic and felsic volcanic rocks, and upper quartz sandstone, conglomerate and dolomites (Thelon Formation) (Blake 1980a). Many hundreds of co-magmatic K-rich dykes, mainly minettes, have been identified in the region.

The disconformably overlying Thelon Formation forms a 350 m-thick blanket sandstone deposit, itself conformably overlain by stromatolitic dolostones.

*Age*: Dubawnt accumulation occurred shortly after the ~1.85 Ga Hudsonian Orogeny and was associated with large-scale regional late to post orogenic faulting. The most reliable data (U–Pb zircon) for Dubawnt Lake alkaline magmatism is 1850 Ma (LeCheminant et al 1987). The age of Thelon sandstone deposition is bracketed in the 1753–1720 Ma range (Miller et al 1989).

### Bathurst Inlet Basin, Coronation Gulf

Helikian strata are exposed on many of the islands in Bathurst Inlet, on the mainland south and west of the inlet, and on Kent Peninsula to the northeast (Fig. 4-8, Table 2-8, column 3). The area underlain by these rocks (13 000 km$^2$), termed the Bathurst Basin, is bounded on the east by Archean basement and Aphebian metasediments (Goulburn Group), and on the west by the Bathurst Trench, a prominent linear graben-like structure that extends 350 km southeastward from the Arctic coast and the Kilohigok Basin. The lowermost Helikian strata occur in and along the trench as far as 140 km from Bathurst Inlet (Fraser et al 1970).

Helikian strata in the Bathurst Basin comprise four mildly deformed sandstone–mudstone–carbonate sequences with an aggregate thickness of more than 5000 m (Campbell 1978).

### Coppermine River to Great Bear Lake

The Paleohelikian Hornby Bay and Dismal Lakes siliciclastic to marine carbonate groups are the lowermost stratigraphic units in the comparatively undeformed, N-dipping Coppermine Homocline of the Amundsen Embayment, an elliptical area 130 km by 80–100 km, located on the northwest flank of the Wopmay Orogen (Fig. 4-8). These Paleohilikian strata are successively overlain by basalt lavas of the Coppermine River Group and sediments of the Rae Group; they are intruded by the Muskox Intrusion and diabase dykes and sills, all of Neohelikian age (1.3 Ga) (Table 2-8, column 3) (Kearns et al 1981, Hoffman and St Orge 1981, Dupuy et al 1992).

Dates from the Hornby Bay Group permit age correlation with Athabasca and Thelon strata, long correlated on the basis of similarities in sedimentation, stratigraphy and tectonic setting (Miller et al 1989).

## 4.6.2 ELSONIAN DISTURBANCE (OROGENY)

The Western Nain Subprovince of the Canadian Shield is the type area for the late Paleohelikian Elsonian Disturbance (Table 2-8, column 4). This disturbance differs from other orogenies of the Canadian Shield in that anorogenic intrusions (Nain Plutonic Suite) constitute an important part of the period of activity (~1450–1290 Ma) (Stockwell 1982, Emslie and Loveridge 1992). However, the effective end of the disturbance is estimated at about 1400 Ma, marking the Paleo-

Neohelikian boundary. Following a period of post-Elsonian erosion, rocks affected by the Elsonian Disturbance were unconformably overlain by the early Neohelikian Seal Lake Group (1323 Ma) of the Grenville Province (see below).

## 4.6.3 NEOHELIKIAN SUBERA

### Grenville Province

*Introduction*

The Grenville Province, 300–600 km wide and 2000 km long in exposed length, is an eroded belt distinctive for its widespread high grade metamorphism, complex deep-level structures, and abundant meta-anorthosites with associated mangerite-charnockite igneous suites (Wynne-Edwards 1972, 1976, Emslie 1978, 1991, Davidson 1985, 1986, Easton 1986, Moore 1986, Roy et al 1986, Green et al 1988, Rivers et al 1989, Emslie and Hunt 1990) (Figs 4-7, 4-9; Table 2-8, column 4). The subsurface extension is at least as long again, reaching to Texas and even Mexico (Fig. 1-5d(ii)) (Rankin et al 1993, Svegard and Callahan 1994). The complete >4000 km-long Grenville Belt evolved over a period of more than 650 Ma (1700–1050 Ma), includes substantial Archean and pre-Grenville Proterozoic crust and is especially marked by the culminating Grenvillian Orogeny (Event) (1.1–1.0 Ga). In contrast to standard orogenic sequences, calc-alkalic igneous suites are rare to absent from Grenvillian magatism (Emslie and Hunt 1990). In addition to the dominant (~60%) quartzofeldspathic grey gneiss, which represents in large part rejuvenated basement, local authochthonous metasedimentary assemblages, 3–9 km thick, of shallow-marine to continental affinity (marbles, calcsilicates, metapelites) are important (e.g. Grenville Supergroup). The Grenville Front Tectonic Zone on the northwest marking one of Earth's great structural discontinuities, is attributed to continental collisional tectonics.

Vast sheet-like anorthositic intrusions (massifs) are distributed in generally N-trending alignments. The anorthosites are typically but not exclusively associated with mangerite-charnockite igneous rocks to form the AMCG (see above) magmatic suites, the most characteristic magmatic association throughout much of the province. AMCG suites were intruded during at least three distinct episodes: ~1.64 Ga, ~1.36 Ga and ~1.15 Ga (Emslie and Hunt 1990, Higgins and van Breemen 1992, McLelland and Chiarenzelli 1990, Doig 1991). The last episode is the most widespread, and most directly related to the Grenvillian Event (1.1–1.0 Ga) in timing, and also marks the culmination of igneous activity in the province (Emslie and Hunt 1989).

Deformation across the entire province is characterized by a pattern of northwestward imbrication/verging of NE-trending folds and thrust (Wynne-Edwards 1976). This deformation is principally manifested in the northwest displacement of SE-dipping crystalline thrust sheets, separated by granulite-upper amphibolite facies ductile thrust zones (Hammer and McEachern 1992). This major overall northwest tectonic transport has resulted in widespread exposure of catazonal granulite facies of metamorphism. The prevailing level of metamorphism in Grenville crust indicates an average depth of erosion of 15 km across the province (Wynne-Edwards 1976) with local removal of as much as 30 km of upper crust (Annovitz and Essene 1990), and with burial depths > 45 km attested to by the presence of well-preserved eclogites (Indares 1993). Gibb and Walcott (1971) proposed a continental rifting–collision model, Wynne-Edwards (1976) a ductile spreading ensialic model, and Rivers et al (1989) a diachronous or oblique continental collisional model.

*Subdivisions*

Following Wynne-Edwards (1972), the Grenville Province itself is divided into six major subdivisions (Fig. 4-9). The boundaries of these subdivisions either outline major lithological changes or cut across geological contacts to separate areas of different structural and/or metamorphic style.

(1) *Grenville Front Tectonic Zone*, a NE-trending, SE-dipping 1200 km long and 20–100 km-wide thrust zone, containing numerous dip–slip ductile shears, forms the northwestern boundary of the province. The dominant rocks are quartzofeldspathic gneiss, commonly in granulite facies, with strong NE-trending foliation and numerous parallel zones of cataclasis and mylonitization. The grade of metamorphism is generally high and kyanite is the dominant aluminosilicate. Seismic reflection lines reveal the deep structures (Davidson 1986, Cannon et al 1987, Green et al 1988, Rivers et al 1989, Corrigan et al 1994, Doigneault and Allard 1994, Kellet et al 1994).

(2) Recent work in the *Central Gneiss Belt* (Davidson 1986) has led to recognition of a number

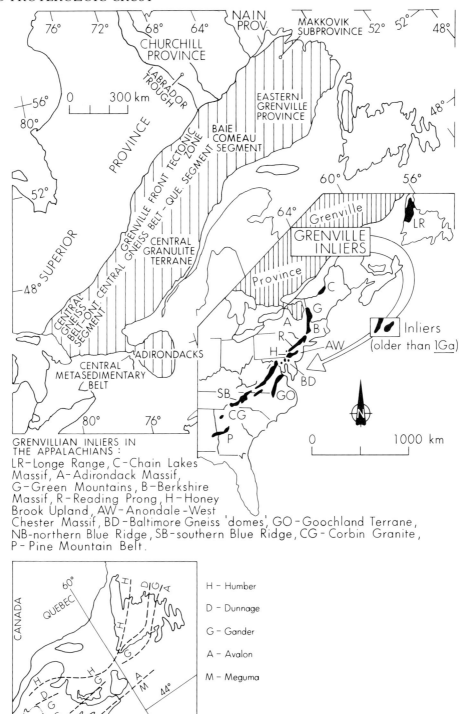

**Fig. 4-9.** Outline of the Grenville Province and its main subdivisions: ONT – Ontario; QUE – Quebec. Inset shows Grenvillian inliers in the Appalachians. (After Easton 1986, Fig. 1, and reproduced with permission of the Geological Association of Canada). Inset below of northeastern Appalachia showing main subdivisions (From Williams et al (1972) and published with permission of the Geological Association of Canada).

of lensoid to lobate lithotectonic regions, referred to as domains and subdomains. Although component rocks are thoroughly deformed, the inter-domain boundary zones contain particularly highly trained rocks, including mylonites, and are undoubtedly tectonic. These continuously tectonite-bounded domains are interpreted as individual nappes and klippes, products of northwestward directed ductile flow, and components of a segment of crust thickened by a process of northwestward directed stacking of large blocks and slices along inclined ductile shear zones.

(3) The metasupracrustal-rich *Central Metasedimentary Belt* encompasses the main exposures of the Grenville Supergroup, a suite of marbles, calcsilicates, quartzites, paragneiss, amphibolite and metavolcanic rocks intruded by gabbro–granite–syenite plutons. Its structural base on the northwest is marked by a gently SE-dipping thrust zone of planar, highly strained granoblastic tectonites, 8–10 km thick, representing thoroughly annealed, granoblastic mylonites derived by cumulative deformation, mechanical degradation, and transposition of megacrystic and pegmatitic granitoid protoliths, intrusive into or intruded by mafic protoliths. The southeast boundary is a prominent lineament marked by cataclasis, mylonite and an abrupt change from amphibolite to granulite facies. In northern New York State it is termed the Adirondak Highland line.

The Grenville Supergroup itself is probably entirely allochthonous, no longer in contact with original basement, except perhaps on the backs of orthogneiss thrust sheets such as the Glamorgan Complex.

(4) *Central Granulite Terrain*, which includes the Adirondak Highlands in New York, is characterized by granulite facies metamorphism including pre-Grenvillian high grade metamorphism at 1450–1430 Ma (Ketchum et al 1994) and the presence of large massifs of anorthosite. The plutonic anorthosite massifs are deformed by cataclasis and recrystallization of variable intensity.

(5) *Baie Comeau Segment* extends along the St Lawrence River from La Malbaie to Sept-Iles, a distance of 350 km, and northward from the river to the Grenville Front Tectonic Zone south of the Labrador Trough. Its northwestern boundary is the granulite facies of the Central Granulite Terrain. The eastern boundary separates the more homogeneous metamorphism and complex reworking of the Baie Comean Segment from anorthosites and basement gneisses with contrasting grades of metamorphism to the east.

(6) *Eastern Grenville Province* is underlain by dominant quartzofeldspathic gneiss in upper amphibolite facies, together with many large bodies of meta-anorthosite. Supracrustal rocks, mainly mature arenites, pelitic schist and marble with intercalated gabbro sills, felsic volcanic rocks and subordinate basalt are assigned to the Wakeham Bay Supergroup, a thick terrestrial sequence, products of rifting at 1280–1180 Ma (Martignole et al 1994).

Tectonostratigraphy

The principal effect of the Grenvillian Orogeny (~1000 Ma) was the structural telescoping within the province of Archean and lower to mid-Proterozoic units, and the development of a low to high grade metamorphic overprinting of all pre-Grenvillian crust.

The Grenville Province structure is now seen as largely consisting of southeastward dipping crustal-scale imbricates bounded by northwestward directed, broad ductile thrust zones (Rivers et al 1989). In brief summary, the province is accordingly divided into three first-order longitudinal belts by three principal tectonic boundaries; from northwest to southeast, the Grenville Front, the Allochthonous Boundary Thrust, and the more restricted Monocyclic Belt Boundary Zone (Fig. 4-10, inset). This surface pattern is reinforced at depth by widespread seismic reflection data (Davidson and Van Breemen 1987, Corrigau et al 1994, Daigneault et al 1994, White et al 1994a).

Age domains

Available age data allow a preliminary division of Grenville Province into chronologic domains (provinces) (Fig. 4-10, northeast half of figure), in which tentative geologic histories have been deduced. Seven areas have been defined, in each of which the crust has had common geologic and chronologic histories. In some areas the age of the basement is the common chronologic factor; in others a magmatic episode is predominant. All seven areas have been affected by the ~1100–1000 Ma Grenvillian overprint, but this is the only event common to all (Easton 1986). In general, maximum ages of these domains decrease (young) southeastward away from the Grenville Front. This is in accord with corresponding southward younging of midcontinent belts in central USA (Fig. 4-10,

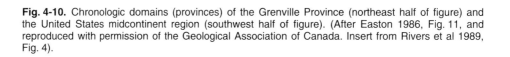

**Fig. 4-10.** Chronologic domains (provinces) of the Grenville Province (northeast half of figure) and the United States midcontinent region (southwest half of figure). (After Easton 1986, Fig. 11, and reproduced with permission of the Geological Association of Canada. Insert from Rivers et al 1989, Fig. 4).

southwest half of figure), leading to the inference that the Grenville Province represents a melange of tectonically distorted and deformed ('accordioned') extensions of pre-Grenvillian crust, formerly distributed to the south and west–southwest. Co-ordinated geochemical data have led to the inference that much of the common grey gneiss, at least in Central Grenville Province, is the product of a subduction-related (i.e. orogenic) major crustal extraction event at 1.53 Ga (Dickin and Higgins 1992). However, the corresponding grey gneiss in the Eastern Grenville Province is mostly attributed to as yet unspecified processes during the 1710–1620 Ma Labradorian Orogeny (Gower et al 1991), Philippe et al 1993).

In conclusion, all Grenville Province domains have complex histories involving multi-aged crust. Complex telescoping involving northwestward transport of allochthonous slices, the culminating multi-pulsed Grenvillian overprint (1035–970 Ma) is common throughout (Easton 1986, Gower et al 1991, Dickin and Higgins 1992, Connelly and Heaman 1993, Scott et al 1993, McEachern and Van Breeman 1993, Sager-Kinsman and Parrish 1993, Van Breemen and Higgins 1993, Smith et al 1994).

### Arctic platform rifting

A 3200 km long trans-Arctic rift zone along the northern edge of the Canadian–Greenlandic shields developed concurrently with the opening, 1.25–1.2 Ga ego, of Poseidon Ocean, the earliest Proto–Arctic Ocean (Jackson and Iannelli 1981). Poseidon Ocean is interpreted to have closed by 1.0 Ga. In accord with the Wilson Cycle rhythm, the subsequent opening–closing of the second Proto–Atlantic (Iapetus) Ocean in early Paleozoic time (0.6–0.4 Ga), was followed by opening of the present Atlantic Ocean, beginning ~0.2 Ga.

During the Poseidon and associated trans-Arctic

rift zone openings (~1.2 Ga), several Arctic basins, most of which were initiated and evolved as grabens or aulacogens, developed along the present E-trending platform margin that extends from the Cordilleran Fold Belt in the west to the Caledonian Fold Belt in eastern Greenland. In the eastern Arctic, sedimentary sequences are preserved in five similar, penecontemporaneous, formerly interconnected rift basins, including, from west to east: (1) Somerset Island; (2) North Baffin Island; (3) Fury and Hecla Strait; (4) southeast Ellesmere Island; and (5) northwest Greenland (Table 2-8, column 2).

In example, as much as 6100 m of Neohelikian strata are spectacularly exposed in towering castellated cliffs along inlets and fjords of the rugged, mountainous northern Baffin and Bylot Islands (Jackson and Iannelli 1981, fig. 16.33). Bylot Supergroup includes the Society Cliffs dolostones which host important Pb–Zn ore deposits at Nanisivik. Lithostratigraphic details are provided by Jackson and Iannelli (1981).

Repeated faulting along NW-trending planes has occurred in northwest Baffin Island from the end of Aphebian (1.8 Ga) to recent times, to which the gentle folds and local vertical dips in Bylot strata are attributed. Several small northwestward dipping, low-angle thrusts may be syn-depositional but are probably related to post-depositional compression from the northwest.

Although much faulting postdates Bylot deposition, the present fault distribution and nature of the North Baffin Rift Zone are considered by Jackson and Iannelli (1981) to approximate the nature of horsts and grabens at the close of Bylot time. Borden Basin, in brief, probably developed along a failed arm or aulacogen during a 1.2 Ga ocean opening to the northwest (Jackson and Iannelli 1981).

Neohelikian strata in the Coppermine Homocline of the western Arctic (Fig 4-8) (Baragar 1972) contain basalt lava flows each 17–700 m and totalling 3000–4000 m in thickness. The lavas dip very gently to the north at angles of 10° or less.

The conformably overlying Rae sediments, 5000 m thick, comprise sandstone, siltstone, shale and stromatolitic carbonates. Gypsum beds up to 20 m thick occur in carbonate units on the Arctic coast.

The closely associated Muskox Intrusion is dyke-like in plan and funnel-shaped in cross-section (Kearns et al 1981, fig. 9-23) (Table 2-8, column 3). It cuts both Aphebian basement and the overlying Hornby Bay quartzites. The intrusion is not in direct contact with the Coppermine volcanic rocks, but the lavas and possibly some of the overlying Rae sediments probably formed cover rocks to the intrusion.

Bronzite, gabbro and norite are the most abundant rock types but, as the dyke widens, lenses of picrite appear in the centre. The central layered sequence is about 3000 m thick and is composed of 38 main layers which dip northward at 10°, about parallel with the dip of the enclosing Coppermine flows.

Radiometric and paleomagnetic data indicate that the Muskox Intrusion, Coppermine lavas and nearby Mackenzie diabase dyke swarm are all 1280–1270 Ma old (Heaman and LeCheminant 1988).

## Cordilleran Fold Belt

Mid-Proterozoic supracrustal rocks accumulated in a number of scattered, fault-bounded troughs and basins distributed along the western Cordillera. The main depocentres and contained strata, including some overlying late Proterozoic sequences (in brackets), include, from south to north: (1) in southwestern USA, (a) the Apache Group and Troy quartzite of east–central Arizona; (b) the Unkar (and Chuar) groups, Grand Canyon Supergroup, Arizona; (c) Crystal Spring, Beck Spring and (Kingston Peak) formations, Pahrump Group, Nevada and California; and (d) Uinta Mountain Group, (Big Cottonwood and Pocatello formations and Brigham Group) Idaho–Utah; (2) Belt–Purcell supergroups in Idaho, Montana, and southern British Columbia; (3) Wernecke Supergroup, and lower Mackenzie Mountains Supergroup in northern British Columbia–Yukon (Harrison 1972, Stewart 1976, Labotka et al 1980, Delaney 1981, Young 1981a, Horodyski 1983, Mitchelmore and Cook 1994, Larson et al 1994 (Figs. 4–11; 5–8; Table 2–8, columns 3, 5). For conflicting mid-late Proterozoic correlations see Harrison and Peterman (1984), Hedge et al (1986) and Link et al (1993).

Common rock types are sandstones, siltstones/shales, locally stromatolitic carbonates and mafic intrusive and extrusive igneous rocks; the association is suggestive of widespread extensional tectonic (taphrogenic) events. These events have been interpreted in three ways: (1) as major epicratonic rift features related to a time of continental fragmentation that shaped the North American Craton about 1200 Ma (Burke and Dewey 1973);

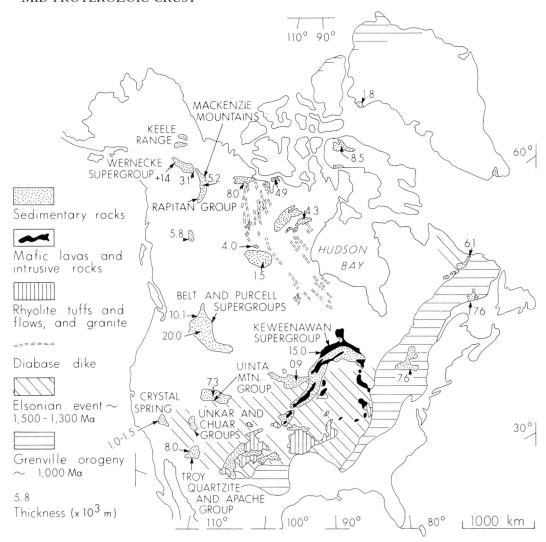

**Fig. 4-11.** Generalized distribution of mid to late Proterozoic rocks (1700–850 Ma) in the North American Craton. (From Stewart 1976, Fig. 1, and published with permission of the author).

(2) as disconnected epicratonic troughs possibly connected to a hypothetical ocean basin to the west (Stewart 1976); and (3) as diachronous intracratonic basins in a dominantly lacustrine environment (Link et al 1993).

Of these, the Belt–Purcell supergroups, extending over an area of about 200 000 km², constitutes a thick prism of dominantly siliciclastic and carbonate sediments deposited at ~1.45–1.2 Ga in a large epicratonic re-entrant of the mid-Proterozoic sea at the western margin of the North American Craton. It forms the most extensive mid-Proterozoic sedimentary sequence in the Cordillera. The dominantly fine-grained sequence comprises a monotonous assemblage of drab-coloured siltstones and mudstones, with intercalated units of quartz arenite, dolomite and limestone and with notably sparse coarser-grained lithologies.

Virtually the entire Belt–Purcell succession is allochthonous, having been thrust eastward substantial distances during the Cretaceous–early Tertiary. As a result, stratigraphic thicknesses are elusive and may have been overestimated.

The sequence in Glacier National Park, Montana, however, is ~2900 m thick, extremely well exposed and, for the most part, structurally simple. It is in lowermost greenschist facies metamorphism, and primary sedimentary structures are exceptionally well preserved (Harrison 1972, Horodyski 1983, Hedge et al 1986).

Lower Belt–Purcell sediments were deposited in a WNW-trending trough, and upper sediments in a NW-trending trough (Stewart 1972, McMechan 1981). Recent stratigraphic evidence provides strong supporting evidence for an overall enclosed basin with a tectonically active western margin (Ross et al 1992). This presumed westerly terrain is inferred to be different from the established North American basement terrains now bordering the basin. In explanation, one set of continental paleo-reconstructions favours Siberia (Sears and Price 1978) as the mid-Proterozoic western counterpart, whereas others favour Australia on the north (Bell and Jefferson 1987, Borg and De Paolo 1994) and Antarctica on the south (see Link et al 1993).

In southeast British Columbia, Canada, the similar coeval Purcell Supergroup, which contains the massive sediment-hosted Sullivan Pb–Zn–Ag ore bodies, is unconformably overlain by sedimentary rocks of the late Proterozoic Windermere Supergroup (Fig. 5-8, Table 2-8, column 3, McMechan 1981, Hamilton et al 1982). Other significant Belt–Purcell mineralization includes Cu–Ag, Co–Cu–Au, and minor magnesite, talc, asbestos and platinum.

## Midcontinent Rift System

In the general Lake Superior Region of central North America, Keweenawan rocks accumulated within the > 2000 km long Midcontinent Rift System (Figs 4-11 Table 2-8, column 1) (Green 1977, Halls 1978, Wallace 1981, Hinze and Wold 1982, Klasner et al 1982, Cannon et al 1991, 1993, 1994, Sims et al 1993, Cannon 1994). Especially between 1109 Ma and 1094 Ma (Cannon 1994), active rifting in the Lake Superior region in response to plate rotations and involving 60–90 km of crustal separation, initiated rapid extrusion of enormous volumes (> 400 000 km$^3$) of primitive, mid-ocean ridge-type, subaerial basalts including giant flows, with andesitic and rhyolitic associates, and subordinate (3–10%) red fluvial to marine interflow lithic sandstones and polymictic conglomerates. Al-rich olivine tholeiite predominates followed by alkaline olivine basalt and a large proportion of high-Fe tholeiite that grades into tholeiitic (basaltic) andesite. Intermediate compositions are sufficiently widespread that the traditional view of bimodal basalt rhyolite volcanism is no longer valid (Sims et al 1993). Volcanism was dominated by localized fissure-fed flood eruptions, overlain by intermediate to felsic stratovolcanic centres (Hamilton 1989). Major Cu(–Ag) ore deposits lie in the volcanic sequence. Coeval mid-Keweenawan intrusions, mainly huge layered gabbroic masses, are dominated by the very large Duluth and Mellen complexes with anorthosite, gabbro and troctolite fractionates and associated uppermost granophyres. Low-grade Cu–Ni mineralization is ubiquitous at the base of the Duluth Complex in particular (Foose and Weiblin 1986). Later (1080–1060 Ma) moderate regional compression steepened the sides of the Keweenawan basin to form the modern Lake Superior Syncline. Following the magmatism, the volcanic pile was deeply weathered, prior to burial by voluminous sediments mainly from exterior sources. Conformably overlying upper Keweenawan mainly sandstones (Oronto and Bayfield groups), including the native copper–chalcocite-bearing, argillite-rich Nonesuch Formation, are themselves unconformably overlain by Cambrian sediments.

Interpretation of recent seismic reflection profiles obtained by GLIMPCE (Great Lakes International Multidisciplinary Program on Crustal Evolution) (Cannon et al 1989, 1991, Shay and Tréhu 1993, Mariano and Hinze 1994, Samson and West 1994) shows an upper crust up to 32 km thick of layered rocks and an underlying basal crust extending to about 6 km above the reflection Moho at a depth of 45 to 55 km. The now-infilled Midcontinent Rift System, up to 32 km deep, probably represents Earth's greatest accumulation of intracratonic rift deposits. It is not known with certainty whether any part of the Archean–early Proterozoic basement completely separated during Keweenawan rifting. However, Keweenawan igneous rocks show little evidence of interaction with continental crust implying limited, if any, contact with older crust, hence probable basement separation during rifting.

Keweenawan strata in the Lake Superior Region continue under Paleozoic cover in a narrow continuous arc at least 2000 km long (Fig. 4-11; Table 2-8, column 1). Southwest of Lake Superior, the Midcontinent Gravity High extends for at least 1000 km to central Kansas. Southeast of the same lake, a similar though weaker linear anomaly extends to Ohio, where it is in contact with the trace of the Grenville Front. It is likely that the rift extends east of the Grenville Front and thus is older than the final Grenvillian thrusting and uplift (Rankin et al 1993). The resulting irregular U-shaped, southward opening, basalt–sediment-filled zone thereby marks a unique Precambrian midcontinental rift. Opinion is divided as to whether the continental rifting was induced by crustal heating and doming, or crustal heating by prior rifting (Hamilton 1989).

Interestingly, the close correspondence in time between Keweenawan magmatism and Grenvillian AMCG (anorthosite–granite) igneous activity (see above) points to continental collision as the ultimate control for both (Gordon and Hempton 1986, Emslie and Hunt 1990, Cannon 1994).

*Bruce River and Seal Lake groups (Central Labrador)*

Folded and metamorphosed Helikian supracrustal rocks in central Labrador are exposed in a cuspate area along the northern foreland zone of the Grenville Province and immediately west of the Nain Province (Fig. 2-7; Table 2-8, column 4). They comprise the Paleohilikian Letitia Lake (formerly upper Croteau) and Bruce River groups, both characterized by the presence of felsic volcanic porphyries with conglomerate–sandstone and the Neohelikian Seal Lake Group composed of arenite, arkose, shale, limestone and copper-bearing basalt lava flows, with associated Harp Lake dyke swarm, all part of what is termed the Central Mineral Belt (Ryan 1981, Wardle et al 1986, Cadman et al 1994).

## 4.7 SOUTH AMERICAN CRATON

Mid-Proterozoic (post-Transamazonian Orogeny) rocks are widespread in South America, forming—as in North America—both extensive epicratonic cover and coeval pericratonic mobile belts (Figs 1-3e, 1-5e). In the central Amazonian region (Amazonian Craton), unusually widespread volcanic–sedimentary platform cover with co-magmatic intrusions, products of craton-wide taphrogenesis, overlies the older Guiana and Central Brazil basements. To the southwest two N–NW-trending, outward younging mobile belts: (1) the substantially exposed Rio Negro–Juruena Belt (1750–1400 Ma), and (2) the poorly exposed Rondonian Belt, (1.4–1.6 Ga), itself comprising in full sub-Andean development both older (~1.3 Ga) San Ignacio parts and younger (~1.0 Ga) Sunsas–Aguapei units. To the east, in the São Francisco Craton of the Atlantic Shield, mid-Proterozoic rocks (1.7–1.2 Ga) form both narrow, NNW-trending, highly deformed, stratigraphically thick mobile belts (Espinhaço) and thinner undeformed coeval platform cover (Chapada Diamantino) on the stable craton to the east. Finally, in the intervening Tocantins Province of the Central Brazil Shield, the mid-Proterozoic Uruaçu Fold Belt (1.7–1.2 Ga) extends mainly south–southeastward for 1000 km from Goias Massif to Minas Gerais, (Almeida et al 1981, Gibbs and Barron 1983, Tassinari et al 1984, Hasui and Almeida 1985, Litherland et al 1985, Brito Neves 1986, Uhlein et al 1986).

### 4.7.1 AMAZONIAN CRATON: GUIANA AND CENTRAL BRAZIL SHIELDS (LESS TOCANTINS PROVINCE)

*Epicratonic cover*

Immediately following the main Transamazonian Orogeny (~2.0 Ga), extraordinarily widespread, rift-induced felsic–intermediate volcanic–plutonic activities, accompanied and followed by equally widespread continental sedimentation, was initiated across the breadth of the newly consolidated Amazonian Craton (Figs 2-12, 2-13; Table 2-9, column 1). The remains of the resulting volcanic–sedimentary cover are distributed craton-wide but their detailed correlations and depositional environments have yet to be fully established (Hasui and Almeida 1985).

*Uatuma Volcanic–Plutonic Complex and equivalents*

An unusually extensive and diverse association of rhyolite–rhyodacite–dacite–andesite tuffs, breccias, agglomerates, ignimbrites and volcaniclastic rocks, together with associated co-magmatic granitoid intrusions, commonly granite, granodiorite, diorite, syenite and diabase, form the Uatuma Volcanic-Plutonic Complex (Fig. 2-13) (Barbosa 1966, Basei 1975, Hasui and Almeida 1985). The 1.85–1.55 Ga felsic to intermediate volcanism of the Uatuma Supergroup, up to several thousand metres in thickness, probably affected no less an area than 2 000 000 km$^2$. The volcanic rocks typically occur in near-fracture depressions or grabens and rest with pronounced unconformity on the basement. In many places they are cut by subvolcanic porphyries, granophyric granites and diorites as well as by alaskites and microcline granites, including rapakivi types. The volcanic rocks are commonly interbedded with arkose and red polymictic sandstone, conglomerate, siltstone and tuffites. The same type of sediments typically overlie the Uatuma volcanic–plutonic rocks. These late, redbed sequences vary greatly in thickness, in some places reaching 5000–6000 m.

In the Guiana Shield to the north, correlatable felsic–intermediate volcanic-rich sequences intermittently cover an area of about 1 000 000 km² (Fig. 2-12).

*Roraima and Goritore groups and equivalents*

The conformably to disconformably overlying, terrigenous clastic sequences, represented by the Roraima and Goritore groups, are products of widespread horst-and-graben movements. They rest like a mantle upon the older predominantly volcanic formations and form a platform cover, commonly in the form of table mountains. The groups are composed of red to grey arkose and sandstone, partly red and green shales, some banded cherts and jaspilitic tuffs, and local basal and intraformational conglomerates. Cross-bedding, ripple marks, suncracks and rain-drop impressions abound. Volcanic tuffs and ignimbrites are locally present. Roraima tuffs yield ages of 1.75–1.66 Ga (Teixeira et al 1989). Goritore sediments, which directly overlie Uatuma volcanic rocks, are cut by basic dykes dated at 1475 Ma (Amaral 1974).

## Mobile belts

The major, ~500 km wide, NW–N-trending *Rio Negro-Juruena Mobile Belt* forms the western margin of the Guiana Shield and the centre–west part of the neighbouring Central Brazil Shield to the south. Together with buried extensions, it probably extends sinuously for at least 2500 km from the northern extremity of the Rio Apa Massif at the Brazil–Paraguay–Bolivia boundary northward to Venezuela (Fig. 1-5e; Table 2-9, column 1).

In both shields, the mobile belt is mainly composed of older partly regenerated crystalline basement, including recognizable granite–granodiorite gneiss and pluton, migmatites and numerous older wispy metasupracrustal (greenstone) belts (Figs 2-12, 2-13). Basement rocks are typically intensively faulted with widespread mylonites and cataclasites, and widely intruded by alkali syenites. Amphibolite facies metamorphism prevails (Cordani and Brito Neves 1982, Tassinari 1988). Available dates fall in the range 1.75–1.60 Ga (Teixeira et al 1989).

Of note is the 1550 Ma, unusually large (> 30 000 km²) Parguaza rapakivi granite mass in the northwestern Guiana Shield (Fig. 2-12), Venezuela, possibly Earth's largest anorogenic rapakivi intrusion.

A final mid-Proterozoic series of thermo-tectonic events which affected especially the southwestern margin of the Amazonian Craton formed the NW-trending *Rondonian (San Ignacio–Sunsas–Aguapei)* belt complex; this complex comprises the older 'interior' (to the northeast) San Ignacio Belt (~1.3 Ga) itself divided into northern and southern segments, and younger 'exterior' (to the southwest) parallel Sunsas (–Aguapei) Belt (~1.0 Ga), also divided into northern (Brazil) and southern (Argentina–Paraguay–Bolivia) segments (Fig. 1-5e; Table 2-9, column 1). Including presumed buried extensions, this belt complex, also up to 500 km wide, is considered by Litherland et al (1989) to extend sinuously from Rio de la Plata, Uruguay, in the south to Venezuela in the north. Other interpretations restrict it to the Amazonian Craton (Teixeira et al 1989). Where exposed in the Bolivia–Paraguay–Brazil junction (Brazilian Precambrian shield) (Fig. 2-13), the belt is marked by widespread faulting of older basement including granulites, gneisses and schist belts, and extensive younger granitoid intrusions, including ring-structured syenites, and widespread trachyte, rhyolite, andesite and basalt extrusions. Cassiterite mineralization is common, culminating in the largest Brazilian tin belt of the Jamari–Alta Candeias–São Miquel region. Scattered Rondonian dates fall in the range 1400–1100 Ma with, however, the main tin-bearing granites in the approximate range 1260–970 Ma (Bettencourt et al 1987, Litherland et al 1989).

The Rondonian (–Sunsas) Belt has been ascribed mainly to the reworking of pre-existing continental material (Teixeira et al 1989). However, this petrogenetic interpretation as well as its geographic distribution require further studies.

### 4.7.2 SÃO FRANCISCO CRATON

Mid-Proterozoic rocks are preserved in two narrow, N–NW-trending, thrust-bounded Espinhaço fold bents, which bisect the north–central part of the craton and in thin, tabular, coeval, diamond-bearing platform cover (Chapada Diamantino) overlying the stable craton to the east. The two narrow but stratigraphically thick (3–4 km) Espinhaço fold belts amalgamate southward to form the Serro do Espinhaço Range, an impressive dominantly quartzitic ridge extending more than 1400 km from Bahia in the north to Minas Gerais in the south, almost parallel to the east Brazilian coast (Fig. 2-15; Table 2-9, column 3) (Almeida et al 1981, Schorscher et al 1982, Dossin et al 1984, Uhlein et al 1986).

### 4.7.3 TOCANTINS PROVINCE (CENTRAL BRAZIL SHIELD)

Mid-Proterozoic strata form the Uruacu Belt extending to the east and southeast of the Goias Massif in Tocantins Province (Fig. 2-14; Table 2-9, column 2). Coeval strata may lie in both the Paraguay–Araguaia Belt bordering Guapore Shield to the west and the Brasilia Belt bordering the São Francisco Province to the east; the latter two belts are considered in Chapter 5. Recent studies have provided a tentative date of ~1565 Ma for the Niquelândia layered intrusion, one of a string of disconnected mafic–ultramafic complexes in the Goias Massif (Ferreira-Filho et al 1994).

#### Uruaçu Belt

The Uruaço Belt, located in the centre–east part of the Province, extends mainly southeastward for more than 1000 km, from Natividade (Goias) on the north to the Guaxupa region (Minas Gerais), from where it continues eastward to be limited by a major transcurrent fault adjoining the N-trending Brasíliano (500 Ma) Belt (Fig. 2-14) (Almeida and Hasui 1984). The mainly arenaceous–pelitic strata of the Uruaçu Belt were metamorphosed to prevailing greenschist facies during the Uruaçu event dated at ~1000 Ma. South of the Pirineus (Pirinopolis) Megaflexure (see above) Uruaçu folds trend SE and verge NE towards the São Francisco Craton, whereas to the north of the flexure they trend NNE.

## 4.8 AFRICAN CRATON: SOUTHERN AFRICA

Mid-Proterozoic rocks in southern Africa comprise the pericratonic Namaqua–Natal mobile belt to the south of the Kalahari Craton, and an arcuate set of coeval, disconnected rift-induced basins in the Koras–Ghanzi–Rehoboth region of northern Namaqualand (Figs 1-3f (i), 1-5f (i,ii)).

### 4.8.1 NAMAQUA–NATAL MOBILE BELT

The mid-Proterozoic Namaqua–Natal Mobile Belt is a 200–500 km wide, high grade domain extending in arcuate form for 2000 km across the southern flank of the Kaapvaal Craton and beyond, from the Atlantic coast in South Africa and southern Namibia to Natal on the Indian ocean. It is divided into the Namaqua Province in the west and the Natal Province in the east, with the intervening area buried beneath thick Mesozoic cover. The buried boundaries of the Belt have been mainly defined by geophysical surveys. The southern boundary is expressed by an ESE-trending, highly conductive, magnetic mass of dense rock located at shallow crustal depth and called the Southern Cape Conductive Belt; it may represent partially serpentinized mafic rocks marking oceanic crust subducted from the south product of collisional tectonics (de Beer and Meyer 1983) (Fig. 4-12).

#### Namaqua Province

The exposed Namaqua Domain occupies the Orange River Basin, an irregular triangle about 800 km on the side extending west from the southern extension of the Kheis Belt to the Atlantic coast and northwest therefrom to Lüderitz vicinity (26°38′S) in Namibia (Joubert 1981, Blignault et al 1983, Hartnady et al 1985, Van Aswegen et al 1987, Colliston et al 1991, Thomas et al 1994). Its eastern boundary is the Namaqua Front, a marginal zone characterized by faulting, shearing and metamorphic transition which separates high grade Namaqua infrastructure on the west from low to intermediate grade, early Proterozoic metasupracrustal rocks of the Kheis Belt in the east. The western boundary of the Namaqua Province is the Atlantic coast, except for the NW-trending late Proterozoic Gariep Basin. The Namaqua Province itself forms a regional gravity high, suggesting high density lower crust different from that of the adjoining Kaapvaal Craton (De Beer and Meyer 1983, De Beer and Stettler 1988).

The Namaqualand Province is divided into three main SE-trending subprovinces: (1) Richtersveld in the centre; (2) Bushmanland in the south; and (3) Gordonia in the north (Fig. 4-12).

#### Subprovinces

(1) The *Richtersveld Subprovince* occupies a comparatively small, wedge-shaped area about 150 km long (E–W) at the western margin of the province, separating NW-trending Gordonia rocks on the northeast from E-trending higher grade Bushmanland gneisses in the south (Blignault 1977). The western boundary with the late Proterozoic Gariep Domain is marked by deformation and metamorphism (Kröner 1978). The Richtersveld Domain is underlain by (1) grey biotite gneiss, of uncertain origin which is at least 1.9 Ga old (Kröner 1978); and by (2) mafic and felsic lavas, ignimbrites and

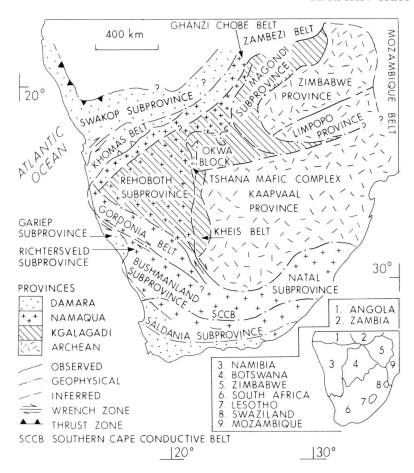

**Fig. 4-12.** Major tectonic provinces and subprovinces of southern Africa. Southern Cape Conductive Belt is shown as a geophysical line bisecting (E–W) the Namaqua (-Natal) Province (Mobile Belt). (From Hartnady et al 1985, Fig. 1, and reproduced with permission of the authors).

minor metasediments, including quartzites, pelitic schist and BIF of the 2.0 Ga old Orange River Group; with (3) the associated 1.9–1.7 Ga old Vioolsdrif plutonic suite. A very large, low grade porphyry copper deposit occurs at Haib in Orange River rocks (Joubert 1981).

Along the southern boundary the lower metamorphic grade Richtersveld rocks were thrust southwards over the higher grade Bushmanland Subprovince.

The currently preferred model for the Richtersveld association involves the development of an island-arc complex at 2.0–1.9 Ga, and its late mobilization at depth along an active continental margin (Reid, quoted in Hartnady et al 1985).

(2) The *Bushmanland Subprovince* to the south is dominated by intensely deformed, high grade gneisses, granulites and massive granitoids. The distinction between pre-Namaqua 'basement' rocks and Namaqua supracrustal cover rocks is difficult and controversial. Unanswered questions concern (a) the sedimentary or volcanic origin of certain paragneisses and their distinction from syn-tectonic orthogneisses; and (b) the identification of certain rocks either as tectonically 'reworked' Archean basement, as proposed by Kröner (1978), or as reworked early Proterozoic cover of mixed volcanoplutonic origin (Hartnady et al 1985, Thomas et al 1994). It is generally agreed, however, that certain heterogeneous gneiss–metapelite–quartzite assemblages could well represent vestigial basement-cover relationships (Hartnady et al 1985).

The dominant, mainly east-to-west fabric in the Bushmanland Subprovince has been related to a regional metamorphism dated at ~1.2 Ga (Clifford et al 1981). The main isoclinal folds are southward vergent and the large open, E-trending folds that extend across Bushmanland are a later phase. Extensive megacrystic and augen gneisses, dated at ~1.2 Ga and younger, intruded the paragneisses as subhorizontal concordant sheets of batholithic dimensions. Their granitoid precursors are attributed to crustal anatexis, reworking a 2.0–1.9 Ga old crustal protolith similar to the Vioolsdrif suite of the Richtersveld Subprovince (Reid and Barton 1983).

The Bushmanland Subprovince is unusually rich in mineral deposits. The Okiep copper sulphide ore bodies, which represent the oldest mining district in South Africa, are enclosed in norite, diorite and anorthosite and contain about $24 \times 10^6$ t of ore with an average grade of 1.58% Cu (Boer 1991, Cawthorn and Meyer 1993, Andreoli et al 1994). The nearby Aggeneys (Cu–Zn–Pb–Ag) and Gamsberg (Zn) deposits, associated with quartzites and aluminous schist, are calculated to have collective reserves exceeding $500 \times 10^6$ t at grades of the order of 0.55% Cu, 2.81% Pb and 1.82% Zn (Joubert 1981, Moore et al 1990). The Copperton Cu–Zn deposit associated with metavolcanic rocks contains $47 \times 10^6$ averaging 1.7% Cu and 3.8% Zn (Joubert 1981). Pegmatites yield Sn, W, Be, Li and REE.

(3) The eastern contact of the WNW-trending up to 120 km wide *Gordonia Subprovince* (Belt) represents a complex zone of convergent thrusting and folding against the adjoining Kheis Belt, followed by dextral shearing with an estimated minimum length of 140 km. The subprovince comprises three parallel domains: (1) the Upington Terrain, a narrow (up to 35 km) belt of quartzite–pelite adjoining the Kheis Subprovince to the east; (2) the still narrower (10–15 km) Jannelsepan volcanoplutonic amphibolite complex; and (3) the broader (75 km) high-grade metasupracrustal Kakamas Terrain in the west and southwest (for details see Hartnady et al 1985). The southwest contact of the subprovince is characterized by southerly vergent overthrusts and associated wrench zones.

*Tectonic model*

The relations in the Namaqua Province support some form of plate convergence involving relative movements of suspect terranes. So far, no crust older than 2.0 Ga has been identified. Certain geochemical signatures point to a major juvenile crust-forming event in the 1.3–1.2 Ga interval (Barton and Burger 1983). The presence of southwestward vergent overthrusts at the southerly margin of the Gordonia Subprovince suggests that the Bushmanland and Richtersveld subprovinces may be part of an older, exotic (2.0–1.6 Ga) microcontinent, accreted about 1.3 Ga ago (Hartnady et al 1985), part of a continent–continent convergence during the Namaqua tectogenesis. However, all observed structures can also be explained by a single long-lived compressional event leading to a progressive ductile shear model (Colliston et al 1991).

## Natal Province

About 1800 km to the east in the Natal Domain (Province), basement crystalline rocks form a zone extending north along the Indian Ocean coast for 350 km where they are thrust northward on to the Kaapvaal craton. The domain contains a complex of granitic and migmatitic gneiss with general E–ENE structural trend and prevailing medium to high grade regional metamorphism indicative of erosion to a deep crustal level, in the order of 20–25 km (Matthews 1972, 1981). At the northern contact, the Natal Thrust Belt, a narrow (2–5 km wide) southward dipping, imbricate complex of 1.2 Ga old lavas, tuffs and metasediments occurs in a tectonic melange beneath an amphibolitic nappe complex. Available dates fall in the range 1200–900 Ma, which is regarded as the time span of the main tectono-thermal event (Matthews 1981).

*Zones*

Based on contrasting lithologic, structural and metamorphic features, the Natal Subprovince is divided by Matthews (1981) into four major E-trending zones, from north to south:

(1) *Northern Frontal Zone*, 15–30 km wide, is characterized by extensive thrust sheets of predominantly amphibolitic gneiss including metaserpentinites that have been transported northwards on to the southern flank of the Kaapvaal Craton (see Matthews 1981, fig. 10-11).

(2) *Migmatite and Granite Gneiss Zone*, about 60 km wide, is composed of four major alternating, ENE-trending linear belts of highly deformed, subvertically dipping migmatitic gneiss, with intervening belts of well foliated granitoids, including porphyroblastic and augen gneiss.

(3) *Granitic Zone*, about 120 km wide, comprises extensive (batholithic) areas of a remarkably homogeneous, generally megacrystic granitoid within a framework of granitic and migmatitic gneiss.

(4) *Southern Granulite Zone*, 20–30 km wide, is composed of metasedimentary granulites and charnockite intrusions.

*Tectonic model*

Matthews (1981) interprets the northerly frontal zone as an ophiolite complex and suggests that it contains tectonic slices of transformed oceanic crust obducted from a marginal basin on to the southern flank of the Kaapvaal Craton. According to this model, the Natal granite–gneiss complex and infolded metasupracrustal sequence represent the northern margin of a continental plate that was deformed above a subduction zone, as the leading edge moved into the collision zone with the Kaapvaal Craton.

Based on limited dating, the members of this basement are characterized by ages of ~1000 Ma. Intrusion of charnockite continued until ~950 Ma ago, following the 1.2 Ga peak of high grade metamorphism in Namaqua Province to the west (Matthews 1972).

## 4.8.2 KORAS–SINCLAIR–GHANZI RIFTS

In northern Namaqualand–Rehoboth (Fig. 4-12) a number of disconnected, relatively undeformed, rift-induced, 1.3–1.0 Ga volcanosedimentary basins form an arcuate loop, concave to the east, around the western and northern margins of the Kalahari Craton (Fig. 1-5f(i)) (Borg 1988). The depositional troughs represent narrow, fault-bounded continental rift grabens with a structural style and sedimentation pattern controlled by strong vertical tectonism. The volcanic–sedimentary infill represents a succession of immature coarse clastic continental redbed sediments and distinctly bimodal volcanic rocks with some intermediate volcanic rocks. The stratigraphic sequences are commonly 8000–15 000 m thick. Sediments were deposited in narrow yoked and graben-like fans or alluvial aprons along the flanks of the fault scarps. The faults bounding the grabens and horst blocks controlled the extrusion and distribution of the volcanic rocks.

The basins and their volcanosedimentary infill formed diachronously, in general younging from an age of 1300–1200 Ma for the NW-trending Koras–Sinclair branch to 1050–900 Ma for the ENE-trending Ghanzi–Lake N'Gami branch. The two branches are interpreted either as sequential parts of a mantle plume-induced propagating continental rift system, parent to the ensuing early Damara Rift in Namibia (Borg 1988), or to the initiation of subduction in response to continental collision at 1.3–1.2 Ga (Hoal 1993).

## 4.9 AFRICAN CRATON: CENTRAL AFRICA

Additional intracratonic extensional structures are present in east–central Africa in the form of the essentially coeval Kibaran, Irumide and Lurio belts (Figs 1-3f(ii), 1-5f(i,ii)).

## 4.9.1 KIBARAN BELT

### Setting

The NE–NNE-trending Kibaran Belt (Kibarides) can be followed practically continuously for more than 1500 km from southern Zambia through Tanzania and Burundi to Rwanda, where it swings to the northwest, ending in Uganda and northern Zaire (Fig. 3-16). This predominantly low grade pelitic–arenaceous intracratonic belt, which occupies a restricted fault-bounded zone, ranges from 100 to 500 km in width. Kibaran rocks are also known as Burundian in the three countries of northern Zaire, Burundi and Rwanda, and as Karagwe–Ankolean in Uganda and Tanzania (Mendelsohn 1981, Cahen et al 1984, Klerkx et al 1987).

### Lithostratigraphy–structure

The *Kibaran Supergroup* is about 10 500 m thick in the type area (Kibara Mountains of central Shaba). It comprises a rather monotonous sequence of phyllites and schists with some intercalated quartzites, conglomerates, mafic–felsic volcanic rocks, carbonaceous shale, and stromatolitic carbonates. Estimated lithic proportions in the complete Kibaran section are fine clastics plus chemical sediments: coarse clastics: volcanics = 6:3:1 (Cahen et al 1984). Deformed mafic–ultramafic rocks locally present along the 350 km-long Kibaran belt, including peridotites, norites and anorthosites (Tack et al 1990, 1992) are tentatively interpreted as ophiolites (Moores 1993).

Northeastward along strike, the corresponding *Burundian Supergroup*, about 12 000 m thick, is also dominated by pelitic rocks with quartzitic intercalations, which are mature and well-sorted in lower levels, but progressively more immature and poorly sorted above (Klerkx et al 1987).

Granitic rocks, emplaced at 1330–1260 Ma and again at ~1249 Ma, mainly in lower stratigraphic levels, form extensive and complex intrusions that are typically associated with smaller mafic intrusions (Tack et al 1994). The late granites are allied with cassiterite, tantalite and wolframite mineralization.

The entire succession has been folded along NNE trends with a marked westerly vergence, the folding dying out to the foreland. Greenschist-facies metamorphism prevails, with local amphibolite-facies present in the lowermost group and in proximity to granitoid intrusions.

## Age

Kibaran–Burundian sedimentation started around 1400 Ma, followed by bimodal magmatism including large granitoid intrusions accompanied by progressive horizontal deformation, in the interval 1350–1260 Ma, when the main magmatic-structural character of the belt was acquired. Subsequent upright folding in the period 1180–1100 Ma has not dramatically modified the earlier features (Klerkx et al 1987).

## 4.9.2 IRUMIDE BELT

Five hundred kilometres to the south, the parallel, lithologically similar, pelitic–arenaceous Irumide Belt (Province) extends southwestward for 1000 km from Lake Malawi on the northeast to pre-Katanga basement inliers with Irumide trend situated northwest of Lake Kariba on the southwest. The Irumide belt occupies the triangular area lying between both the Bangweulu (Zambian) Block and Lufilian Arc on the northwest, the Zimbabwe Craton with flanking Zambezi Belt on the south, and the Mozambique Belt on the east. To the northeast the Irumide Belt adjoins the SE–ENE-trending early Proterozoic Ubendian–Usagaran belt systems (Figs 3-16, 4-13) (Cahen and Snelling 1966, Fitches 1971, Daly 1986a,b, Ring 1993).

The Irumide Belt is related by Cahen et al (1984) to the Kibaran Belt system to the north and, alternatively, by others to the Mozambique Belt system to the east (Daly 1986a, Ring 1993). The Irumide Belt has been widely subjected to Pan-African thermal effects (Vail et al 1968) which complicate the chronologic correlation of structures within the belt (Daly 1986a). As considered further below, the contrary views expressed above are readily reconciled in terms of a major continental-scale plate collision in eastern Africa with which all three named belts were associated. The partly coeval Lurio Belt, 600 km to the southeast within the Mozambique Belt (Fig. 4-13) is considered below (Chapter 5).

## Lithostratigraphy and tectonic development

The northwestern margin of the Irumide Belt is well defined in northern Zambia (Fig. 4-13). Here it comprises a northeastward striking, up to 10 km thick, conformable sedimentary sequence, the Muva Supergroup (Daly and Unrug 1982), composed of alternating layers of quartzites and pelites, each layer typically hundreds of metres thick. These pass eastward into a granitoid-migmatite terrain where only remnants of metasediments are preserved (Daly and Unrug 1982).

Muva sediments can be traced southeastward across strike towards the internal zone of the belt, where they appear to thicken markedly and take on an increasingly marine character. The subsequent Irumide Orogeny resulted in the formation of a distinct northern foreland region and, to the southeast, an internal zone, the Irumide Belt proper, involving granitoid magmatism and amphibolite facies metamorphism.

The Irumide structures pass eastward without interruption into gneisses of the Mozambique Belt. These gneisses locally form extensive basement to Muva metasediments and include large tracts of granulite-facies rocks. The granulites at least partly represent reworked early Proterozoic (~2300 Ma) continental crust. However, in further complication, to the southeast in southern Malawi occur granulites of Irumide age (i.e. ~1100 Ma) (Haslam et al 1983). Moreover, large southeastward-transported klippes of corresponding Irumide age are reported in Mozambique (Daly 1986a). In brief, continuity of structure and movement directions between the Irumide Belt and the adjoining higher grade Mozambique Belt suggests that the two belts were formed during the same orogenic cycle, about 1100 Ma ago (Daly 1986a).

According to Daly (1986a), the Irumide Belt is best viewed as the western margin of the Southern Mozambique Belt, the two having been developing during the ~1100 Ma old orogenesis. This places the Southern Mozambique Belt into the Kibaran age bracket, as originally proposed by Holmes (1951). Hence the northward extensions of the Mozambique Belt, representing the Pan-African of northeast Africa with an older age limit ~900 Ma, is distinctly younger (see below).

The horizontal nature of the tectonics, the syntectonic granulite-facies metamorphism, the presence of possible suture zones in the Mozambique Belt and the associated volcanic arcs (see Chapter 5), all suggest that the Kibaran Orogen resulted from collisional tectonics involving accretion on to the eastern flank of Africa. Within such a continental collisional system, the Irumide Belt in Zambia would represent an intracontinental basin (Daly 1986a).

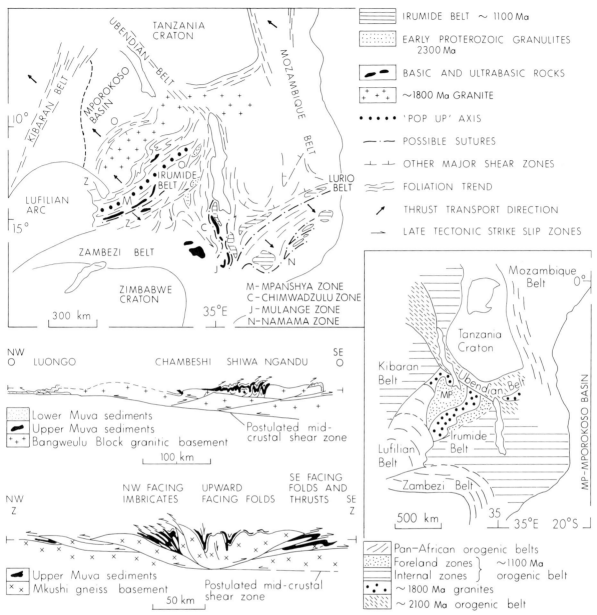

**Fig. 4-13.** Geologic sketch showing the regional setting of the Proterozoic mobile belts of south Central Africa. The Irumide foreland fold-and-thrust zones are shown in relation to the internal zone of the orogen. The locations of two sections presented in the figure, i.e. O–O and Z–Z, are shown. Section O–O—A sketch cross-section through the Northern Irumide Belt margin. Note locations of Luongo fold-and-thrust zone, Chambeshi fold-and-thrust zone and Shiwa Ngandu fold zone. Note the foreland fold-and-thrust zones facing to the northwest linked to a postulated mid-crustal shear zone. Section Z-Z—A sketch cross-section through the Southern Irumide Belt. Note how the foreland zones of the northeast have been replaced by large scale imbricates in the southwest. The facing direction of the major structures clearly changes across the 'pop-up' structure which is thought to root in a mid-crustal shear zone. (From Daly 1986a, Fig. 1, and reproduced with permission of the author).

## Mporokoso Basin, Irumide Foreland Zone

The preserved Muva sediments on the northern range of the Irumide Belt are separated by 50–100 km of intervening early Proterozoic Ubendian (2.2–1.8 Ga) basement from a northern mid-Proterozoic depocentre, the Mporokoso Basin (Fig. 4-13). This broadly triangular unit, some 200–400 km on the side, occupies much of the northern half of the Bangweulu Block.

The Mporokoso Basin is filled with up to 5060 m

thick of continental and marine clastic sediments covering an area of 50 000 km². The comparable siliclastic metasediments of the Muva Group to the south either postdate the Mporokoso Group or are equivalent to the upper part of it (Daly and Unrug 1983). The Mporokoso Group lies in the foreland region of the Irumide Belt and was locally overthrust by Ubendian basement. This deformation generated a narrow, arcuate fold-and-thrust belt (Luongo) in the southern part of the basin. Greenschist facies metamorphism accompanied this deformation, with a minimum age of 1020 Ma (Daly 1986a).

The development of the Mporokoso Basin is attributed to crustal extension and subsidence representing the gravitational collapse of a previously thickened (Ubendian) crust by tectonic and magmatic processes, which is considered to be a common feature of mid-Proterozoic basins of the world (Andrews-Speed, 1989).

## 4.10 AFRICAN CRATON: NORTHERN AFRICA

Mid-Proterozoic rocks are locally exposed in the Darfur Province of northeast Africa, and in the Tuareg Shield of West Africa (Figs 1-3f(ii), 1-5f(i,ii)).

### 4.10.1 DARFUR AND EASTERN TCHAD BASEMENT

A large area of basement rocks exposed in eastern Tchad extends across the Sudan border into Darfur Province (Fig. 3-17) (Vail 1978). The main rock types are granitic gneiss, and quartzofeldspathic, graphitic and pelitic schists with occasional calcareous and volcanic units. These rocks are all deformed about NE-trending axes. Amphibolite facies metamorphism prevails. Numerous late orogenic granitoid batholiths occur, some with Sn–W-bearing pegmatites and quartz veins. On the basis of limited data, Vail (1976) tentatively proposed the presence of a ~1 Ga linear mobile belt, incorporating Darfur rocks, which would link the Kibaran Belt in the south through the Central African Republic to Darfour in the north. However, this interpretation requires corroboration.

### 4.10.2 TUAREG SHIELD

In the Pharusian belt of the Tuareg Shield (Fig. 2-23) local Al-rich quartzites and pelitic schists have been dated at 1742 and 1843 Ma (Andreopoulos-Renaud, quoted in Caby et al 1981) placing them at the early–mid-Proterozoic transition.

Also to the east, in Central Hoggar, Bertrand (1974) confirmed the presence of the ~100 m-thick Aleksod series (Table 2-14, columns 2–4), composed of amphibolite, kyanite-bearing schists, quartzite, marble and calcsilicate gneiss. Deformation and metamorphism of Aleksod rocks have been dated at 1130 Ma (Bertrand and Laserre 1976), thereby suggesting a Kibaran (~1.0 Ga) presence in the Tuareg Shield. However, recent U–Pb zircon dating on plutonic rocks in Central Hoggar has failed to find any evidence of ~1.0 Ga ages (Bertrand et al 1986) raising in question whether there is a Kibaran presence in central Hoggar. Clearly, further studies are required to resolve this point.

## 4.11 INDIAN CRATON

Mid-Proterozoic activities in the Indian Subcontinent are mainly represented by several large epicontinental sedimentary basins, troughs, grabens or aulacogens located primarily in the central (Cuddapah, Kaladgi, Godavari) and northern (Vindhyan) parts (Figs 1-3g, 1-5g, 5-17; Table 2-15, column 1). In addition, some mid-Proterozoic events are documented in both the Eastern Ghats Belt and neighbouring Sri Lanka (see Chapter 5). Diamondiferous kimberlites of age range 1.5–1.0 Ga, source of the famous Koh-i-Nur, Great Mogul and Taverner blue diamonds, are found in the northern (Panna and Jungel) and southern (Wajrakarur) areas (Miller and Hargraves 1994).

### 4.11.1 CUDDAPAH–KALADGI–GODAVARI REGION

#### Cuddapah Basin

The Cuddapah Basin in the south and the great Vindhyan Basin in the north are two mega-depositories developed in mid- to late Proterozoic time (Fig. 5-17). These and other contemporary basins, all of passive continental margin type associated with extensional tectonic regimes (Kale 1991), are characterized by arenaceous–pelitic redbed successions interspersed with local carbonates and volcanic rocks. Cuddapah strata are primarily mid-Proterozoic, whereas the Vindhyan Basin is mainly, if not exclusively, late Proterozoic in age.

*Cuddapah Basin*, in south–central India, covers an area of 21 600 km². It has a crescent shape, convex to the west, with a length of 400 km and a maximum width of 145 km (Fig. 5-17) (Pichamuthu 1967, Meijerink et al 1984, Nagaraja Rao et al 1987). Sedimentary thickness is about 600 m. Cuddapah basal sediments clearly unconformably overlie Peninsula gneiss (~3.0 Ga). Sediments in the western portion of the basin are essentially unfolded with shallow eastward dips towards the basin centre. The eastern portion of the basin, in contrast, is a 50 km wide arcuate zone of steeply dipping, isoclinally folded, inverted strata which form the Nallamalai hills. The Cuddapah Basin is tectonically overlain to the east by an Archean thrust mass, which includes the Dharwar-type (3.0–2.6 Ga) Nellore schist belt. Deep seismic soundings show that this thrust and several small imbrications within the basin are steeply inclined and penetrate to the base of the crust (Drury and Holt 1983). Clearly the present Cuddapah Basin is but the remnant of an original depository that extended a considerable distance to the east.

The Cuddapah Basin is of composite structure, being divided into three separate units of deposition, each forming a stratigraphic group, as follows, with approximate ages in brackets (Meijerink et al 1984): Cuddapah (2000–1470 Ma), Nallamalai (1470–1000 Ma) and Kurnool (1000–500 Ma).

Kaladgi–Godavari Successions

Two hundred kilometres to the northwest, rocks of the Kaladgi and nearby Bhima successions are intermittently exposed along the southern border of the Deccan Trap for more than 250 km (Fig. 5-17). Kaladgi and Bhima strata, each ~2000 m thick, closely resemble those of the Cuddapah Basin and are thought to represent either an original extension or separate coeval basins (Viswanathian 1977).

Pakhal sediments extend along the NW-trending Godavari graben for 440 km where they are exposed in two parallel belts, separated by about 50 km of coal-bearing Gondwana sediments. This distribution pattern is due to simple steep normal faulting of eastward-dipping Pakhal–Gondwana strata (Mathur 1982, Chaudhuri and Hower 1985, Naqvi and Rogers 1987, Sreenivasa Rao 1987).

In the Bastar and Chattisgarh basins to the east (Fig. 5-17), the Indravati Group and Raipur Limestone respectively are correlated by Raha and Sastry (1982) with Cuddapah strata.

Based on biostrome content, certain Himalayan inliers, as at Jammu, Shali, Deoban, Pithoragarh (Nepal), Rangit and Bhutan Buxa, are of lower to mid-Riphean age and are generally correlated with the Chatisgarh and Bastar sequences (Raha and Sastry 1982).

### 4.11.2 EASTERN GHATS BELT

Possible mid-Proterozoic rocks in the Eastern Ghats Belt (see above) are represented by widespread charnockites, khondalites, kodurites and related rocks, including calcsilicates, mafic granulites, anorthosites and alkaline rocks (Radhakrishna and Naqvi 1986). Although the map pattern of the principal Eastern Ghats lithologies is very complex and few reliable age dates are available, it has been noted that charnockites are more common in central–western parts of the belt, whereas khondalites and other pyroxene-free rocks are more common in the east (Fig. 5-17).

The Eastern Ghats Belt has a long, complex history. The Chilka Lake anorthosite is dated at 1400–1300 Ma (Sarkar et al 1981) and the Kunavaram alkaline complex at ~1300 Ma (Clark and Subbarao 1971). Granites intruding the Nellore Belt provide ages in the range 1615–995 Ma (Gupta et al 1984b). The timing of the main metamorphism producing the bulk of the charnockites is not well established. In this regard, Eastern Ghats granulites at Anakapalle are, in a super continental paleo-reconstruction, juxtaposed directly against the 1000 Ma granulite belt of East Antarctica (Dasgupta et al 1994). Eastern Ghats granulites may well have a corresponding age. Formation of the western margin of the Eastern Ghats was probably caused by some type of continent–continent collision (Naqvi and Rogers 1987).

## 4.12 AUSTRALIAN CRATON

Mid-Proterozoic events in this craton are conveniently divided into older Carpentarian (1.8–1.4 Ga) and younger Musgravian (1.4–1.0 Ga) divisions. The older Carpentarian events are dominated in North Australia by the concurrent development of (1) widespread bimodal anorogenic magmatism; and (2) major sedimentary basins, including (a) McArthur Basin; (b) Lawn Hill Platform and Mount Isa Orogen; (c) Birrindudu Basin; and (d) Kimberley Basin; and in South Australia, by accumulation of (e) Gawler Range Volcanics. The younger Musgravian events include development of the (a) Victoria River; and (b) South Nicholson basins in North Australia; (c) Bangemall Basin in West Australia; and (d) Amadeus–Albany–Fraser tectonic zone in Central Australia (Figs 1-3h, 1-5h; Table 2-17).

### 4.12.1 CARPENTARIAN DIVISION

#### Anorogenic magmatism

Following the 1870–1850 Ma Barramundi igneous activity and a further depositional break of 50–70 Ma, widespread crustal rifting, with laterally extensive thermal subsidence, marked the development of several large sedimentary basins with concurrent episodic bimodal magmatism in the form of intrusive–extrusive associations, chiefly in the period 1800–1600 Ma but extending to 1500 Ma. These bimodal magmatic associations are in anorogenic settings, have dominantly A-type chemical affinities (Collins et al 1982) and can be distinguished from the older Barramundi suite on the basis of their less coherent age and contrasting geochemistry, notably enrichment in alkalis, Zr, Nb, Y and REE (Page 1988). Most fall in four age ranges: 1800–1780 Ma, 1760–1740 Ma, 1670–1640 Ma and ~1500 Ma.

The granitoid intrusions are of two types: large anorogenic batholiths including rapakivi granites, and small microgranite plutons. Bimodal rhyolite–basalt extrusive suites are closely associated. Examples are widespread in the north in Mount Isa, the Granites–Tanami, Tennant Creek and Arunta inliers and in the south, in Gawler Craton (Gulson et al 1983, Wyborn et al 1987, Page 1988). The Charleston Granite, Gawler Craton, which crystallized at 1585 Ma, provides inherited zircon core dates of ~1780, ~1970, and >3150 Ma, thereby demonstrating diverse crustal heritage, and the presence of ancient crustal material at depth in the craton (Creaser and Fanning 1993).

#### McArthur Basin

The SE-trending McArthur Basin, about 200 000 km$^2$ in exposed area, is the principal depositional element of the North Australian platform cover (Figs 1-5h; Table 2-16, column 2). The stratigraphic succession is characterized by vast thicknesses (up to 12 km) of shallow marine–tidal–fluviatile deposits, widespread evaporitic–carbonate (sabkha) sequences, abundant stromatolites and microfossils, exceptionally low grade of metamorphism, and the huge McArthur River Pb–Zn–Ag deposit, considered a model for stratiform ore formation (Oehler 1978, Plumb et al 1979, Williams 1980, Plumb et al 1981, Jackson et al 1987).

The McArthur Basin lies towards the eastern edge of the North Australian Craton (Fig. 1-2) and next to the Mount Isa continental margin belt. The McArthur Basin, Lawn Hill Platform and Mount Isa Orogen, as well as the Birrindudu Basin located 800 km to the west, all have similar stratigraphic successions and parallel evolutionary histories (Fig. 3-22). The McArthur Basin is bounded by and/or unconformably overlies the early Proterozoic Pine Creek (west), Arnhem (northeast), and Murphy (south) inliers (Figs 3-21, 3-22). Paleogeographic trends indicate that a basin margin existed not very far east of the presently exposed McArthur Basin rocks. Thus the setting was that of a linear marine embayment, comparable either to the Persian Gulf or Red Sea, or to the Gulf of California (Plumb et al 1981).

The paleogeography of the basin at large was dominated by the N-trending, 600 km long, 75–100 km broad centrally disposed Batten Trough (Fig. 3-21) (Plumb et al 1981, fig. 4.10), a tectonic feature controlled by syn-depositional faults, in which up to 12 km of shallow-water sediments accumulated compared to only 1.5–4 km on adjoining stable shelves (e.g. Arnhem shelf). Large vertical uplifts subsequently reversed the Batten Trough, into a horst-like feature, with basement rocks locally exposed within it.

#### Structure

The McArthur Basin was deformed in response to complex block faulting, particularly along the Batten, Walker and Urapunga fault zones. The major fracture systems were inherited from at least early Proterozoic time and were repeatedly reactivated throughout the basin's history. At least two sets of conjugate faults—north–northwest to north, and northeast—of uncertain relative ages, have resulted in displacements up to 7.5 km.

#### Mineralization

Mineralization in the McArthur Basin is dominated by the exceptionally large, shale-hosted, stratiform HYC Pb–Zn–Ag deposit and similar smaller deposits nearby. Reserves of the single HYC deposit are about 190 × 10$^6$ t, assaying 9.5% Zn, 4.1% Pb, 0.2% Cu and 44 g Ag/t. Finely banded galena and sphalerite is hosted by laminated, turbiditic, pyritic–carbonaceous–dolomitic shales.

The major copper resource of the McArthur Basin is contained in several breccia pipes and veins in the Redbank area (Plumb et al 1990). Small tonnages have been produced so far from substantial ore reserves.

Low grade iron deposits at Roper lie in several beds of the Serwin Ironstone Member. The ironstones comprise combinations of oolitic–pisolitic hematite, goethite, greenalite and minor chamosite in a matrix of siderite. Small showings of uranium, gold and diamonds are reported.

### Age

Available data support the conclusion that McArthur Basin deposition occurred in the 1700–1400 Ma interval (Kralik 1982, Plumb 1985, Page 1988). The main mineralized member of the McArthur Group is dated at 1690 Ma (Page 1988).

### Lawn Hill Platform and Mount Isa Orogen (Inlier)

The Mount Isa Orogen (Inlier) and undeformed correlative Lawn Hill Platform immediately adjoining to the north, are exposed over about 80 000 km$^2$ (Figs 3-21, 3-22) and, on geophysical evidence, the orogen extends southward for at least another 70 000 km$^2$ and northward to the western edge of the Cape York Peninsula (Fig. 3-21) (Williams 1980, Plumb et al 1981, 1990, Plumb 1985, Blake 1987, Page 1988).

### Setting and Lithostratigraphy

Mount Isa Orogen and Lawn Hill Platform include seven major stratigraphic subdivisions (Plumb et al 1981, table 4.VII) totalling 6–16 km thick and dominated by transgressive clastic–carbonate sheets, and assorted volcanic rocks, including conspicuous felsic tuffs and flows. Of these, the *Mount Isa Group*, up to 7.5 km thick, is composed of dolomitic and pyritic siltstone, black shale, dolomite, K- and Na-rich tuffites, chert, quartzite and siliceous and dolomitic breccia, which host the major stratiform Pb–Zn–Ag deposits.

Paleogeographic studies suggest that the depositional phase of the Mount Isa Orogen repeats a rifted continental margin similar to parts of the Red Sea marine embayment or the Appalachian–Ouachita structural system of eastern North America (Plumb et al 1990). However, recent lithostratigraphic studies point to several stages of intracontinental rifting, all without formation of oceanic crust (Loosveld 1989).

The McArthur and Mount Isa groups are well established chronostratigraphic equivalents (1690–1670 Ma) which, together with subsequent deformation and metamorphism in the Mount Isa Inlier, fall in the Carpentarian age range (1800–1400 Ma) (Page 1979, 1983, 1988, Gulson et al 1983, Page and Bell 1986, Beardsmore et al 1988).

Massive Pb–Zn–Ag deposits formed during deposition of the Mount Isa Group. The Mount Isa deposit, with reserves of at least $56 \times 10^6$ t, at an average grade of 7% Pb, 6% Zn, 149 g Ag/t, and the nearby Hilton deposit, with reserves of $36 \times 10^6$ t at an average grade of 7.7% Pb with associated subeconomic Cu mineralization, consist of well-bedded sulphide ore layers hosted by pyritic and dolomitic siltstone and shale, lacking obvious shallow-water features, and deposited adjacent to possible growth faults (Valenta 1994).

The McArthur and Mount Isa deposits show many similarities. They are the same age, occur in similar host rocks and are located within or at the margins of major intracontinental or continent margin rift zones. At Mount Isa, greenschist facies metamorphism has caused coarsening of galena and sphalerite, remobilization of sulphides, and the crystallization of pyrrhotite.

At Lady Loretta, on the Lawn Hill Platform, bedded Pb–Zn–Ag sulphides ($8.7\% \times 10^6$ t at 6.7% Pb, 18.1% Zn, and 109 g Ag/t) are contained within a massive pyrite body transitional vertically and laterally to carbonaceous shale and dolomitic siltstone (Plumb et al 1981).

### Other basins

Broadly equivalent carbonate-rich sequences of the *Birrindudu Basin* lie 800 km to the west of Mount Isa (Figs 1-5h, 3-22) (Plumb et al 1981, 1990). This mildly deformed basin, about 120 000 km$^2$ in area, overlies the Sturt Block including the Halls Creek Inlier. The Birrindudu sequence, which is up to 6000 m thick, comprises basal sandstone–conglomerate units passing up into a cyclic hypersaline carbonate sequence which broadly resembles the McArthur and Mount Isa groups. A wide peritidal shelf and/or lacustrine environment is indicated.

The relatively undisturbed *Kimberley Basin*, about 160 000 km$^2$ in area, overlies the Kimberley Block. It is flanked by the Halls Creek and King Leopold mobile zone (inliers) to the southeast and southwest respectively (Fig. 3-21). Following the 1850 Ma Barramundi orogeny, up to 5 km thick of quartz-rich arenite, lutite, flood basalts and minor carbonate rocks were unconformably deposited upon Halls Creek Inlier rocks (Fig. 3-22; Table 2-16, column 2). A feature of the Kimberley Basin is its very uniform stratigraphy and paleocurrents (Plumb et al 1981, 1990). Kimberley deposition (1850–1760 Ma) (Page and Hancock 1988) straddles the early–mid Proterozoic boundary.

In the *Gawler Domain* (Craton) to the south (Fig. 3-23; Table 2-17, column 4) the eruption of the Gawler Range Volcanics (1.6 Ga), the synchronous deposition of predominantly clastic sediments, and subsequent co-magmatic post-tectonic granitoid intrusions define a major tectonic episode which marks the final cratonization by 1.5 Ga of this domain with subsequent late Proterozoic Adelaidean sedimentation (Rutland et al 1981, Fanning et al 1988, Giles 1988).

The Gawler Range Volcanics form part of a vast post-orogenic bimodal magmatic province that developed in South Australia between 1600 and 1500 Ma. The volcanic pile comprises thick extensive units of subaerial rhyolitic to dacite ashflow tuff with subordinate andesite and basalt, and is intruded by comagmatic granite. The volcanic rocks are virtually unmetamorphosed and only gently warped, but they have undergone variable deuteric alteration. The voluminous felsic volcanic rocks were erupted over a short interval at 1592 Ma and were intruded by anorogenic granites at ~1575 Ma (Fanning et al 1988).

Cropping out 130 km to the northeast of the Gawler Range Volcanics is the coeval, lithologically similar Olympic Dam suite of volcanic–plutonic rocks. Here the Olympic Dam Cu–U–Au–Ag–(REE) ore deposit represents an unusual type of large hematite-rich breccia-hosted ore concentration. The deposit, which occupies a fault-bounded depression at or near the paleosurface, contains in excess of $2 \times 10^9$ t of mineralized material with an average grade of 1.6% Cu, 0.06% $U_3O_8$ and 0.6 g Au/t (Mortimer et al 1988, Oreskes and Einaudi 1990, 1992, Creasec and Cooper 1993). The Olympic Dam deposit is directly associated with variably altered felsic volcanic rocks and granitoids, including quartz monzodiorite and quartz syenite. The ore-bearing granitoids range in age from 1576 Ma to 1613 Ma, comparable in age to the Gawler Range Volcanics. Close to the Olympic Dam deposit, the zircon U–Pb system of Olympic Dam suite quartz syenite was contaminated by barite precipitated from hydrothermal fluids responsible for development of this huge ore deposit. The data, together with those from Emmie Bluff deposit, 50 km to the south, support a genetic model involving two temporally distinct fluids (Gaw et al 1994).

Cu mineralization at nearby Mount Gunson is attributed to hydrothermal alteration of Gawler Range Volcanic and subvolcanic rocks which are the likely regional source of Cu (U, Pb, Zn) mineralization in South Australian Proterozoic rocks (Knutson et al 1992).

### 4.12.2 MUSGRAVIAN DIVISION

#### Victoria River Basin

Victoria River Basin of North Australia covers an area of about 160 000 km² on the Sturt Block to the northeast of the Hall Creek Inlier (Figs 1-5h, 3-22). Up to 3.5 km thick of virtually undisturbed, stable shelf, carbonate–terrigenous strata accumulated in the Victoria River Basin. Coeval, very much thicker (up to 9 km), unstable shelf, terrigenous sequences accumulated in the comparatively narrow, linear, now moderately to intensely deformed Halls Creek (Carr Boyd Group) and nearby Fitzmaurice (Fitzmaurice Group) mobile zones immediately adjoining to the west (Plumb et al 1981, Plumb 1985). The successions are unconformably overlain by the late Proterozoic 'glacial' Duerdin Group, a correlative of the late Proterozoic–Paleozoic Adelaidean sequence.

#### Bangemall Basin

The Bangemall Basin is a generally elongate concave unit of approximately 150 000 km² covering the southerly and northerly flanks respectively of the Hamersley and Nabberu basins in Western Australia; it lies between older rocks of Gascoyne Province (Block) to the west and Paterson Province to the northeast and Phanerozoic cover of Canning (Officer) Basin to the east (Figs 1-5h, 3-18). It is an intracratonic sedimentary basin containing a marine shelf sequence, the Bangemall Group, of age ~1.1 Ga (Table 2-16, column 1). Bangemall strata vary from flat-lying and unmetamorphosed in the east to folded and metamorphosed in the west. They unconformably overlie older Hammersley, Nabberu, Bresnahan and Mount Minnie successions. The basin is divided into western and eastern parts (Gee 1979, Goode 1981, Muhling and Brakel 1985).

The Western Bangemall Basin, composed of arenites, shales and dolomite, is broadly synclinorial. The northern flank dips gently off the Ashburton Fold Belt (Trough). The southern flank is mildly to moderately folded with tight, SE-trending folds, commonly overturned to the north, in which slaty cleavage is common. The Eastern Bangemall Basin (Goode 1981) is dominated by generally flat-lying or gently dipping fine- to coarse-grained sandstones with minor carbonates. Structural features characteristic of shallow-water deposition are abundant. Along the extreme southern margin of the basin, basal sandstone exhibits spectacular, large-scale, trough cross-bedding of possible coastal aeolian origin. Thin stromatolitic carbonate units are present in the lower part of the succession.

### Amadeus–Albany–Fraser Tectonic Zone

The Musgrave–Fraser Province, as used here, includes the broad orogenic zones situated both north (Arunta Block) and south (Musgrave Block) of the Amadeus Basin, together with those of the Albany–Fraser Province adjoining the Yilgarn Block to the southwest (Fig. 1-5h). Also, the Naturaliste and Northampton blocks, two small granulite facies inliers situated west of the Yilgarn Block, and the thin NW-trending Paterson Province to the northeast of the Bangemall Basin (see above), are included for convenience. All these domains experienced repeated mid-Proterozoic taphrogenic reworking and intrusions with the development of various shear trends and reworkings, including granulite facies metamorphism (Rutland 1981).

Rutland (1981), in noting that the Musgrave, Redbank (southern Arunta) and Albany–Fraser domains are the most striking manifestations of the Musgravian Division, emphasizes that the 1400–1000 Ma period was also notable for widespread faulting and shearing throughout the Australian Craton; indeed the resulting pattern is responsible for the present continental 'block' structure of Australia, the more deformed zones developing into thick late Proterozoic (Adelaidean) to Paleozoic sedimentary basins, which now separate the several basement 'blocks' of the Australian Craton.

### Albany-Fraser Province

The high grade Albany–Fraser Province, some 800 000 km² in area, forms an arcuate metamorphic belt that sweeps around the southern and southeastern margins of the Yilgarn Block, truncating Archean trends therein at high angles (Gee 1979, Goode 1981, Beeson et al 1988). The NE-trending Fraser Province is broadly similar to the contiguous E-trending Albany Province. Both are characterized by high grade metamorphic and plutonic activities and strong tectonism that involved both Archean basement and Proterozoic cover rocks.

The Albany–Fraser Province is interpreted as a linear zone of crustal downwarp, involving (1) basement reworking, metamorphism and granitoid plutonism; and (2) shallow-water, stable shelf facies sediments that may have extended across the province and on to the adjoining Yilgarn Block. Granulite-facies metamorphisms have been deformed both at 1.8–1.6 Ga and 1.3–1.1 Ga in some zones. Steep lateral thermal gradients must have existed, suggesting a zone of high heat flow (Gee 1979).

### Arunta and Musgrave domains

The mainly E-trending Arunta and Musgrave domains, the main components of the Amadeus Transverse Zone (Rutland 1981), are characterized by granulite facies terrains, reactivated shear zones (lineaments) with mylonites, layered mafic–ultramafic intrusions and a bimodal volcanic suite dated at about 1050 Ma. Taken together with gravity evidence, the Amadeus Transverse Zone constitutes a 500 km broad zone separating two stable shield areas, situated respectively to the north (North Australian) and south (Gawler).

The Arunta Inlier (Block), covering about 200 000 km², along the southern edge of the North Australian Craton, typifies the intracontinental polymetamorphic mobile belts which have experienced a long, complex history (Foden et al 1988, Oliver et al 1988). The structure of the inlier is dominated by large W–NW-trending belts of overthrusting, shearing, mylonitization and retrogression which commonly separate the principal tectonic zones of markedly different metamorphic grade. These were the loci for most of the overthrusting and retrogression which uplifted the inlier and cover to its present attitude.

The northwestward (Paterson–Musgrave Belt) and southeastward (towards Tasmania) extensions of this broad transcontinental taphrogenic–deformation zone are largely buried and little known. The Paterson Province consists of a Proterozoic (?) metamorphic basement and a cover sequence that exhibits asymmetric folding and high grade thrusting to the southwest (Williams et al 1976).

## 4.13 ANTARCTIC CRATON

Mid-Proterozoic metasupracrustal rocks have been located in three main parts of the East Antarctic Metamorphic Shield (Figs 1-3i, 1-5i) (James and Tingey 1983). In southern Prince Charles Mountains (70°E) there occurs a sequence of recumbently folded greenschist facies calcareous phyllites, sandstone, and conglomerate with BIF clasts. These rocks are considered to have been folded at 891 Ma, which is the age of nearby recumbently folded, high grade rocks (see below). The Shackleton Range (30°W) includes some lower amphibolite facies metasediments which unconformably overlie metamorphic basement rocks and are dated at 1446 Ma. At nearby Coats Land, coastal rhyolite porphyries of Littlewood Nunataks are dated at 1001 Ma.

Metamorphic rocks of corresponding Grenvillian age are widespread though discontinuous from Windmill Islands (110°E) to Dronning Maud Land (0°). Similar rocks in the Bunger Hills (100°E)–Vestfold Hills–Prydz Bay area (80°E) give corresponding K–Ar dates, but these are now difficult to interpret. In the Prydz Bay–Prince Charles Mountains–Mawson area (60–70°E), high grade metamorphic rocks underlie an E-trending belt about 500 km wide that is probably continuous with similar rocks in all three of the Rayner Complex of Enderby Land (50°E), the Lutzow–Holm Bay area (40°E), and the Dronning Maud Land (0–15°E). This appears to represent a major zone of Grenvillian age metamorphism, one of the most important thermo-tectonic events in the East Antarctic Metamorphic Shield. Dyke swarms in the Vestfold Block (75°E) show a very long and complicated Proterozoic deformational history, including megaemplacements at 1380 Ma and 1250 Ma, followed by a major 1000 Ma compressional event (Dirks et al 1994).

Enderby Land, Kemp Coast and Mawson Coast are underlain by high grade, mostly granulitic facies rocks some with a history extending back to almost 4.0 Ga (see above). Archean rocks of the Napier Complex were little affected by events after 2.5 Ga (see above), whereas the Rayner Complex, adjoining to the south, was affected by pervasive 1.0 Ga old deformation, metamorphism, and charnockitic activities (Rayner event). Rocks similar to those of the Rayner Complex extend eastward (52–68°E) along, successively, Kemp Coast and Mawson Coast. Much of the Kemp Coast is underlain by quartzofeldspathic gneisses (charnockitic and enderbitic) and pyroxene granulite, with subordinate interleaved garnet–biotite–sillimanite gneiss, garnet–clino–pyroxene–plagioclase–calcsilicate rocks, magnetite-bearing rocks, and ultramafic rocks (Grew et al 1988). Available zircon dates record a long and complex history for this part of Antarctica, with new zircon growth documented episodically (at least sixfold) over an interval of ~2300 Ma (2825–469 Ma). This study thereby documents the first conclusive evidence for Archean geological activities from anywhere near the eastern margin of the Arctic shield (Grew et al 1988). However, the main conclusions to be drawn pertain to the pervasive deformation, granulite facies metamorphism and pegmatite emplacements at ~1.0 Ga. A similar sequence of events has been reported for parts of the Eastern Ghats Belt in India which adjoins the Kemp–Mawson Coastline of Antarctica in a reassembled Gondwanaland (Grew et al 1988).

# Chapter 5

# Late Proterozoic Crust

## 5.1 INTRODUCTION

### 5.1.1 DISTRIBUTION

The area of exposed plus buried (total) late Proterozoic crust is $45.3 \times 10^6$ km² (Table 5-1), or 43% of Earth's total Precambrian crust (Table 1-2). Of this total, Africa alone accounts for 51%; South America 15%; Asia, Europe and Antarctica each 8–9% ($\Sigma 25\%$); Australia, India and North America each 1–5% ($\Sigma 8\%$). In this case, the four Atlantic continents account for 75%; the three Laurasian continents 19%; and the five Gonwanaland continents 81% of the total.

The area of exposed late Proterozoic crust is $10.1 \times 10^6$ km² or 33% of Earth's exposed Precambrian crust. Africa alone represents 56% of this total; south American 17%; Asia, Europe and North America each 5–7% ($\Sigma 18\%$); India, Australia and Antarctica each <4% ($\Sigma 8\%$).

In summary, Africa is the dominant late Proterozoic continent, with South America second, the two combining to provide a western Gondwanaland predominance (Fig. 5-1).

### 5.1.2 SALIENT CHARACTERISTICS (Table 5-2)

(1) *Large interior basins*. Following widespread consolidation of the Precambrian cratons including implementing thermo-tectonic events at ~1.0 Ga, the central parts of several cratons were gently downwarped to form vast central basins, up to 1500 km in diameter, and smaller marginal basins, foredeeps and troughs. The resulting sandstone–pelite–stromatolitic carbonate platform cover, typically 1000–4000 m thick, is characterized by wide extent and great homogeneity of facies, and an average subsidence rate of ~5 m/Ma; for the most part, it represents deposition on remarkably flat surfaces. Lithostratigraphic successions are typically divided into discordant subgroups, each representing a transgressive–regressive cycle, the successions unconformably to conformably overlain by Phanerozoic strata. These late Proterozoic sequences are, for the first time, characterized by the presence of tillites, redbeds and metazoan fossils, a mark of the evolving continental crust and environments.

In some cratons, basement subsidence involved linear array of developing intracratonic extensional troughs with adjoining continental shelves, sites of mainly shallow-water sediments accumulations up to 15 km thick. These successions typically include lower blanket sands, shelf carbonates and basaltic flows, median thick local deltaic units, and upper evaporite-bearing stromatolitic platform carbonates.

(2) *Mobile belt network*. In sympathetic relation to intracratonic basin–trough development, late Proterozoic tectonism, culminating in the Pan-

*Table 5-1.* Distribution of preserved late Proterozoic crust by continent.

| | Total crust (%) | Exposed crust (%) |
|---|---|---|
| Asia | 9.1 | 6.6 |
| Europe | 8.4 | 5.5 |
| North America | 1.6 | 6.0 |
| South America | 14.9 | 17.2 |
| Africa | 50.6 | 56.4 |
| India | 4.3 | 1.8 |
| Australia | 2.8 | 2.6 |
| Antarctica | 8.3 | 3.9 |
| | 100.0 | 100.0 |

Total (exposed + buried) crust = 45 256 000 km²
Exposed crust = 10 085 000 km² (24% of total)

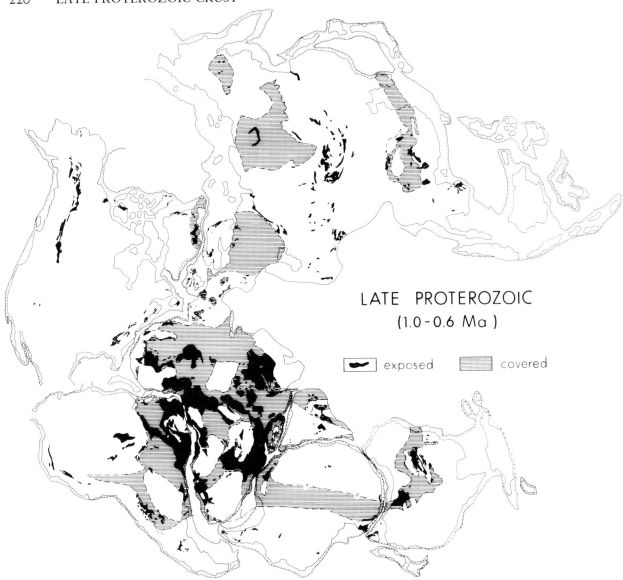

**Fig. 5-1.** Distribution of exposed and covered (i.e. sub-Phanerozoic) late Proterozoic crust in the Briden et al (1971) pre-drift Pangean reconstruction of the continents.

African thermo-tectonic events, affected large parts of the African Craton in particular, as well as, in the pre-Mesozoic paleoreconstruction, adjoining parts of South America, Madagascar and Antarctica. In Africa, in particular, this tectonism produced an impressive network of supracrustal wedges aligned in broadly latitudinal and longitudinal directions, subparallel to the present continental coastlines. In the larger Gondwanaland setting (i.e. southern supercontinent), the Pan-African mobile belt system, which stabilized in latest Precambrian to early Paleozoic time, includes amongst others the Brasilide–Pharuside, West Congo, Damaran–Katangan–Zambezi, Mozambique, Arabian/Nubian, Cameroon–Central Africa, Transantarctican and Adelaidean belts. In Laurasia, or the northern supercontinent, the corresponding but significantly younger belt system, which includes the Appalachian–Caledonian, Innuitian, Cordilleran, Variscan and Uralides, although partly initiated in late Proterozoic time, did not stabilize until the late Paleozoic (300 Ma ago) and even later.

(3) *Ophiolites.* Convincing 900–800 Ma ophiolites, seemingly representing pieces of obducted oceanic crust, have been identified at Bou Azzer, Morocco, and at many localities in Saudi Arabia, Egypt, Sudan and southward through Kenya to Mozambique. These, together with associated

*Table 5-2.* Salient late Proterozoic characteristics.

| | |
|---|---|
| Distribution | Dominant in Africa; moderate in Australia, South America, Asia and Antarctica; paucity in North America. Gondwanaland continents predominate. |
| Composition | Large interior basins with widespread, homogeneous shallow-marine-fluviatile red-bed arenite-pelite-carbonate platform cover upon exceptionally flat surfaces, products of craton-wide peneplanation and gentle platformal subsidence. Aulacogens and continental rifts. |
| | Mobile belt network involving rift-induced, crudely orthogonal latitudinal-sublongitudinal, intra- and peri-platformal wedges composed of (1) subtabular exterior shelf facies and (2) variably deformed interior trough facies. Common subgreenschist facies with increasing amphibolite-granulite facies towards the hinterland. |
| | Undoubted ophiolites and magmatic arc volcanics-plutonites scattered about the mobile belt network. |
| | Plethora of end-of-era diamictites including olistostrome and tillites, the latter products of globally intermittent glaciogenic, high altitude environments. |
| | Common continental red-beds, phosphorites and evaporites including gypsum-halite casts. Sparse but extensive BIF. |
| Deformation | Platform cover little deformed. Periplatformal wedges increasingly deformed towards the hinterland with pronounced foreland vergence including prominent overthrusting, nappe-stacking and decollements. Terminal orogeny achieved platform stability. |
| Mineralization | Cu(-Co-U-Pb-Zn-V-Cd-Ag) stratiform shale hosted deposits; Pb-Zn carbonate deposits; Cu-U deposits. |
| Paleoenvironment | Steady-state oxidizing environment (~12% PAL); local reducing sinks, e.g. volcanic fumaroles; predominant continental flux signature. |
| Paleobiology | Ubiquitous stromatolites; widespread cyanobacteria. Further diversification of eucaryotic cell. End-of-era origin of sexuality, expressed in soft-bodied metazoa. |
| Paleotectonics | Subduction-involved, Wilson-type plate motions, involving plate fragmentation and reorganizations. Final periplatformal arc and exotic terrain accretions to consolidate the Precambrian platforms. |

island-arc-type calc-alkalic volcanic rocks and plutonites, leave little doubt that actualistic plate tectonic processes were operating in the late Proterozoic.

(4) *Tillites*. There is a plethora of diamictites in the 950–550 Ma age range. Many are well established tillites according to a wide range of criteria, but only a few are well enough dated to distinguish glacial periods or epochs. Four such glacial periods are named: Lower Congo, 950–850 Ma; Sturtian, 670–660 Ma; Varangian (Marinoan–Vendian), 620–590 Ma; and Late Sinian, 600–555 Ma. These late Proterozoic tillites are recognized in practically all Precambrian cratons of the world. They demonstrate widespread, albeit intermittent, glaciogenic environments. Paleomagnetic evidence suggests that many of these glaciation occurred at near-tropical latitudes ($<25°$).

(5) *Metazoans*. In the late Proterozoic sequences of the world, stromatolites are ubiquitous, microfossils and trace fossils are widespread, and new finds of fossil soft-bodied invertebrates are increasingly common. Rocks in the 750–570 Ma age range, particularly in Australia and Russia, are remarkable for their record of soft-bodied metazoans. The oldest metazoan soft-body fossils typically occur in Precambrian rocks less than 100 Ma older than the Precambrian–Cambrian boundary. Moreover, the rocks containing them are closely associated with, though commonly somewhat younger than, the youngest Precambrian (Eocambrian) tillites, which collectively document the latest Precambrian global glaciation. Accordingly, a causal relation between glaciations and the first differentiation of marine metazoan faunas has been suggested. In this regard, the host environment and composition of the Ediacara fauna, the oldest and most varied metazoan fauna in Australia, are particularly well known.

## 5.2 CATHAYSIAN CRATON

The Yangtze and Tarim component cratons were widely consolidated by the Jinningian Orogeny at 850 Ma, such that the resulting unconformity provided widespread basement for the Sinian System (850–615 Ma) (Figs 1-3a, 1-5a). Whereas Sinian successions are rare to incomplete in the Sino–Korean and Tarim cratons, they are fully developed in South China, notably the Yangtze Craton and its margins (Fig. 4-2). Thus, the Sinian is notably exposed at the Yangtze Gorges near Yichang, Hubei Province (Fig. 4-2, C10); this, the stratotype of the Sinian system and 693 Ma, is characterized by a complete succession some 1095 m thick, continuous exposure, simple structure, low metamorphic grade including blueschist facies, the presence of renowned tillites, and abundant stromatolites, plant micro-fossils and metazoans, all transitional up to abundant shelly fauna of the lower Cambrian member (Gao et al 1980, Zhao et al 1980, Chen et al 1981, Yang et al 1986, Liou et al 1989).

Sinian strata of stable craton-type, commonly 3000–5000 m thick, are typically divided into lower tillite-bearing clastic sequences, and upper carbonaceous pelite–stromatolitic carbonate with phosphorite assemblages. Over most of the Yangtze craton, one glacial unit overlies early Sinian basement and passes up into metazoan-bearing carbonates overlain by earliest Cambrian phosphorites, as in the Yangtze gorges. In marked contrast, Sinian sequences at the eastern and western margins of the same Yangtze Craton, up to 10 km thick, are distinguished by thick volcanic–turbiditic accumulations lacking glacial deposits, thereby marking impressive pericratonic instability. For biostratigraphic reasons, the lower limit of the Cambrian is placed at 615 ± 20 Ma in China (Zhao et al 1980), rather than at the conventional 570 Ma. Representative Sinian lithostratigraphic sections are illustrated in Yang et al (1986, fig. 6.2).

Sinian strata are particularly important for their stromatolite and tillite contents.

Six distinctive stromatolite assemblages, each of specific age range, have been identified in mid- to late-Proterozoic strata of China. These stromalite assemblages, in combination with isotopic dates, provide regional stratigraphic correlations over the Cathaysian Craton (Chen et al 1981).

Numerous discoveries of metazoans and worms, or their traces, are reported from the Sinian System. Exceptional Precambrian fossils, recently discovered in upper Sinian strata, include Cyclomedusa, Medusinites, Lorenzites, Eosomedusa and Bohaimedusa, as well as certain trace fossils (Chen et al 1981, Zang and Walter 1992). Their reliability for widescale stratigraphic correlations is under constant study.

*Tillites* are also of great interest for their potential use as regional stratigraphic markers (Fig. 4-2). Three main periods of glaciation in the Sinian System of China are recognized (Yang et al 1986).

(1) *Chang'an Glacial Epoch (800–760 Ma)*: tillites of this age are present in southern China, including Yangtze Provinces, where they are represented by boulder-bearing slate and sandstone measuring 15–2000 m in thickness (Chang'an Formation), and in Tarim–Tianshan Province to the west where the Beiyixi Formation consists of interbedded marine clastics, lavas, tuffs and glacial conglomerates totalling 1490 m in thickness. The Chang'an tillite corresponds to the Sturtian glacial period of the East European Craton (see below).

(2) *Nantuo Glacial Epoch (740–700 Ma)*: Nantuo tillites, widespread in southern China, are the most extensive of all Sinian glacial deposits, the glacial environments illustrated varying considerably from continental to glaciomarine.

(3) *Luoquan Glacial Epoch (640–600 Ma)*: these deposits are mainly continental glacial sediments. Those distributed on the northern slope of Qinling mountains and nearby to the west (Luoquan Formation), about 200 m thick, commonly overlie well-exposed glacial pavement. Those still farther northwest in Zinjian Province, are composed almost entirely of glacial conglomerate about 450 m thick, with varied mudstones restricted to the base and to the top, the latter transitional up to phosphate-bearing, fossiliferous lower Cambrian strata. This third glaciation is referred to as the Late Sinian Glacial Epoch.

## 5.3 SIBERIAN CRATON

Late Proterozoic platform cover on the Siberian Craton is represented by two late Riphean systems—Karatavian (1.0–0.68 Ga) and Kudashian (0.68–0.65 Ga)—together with the Vendian (0.65–0.57 Ga) system (Figs 1-3b, 4-3; Table 4-3). Collectively these are equivalent to the Baikalian system (0.9–0.57 Ga) of the Baltic Shield to the west (Fig. 1-3c(i)). These mainly siltstone–shale–carbonate sequences, products of repeated widespread marine transgressions across the now-stable craton, are exposed on the relevant flanks of the Aldan and Anabar shields, and in the pericratonic mobile belts (Fig. 4-3). Additional strata are present in the buried interior platform.

### 5.3.1 KARATAVIAN (LATE RIPHEAN) DIVISION

Karatavian sequences are moderately well distributed in Siberia (Chumakov and Semikhatov 1981) (Fig. 4-3). The lower boundary is based on a paleontologic change in stromatolite assemblages at 1000 ± Ma (Table 4-3). Karatavian strata in Siberia transitionally overlie Yurmatinian (mid-Riphean) strata. They are conformably overlain by Kudashian strata on the Yenisei Ridge and Patom Highlands but are elsewhere separated by a hiatus from overlying Kudashian–Vendian strata. Representative sections comprise lower terrigenous-carbonates, middle shales, and upper shale carbonates (Chumakov and Semikhatov 1981, fig. 2).

## 5.3.2 KUDASHIAN (LATEST RIPHEAN) DIVISION

The lower boundary of the Kudashian is set where the upper Riphean ($R_3$) assemblage of stromatolites and microphytolites is replaced by a subsequent assemblage known as the Yudomian ($R_4$) assemblage. The isotopic age of the lower boundary is tentatively placed at 680 Ma.

This change can be observed especially in the partly conformable sequences of the Patom Highlands, Baikal Region; elsewhere in Siberia and middle Asia it coincides with a disconformity.

In the Uchur–Maya Region, the Kudashian sequence, about 400 m thick, is composed of terrigenous and carbonate rocks. Correlatable strata of both the Patom Highlands and the Yenisei Ridge to the west are composed of sandstone–siltstone–shale transitions. On the Olenek Uplift and Anabar Shield, the equivalent sequences, about 200 m thick, comprise lower sandstones and upper dolomites.

## 5.3.3 VENDIAN (LATEST PROTEROZOIC) DIVISION

In Siberia, the Laplandian glacial horizon, which elsewhere defined the Vendian (see below), has not been recognized, thereby preventing unequivocal identification. Instead, the Yudomian (incorporating both Kudashian and Vendian) constitutes the upper Proterozoic division. The type Yodoma Group of the Uchur–Maya Region (Komar et al 1977), which unconformably overlies Karatavian strata, is about 1000 m thick. It is divided into a lower terrigenous-carbonate formation and an equally thick, upper, largely dolomitic formation. It contains plentiful 'fourth assemblage' ($R_4$) stromatolites and microphytolites throughout the section, and rare soft-bodied metazoans in the middle part (Sokolov and Fedonkin 1990).

The most complete Vendian section, up to 1500 m thick, occurs in the Bajkonyr synclinorium of Tien Shan, including the Ulutau area, located midway between the Siberian and East European cratons (Fig. 4-3). The section comprises lower conglomerate–sandstone and upper sandstone–shale–carbonate sequences with at least two undoubted tillite horizons, respectively 400–2500 m and 15–100+ m in thickness. These tillite horizons have been traced from Kazakhstan in Siberia to the Kuruktag Ridge, Tarim Craton, China (Salop 1983).

## 5.4 EAST EUROPEAN CRATON

### 5.4.1 INTERIOR PLATFORM AND PERICRATONIC DOWNWARPS

Later Riphean (<1000 Ma) strata, less common on the East European Craton than in the subjacent, aulacogen-induced, earlier Riphean-rich troughs, reach a maximum thickness of only 600 m in the central part of the platform, and 1000 m in the coeval Kama pericratonic belt to the east (Fig. 5-2). This comparatively modest later Riphean development is attributed to a combination of tectonic decay of the graben (aulacogen) system and to craton-wide pre-Vendian erosion (Postnikova 1976, Sokolov and Fedonkin 1990). Preserved later Riphean strata are almost exclusively marine siltstones and carbonates, the latter commonly glauconitic, with barrier reefs commonly developed in the Kama pericratonic downwarp (Khain 1985). Terminal Riphean (650 Ma) was marked by renewed epeirogenic uplift and gabbro-diabase intrusions.

By early Vendian times the preceding aulacogen stage of craton development was completely superseded by the platform stage, in which sedimentation expanded beyond the aulacogens to cover much of the platform, although in reduced thickness (Sokolov and Fedonkin 1990). Thus by the end of the early Vendian, the huge NE-trending Moscow Syneclise had formed, stretching for 1800 km from southern Belorussia on the west to Timan on the east (Fig. 5-2). It was bounded on the northwest by the Baltic Shield slope. Lower Vendian sediments are represented by marine-terrigenous strata, which are red or variegated below but grey-coloured and glauconitic above. In mid-Vendian time, the East European Craton underwent intense glaciation, now represented by tillites and tilloids (marine-glacial deposits) (see below). This coincided with basaltic volcanism, especially at the western margins of the craton. In the late Vendian, sedimen-

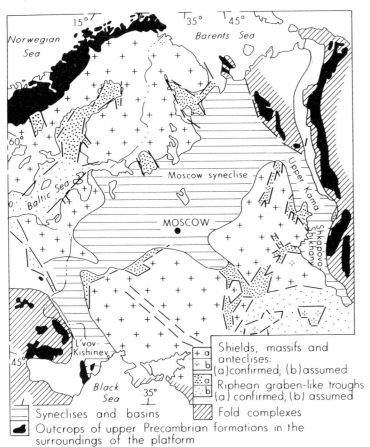

Fig. 5-2. Vendian structures of the East European Craton. (From Aksenov et al 1978, Fig. 2, and reproduced with permission of the authors).

tation spread to the Timan, cis-Uralian and Visla–Dneister pericratonic downwarps, thereby forming a vast gulf in the direction of Volgograd to the south. Uniform grey, marine-terrigenous strata with glauconite predominate in the central platform, but are replaced peripherally with increasingly variegated strata.

Periodic volcanism waxed and waned, synchronous with tectonism. The upward transition to the Cambrian is gradual but with a notable westward expansion of the Baltic–Moscow Syneclise to absorb the southern part of the Baltic Shield, thereby merging with the Grampian Geosyncline (UK) to the west. Conversely, the area of the syneclise was reduced in the northeast as a result of Timan uplift. The cis-Uralian pericratonic downwarp effectively ceased to exist by the beginning of Cambrian time (Khain 1985).

In late Vendian to early Cambrian (i.e. Baikalian) time, the principal pericratonic downwarps were the Timan (northeastward), Uralian (eastward) and cis-Carpathian (westward); these gradually spread and merged by inheritance with the main interior Baltic–Moscow (east to west) and Pachelma (north to south) depressions.

The Timan Geosyncline, a representative pericratonic downwarp, underwent intense subsidence in mid- to late-Riphean time, and especially along offshore Kola Peninsula to the northwest, in early Vendian time. Sedimentary sequences, with local mafic volcanic rocks, range to 10 km thick, with the clastics derived from platform provenances to the southwest. The final diastrophism occurred in late Vendian time with intense southwestward fold-thrust vergence in the direction of the parent craton. Deformation was accompanied by regional metamorphism up to amphibolite facies and by widespread granitoid intrusion (789 Ma) (Khain 1985). Uplift of the resulting Kanin–Timan Ridge was accompanied by compensatory subsidence of the fore-Timan Trough. These and related pericratonic activities completed the consolidation of the East European Craton.

## 5.4.2 BALTIC SHIELD

Baikalian (<0.9 Ga) events in the Baltic Shield, mainly restricted to the west, include limited tectonic reworking of the Southwest Scandinavian

Domain and extensive glaciogenic deposition along the northwestern boundary of the shield (Figs 2-3, 5-2).

Along the eastern border of the Caledonian Fold Belt there is an abundance of chiefly arenaceous Late Proterozoic metasediments which are divided into the older Sparagmite and younger Varegian formations, both including products of the Varangian Glacial Period.

These formations are included in a series of imbricated nappes that characterize the eastern part of the fold belt. The older Sparagmite Formation of southern Norway, which constitutes the Eocambrian type section, includes the 15–30 m thick Mirelv tillite (Harland 1983). In northern Norway, thick Finnmark sandstones include the two famous tillites at Varanger Fjord (Siedlecka et al 1989), which are separated by sandstone–slate units dated at 654 Ma (Harland 1983).

## 5.4.3 MEDIAN MASSIFS

Numerous Precambrian massifs (inliers) contained in the enclosing Caledonian and Variscan fold belts are considered here, including those in Spitzbergen and the British Isles to the west, and the Variscan massifs of middle Europe to the south.

### Spitzbergen

In the Spitzbergen Archipelago of the North Atlantic, 1000 km north of Narvik, Norway (Fig. 1-5d(i)), Vendian rocks are present both in the Hecla Hoek Geosyncline of the Eastern Province and in the Western Province (Winsnes 1965, Harland 1983). The contained Vendian tillites, commonly 60–100 m thick, belong to the Varangian Glacial Period. They are intercalated with variegated shale–siltstone–carbonate assemblages with occasional volcanic facies.

### British Isles

With the exception of the older Lewisian and associates in the Northwest Highlands and Islands of Scotland, which collectively form the Scottish Shield Fragment or Hebridean Craton (NW Kratogen) of the former North Atlantic Craton (see above), Precambrian rocks of the British Isles, broadly though sparsely distributed, are typically late Proterozoic in age, commonly transitional up to early Paleozoic strata (Figs 1-3c(i), 5-3, 5-4). In this respect they closely resemble the late Riphean to Vendian division of the East European Craton. The broad British tectonic divisions, in addition to the Hebridean Craton, are, from north to south, Metamorphic and Non-Metamorphic Caledonides, Midland Craton and Hercynides (Fig. 5-4).

The Northwest Caledonian Front of the Caledonian Belt is defined by a border thrust zone, including the well known Moine Thrust (MTP). This front probably passes still further southwestward, beyond the northwest coast of Ireland (Fig. 5-3).

To the northwest of the Moine Thrust, the rocky moorlands and mountains of the foreland (Hebridean Craton) consist of a Lewisian basement unconformably overlain by mainly unaltered late Precambrian (Torridonian) and early Paleozoic (Cambrian) sediments (Fig. 5-4).

To the southeast of this front, the Caledonian Fold Belt underlies nearly 75% of the total area of the British Isles (Anderson 1965). The fold belt is split by the SW-trending, >610 km-long Highland Boundary Zone into Metamorphic (NW) and Non-Metamorphic (SE) divisions (Fig. 5-3). The Metamorphic Caledonides on the northwest, split by the Great Glen Fault, consist of two major groups of complexly folded and metamorphosed metasedimentary assemblages, the Moinian (Grampian) and Dalradian (Fig. 5-4). These metasediments are transitional up to Cambrian strata.

In the parts of the Caledonian Fold Belt to the southeast of the Highland Boundary Zone (i.e. Non-Metamorphic Caledonides and Midland Craton), outcrops of Precambrian strata are comparatively isolated and of limited extent (Fig. 5-3). They are mainly grouped into Monian, Uriconian, Charnian and some other assemblages of uncertain age, commonly transitional up to Cambrian strata (Figs 1-3c(i), 5-4).

#### Torridonian Supergroup

To the west of the Moine Thrust (North-West Caledonian Front), a thick Torridonian succession of dominantly redbed Facies and associated marine deposits unconformably overlies Lewisian basement (Park et al 1994). The Torridonian Supergroup (Piasecki et al 1981) contains two groups, the lower Stoer (~2 km) and upper Torridon (up to 7 km) composed of red conglomerates and sandstones with southward thickening shales. On the mainland the Torridonian is largely unmetamor-

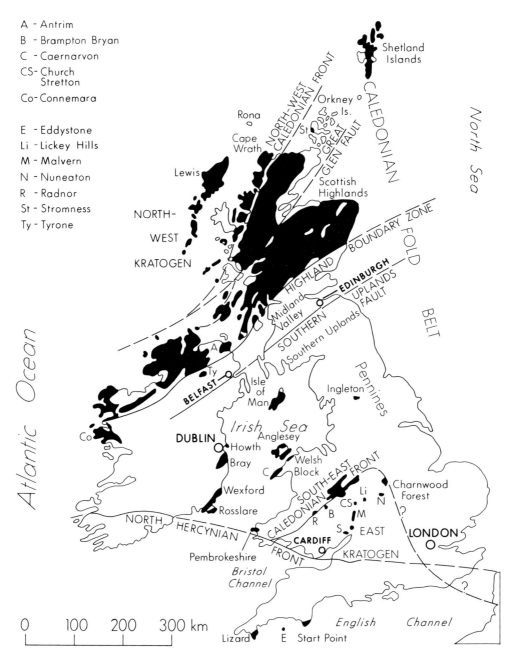

**Fig. 5-3.** Structural map of the British Isles showing Precambrian outcrops. (From Anderson 1978, Fig. 1, and reproduced with permission of the author and of Pergamon Press.)

phosed, but on some of the Hebridean Islands the rocks are deformed and metamorphosed in low greenschist facies.

Within the Moine Thrust Zone itself, Lewisian and Torridonian rocks occur in autochthonous nappes beneath the main Moine Thrust, which itself brings forward an exotic nappe consisting of Moinian metasediments of the fold belt to the southeast (see below). Thrusting, accompanied by the development of mylonites, has taken place on a grand scale, the general direction of tectonic transport being towards the foreland in the northwest. Near the southern end of Loch Maree, beneath the Moine nappe, spectacular exposures reveal the Kinlochewe nappe, consisting of Lewisian and Torridonian units thrust over the Cambrian (see Anderson 1965, fig. 3). Dates of 970 and 790 Ma were obtained for deposits of the older and younger Torridonian redbed assemblages respectively (summarized in Piasecki et al 1981).

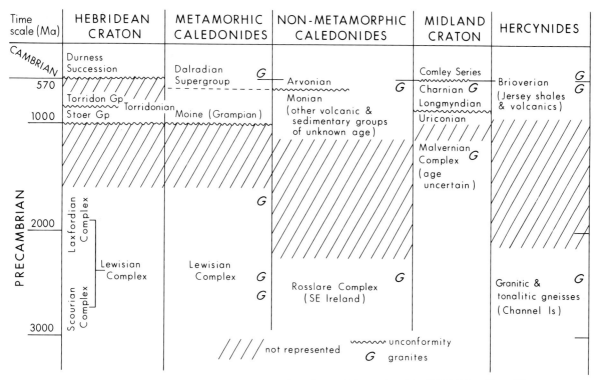

Fig. 5-4. Regional stratigraphic arrangement of the Precambrian rocks of the British Isles. (From Watson 1975, Fig. 1, and reproduced with permission of the Geological Society of London.)

Dalradian Supergroup

The Northern Highlands, which extend for 50–100 km between the Northwest Caledonian Front and the Great Glen Fault, as well as the Grampian (Scottish) Highlands, which extend at least an equivalent distance further southeastward to the Highland Boundary Zone (Fig. 5-3), are both underlain by metasedimentary assemblages of uncertain relationship, collectively referred to as the Dalradian Supergroup, and closely associated Moinian (Grampian) assemblages (Fig. 5-4).

The *Moinian assemblage* is dominated by meta-arkosic sandstones, commonly banded or flaggy due to the presence of intercalated micaceous laminae and semi-pelitic mica schists. A few narrow bands of marble and calcsilicates are present. Migmatites and injection complexes abound. Pegmatite sills and dykes, and irregular lenticular masses of lit-par-lit injection, are also common (Piasecki et al 1981).

Moinian rocks are characterized by sharp, often isoclinal, folding along N–NNE-trending lines. Three fold phases prior to the formation of the border thrust zone are recognized. Moinian structures include abundant lineations, rodding and mullion.

The *Dalradian Supergroup* proper consists of a metasupracrustal sequence, up to 25 km thick, dominated by metasediments. The main sequence is divided into the lower Appin and upper Argyll groups (Harris et al 1978).

The lower *Appin Group*, an interbedded sandstone–mudstone sequence up to 6 km thick, represents a complete basin-deepening to basin-shallowing cycle.

The overlying *Argyll Group* is introduced by the Port Askaig tillites, about 750 m thick, comprising several complex glacial facies including massive tillite, reworked marine tillite and outwash sands (Spencer 1978, Harland 1983). Deposition was essentially continental or marginal marine. The tillite lies 10 km below rocks with early Cambrian trilobites, yet the age of the tillite has not been accurately established.

Non-Metamorphic Caledonides and Midland Craton

Volcanic and sedimentary rocks roughly coeval with the Moine and Dalradian are known from scattered exposures and drill holes over a considerable area in the Midland Craton as well as in fold cores and fault blocks in the adjoining Non-Metamorphic Caledonides to the north including glaucophone-bearing Monian quartzites, turbidites, breccia and ophiolites and Charnian pyroclastic rocks and granophytes (Figs 1-3c(i), 5-3, 5-4) (Watson 1975b, Piasecki et al 1981).

In southwest England, the Lizard Peninsula, the Dodman and Start Points lie to the south of a major tectonic line bringing metamorphic rocks against Lower Paleozoic and Devonian rocks of Devon and Cornwall. Low grade phyllites of unknown age occur on the headlands of Dodman and Start Points. The rocks of the Lizard Peninsula show complex sequences of events punctuated by the emplacement of a large mass of now entirely serpentinized peridotite, possibly representing an obducted slice of oceanic crust and mantle (Piasecki et al 1981).

### Variscan massifs

The Variscan Fold Belt (Variscides) refers to that generally E-trending middle division of Europe characterized by flat-lying Mesozoic and Tertiary sediments resting on strongly folded rocks up to and including late Paleozoic systems which were affected by the late Paleozoic Variscan Orogeny (345–280 Ma). The dividing line between this fold belt (meso-Europa) and the older divisions to the north, including the Caledonides (early Paleozoic) and East European Craton (Precambrian) (Palaeo- and Eo-Europa), passes consecutively eastward through southern Ireland, southern England, northeast France and Belgium, Germany and northern Czechoslovakia (Fig. 5-5). South of this line lie all but one (Holy Cross Mountain) of 10 Variscan-enclosed Precambrian-bearing massifs, which extend from the Iberian Massif on the southwest, around the outer (northern) rim of the Alps and Carpathians, as far as Dobrogea at the mouth of the Danube on the Black Sea. The Ural Mountains and their northerly extension in Novaya Zemlya with contained massifs (see above) form an equivalent fold belt to the east of the East European Platform (Fig. 5-5).

All 10 Variscan massives probably include some Precambrian rocks, although basement ages in some have yet to be confirmed. The massifs are characterized by thick, late Proterozoic, 'geosynclinal' cover (e.g. Brioverian) containing diamictites (either tillites or olistostromes) and stromatolitic carbonates, together with variable older Precambrian basement (e.g. Pentevrian) (Figs 1-3c(i), 5-4).

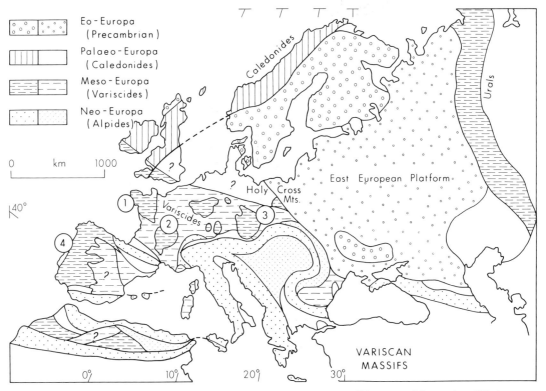

**Fig. 5-5.** Geologic sketch map showing the main tectonic divisions of Europe and the enclosed Variscan massifs including: 1 – Armorican; 2 – Central; 3 – Bohemian; 4 – Iberian. (From Ager 1980, Fig. 0.2, and reproduced with permission of McGraw-Hill (U.K.)

The Armorican Massif (Bishop et al 1975, Auvray 1979, Dupret 1984, Shufflebotham 1989, Brown and D'Lemos 1991) has been dated locally at 2650 Ma. The Central Massif (Anderson 1978, Ager 1980) provides nearby detrital zircons dating to 3.7 Ga. The Bohemian Massif (Svoboda 1966, Anderson 1978, Ellenburger and Tamain 1980, Wendt et al 1987) has been dated in the 2.0–2.3 Ga range (Grauert et al 1974). The Iberian Massif (Anderson 1978, Ager 1980, Aubouin 1980, Dallmeyer and Garcia 1990) and adjoining Pyrenees Belt provide an age of 1887 Ma (Guenot et al 1987).

## 5.5 GREENLAND SHIELD (NORTH AMERICAN CRATON)

### 5.5.1 EAST GREENLAND

Unusually thick, late Precambrian to early Paleozoic sedimentary accumulations are a spectacular feature of the East Greenland Caledonian Fold Belt (Figs 1-3d(i), 1-5d(i)) (Henriksen and Higgins 1976). The most complete sections are found in the outer fjord zone of central East Greenland where occur cumulative thicknesses of up to 17 000 m thick of shallow-water sediments characteristic of shelf environments. In central East Greenland, three main divisions are recognized: Eleonore Bay Group, Tillite Group and Cambro-Ordovician. The broadly coeval Hagen Fjord Group, up to 5000 km thick, is preserved in several thrust sheets within the fold belt 500–700 km to the north (Fig. 4-6).

Non-metamorphosed or mildly metamorphosed *Eleonore Bay sediments*, typical of the three divisions, outcrop over a 450 km by 200 km region. The bulk of the Lower Eleonore Bay Group has a simple lithology of alternating argillaceous (fine-grained greywacke) and arenaceous layers. The Upper Eleonore Bay Group is noted for its spectacular and distinctive lithology featuring quartzite–argillite–carbonate alternations and great uniformity of development over an extensive region (Henriksen and Higgins 1976).

The unconformably overlying Tillite Group, 300–1300 m thick and confined to a narrow belt at least 300 km long, is rich in tillites separated by shale–sandstone intertillite beds (Haller 1971, Harland 1983).

Basement gneisses in the Caledonian Fold Belt of north–east Greenland show that most of these rocks are related to an early Proterozoic event of crust formation at ~2.0 Ga (Kalsbeek et al 1993b). Archean rocks are also present, but do not appear to be common. The data suggest the presence in this fold belt of a major hitherto unrecognized mainly juvenile early Proterozoic belt.

### 5.5.2 NORTHERN GREENLAND

In the platform region of northern Greenland, sedimentary strata which unconformably overlie the crystalline basement range in age from late mid-Proterozoic to Silurian (Fig. 1-5d(i)). Unconformities and glacial deposits mark the Proterozoic sedimentary sequences which are disposed in a number of sedimentary basins formed between tectonic highs or arches in the basement. Of three such Proterozoic depositories across northern Greenland, Thule Basin, facing Ellesmere Island to the northwest, is the thickest and most characteristic (Dawes 1976).

Thule Basin has a composite thickness of >4500 m. Horizontal or gently inclined strata with broad flexures are preserved mainly in downfaulted blocks on the Rinkian Province crystalline basement. The sequence comprises sandstone–shale alternations, calcareous shales, and locally sulphidic and stromatolitic dolostones with associated gypsum layers. Available age data indicate the presence of both mid- and late-Proterozoic strata (Henrikson and Jepsen 1970).

## 5.6 NORTH AMERICAN CRATON (LESS GREENLAND SHIELD)

With the exception of the Lake Superior site previously considered, late Proterozoic (Hadrynian) supracrustal assemblages are confined to the perimeter of the North American Craton, notably the Arctic, Cordilleran, Quachitan–Appalachian, and, in the reconstructed North America, the Caledonide belts of eastern and northern Greenland (Fig. 5-6). These pericratonic assemblages, characteristically composed of quartzite–siltstone–carbonate sequences with minor conglomerate, are remarkably uniform around the rim of the craton, typically grading from thin sandstone shelf units to thicker off-shelf miogeoclinal sequences. Eugeoclinal assemblages, notably rare in the Cordilleran Belt, are present in the Appalachian Belt. Basement rocks take the form of crystalline complexes, with rare intervening unmetamorphosed earlier Proterozoic platform cover. A three-part tillite–mafic volcanic–

# LATE PROTEROZOIC CRUST

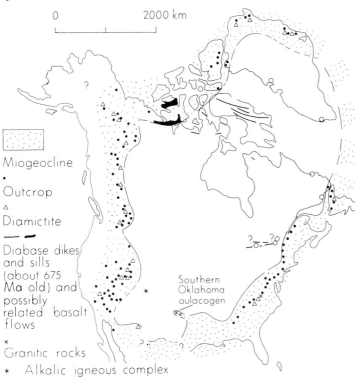

**Fig. 5-6.** Geologic sketch map of the broad distribution of late Proterozoic to early Cambrian rocks, ranging in age from 850 to 540 Ma in the reconstructed North American Craton. (From Stewart 1976, Fig. 2, and reproduced with permission of the author).

BIF association is widespread near or at the base of the miogeoclinal wedges.

## 5.6.1 ARCTIC PROVINCE

In the pericratonic *Franklinian Fold Belt* of the eastern Arctic, late Proterozoic strata are mainly limited to east–central Ellesmere Island (Fig. 5-6; Table 2-8, column 2), where they form the lower part of a thick succession of clastic and carbonate sediments (Kennedy Channel, Ella Bay and Ellesmere groups) deposited upon Precambrian basement in paralic and shelf environments (Trettin 1972, Kerr 1980, Young 1981a,b). Included are the Kennedy Channel–Ella Bay groups (1.0–0.7 Ga) and disconformably overlying Ellesmere Group (~0.5 Ga). The Boothia Uplift, 600 km to the west, includes local, thin Hadrynian cover, mainly composed of arkosic and quartzitic sandstones.

The *Amundsen Embayment* of the western Arctic includes Hadrynian supracrustal sequences of the Shaler Group on Victoria (Minto Arch) and Banks islands, and the correlative Rae Group in the Brock Inlier and Coppermine Area on the mainland to the south (Fig. 5-7; Table 2-8, column 3). Additional small exposures are located at southern Victoria Island. Rocks of these groups consist of shallow-marine and fluviodeltaic sediments. Paleocurrents suggest northwesterly transport. The sediments locally unconformably overlie Archean crystalline basement; elsewhere, they conformably overlie ~1.2 Ga-old Coppermine River lavas. They are intruded by 723–718 Ma Franklinian mafic sills and dykes, and are disconformably overlain by comagmatic (Franklinian) Natkusiak lavas (Young 1981b, Heaman et al 1992).

## 5.6.2 CORDILLERAN OROGEN

Hadrynian strata of the Windermere Supergroup and equivalents were deposited along the western margin of the developing North American craton at some time between 900 and 570 Ma. They are now exposed discontinuously for 4000 km from Alaska–Yukon, through British Columbia, Washington, Idaho, Utah and Nevada, to California and northern Mexico (Figs 5-6, 5-7, 5-8; Table 2-8, column 3). Within this discontinuous belt, the strata thicken from a zero edge in the east to over 10 000 m locally, some 300–500 km to the west. Thrusts, strike-slip faults and associated folds, and large tectonic discontinuities have affected various parts of the belt. The late Proterozoic sequences are characterized by lower diamictite–mafic volcanic–

**Fig. 5-7.** Geologic sketch map of the Amundsen Embayment and Mackenzie Fold Belt in northwest Canada. Major areas of Shaler Group and Mackenzie Mountains Supergroup are shown by the dotted ornament. (From Young 1981a, Fig. 12.2, and reproduced with permission of the Geological Survey of Canada and of the Minister of Supply and Services Canada).

BIF associates overlain by generally thick unfossiliferous sandstone–mudstone assemblages (Harrison and Reynolds 1976, Stewart 1976, Link et al 1993). They were deposited unconformably upon and 'seaward' of the mid-Proterozoic sequences (e.g. Belt–Purcell).

Windermere rocks are considered to be mainly shallow-marine deposits (Stewart 1972). The primary source area, based on paleocurrent indicators and facies changes, lay to the east. Lateral continuity of individual sedimentary units for several hundred kilometres along sedimentary strike indicates a uniform tectonic and sedimentary environment, probably on a continental shelf (Stewart 1972, 1976, Wheeler et al 1972, Young 1981a).

Deposition of the Windermere assemblage followed a thermo-tectonic activity involving uplift and mild folding of subjacent strata including the Belt–Purcell assemblage. The event (~1.2 Ga) is called the East Kootenay Orogeny and Racklan Orogeny in British Columbia and Yukon respectively; southward in the USA the corresponding strata relations vary from unconformable to conformable (Stewart 1972).

Yukon, Alaska and Northwest Territories

The Mackenzie Mountain Supergroup, an epicratonic succession, 4–6 km thick, of mainly shallow-water siliciclastic and carbonate strata, was deposited in the interval 1.0–0.78 Ga. Correlatives of this supergroup lie in the Coppermine Homocline to the north (Fig. 5-7) (Narbonne and Aitken 1995).

Adjoing to the west the disconformably overlying Windermere (Ekwi) Supergroup a 5–7 km-thick siliciclastic-dominated assemblage including deep-

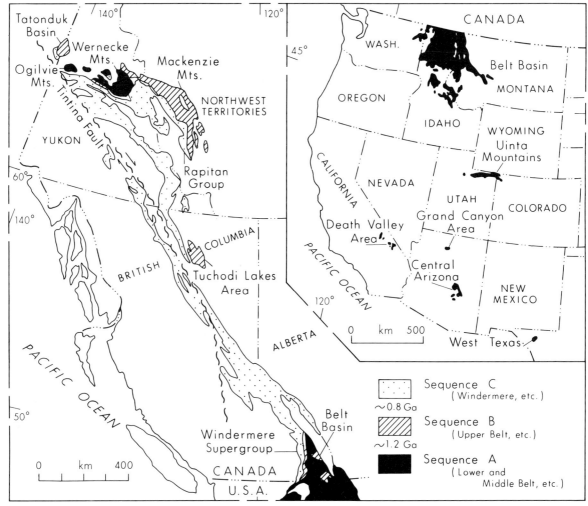

**Fig. 5-8.** Sketch map of the Canadian Cordillera showing the distribution of well-preserved mid and late Proterozoic supracrustal sequences. The inset map shows the corresponding distribution in the Western United States. (From Young 1981b, Fig. 2, and reproduced with permission of the author).

water slope carbonates, is characterized by local diamictites –BIF in the lower Rapitan Group, and recessive mudstones in the upper Hay Creek Group. Windermere strata contain Ediacaran megafossils, ichnofossils and small shelly fossils. The age of Windermere deposition is 780–~730 Ma. The Windermere Supergroup is unconformably overlain by newly defined Ingta formation, (Yeo 1981, Narbonne et al 1994).

The Rapitan Group, which outcrops discontinuously for about 630 km in the arcuate Mackenzie Mountains, comprises a 600–700 m-thick sequence of marine, glaciomarine and glacial sediments that were deposited in the 755–730 Ma interval (Klein and Beukes 1993). These have been correlated with the Sturtian glaciation of Australia (Young 1992). Mixtite beds up to 7 km thick include numerous intraformational fragments up to 25 cm in diameter. Jasper–hematite iron-formation is present in the red mixtite. Two major basins of deposition are recognized: Snake River Basin in the northwest and Mountain River–Redstone River Basin in the south.

Iron formation in the Rapitan Group (Gross 1965) contains a reported reserve in the Crest deposit alone in excess of $20 \times 10^9$ t, with an average grade of about 50% Fe. A hydrothermal origin by which iron and silica precipitated from fumarolic waters discharged along submarine faults is preferred (Gross 1965, Yeo 1981).

In the western Ogilvie Mountains of east–central Alaska and Canada, the Tindir Group (Chuskin 1973) contains a BIF-bearing sequence strongly resembling the Rapitan. On the basis of microfossils and isotopic tracers, the upper Tindir Group is assigned a likely age of 780–620 Ma (Kaufman et al 1992).

## British Columbia and Alberta

In south–central British Columbia the *Windermere Supergroup*, which disconformably overlies the Purcell Supergroup, is mainly represented by two grit–shale–carbonate packages (Horsethief Creek and Kaza–Cariboo), each a southward tapering wedge, with the second wedge deposited above and outboard (westward) of the first (Pell and Simony 1987).

In each grit–shale–carbonate wedge, the lower grit facies is represented by diamictites with local volcanic rocks, coarse clastic detritus shed off fault scarps, and turbidites and grain flows deposited in submarine fans. Medium- and fine-grained metasediments of the subsequent shale facies infilled the basin; the ensuing period of quiescence culminated in the accumulation of carbonates in a stable platform environment.

## Western United States

Late Proterozoic rocks are decidedly discontinuous in the southern Cordillera, a pattern attributable in large part to tectonic dislocations (Fig. 5-8, inset); Table 2-8, column 5) (Stewart 1976, Young 1981a, Hedge et al 1986, Link et al 1993). Representative sections occur in the Uinta Mountains of northeastern Utah–southeast Idaho where all rock units are allochthonous on older Precambrian rocks. Farther south, the Chuar Group, forming the upper part of the Grand Canyon Supergroup as well as correlatives in southern California including the upper part of the Pahrump Group, fall in the age range 1084–600 Ma (Heaman and Grotzinger 1992).

Clearly, the North American Craton had been broadly outlined by rifting of some pre-existing (1.0–0.6 Ga?) supercontinent (Rodinia?), the ~625–555 Ma break-up resulting in the opening of surrounding ocean basins, including the Pacific, as the fissured plates moved away from the craton on all sides to produce the miogeoclinal wedges (Hamilton 1989).

In general, the earlier tectonic picture for the Cordillera is interpreted to be a series of diachronous intracratonic basins featuring lacustrine environments rather than a single passive margin with marine-marginal marine environments. Thus, multiple rift events, the earliest of which did not produce continental separation, are required (Link et al 1993, Hansen et al 1993). Passive marine environments were eventually established in latest Proterozoic–Cambrian time.

## 5.6.3 APPALACHIAN OROGEN

Throughout the entire Appalachian region, extending northeastward for 3500 km from Alabama to Newfoundland, with some additional buried extensions, the general Precambrian history can be considered in terms of a crystalline (gneiss–schist–anorthosite) basement of dominant Grenvillain (~1.0 Ga) character now represented by Grenville inliers (Figs 5-6, 4-9, insets; Table 2-8, column 4), overlain unconformably by an eastward thickening (up to 8000 m) continental terrace wedge of sedimentary and volcanic assemblages with local prominent diamictites. In the northeast, the five parallel zones of the Appalachian Orogen are arranged in terms of the Iapetus (late Proterozoic to early Paleozoic) ocean, predecessor of the Atlantic (Stewart 1976, Williams 1979, Young 1981a, Rankin et al 1983, 1993, Bond et al 1984, Doig et al 1993).

The *Canadian Appalachians* constitute a NE-trending belt of late Proterozoic to Paleozoic strata, together with Grenville inliers that were successively deformed in late Proterozoic (Avalonian), mid-Ordovician (Taconian), Devonian (Acadian) and Permo-Carboniferous (Alleghanian) times. The late Proterozoic to early Paleozoic rocks show sharp contrasts in thickness, facies and structural style across the Canadian Appalachians, the basis of the distinctive five-part stratigraphic–tectonic zonation mentioned above (Williams 1979, Nance and Murphy 1994, Dostal et al 1994).

The *US Appalachians*, also dominated by the late Proterozoic–early Paleozoic orogen, represent the product of collisional events between at least two continental masses. Late Precambrian rocks are recognizable on each flank of this orogen. A pre-Appalachian crystalline basement exposed in the Adirondack Massif and discontinuously along the Blue–Green–Long axis of the Western Flank, marked by the line of basement inliers (Fig. 4-9, upper inset) provides isotopic ages of 1250–1000 Ma (Rankin et al 1983, 1993).

The main belt of late Proterozoic rocks extends sinuously for 1200 km from Alabama to Pennsylvania, ranging in width from 50 to 110 km in the southern part to less than 10 km at the north end. Additional thin units are present in New York and Vermont to the north. These belts are divided along strike into western and eastern limbs (Rodgers 1975, Cook et al 1979, Rankin et al 1983, 1993).

The western limb of the Blue–Green–Long axis (see above) includes thick sections of late Proterozoic predominantly clastic rocks. The Ocoee Supergroup, a representative assemblage at least 8000 m thick, takes the form of great clastic wedges, mainly sandstone–siltstone–shale with some bimodal volcanic rocks. Abrupt eastward Ocoee thickening probably represents the hinge of a major late Proterozoic crustal downwarp. Late Proterozoic sequences along the eastern limb of the Blue–Green–Long axis are thicker and comparatively rich in greywacke, shale and volcanic rocks. The eugeoclinal-type Carolina Slate Belt, at least 9000 m thick, is rich in felsic–mafic volcanic rocks including tuff, lapillistone, pyroclastic breccia and lava flows (Rankin et al 1983, 1993).

## 5.7 SOUTH AMERICAN CRATON

Late Proterozoic rocks occur (1) in the Paraguay–Araguaia and Brasilia fold belts located along the western and eastern margins respectively of Tocantins Province; (2) as cover remnants on the nearby São Francisco Craton; and (3) as dispersed elements within two polycyclic provinces—Mantiqueira and Barborema—of the broad, complex NE-trending Brasílano Fold Belt bordering the Atlantic (Figs 1-3e, 1-5e).

### 5.7.1 TOCANTINS PROVINCE

The *Paraguay–Araguaia Fold Belt* follows the western border of Tocantins Province, Central Brazil Shield, for at least 3200 km (Fig. 2-14; Table 2-9, column 2). Typically composed of quartzite, schist, metagreywacke and volcanic rocks, it describes a sinuous, generally S-trending course with a pronounced westerly indentation at mid-length in conformity with the pre-Brasiliano outline of the adjoining Guapore Craton. The northern and southern extremities of this belt are covered by Phanerozoic sediments of the Parnaiba and Paraná basins respectively (Fig. 1-5e). In addition, a great part of the belt is covered by modern sediments of the Araguaia Plain. There has been considerable uncertainty about the age of some lower stratigraphic beds in the belt, whether mid- or late-Proterozoic. However, following Almeida and Hasui (1984), all Precambrian strata are here designated as late Proterozoic (1.0–0.57 Ga). With the exception of undeformed to mildly deformed shelf facies on the Amazonian foreland to the west, all parts of the Paraguay–Araguaia sequence are deformed increasingly so eastward towards the hinterland, with pronounced structural vergence westward to the craton (Almeida and Hasui 1984, Herz et al 1989).

The Urucum district, State of Mato Grasso do Sul near the city of Corumbá (Fig. 2-13) is famous for its unusually large iron and manganese ore deposits. Intercalated hematite–jaspilites and manganese ore horizons of the Santa Cruz Formation, Jacadigo Group, which directly underlie the latest Precambrian (~600 Ma) Corumbá Group, contain an estimated $36 \times 10^9$ t iron ore and $608 \times 10^6$ t manganese ore. Most of the iron and manganese ores precipitated as chemical sediments in a partly ice-covered, fjordlike basin in this latest Precambrian period (Urban et al 1992).

The N-trending *Brasilia Fold Belt*, at the eastern margin of Tocantins Province, extends on strike for more than 1100 km from the Minas Gerais on the south to Parnaiba Basin cover on the north, and is up to 300 km wide extending from the west edge of the São Francisco Craton on the east, to the SE-trending early Proterozoic Uruaçu Fold Belt on the west (Fig. 2-14; Table 2-9, column 2). The sandstone–pelite–carbonate assemblage is of general miogeoclinal character; the upper part includes prominent tillites. The strata are considerably deformed with structural vergence eastward to the São Francisco Craton. Greenschist facies metamorphism prevails in the main part of the belt and decreases to the east.

### 5.7.2 SÃO FRANCISCO CRATON

Late Proterozoic cover on the São Francisco Craton represents the extensions upon it of thicker coeval pericratonic sequences including those of the Brasilia Fold Belt to the west. This epicratonic cover, which extends over a great area of central Brazil (Fig. 2-15; Table 2-9, column 3), is included in the São Francisco Supergroup which comprises, up-section, the Macaubas and Bambuí groups, (Dardenne and Waldo 1979, Almeida and Hasui 1984).

The *Macaubas Group* (about 1250 m thick) includes basal tillite overlain by a thick sequence of quartzites, arkose and siltstones with some greywacke. Correlated tillites and tilloids have been identified in Minas Gerais, Bahia and Mato Grosso.

The overlying *Bambuí Group* (820 m), which contains basal conglomerate (Jequitai tillite), carbonates, siltstones and arkoses, reflects a remarkably constant shallow-marine environment of deposition.

## 5.7.3 ATLANTIC SHIELD

### Barborema Province

The Barborema Province, which covers an area of about 380 000 km², comprises a complex mosaic of linear Brasíliano fold belts, separated by more or less equidimensional blocks of rejuvenated older basement rocks, commonly in faulted contact. The province is characterized by huge volumes of granitoids, mainly Brasíliano in age but also older (Fig. 1-5e; Table 2-9, column 4) (Almeida et al 1981, Brito Neves et al 1982, Santos and Brito Neves 1984, Brito Neves 1986, Sial et al 1987, Davison and Santos 1989, Janasi and Ulbrich 1991, DaSilva Filho et al 1993).

A complex system of faults divides the province into sections, separating the fold systems or cutting through them. These faults seem to be very old, deep and repeatedly reactivated, with varied character. The structural trend in the provinces is fan-like, opening towards the northeast, diagonally to the arcuate-shaped coast. In the pre-Mesozoic Brazil–Africa supercontinental reconstruction, the E-trending Barborema belts in the south align with the North Equatorial (Central African) belt in Cameroon–Central African Republic, whereas the NNE-trending belts in the north align with the Benin Nigeria and Tuareg Shields of West Africa.

The principal mineral deposits of the province are Be, Ta, and Li in pegmatites, W and Mo in scarns, and Fe, magnesite, marble and graphite in sedimentary rocks.

### Mantiqueira Province

The essentially similar Mantiqueira Province, situated along the Atlantic coast south of 15°S, occupies an area of about 450 000 km² (Fig.1-5e).

The province, also largely defined by the Brasíliano Orogeny, is underlain by NE-trending, alternating Brasíliano fold belts, reworked pre-Brasíliano basement, and, in the southern sector, cratonic nuclei and both interior and marginal massifs, the latter association forming a polycyclic aggregation of granitoid gneiss, migmatite and plutons. Greenschist–amphibolite facies metamorphism prevails in the northwestern parts of the province whereas granulite facies is conspicuous in the southwest.

The northern segment of the province, north of Luis Alves Craton, is commonly called the Ribeira Belt, and the southern segment, the Don Feliciano Belt.

In the northern part of Mantiqueira Province, the arcuate *Aracuai Fold Belt*, some 500 km long by up to 200 km wide, lies at the eastern margin of São Francisco Craton. The belt is composed of pelitic to psammitic sediments, limestone, itabirite and mafic volcanic rocks. Isoclinal folding and large thrust faults affected these rocks, the structures typically verging westward towards the craton.

Amphibolite facies rocks of the Aracuai Belt grade into adjoining gneiss and migmatite. The limit of this belt has not yet been adequately fixed.

In the central part of the province, the so called 'southern' fold belt comprises three fold systems—separated by two median massifs, all with a general NE trend. The fold systems are composed of a lower psammitic–pelitic sequence, with carbonate lenses near the top, and an upper chiefly pelitic to flysch-type sequence. Intense linear, polyphase folding has a poorly developed vergence. Some posttectonic intrusions are dated at 540 Ma (Almeida et al 1981).

The principal mineral resources of the fold belts are Pb–Zn deposits associated with limestone, W and Sn in gneiss, feldspar- and Li-bearing pegmatites, Mn and Ti associated with basement rocks, and Au and Cu associated with rare molassic deposits.

In the central part of the Mantiqueira Province the *Luis Alves cratonic fragment* is differentiated from the adjoining Curitiba Marginal Massif on the north. In the southernmost part of the same province (30°S to Rio de la Plata) the southwestward extension of this same narrow Brasíliano fold belt is bordered by the *Rio de La Plata cratonic nucleus* on the northwest and the *Pelotas Marginal Massif* on the southeast.

Finally, the São Luis Craton is exposed in the northern part of Parnaiba Province along the Atlantic at 1–3°S, 45–47°W (Fig. 1-5e). The Precambrian exposure measuring 250 km by 200 km, is composed by NW-trending zones of granitoid gneiss–migmatite of possible Archean age, with intercalated zones of Proterozoic paraschists including biotite–garnet assemblages, quartzites and metagreywacke. These are associated with widespread granitoid intrusive–extrusive suites.

## 5.8 AFRICAN CRATON: SOUTHERN AFRICA

### 5.8.1 INTRODUCTION

Late Proterozoic tectonism, culminating in the Pan-African thermo-tectonic event (~0.6 Ga), affected large parts of the African Craton, including offshore Madagascar and Seychelle Islands (Figs 1-3f(i,ii), 1-5f(i,ii)). In southern Africa, used broadly, this tectonism produced: (1) the NNW-trending Atlantic coastal chain of geosynclinal wedges, more than 300 km long and incorporating, in southward succession (a) northern and southern coastal branches of the Damara Province; (b) Gariep Province; and (c) southernmost Saldanian province; and, also, disposed almost perpendicular thereto, (2) the main 500-km long, ENE-trending Intracontinental Branch of the Damara Province, in turn succeeded eastward in highly diachronous fashion (Hanson et al 1993), by (a) the Katanga Belt, including the Lufilian Arc and (b) the Zambezi Belt (Figs 5-9, 5-10). At and near the eastern seaboard (Indian Ocean) the Pan-African belt system is marked by considerable tectonic–thermal reworking of basement rocks, widespread isotopic resetting and the presence of scattered ophiolitic melanges, all part of the major N-trending Mozambique Province. In the pre-Mesozoic South America–Africa supercontinental reconstruction, the Damara belt system correlates with the Ribeira fold belt of the Mantiqueira Province (Fig. 5-9).

### 5.8.2 DAMARA PROVINCE

The Damara Province represents an integral part of the Pan-African (~950–450 Ma) structural framework. It extends both eastward between (1) the Kalahari and Congo cratons to form the Intracontinental Branch (Central Zone); and (2) northward along the Atlantic continental margins to form the two contiguous arms, respectively, Northern Coastal Branch (Kaokoveld Orogen) and Southern Coastal Branch (Fig. 5-10). The Intracontinental and Coastal branches collectively constitute an asymmetric triple-junction with lithologic and structural continuity between the component arms. The geologic relations have been studied and documented in detail by, amongst others: Martin and Porada (1977), Kröner (1980), Mason, in Hunter (1981), Tankard et al (1982), Coward (1983), Miller (1983), Hawkesworth et al (1986), Porada (1989) and de Kock (1992).

#### Basement

Relicts of a continental basement occur on the northern and southern flanks of, as well as within, the Damaran Orogen. Radiometric ages indicate that those in the northern and central parts are as old as 2.0 Ga (Burger et al 1976, Tegtmeyer and Kröner 1985), whereas those in the south reflect a significant magmatic event at 1.1 Ga (Watters 1976).

#### Damara Supergroup

The dominantly sedimentary Damara Supergroup comprises three main facies (Fig. 5-11a): (1) the early Nosib graben stage of highly variable thickness; (2) to the north, in the acute angle between the Northern Coastal (Kaokoveld) Branch and the core of the Intracontinental Branch, the platform or miogeoclinal Otavi Group, which is up to 7 km thick; (3) the eugeoclinal Swakop Group to the south, with a maximum preserved thickness of 10–17 km.

(1) Sedimentation in the early *Nosib stage* began 1000–900 Ma ago (Kröner 1980a) in four main

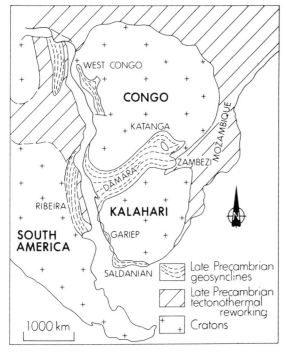

**Fig. 5-9.** Map showing the late Precambrian structural framework of central and southern Africa and eastern South America on a pre-drift reconstruction of the continents. (From Tankard et al 1982, Fig. 9.1, and reproduced with permission of the authors and of Springer-Verlag Publishers).

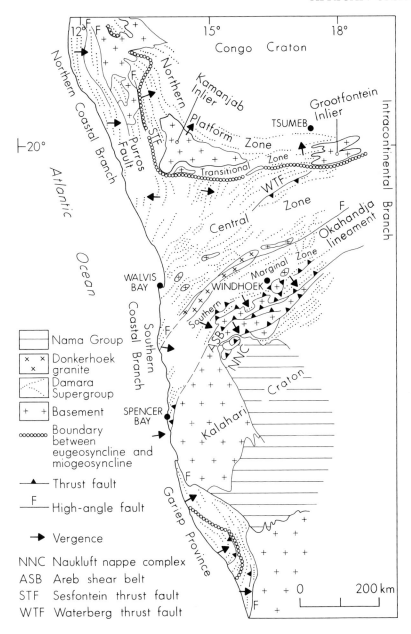

**Fig. 5-10.** Tectonic map of the Damara and Gariep structural provinces, southern Africa. (From Tankard et al 1982, Fig. 9.14, and reproduced with permission of the authors and of Springer-Verlag Publishers).

discontinuous, restricted, fault-bounded troughs and basins. Each trough was approximately 50–70 km wide and 200 km long, with infillings up to 6 km thick.

(2) Following Nosib time, about 830 Ma ago, the area of deposition widened so that the original four narrow troughs coalesced and sediments overstepped the basement highs. Furthermore, the overall shelf–eugeoclinal dichotomy developed. The shelf facies (*Otavi Group*)—up to 3000 m thick of dolomitic limestone including Pb–Zn deposits, with subordinate marl near the top—accumulated in a stable, largely shallow-marine environment molasse (up to 2000 m) composed of quartzite, arkose, greywacke and shale, deposited in discontinuous basins during the interval ~660–570 Ma (Kröner 1980a), and ended with accumulation of fine-grained terrigenous clastics of the Owambo Formation (Fig. 5-11a).

In the central and southern part of the Damaran Orogen, infilling of the Nosib grabens was followed by unconformable or paraconformable deposition of the essentially coeval *Swakop Group* in the deepening Khomas Trough (Fig. 5-11a). The extremely thick Khomas Subgroup now consists largely of micaschist, representing a monotonous succession

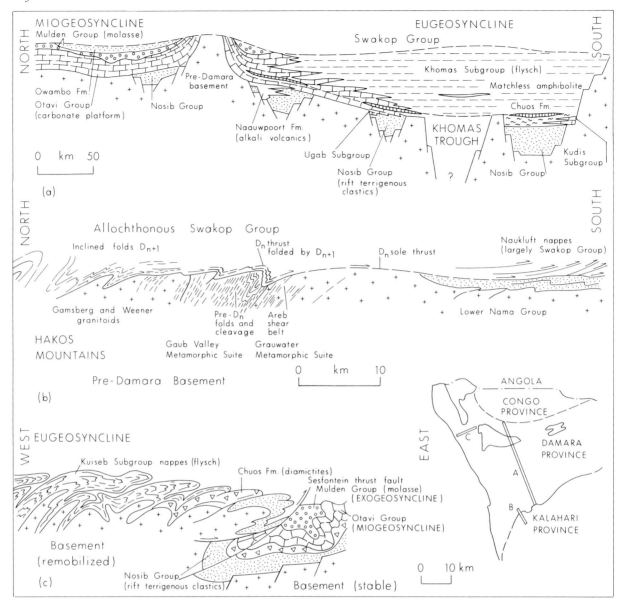

**Fig. 5-11.** (a) Schematic stratigraphic cross-section across the Intracontinental Branch of the Damara Belt; (b) Schematic structural cross-section across the southern marginal thrust zone of the Damara Belt, the Kalahari Foreland and the Naukluft nappe complex; (c) Schematic geologic cross-section across the coastal branch of the Damara province. (From Tankard et al 1982, Fig. 9.16, and reproduced with permission of the authors and of Springer-Verlag Publishers).

of argillite, which overlies local calcareous and quartzose turbidites and includes some lower, discontinuous beds of amphibolites, notably the Matchless amphibolites. This flysch sequence is overlain by a thick monotonous sequence of pelitic–quartzitic mica schist with minor but characteristic lenses of calcsilicate rocks. Widespread manganiferous BIF is intercalated with mostly glaciomarine sediments within the Chuos Formation (Bühn et al 1992). The whole sequence in the Khomas Trough is overturned to the southeast; a stratigraphic thickness exceeding 10 km has been suggested (Martin and Porada 1977). The Khomas Zone represents a fore-arc basin which evolved complexly with time. It was presumably initiated when the Kalahari plate to the south subducted oceanic crust underneath the passive continental margin of the Congo Craton to the north (de Kock 1992).

## Coastal branches

The Northern and Southern Coastal branches of the Damara Province represent slight modifications of the main ENE-trending Intracontinental Branch (Figs 5-10, 5-11c). The >500 km long Northern Coastal Branch (Kaoko Belt), with a maximum width of 150–200 km, displays a strong miogeo–eugeoclinal facies variation from east to west (Porada 1989). From a weakly deformed succession (up to 5000 m) of conglomerate–quartzite (Nosib) and carbonates (Swakop) in the east, deformation and metamorphism increase westward and the sediments become more clastic. Essentially similar, though less well developed, east-to-west relations prevail in the shorter Southern Coast Branch. Well developed mixtites including probable tillites are present in both the Kaokoveld and main Damara orogens.

## 5.8.3 GARIEP BELT

The coastal Gariep Belt is a crescent-shaped zone extending southeast of Lüderitz for nearly 400 km with a maximum width of 80 km (Fig. 5-10). Within this belt, rocks of the Gariep Group, correlated with the Damaran Nosib Group, unconformably overlie a basement composed mainly of the Namaqua Metamorphic Complex, in which late-stage pegmatites provide an age of ~965 Ma, the maximum Gariep age.

As in the Northern Coastal Branch (Kaokoveld Orogen) to the north, there is a strong facies variation in the Gariep Group (Fig. 5-10) from relatively thin, rift-induced miogeoclinal deposits in the east to thick eugeoclinal clastic–volcanic sequences to the west (Kröner 1974, Kröner and Blignault 1976, Tankard et al 1982, Porada 1989).

Tectogenesis under prevailing low-grade metamorphism is characterized by large-scale eastward thrusting of cover sediments, ocean-floor components (ophiolites?) and basement gneisses as flat-lying sheets emplaced to the southeast (Davies and Coward 1982). The presence of ophiolitic rocks in the nappes and the reported occurrence of glaucophane-bearing schists (Kröner 1974) suggest that a plate collision event may have occurred.

## 5.8.4 MALMESBURY GROUP (SALDANIAN BELT)

A thick, highly folded, late Proterozoic turbidite plus quartzite–carbonate assemblage is unconformably overlain by Phanerozoic rocks of the Cape (Saldanian) Province at the southern tip of Africa (Fig. 5-9) (Hartnady et al 1974, Tankard et al 1982). Two narrow, linear NE-striking belts are present. Poor exposure, strong deformation and lack of distinctive marker beds have hindered stratigraphic elucidation.

## 5.8.5 NAMA GROUP

The Nama Group, developed over much of the interior plateau of southern Namibia (Figs 1-5f(i,ii), 5-10), comprises limestone, shale, quartzite and sandstone, together with two mixtites and occasional conglomerates. Except in the north, Nama strata are virtually undisturbed. The sequence varies in thickness from 200 m in the east to about 2000 m in the west and is distributed in three sub-basins of somewhat contrasting lithology (Tankard et al 1982, Gresse and Germs 1993). Lower Nama strata contain Ediacara-type fauna (650–550 Ma range); upper Nama strata contain early Cambrian trace fossils; this represents an exceptional record of early invertebrate evolution. The Nama Basin developed as a peripheral foreland basin between the Kalahari Craton in the east and the partly contemporaneous Damara and Gariep belts in the north and west; the formative transpression probably involved NW-directed subduction of the Kalahari plate under the Congo–South American plates during the sequential closing of the proto-Atlantic Ocean (650–500 Ma) (Gresse and Germs 1993).

## 5.8.6 ZAMBEZI BELT

The structural feature known as the Zambezi Belt, first recognized by Macgregor (1951), comprises a zone of intensely sheared high grade metamorphic rocks including metasediments, intercalated basement gneisses and granitoid intrusions, forming a narrow arcuate terrane between the northern margin of the Zimbabwe Craton on the south and the Karoo (Permo–Triassic) rocks of the Zambezi Rift Valley on the north (Vail and Snelling 1971, Hanson et al 1988b, Porada 1989). The belt strikes east to west for 400 km and is up to 60 km wide (Fig. 5-12). It contains at about mid-length several faulted segments of the Great Dyke (2.49 Ga). At its eastern extremity the southern portion of the belt swings southward and merges with the western margin of the Mozambique Belt. At its western extremity, the Zambezi Belt disappears under

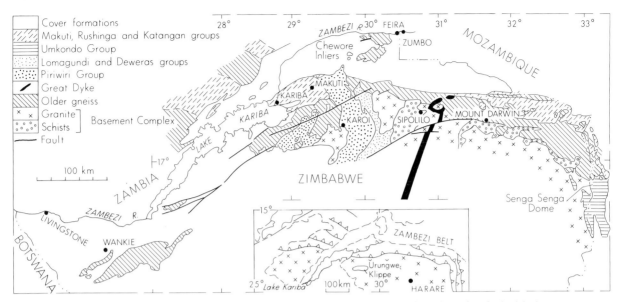

**Fig. 5-12.** Simplified geologic map of the Zambezi Metamorphic Belt. (Modified after Brokerick, in Hunter et al 1981, Fig. 10.16, and reproduced with permission of Elsevier Science Publishers; inset based on Porada 1989, Fig. 7 and reproduced with permission of the author).

Karoo sedimentary rocks of the Zambezi Valley. The belt may link with the main Intracontinental Branch of the Damara Belt in Namibia to the west, and with the Lufilian Arc of the Katangan Belt in Zambia and Zaire to the north. However, the Zambezi belt is separated from the latter by the ENE-trending, sinistral Mwembeshi dislocation, a regionally significant Pan-African transcurrent shear zone (550–500 Ma), which forms the major dividing line between the Kalahari and Congo shields (Daly 1986b, Hanson et al 1993).

The arcuate boundary between the Zambezi Belt and the Zimbabwe Craton is a major thrust zone, along which the metasediments and gneisses are thrust southward on to the craton. The Urungwe Klippe, situated on the craton some 40 km south of this E-trending thrust, is regarded as an outlier of the belt brought into its present position during the thrusting event (Fig. 5-12, inset) (Porada 1989, Coward and Daly 1984).

The Zambezi Belt has been shown to consist of a number of units of different age and composition, overprinted by the thermo-tectonic event which gave rise to the belt in its present form. Six units are specified within the belt (Fig. 5-12), all strongly deformed and in places metamorphosed to the sillimanite grade.

Daly (1986b) correlates the main thrusting event of the Zambezi Belt with the east–northeastward directed thrusting in the Lufilian Arc of the Katanga Belt (see below) and hence interprets the intervening ENE-trending, sinistral Mwembeshi dislocation as a transform fault between the two belts with opposite movement directions. A further consequence of this correlation would be that the age of Zambezi deformation is in the 950–850 Ma range (Porada 1989).

## 5.9 AFRICAN CRATON: CENTRAL AFRICA

### 5.9.1 INTRODUCTION

Little folded or unfolded late Proterozoic arenaceous–pelitic–carbonate sedimentary rocks of the Katangan Supergroup and stratigraphic equivalents (Mbuji Mayi (Bushimay), West Congolian, Lindian–Ubangian, Itombwe and Malagarasian–Bukoban supergroups) crop out around much of the periphery of the vast central Congo Basin and appear to underlie virtually all of it (Fig. 5-13; Table 5-3). The sedimentary rocks lie with marked unconformity on a variety of older Precambrian rocks and are unconformably overlain by Phanerozoic cover. The Katangan Supergroup is the key stratigraphic succession in the region. In southern Zaire and Zambia of the south–central sector, Katangan folding is more marked than elsewhere, where it defines the Lufilian Arc. Similarly, deformed coeval strata occupy the more peripheral mobile (internal) zones surrounding much of the Congo Basin.

Considered in general, the key Katangan suc-

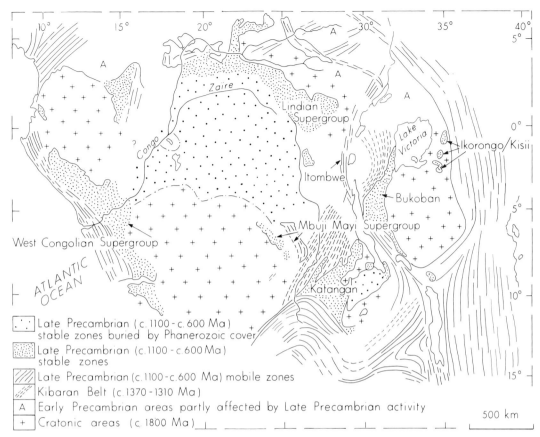

**Fig. 5-13.** Late Proterozoic belt and cover pattern around the Congo Basin, Equatorial Africa. (From Cahen 1982, Fig. 1).

cession is characterized by a threefold division: lower arenaceous (–pelitic) complex; middle calcareous (–pelitic) complex; and upper arenaceous (–pelitic) complex, which is typically red in colour and more or less calcareous. The Katangan sequence proper and practically all the circum-basin correlatives contain one or two tilloid horizons that constitute important marker horizons. Volcanic and intrusive rocks are rare but, where present, occur in the lower to middle parts of the succession. The Shaba–Zambia Copperbelt is famous for its large complex copper-rich ore deposits.

The Katangan sequence and correlatives of the Congo Basin vary in stratigraphic thickness from 3000 to 10 000 m. Deposition occurred in the general interval 1100–660 Ma. The stratigraphic sequences surrounding the basin display lithologic similarities that are in keeping with the sequences having been deposited in various parts of one major basin system. (Key references: Cahen et al 1976, 1984, Mendelsohn 1981, Cahen 1982, Porada 1989, Hanson et al 1993.)

### 5.9.2 CIRCUM-CONGO BASIN SEQUENCES

The circum-Congo late Proterozoic sequences are briefly reviewed in clockwise succession around the basin (Fig. 5-13; Table 5-3).

#### Katangan Supergroup (Shaba–Zambia)

The type Katangan sequence lies in the Copperbelt of southern Shaba province (Zaire) and Zambia. From here Katangan strata extend northeastward into Shaba for 500 km in the form of a complexly deformed synclinorial arc (Lufilian) and for another 500 km as stable platform cover (Golfe du Katanga).

The Katangan Supergroup in the Lufilian Arc area constitutes a more or less concordant pile up to 10 km thick (Table 5-3, column 1). This arc provides the type Katangan section as well as containing the copper and other mineral riches for which this area is so famous. The following stratigraphic subdivisions, adapted from Mendelsohn (1981) and

**Table 5-3. Generalized chronostratigraphy of the major late Proterozoic sequences of equatorial Africa.**

| (1) Shaba–Zambia: Katangan Supergroup (602 Ma) | (2) Kasai–West Shaba: Mbuji Mayi Supergroup | (3) West Congo (Zaire, Angola): West Congolian Supergroup (625±25 Ma) | (4) Northeast Zaire: Lindian Supergroup (>590±20 Ma) | (5) Central Africa Republic | (6) Eastern Kivu (Zaire): Itombwe Supergroup (660±10 Ma) Late shearing and folding | (7) Southeast Burundi and West Tanzania: Malagarasian (Bu) and Bukoban (T) Supergroups |
|---|---|---|---|---|---|---|
| Upper Katangan (Kundelungu): Upper Kundelungu: (I-II-III) | | Schisto-Greseux Sequences: (2) Inkisi Group (1) Mpioka Group (734±10 Ma) | Aruwimi Group: (3) Banalia arkoses (2) Alolo shales (700±9 Ma) (1) Galamboge quartzites | Dialinga Formation (700±9 Ma) Bakouma Formation | | Kibago Group (Bu) (Manyovu redbeds (T)) |
| 'Calcaire Rose' | | Schisto-Calcaire Group | | Bondo 'fluvioglacial' conglomerate | | Uha Group (T) = (Mosso (Bu)) |
| Petit Conglomérat-mixtite | | | (788±15 Ma) | Nakondo quartzite | Alkaline granites-syenites, (774±44 Ma) | Gagwe (T) lavas (Kabuye (Bu)) (815±14 Ma) |
| Lower Kundelungu: Kakontwe Limestone | | Upper Mixtite Haut Shiloango Group | Lokoma Group | Bougboulou Group | Tshibangu Group | Kigonero Flags (T) Musindozi Group (Bu), (>900±24 Ma) |
| Grand Conglomérat and lavas Pegmatites (976±10 Ma) | Lavas (948±20 Ma) | Lower Mixtite and lavas | Akwokwo mixtite (950±50 Ma) | Mixtite | Basal mixtite K-granites, (976±10 Ma) | |
| Lower Katangan (Roan, Mine): Roan Group: Mwashya BII Group Upper Roan BI Group Lower Roan B0 Group | | Sansikwa Group (Terreiro Formation) Mayumbian-Zadinian | Ituri Group (Lenda Formation) | | Nyakasiba Group | Busondo Group (T) = Bukoba Ss (T) (1020±25 Ma) (Kawumwe (Bu)) Masontwa Group (T) Itiaso Group (T) |
| Basement (1110±1200 Ma) (1310±25 Ma) | | Kimezian Basement (~2.0 Ga) | Basement | | Basement (Kibaran) (1204±65 Ma) | Basement and (Burundian) (1310±25 Ma) |

Modified after Cahen et al (1984)

including four key marker units (to the right), are synthesized from a variety of existing subdivisions. The general framework is applicable to Katangan-type sequences of the Congo region at large, with due allowance for local variations:

|  |  |
|---|---|
|  | Upper Kundelungu |
|  | Calcaire Rose |
| Upper Katangan | Petit Conglomérat |
| (Kundelungu) |  |
|  | (tillite) |
|  | Lower Kundelungu |
|  | Kakontwe limestone |
|  | Grand Conglomérat |
|  | (tillite) |
|  |  |
|  | Mwashya |
| Lower Katangan | Upper Roan/or Dipeta |
| (Roan, Mine) | Lower Roan or Mine series |

The lower clastic division (Roan–Mwashya) is composed of arenites and conglomerate with local argillites; the middle division (Lower Kundelungu) of generally dolomitic carbonate rocks, and interstratified argillites (carbonaceous) and glaciogenic mixtites; and the upper clastic division (Upper Kundelungu) of arenites and argillites. Volcanic rocks and intrusions are rare, but where present occur in the lower to middle part of the succession. The detailed correlation is based mainly on the presence of two tilloids (Grand and Petit Conglomérats), which, though not everywhere true tillites, have undoubted glacial affinities and represent a widespread late Proterozoic glacial period (Lower Congo Glacial Period of Harland 1983). Kundelungu deposition occurred in the interval ~950–550 Ma. The younger limit for Katangan deposition in central Zambia is 570–560 Ma. Lufilian arc orogenesis in central Zambia was diachronous in the interval 602–560 Ma (Hanson et al 1993).

Mineral deposits

The copper (Cu–U–Pb–Zn–V–Cd–Ag) deposits of the Shaba–Zambia Copperbelt form one of the world's major mineral fields (Mendelsohn 1981). By far the greatest production is copper. The stratiform ore deposits, concentrated in the lower Roan (Mine) clastic sequences, comprise both Zambian (shale)- and Shaba (dolomitic)-types.

The main copper minerals are chalcopyrite and bornite. There is a well-defined mineral zoning, in places demonstrably parallel to the depositional shoreline. Secondary alteration is variable and many of the northern deposits consist largely of carbonate and oxide minerals. Cobalt is present in significant quantities. Uranium occurs in many deposits. Among the other elements, found as separate stratiform bodies or accompanying the copper, are nickel, molybdenum and tungsten.

The reason for the intense concentration of ore deposits along the Lufilian Arc, in contrast to their comparative sparsity elsewhere, particularly outside the Katanga–Damara 'trough', is not clearly understood. The stratiform deposits formed during sedimentation and diagenesis. They were later modified during the Pan-African events, without, however, any significant metal redistribution. Mineralization is probably related to a primary basin-edge (Mendelsohn 1981, Unrug 1988, Sweeney et al 1991), on which the second-order Lufilian Arc was structurally imposed.

## Mbuji Mayi (Bushimay) Supergroup

The Bushimay Supergroup occupies a separate basin or sub-basin lying 500 km to the northwest in western Shaba and eastern Kasai (Fig. 5-13; Table 5-3, column 2). Mbuji Mayi strata unconformably overlie Kibaran rocks or older granitoid intrusions of the Kasai Craton which are locally dated at 2037 Ma (Cahen et al 1984).

The Mbuji Mayi Supergroup, divided up-section into B0, BI and BII groups, is composed of conglomerate, sandstone, shales and stromatolitic carbonates. Basalt flows, dated at 948 Ma, unconformably overlie the uppermost strata (Cahen et al 1984).

## West Congo Belt

The West Congo Belt, traceable southward for over 1300 km from Gabon through Congo to Angola (Fig. 5-13), constitutes both (1) an internal (westward) mobile zone, 50–150 km wide including eastward verging deformed and metamorphosed strata of the Zadinian–Mayumbian association, in faulted contact to the east; with (2) 100–250-wide median-external stable zones. The latter are composed of subhorizontal (external) to moderately folded (median) strata of the West Congolian Supergroup, overlain to the east by flat-lying platform cover of the Congo Basin (Fig. 5-14).

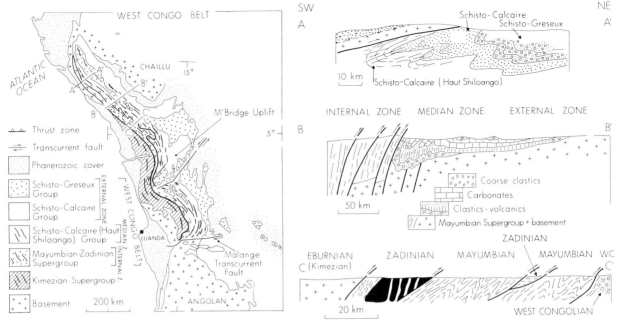

**Fig. 5-14.** Geologic map of the West Congo Fold Belt. A–A′, B–B′, C–C′ represent locations of accompanying geologic sections. Note relative locations of internal zone, median zone and external zone. Section A–A′ – simplified section through basement, internal and median zones, showing effects of $D_1$, thrusting and folding; Section B–B′ – geologic section through the West Congo Belt; Section C–C′ – geologic section through the internal zone, showing position of possible ophiolite complex and effects of $D_1$ thrusting: WC – West Congolian Supergroup. (Modified after Porada 1989, Figs 5, 6 and reproduced with permission of the author).

The stratigraphic scheme of Tack (1983) and Franssen and Andre (1988) corresponds to that developed in Angola (Schermerhorn 1981) and Congo (Vellutini et al 1983) and is the one generally followed here (Table 5-3, column 3; Fig. 5-14).

### Kimezian Basement

Kimezian Supergroup rocks form a westerly basement strip in tectonic contact to the east with Zadinian–Mayumbian rocks.

### Zadinian Group (Supergroup)

In Lower Zaire, the Zadinian sequence (>3500 m), which is separated from Kimezian basement by thin fault gouges, is composed, up-section, of conglomerate, quartzite, metarhyolite, metabasalt, pelitic schist and calcsilicates. Westward, the corresponding, somewhat more metamorphosed (amphibolite facies) sequence is rich in quartzites, amphibolites, staurolite–garnet–mica schists and calcsilicate schists (Cahen et al 1984, Franssen and André 1988).

### Mayumbian Group (Supergroup)

The Mayumbian Group, to the east, is more than 4 km thick (Cahen et al 1984). It comprises a gross twofold division of lower mafic volcanic rocks (1600–2200 m) including possible ophiolites (Vellutini et al 1983), and a thick upper sequence (>3000 m) of metarhyolites and metavolcaniclastics with subordinate quartzites and schists. Mayumbian rocks have been deformed along a dominant NNW trend (Cahen et al 1984).

### West Congolian Supergroup

The slightly to non-deformed West Congolian Supergroup of the combined median-external (eastern) zone is about 4600 thick. It closely resembles the type Katangan section to the east in lithostratigraphy. A broad threefold division is provided by the presence of two mixtites, as in the type Katangan section (Table 5-3, column 3). The uppermost Inkisi redbeds closely resemble and are correlated with those encountered in the two drill intersections that penetrated the cover of the Congo Basin (see above), thereby suggesting a continuous blanket redbed cover.

West Congolian strata are moderately deformed in eastward verging isoclinal to broad open folds that are transitional eastward to tabular beds upon the buried stable craton (Fig. 5-14, Section B-B').

Tectonic models

Vellunti et al (1983) and Franssen and André (1988) interpret the internal zone as a W-directed collisional orogen (Mayumbian Belt) which, although Kibaran (i.e. 1.3 Ga) in age, formed after subduction of ocean floor.

Porada (1989) concludes that some kind of convergence must have occurred to the west of the internal zone. This process may have involved continental separation, ocean floor formation, subduction and eventual continental collision. Even in the absence of established ophiolites and ocean floor associates, some process of crustal stretching and thinning, accompanied by subsidence and sedimentation with volcanism, would be required to allow for the eventual crustal convergence. It is therefore assumed by Porada that a Pan-African rift structure existed west of the internal zone and was subsequently closed during the West Congolian Orogeny.

Lindian Supergroup

The ESE-trending Lindian belt (supergroup) extends continuously along strike for more than 1000 km (Fig. 5-13). To the north, Lindian strata become increasingly deformed and metamorphosed in the adjoining E-trending Central African (Pan-African) Belt.

The type Lindian Supergroup (Table 5-3, column 4) is divided by disconformity or unconformity into three groups—Ituri, Lokoma and Aruwimi—and includes a prominent mixtite of probable glaciogenic origin.

Lindian strata were deposited in an anorogenic basin. The uppermost red Banalia arkoses were slightly folded along NW-trending axes. The Akwokwo mixtite, an important marker zone, although locally containing glaciogenic material, was generally deposited under water.

The corresponding section in the Central African Republic, which includes a presumably coeval mixtite, is illustrated in Table 5-3, column 5.

Itombwe Belt

This narrow N-NE-trending belt, situated in the western rift zone between the Congo and Tanzania cratons, extends a total distance of ~800 km (Fig. 5-13). Formerly considered to belong to the Burundian (Kibaride) sequence, it is now known to unconformably overlie Kibaran rocks (Cahen et al 1984).

The Itombwe Supergroup (Table 5-3, column 6) is divided by a prominent (2500 m thick) mixtite into two groups, composed, respectively, of lower (1000–1500 m) conglomerate–quartzite–phyllite–slates, and upper (~2000 m) sandstone–pelite alternations.

Itombwe rocks are deformed on essentially north–south axes. The most prominent structures are symmetrical open folds of similar type, with axial planes inclined either to the east or west, and accompanied by axial plane flow schistosity (Villeneuve 1977).

Bukoban Supergroup

The Bukoban Supergroup of Tanzania (Malagarasian Supergroup of Burundi) has an arcuate, 800 km by 200 km, outcrop pattern along the western side of the Tanzania Craton (Fig. 5-13). Stratigraphic nomenclature is complex. The Tanzanian conglomerate–sandstone–shales section, about 3000 m thick, is divided into three groups (Table 5-3, column 7) (Cahen et al 1984).

Bukoban folding is restricted to those western areas bordering the strongly folded Itiaso 'geosyncline'. The folds are distributed along simple northeast axes, have eastward vergence, and die out rapidly to the northeast.

Ikorongo Group

On the eastern side of the Tanzania Craton at the latitude of Lake Victoria, the Ikorongo Group, distributed in small scattered outliers, has also been correlated with the Bukoban and shown to correspond to beds involved in the Mozambique Belt to the east (Fig. 5-13). Some uncertain dates fall in the range 1138–1221 Ma (Briden et al 1971a).

5.9.3 MOZAMBIQUE BELT

Setting and composition

The Mozambique Belt of Holmes (1951) is a N-trending polycyclic structural entity extending along the eastern margin of most of southern, central and parts of northern Africa, thereby cutting off a number of older transverse belts. The Mozambique Belt is part of the continent-wide network of Pan-African belts, product of late Precambrian to early Paleozoic activity (Kennedy 1964). The general uniformity of this major transcontinental belt reflects to a great extent the magnitude and intensity of Pan-African superposition upon older rocks of diverse ages. (Key references: Haslam et al 1983, Cahen et al 1984, Daly 1986a,b, Shackleton 1986, Key et al 1989, Porada 1989.)

The southern–central Mozambique Belt is subdivided into four provinces: (1) Malawi; (2) Mid-Zambezi; (3) Mozambique; and (4) Kenya–Tanzania, the last situated to the east of the Tanzania Craton (Figs 5-15, 4-13). The northern extension of the Mozambique Belt, which is still poorly defined due to widespread overburden, extends into Ethiopia and Sudan (Fig. 3-17). Granulite facies rocks together with associated amphibolite facies assemblages make up an important part of the central–eastern Mozambique Belt.

(1) The *Malawi Province* forms a triangular area, 600 km on the side bordered to the northwest by the Irumide Belt and to the south, across the E-trending Mwembeshi dislocation, by the Zambezi Belt.

The basement of this province comprises medium to high grade gneisses, schists, amphibolites and granulitic rocks which have undergone a complex, polycyclic history. These basement gneisses underlie the Muva metasediments (~1.1 Ga) of the Irumide Belt and include large tracts of granulite facies rocks (Daly 1986a) of both Irumide (1.1 Ga) and earlier (~2.3 Ga) ages (Haslam et al 1983, Andreoli 1984). Included with the former serpentinized Alpine-type peridotite interpreted by Andreoli (1984) as a suture zone associated with a metamorphosed island-arc terrain developed during the Kibaran Orogeny (1.1 Ga).

(2) The *Mid-Zambezi Province* is a narrow (100–150 km), S-trending, 400 km-long belt situated at the north eastern margin of the Zimbabwe Craton; it is underlain almost entirely by a complex group of banded quartzofeldspathic micaceous gneisses and migmatites with locally preserved quartzite and marble (Cahen et al 1984). The rocks have been intensely folded and have undergone widespread granulite-facies metamorphism. The western margin of the belt in Zimbabwe is marked by the progressively folded and metamorphosed sedimentary cover of the Umkondo Group (~1.8 Ga), which overlaps the edge of the Zimbabwe Craton. The granitoid plutons and E-striking greenstone belts of the Zimbabwe Craton, as well as the gneisses and granulites of the Limpopo Belt, can be traced for a short distance eastward into the Mozambique gneisses, where they become intensely deformed on north–south axes and quickly indistinguishable from the other gneisses.

(3) The geology of the *Mozambique Province* to the east, a large quasi-rectangular area about 500 km on the side, is complex. Gneisses of several different assemblages predominate which have not yet been clearly unravelled. Metasedimentary micaschist, quartzite and crystalline marble are widespread, and charnockitic gneisses are developed near the Malawi border.

A granulite-facies metamorphism of Irumide age (~1.1 Ga) is described by Sacchi et al (1984) from several places in central and southern Mozambique, and large klippen of Irumide age (~1.1 Ga) occur

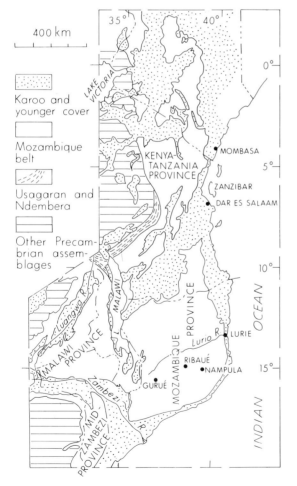

**Fig. 5-15.** Provinces within the Mozambique Belt of Equatorial Africa. (Modified after Cahen et al 1984, Fig. 5.1, and reproduced with permission of the authors and of Oxford University Press).

in northern Mozambique (Jourde 1983, Sacchi et al 1984).

Across the central part of the province a narrow belt of ENE-trending, highly sheared gneisses strike along the valley of the Lurio River to the coast (Fig. 4-13) (Jourde 1983). The Lurio Belt is mainly composed of amphibolites, amphibolitic gneisses and metavolcanic rocks at prevailing amphibolite facies metamorphism, with which are associated serpentinite bodies and granulite facies gneisses. Thus, the Lurio belt clearly represents a zone of major structural significance in the Mozambique Belt (Daly 1986a). Granulite nappes therein are thrust over allochthonous supracrustal units, and with both sequences then thrust over an autochthonous migmatitic foreland (Pinna et al 1993).

(4) The *Kenya–Tanzania Province* extends northward for 1500 km through Tanzania and Kenya to the vicinity of Ethiopia (Shackleton 1986). It represents the principal domain of late Proterozoic deformation east of the Tanzania Craton. The western Mozambique front is well defined at the eastern margin of the craton, where the E-trending shield structures are sharply truncated and transposed to N–S trends along a prominent mylonite zone up to several kilometres wide. To the east there is a rapid rise in the metamorphic grade and development of mainly eastward dipping planar foliations within prevailing high grade metasediments and granitoid gneisses. The eastern front of the Mozambique Belt is more obscure as a result of widespread Phanerozoic cover, including the Tertiary Rift Volcanics (Key et al 1989).

Stratigraphic relations in the province are poorly understood, mainly because of the intense deformation and metamorphism. Three zones of mafic–ultramafic rocks, representing possible ophiolites (Vearncombe 1983, Shackleton 1986), are recognized, two in north–central Kenya and one in northeast Kenya–Ethiopia. The last includes serpentinized peridotite with podiform chromite, metagabbro, amphibolite and pillow basalts associated with mafic gneisses which occur on-strike with extensive outcrops of well-documented ophiolites in southeast Ethiopia (Kazmin 1978). Current evidence supports the suggestion that the low-grade ophiolitic sequences in Kenya are allochthonous and structurally emplaced over the higher-grade gneisses (Mosley 1993).

Tectonic model

In reviewing the Mozambique Belt in Kenya–Tanzania, Shackleton (1986) emphasizes four critical relationships: (1) the general westward vergence; (2) the presence of presumed ophiolites; (3) the tectonic alternation of shelf-facies metasediments and presumed basement gneisses; and (4) the recumbent structures involving thrusting, imbrication and isoclinal folding, all of Alpine type.

On this basis, Shackleton postulates that the Mozambique Belt evolved as a result of plate tectonic processes involving continent–continent collision. The presence of several suture zones suggests repeated collisions by two or more continental fragments (Bonavia and Chorowicz 1993). The scanty geochronologic data suggest that the suture ophiolites and the collisions were late Proterozoic (Pan-African) in age (~600 Ma).

Furthermore, the paleomagnetic record of a major misfit between the apparent polar wander paths for East and West Gondwana (McWilliams 1981) supports the view that the Mozambique Belt resulted from continent–continent collision at ~ 600 Ma.

### 5.9.4 MADAGASCAR (MALAGASY)

The island of Madagascar, a pre-drift Central African component, is similar to the Mozambique Belt in displaying a dominant N–S trend and abundant ages in the ~650–485 Ma range. However, older Precambrian terranes, reworked during later tectono-thermal events, are widespread in this island (Fig. 1-5f(i,ii)) (Besaire 1967, 1971, Vachette 1979a).

The island bears the dominant imprint of the Pan-African events, as in the Mozambique Belt to the west but with widespread occurrences of well-preserved Archean and early- to mid-Proterozoic rocks.

### 5.9.5 SEYCHELLES ISLANDS

The Seychelles Archipelago consists of well over 100 islands within a NE-trending 1200 km-long belt located 1000 km northeast of Madagascar. The principal islet cluster rises from the central part of the submarine Seychelles Bank which is less than 60 m below sea-level and covers 41 000 km$^2$. The principal islands are composed of Precambrian granitoid intrusions of general post-orogenic type (Baker 1967, Suwa et al 1994).

Three types of late Proterozoic granitoids predominate: (1) gneissose tonalite–granodiorite of calc-alkalic, I-type, and age 713 Ma; (2) porphyritic granite of sub-alkaline, A-type, and age 683 Ma;

and (3) grey granite of alkaline, A-type, and age 570 Ma. Seychelles granitoids correlate well in types and ages with those of the Arabian–Nubian shield to the north. Together with paleomagnetic data, the relations indicate that the Seychelles were originally located near the eastern end of the Horn of Africa (Somalia) and have drifted to their present position since continental break-up (Suwa et al 1994).

### 5.9.6 CENTRAL AFRICAN BELT (CAMEROON–CENTRAL AFRICAN REPUBLIC–SOUTHERN SUDAN)

The region extending eastward from northern Cameroon through Central African Republic to southern Sudan occupies a critical intercratonic position separating the Congo Craton on the south, the West African Craton on the northwest and the poorly exposed East Saharan Craton on the north (Fig. 1-5f(i,ii)). This E-trending, ~3200 km long and up to 1000 km broad Pan-African belt is as yet poorly defined, especially in the central and eastern parts, because of widespread cover. It is variably referred to as the Central African or North Equatorial Belt (Bessoles and Lasserre 1978, Lasserre 1978, Bessoles and Trompette 1980, Toteu et al 1987, 1994).

The Precambrian rocks in the western Cameroon part of the belt were classically divided into two groups: (1) an older 'basement complex', composed of metasediments, amphibolites, orthogneisses, and syn- to late-tectonic granitoids, the group largely metamorphosed under amphibolite-facies with widespread development of migmatites; and (2) a younger 'intermediate group' (Bessoles and Lasserre 1978), composed of low–intermediate-facies metavolcanic and metasedimentary rocks of presumed mid-Proterozoic age (Bessoles and Lasserre 1978).

However, recent studies in northern Cameroon (Toteu et al 1987) cast doubt on this classical basement-cover relationship, instead suggesting that practically all the rocks in the belt represent juvenile accretions during the Pan-African cycle of event. The only earlier ages in the belt are provided by local migmatites dated at 1295 Ma and 1940 Ma, possibly relicts of older crust. In brief, Toteu et al (1987, 1994) propose a plate-tectonic model involving opening and closing of an ocean bordering the eastern margin of the adjoining West African Craton. In the larger Africa–Brazil context, they propose that the Congo Craton–São Francisco Craton (Brazil) collective, joined during the Eburnean–Trans-Amazonian Orogeny (~2100 Ma), were collectively bordered to the north by a late Proterozoic rift basin that was closed during the Pan-African–Brasíliano Orogeny (~600 Ma) to produce the Central African Belt and Brazilian counterpart to the west.

## 5.10 AFRICAN CRATON: NORTHWEST AFRICA

Late Proterozoic components in this large tract include: (1) Taoudeni Basin; (2) Anti-Atlas Domain and Tindouf Basin; (3) Trans-Saharan (Pharusian) Mobile Belt, including (a) Tuareg Shield and (b) Benin Nigeria Shield; (4) Togo Belt and Volta Basin; and (5) Gourma Basin (Fig. 1-5f(i,ii)).

### 5.10.1 TAOUDENI BASIN

Taoudeni Basin, the largest single structural unit of the West African Craton, covers some $2 \times 10^6$ km$^2$ (Fig. 2-23). The late Precambrian to Paleozoic (1.0–0.36 Ga) basin fill unconformably overlies older Precambrian (Archean to Eburnean (1.9 Ga)) basement and is in turn conformably overlain by Mesozoic–Cenozoic cover, such that the late Precambrian to Paleozoic strata are mainly exposed at the rim of the basin (Trompette 1973, Affaton et al 1980, 1991, Bronner et al 1981, Villeneuve and Cornée 1994). The platform cover is up to 4000 m in thickness but averages 1000–1500 m.

Following the unusually long mid-Proterozoic (1.9–1.0 Ga) basement peneplanation ('La Grande Lacune'), the central part and eastern and northern margins of the then-existing 'West African' craton were gently downwarped to form the vast central Taoudeni Basin and the smaller Volta and Tindouf basins at the southeastern and northern margins respectively.

The sandstone–stromatolitic carbonate–pelite platform cover in the central Taoudeni Basin includes in the lower part the 700 m thick Atar Group of age 850–700 Ma, which provides an important direct stratigraphic correlation with the Tuareg Shield to the west (see Table 5-4, columns 1 and 6). This platform cover is characterized in general by deposition on a remarkably flat surface, wide extent, great homogeneity of facies, and average subsidence rates of 5 m/Ma (Bronner et al 1980). The earlier dominantly sandstone phase of Taoudeni and Volta sedimentation (Supergroup 1) (Affaton et al 1980, Rocci et al 1991) began about

1035 Ma ago and ended with the pre-tillite unconformity dated at 694–605 Ma (Katangan Unconformity). Supergroup 1 is thus entirely Precambrian in age and the diastrophism which produced the upper angular unconformity is also of Precambrian age. Supergroup 2, which begins with glacial beds older than 605 Ma, ends with Ordovician rocks. An important disconformity separates it from Supergroup 3 which is composed of Upper Ordovician to Silurian strata, but which also begins with extensive glacial deposits. The top of Supergroup 3, at least in the Mauritanian Adrar, represents a Silurian–Devonian disconformity.

Within this general setting, the Taoudeni Basin, considered in more detail, is divided into eight sub-basins each with contrasting structures and sedimentologic evolutions. Of these, the largest and most conspicuous are the Hank Basin in the northeast, and the Gourma aulacogen and Mopti trough in the southeast (Villeneuve and Cornée 1994, fig. 5).

## 5.10.2 ANTI-ATLAS DOMAIN

The northwestern part of Africa, encompassing Morocco, is generally divided into three main structural domains or provinces—Anti-Atlas, Atlas (plus Meseta) and Rif (Fig. 1-5f(i,ii)). Of these, the Anti-Atlas, which borders the Tindouf Basin in the north, contains considerable late Proterozoic to early Paleozoic rocks, including both deformed supracrustal rocks and post-orogenic intrusions. The Anti-Atlas continues southeastward discontinuously in the form of the Ougarta Belt, a link with the Hoggar. The Anti-Atlas Domain, only very slightly affected by Alpine movements, is characterized by inliers of Precambrian formations, cropping out from under a folded or subtabular cover of latest Proterozoic?–Paleozoic strata (Saquaque et al 1989, Naidoo et al 1991, Leblanc and Moussine-Pouchkine 1994).

The important Bou Azzer–El Graara area (Table 5-4, column 5) (Leblanc 1975, 1976, 1981, Leblanc and Moussine-Poushkine 1994) comprises three across-strike terranes, from northeast to southwest: (1) an autochthonous metavolcanic–sediment assemblage intruded by tonalitic plutons; (2) an allochthonous ophiolite-bearing assemblage cut by tonalitic plutons; and (3) a highly allochthonous (thrust plus strike-slip) assemblage of imbricate slices composed of oceanic-type sediments, alkalic mafic volcanic rocks and ophiolites, in a highly sheared serpentinite–metasediment mixture, also cut by gabbro–granodiorite plutons. The three tectonically separated terranes are interpreted as parts of, respectively, a forearc apron–upper slope basin, a highly fractured forearc basement, and a sediment-starved accretionary wedge (Saquaque et al 1989).

The nearby intervening (Reguibat–Anti-Atlas) Tindouf basin, an asymmetric, tectonically distorted, 7–8 km-deep, synclinal structure, some 150 000 km$^2$ in area, contains mainly late Proterozoic volcaniclastic sediments with tillites, molasse and alkalic rhyolites (Supergroups 1 and 2), resting unconformably on Eburnean basement, and unconformably overlain by shallow-marine Cambrian cover (Villeneuve and Cornée 1994).

## 5.10.3 TRANS-SAHARAN MOBILE BELT, TUAREG AND BENIN–NIGERIA SHIELDS

The Pan-African Trans-Saharan (Pharusian–Dahomeyan) Belt apparently formed during collision (~600 Ma) of two continental masses: the West African Craton to the west and the now largely buried East Saharan Craton to the east. The suture zone of this proposed collisional orogen runs southwards from western Hoggar of the Tuareg Shield to the Gulf of Benin, passing, *en route*, through the Gourma Aulacogen and Volta Basin including the Togo Discontinuity, both products of this plate interaction (Affaton et al 1991) (Fig. 2-23, Table 5-4, column 7). Rifting along the eastern margin of the West African Craton and plate extension occurred at ~800 Ma (Black et al 1979b, 1994, Caby et al 1981, 1987, Caby 1987, Bouiller 1991, Black and Liégois 1993, Castaing et al 1993, Dostal et al 1994, Liégois et al 1994).

### Tuareg Shield

The Tuareg Shield is viewed as a collage of displaced terranes (23 now recognized) each of distinctive lithologic, metamorphic, magmatic and tectonic characteristics, and mutually separated by sub-vertical strike-slip meta-shear zones or major thrust faults, some containing ophiolitic assemblages and molassic deposits (Black et al 1994).

In the western *Tuareg Shield*, the late Proterozoic Stromatolite Group, 3 km thick (Tables 2-13, column 1; 5-4, column 6), is correlated in fine detail with the Atar Group of the West African Craton. Marginal to and intruding this platform sequence, occurs a remarkable mafic–ultramafic complex collectively underlying more than 70% of the local Pharuside basement to the overlying Green Group (Caby et al 1981). The central and eastern Tuareg

250  LATE PROTEROZOIC CRUST

Table 5-4. Correlation chart for the late Precambrian geology of the West African Craton and Tuareg Shield Region.

| Time scale (Ma) | (1) Adrar, Mauritania Atar Type Section (Trompette 1983) | (2) Volta Basin (Affaton et al 1980) | (3) Dahomeyides Togo Belt (Affaton et al 1980) | (4) Gourma Basin (Moussine-Pouchkine and Bertrand-Sarfati 1977) | (5) Bou Azzer, Morocco (Anti-Atlas) (Leblanc 1981) | (6) Northwest Hoggar, Tuareg Shield (Black et al 1979a) | (7) Composite West African and Tuareg tectonics |
|---|---|---|---|---|---|---|---|
| | Carboniferous Devonian | | | | | | |
| | ——— disconformity ——— | | | | | | |
| | Oueb Chig Group (Silurian) | | | | | | |
| | Abteilli Group (upper Ordovician) Tillites | Obosum Group molasse (upper Voltaian) Ordovician | Obosum Group Remnants ——— overthrust contact ——— | | | | Annet Purple Group, Tillites (~518±26 Ma) | Glaciation 30 Ma erosional gap |
| 500 | ——— disconformity ——— | | | | | ——— unconformity ——— | ——— unconformity ——— |
| | Oujeft Plateau Group (Ordovician) | | | | Adoudounian Group (534±10 Ma) | | Plate Extension (Hercynian cycle), (563–534 Ma) |
| | Atar Cliff Group (595±43 Ma) | ——— local unconformity ——— | | | Quarzazate Group, (563±20, 578±15 Ma) | | |
| | Bthaat Ergii Group Triad including Jbeliat Tilite (605±13 Ma) | Pendjari (Oti) Group (Middle Voltaian): 2. Pendjari Formation (638±8 Ma) 1. Kodjari Formation with triad including tillite | Tiélé (=Buem) Unit including clastics, cherts, volcanic rocks; important cataclasis | Bandiagara Group (~650 Ma) | Tectonic event (B₂) (600 Ma) Tidline-Ambed Groups (609±15 Ma) Bleida granodiorite (615±12 Ma) | Granitoid plutonism (660–592 Ma) | Plate Collision (620–600 Ma) (Pharusian Orogeny) Glaciation (615–563 Ma) 50 Ma erosional gap |
| 650 | ——— angular unconformity ——— | ——— angular unconformity ——— | ——— overthrust contact ——— | ——— major unconformity ——— | | | ——— unconformity ——— |
| | Assabet el Hassiane Group (694 Ma) | | | Toun-Koutiale Group | B₁ metamorphism (672±12 Ma) | | |
| | ——— disconformity ——— | | | | | | |
| | Tifounke Group | Dapango-Bombouaka Group (Lower Voltaian) | Atacora (=Togo) Unit plus slices of reworked basement | | Tachdamt-Bleida Groups (including El Graara ophiolites) | Green Group ——— unconformity ——— Mafic plutonic rocks (793±30 Ma) | Obduction of oceanic crust (672 Ma) Plate extension (787 Ma) |
| | ——— disconformity ——— | 3. Panabako Formation | | Bobo Group | metamorphism (787±10 Ma) | | |
| | Atar Group Mixite (888±16 Ma) | 2. Poubogou Formation | ——— overthrusts ——— | | | Stromatolite Group | West African-Tuareg Craton |
| | ——— disconformity ——— | 1. Tossiegou Formation | Benin Plain Unit | Sotuba-Sikasso Group (1025 and 960 Ma) | Early Precambrian Basement | | |
| | Char Group | | | | | | ——— unconformity ——— |
| 1050 | ——— angular unconformity ——— (1003±26 Ma) Reguibat Basement | ——— major unconformity ——— Birrimian Basement (Main Shield) | ——— overthrusts ——— Birrimian-Liberian blocks | ——— major unconformity ——— Birrimian Basement (Main Shield) | | In Ouzzal Basement (Archean-Eburnean) | Basement |

Supergroup 3 | Supergroup 2 | Supergroup 1

Shield (Table 2-13, columns 2–4) consists of pre-Pharusian basement reactivated in varying degrees during the closing stages of the Pharusian cycle. A notable feature of all Tuareg domains is the widespread occurrence of syn- to late tectonic and post-tectonic granitoids emplaced during the late stages of the Pharusian cycle.

### Benin Nigeria Shield

The geologic evolution of the *Benin Nigeria Shield* to the south (Fig. 2-23) is not well known. The common rock types include a variety of migmatites and gneiss, granitoid intrusions, a range of metasedimentary schist and gneisses, various amphibolites, pyroxenites and metagabbros, and charnockite (Grant 1973, 1978). The metasedimentary rocks occur in some cases intercalated with gneiss, and in others as substantial, meridionally elongated, epizonal, schist–quartzite units, locally including marble and conglomerate. The contacts between the various rocks types are typically tectonic (Fitches et al 1985).

The *Volta Basin*, 450 000 km² in area and up to 7 km deep, unconformably overlies the Birrimian-bearing basement of the Man (Ivory Coast) Shield (Fig. 2-23, Table 5-4, column 2). Sediments in the south–central Volta Basin make up the Voltaian Supergroup, which has been subdivided into: Lower Voltaian Group (at least 600 m), comprising sandstones, quartzites and shales; the unconformably overlying Middle Voltaian (or Oti) Group, (about 1600 m), composed of tillitic conglomerates, shales, sandstones, limestone and dolomite, products of a stable-shelf marine environment; and the Upper Voltaian (or Obosum) Group, a molasse unit that has no equivalent within the Togo Belt from which it was derived (Grant 1973). A recent alternative classification involves lower Bombuaka and Pendjari groups and upper Tamale Supergroup (Villeneuve and Cornée 1994).

The highly faulted main eastern contact zone (Togo discontinuity) of the Volta Basin has been interpreted as a possible suture zone, product of westward thrust of the Benin–Nigeria basement on to the passive margin of the West African Craton (Black 1980, Castaing et al 1993).

### Gourma Basin (Aulacogen)

In the *Gourma region* of the eastern Taoudeni Basin (Fig. 2-23; Table 5-4, column 4), a major change in sedimentation occurred 800 Ma ago as a result of rifting along the eastern margin of the West African Shield. The Gourma embayment, some 700 km long and up to 400 km wide, is interpreted to be a failed arm, which evolved as an aulacogen (Black et al 1979a, Black 1980, Caby et al 1981, Caby 1987). The shape of the resulting basin is that of a west–southwestward oriented trough which represents a re-entrant within the West African Craton.

In contrast to the thin, flat-lying sediments that cover much of the West African Craton, the Gourma Basin is characterized by deep subsidence with a terrigenous clastic–carbonate accumulation more than 8000 m thick, ending with prograde continental clastic deposits including local eclogites (Reichelt 1972, Moussine-Pouchkine and Bertrand-Sarfati 1977, Villeneuve and Cornée 1994) (Table 5-4, column 6).

## 5.11 AFRICAN CRATON: NORTHEAST AFRICA

Those parts of the Arabian–Nubian Shield straddling the Red Sea contain large proportions of volcanoclastic and extrusive–intrusive magmatic suites, quite unlike the Pan-African mobile belts elsewhere in the continent (Fig. 5-16, inset). The Arabian–Nubian Shield is readily interpretable in terms of actualistic plate tectonics involving plate collisions with accretion of island-arcs, oceanic plateaus and exotic terrains (Shackleton et al 1980, Greenwood et al 1982, Kröner 1985b, Kröner et al 1987a, Pallister et al 1987, 1988).

### 5.11.1 ARABIAN–NUBIAN SHIELD

#### Nubian Shield (Egypt, Sudan, Ethiopia)

This large region some 1800 km long north to south and up to 600 km wide lying to the west of the Red Sea, is divided into the Eastern Desert of Egypt in the north including northern, central and southern sectors, and the Red Sea Hills of Sudan–northern Ethiopia in the south including northern and southern sectors (Fig. 5-16).

The common rock types in this large area are amphibolite facies quartzofeldspathic gneiss, garnet–mica schist, mafic–intermediate–felsic metavolcanic rocks and foliated granitoid bodies and occasional ultramafic masses, all of which are intruded by granodioritic–dioritic masses and cut across by younger granite, gabbro and syenite plutons and dykes (Vail 1976). The rocks are interpreted as representing a mixture of Pan-African

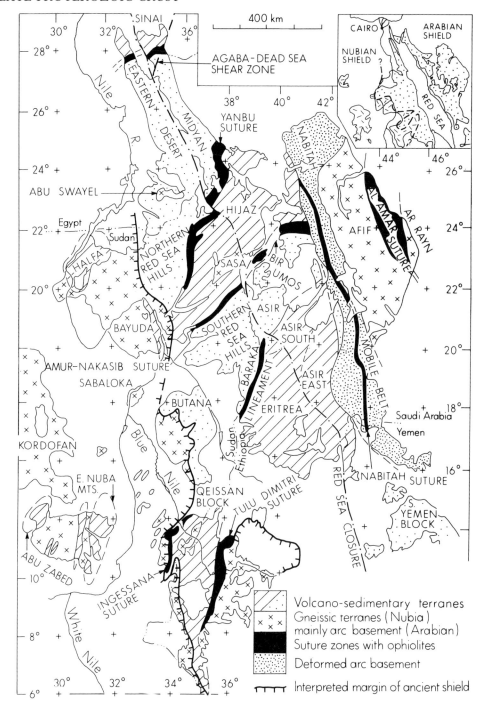

**Fig. 5-16.** Palinspastic sketch map of the Arabian–Nubian Shield showing (1) the approximate boundaries between Pan–African volcanosedimentary terrains, gneissic terrains including a heterogeneity of rock types, not all of continental derivation, and metamorphic belts; and (2) suture zones with ophiolites. (After Vail 1985, Fig. 1, and reproduced with permission of the author). Inset map shows relative positions of the Arabian and Nubian Shields, and the interpreted margin of the ancient Sudan shield, to the west: Stippled – Pan-African; hatched – pre-Pan-African gneisses; dashed line – continental margin of ancient Sudan Shield. (From Kröner et al 1987, Fig. 1A, and reproduced with permission of the authors).

components and pre-Pan-African gneisses, the latter including segments of the ancient 'Sudan Shield' to the west (see Fig. 5-16 for two interpretations).

Between the Nile valley and the Red Sea, a number of stratigraphic units with characteristic lithologies and metamorphic grade were early established as forming the basement complex in Egypt, Sudan and Ethiopia (Vail 1976). The uppermost and generally least metamorphosed of these units is a sequence of basaltic, andesitic and felsic pyroclastic rocks and lava flows, minor limestones, shale, greywacke and conglomerate, with occasional serpentine masses, all usually in the greenschist facies of metamorphism. They crop out intermittently from the Gulf of Suez as far south as Lake Turkana in Kenya (Fig. 5-16).

Following recent intensive studies, Kröner (1985b) and Kröner et al (1993) summarize four principal rock associations that make up the bulk of the Nubian Shield in southeast Egypt and in the Red Sea Hills: (1) an older shelf sequence of dominant metaquartzite; (2) younger arc assemblages rich in tholeiitic to calc-alkalic volcanic and volcaniclastic rocks; (3) tectonically dispersed and dismembered ophiolite complexes, many now serpentinized and carbonatized; and (4) large composite granitic batholiths and gabbroic associates.

Domains previously interpreted as pre Pan-African basement, in view of their high metamorphic grade and poly-deformation, are now known to be no older than ~800 Ma, and almost all the magmatically derived suites have isotopic systematics implying a juvenile origin (Engel et al 1980, Stern 1981, Stern and Hedge 1985). One such island arc terrane (northern Haya) of the Red Sea Hills provides an age of formation of 887–854 Ma (Reischmann et al 1992).

However, at least some of the component metasediments were derived from older sialic basements as shown by detrital zircon ages between 1120 Ma and possibly 2060 Ma (Kröner 1985b) which suggest an old cratonic provenance to the west.

In brief summary, passive margin development occurred contemporaneously with terrane accretion in Arabia to the east (see below). Marginal basin closure between a growing Arabian plate and eastern Nubia, now converted into an active continental margin, began some 750 Ma ago. Part of the Red Sea Hills and the Al-Lith area of Arabia probably formed part of one large arc complex that was accreted to Africa about 720 Ma ago (Kröner et al 1991).

Subduction-related arc volcanism came to an end when the southern Arabian plate finally 'docked' with Nubia, first in the south about 640 Ma ago, then gradually farther north, when activity ceased some 40 Ma later in the northern Eastern Desert of Egypt (Kröner 1985b).

Of interest is a discontinuous belt of ophiolitic nappe remnants in NE Sudan which extends for at least 200 km WSW from the western margin of the Nubian into the gneissic terrains of the Nile Craton. This may represent a previously unrecognized E–W Pan-African suture zone, hence an added complexity to the tectonics of NE Africa (Sultan et al 1987, Schandelmeir et al 1994).

### Arabian Shield

The Arabian Shield, lying east of the Red Sea, consists of Precambrian rocks mainly island arc-type volcanic rocks and sediments, including ophiolites, and widespread younger granitoids, which are unconformably overlain to the north and east by Phanerozoic strata (Fig. 5-16).

The shield is divided into three major tectonic microplates by the NNW-trending Najd fault system: from southwest to northeast, Hijaz–Asir, Nabitah and Afaf–Ar Rayn (Fig. 5-16) (Greenwood et al 1980).

In brief, the Hijaz tectonic cycle responsible for the major folding, metamorphism and plutonism, has been related to a period of island-arc activities reflecting the convergence of oceanic plates and the subduction of oceanic lithosphere (Fleck et al 1980a,b, Stacey and Hedge, 1984, Pallister et al 1988, Kröner et al 1992). Subsequently, crust which had developed during the Hijaz cycle collided with and was sutured to the African plate. The Najd fault system is one aspect of this collision (Greenwood et al 1980, Jackson 1980).

Based on comprehensive studies of the geologic setting, U–Pb geochronology and Pb–isotope characteristics, Pallister et al (1988) draw four conclusions:

(1) The Arabian Shield is composed of a series of late Proterozoic intra-oceanic island arcs that were accreted along with sedimentary aprons and occasional slivers of oceanic lithosphere ophiolites.
(2) Shield formation involved a number of individual episodes of arc magmatism and microplate accretion. This process began 950–900 Ma ago and culminated with final post-orogenic intrusion, uplift and cooling 570–500 Ma ago.

(3) The process of arc-accretion resulted in growth of at least $1 \times 10^6$ km² (Arabian–Nubian Shield) of new continental crust with an average thickness of 40 km adjacent to the African continent.

(4) Early Proterozoic continental fragments which were entrained between the accreted arc complexes of the eastern Arabian Shield may represent rifted fragments of the African continent to the west.

The common post-orogenic plutons, falling in the age range 610–510 Ma, represent products of melting induced by crustal thickening following accretion.

Pallister et al (1988) conclude that an arc and exotic-terrane accretion model best accounts for the growth of the Arabian Shield. This included accretion of a complex region of island arc, back-arc basins, rifted island arcs, and small fragments of older continental crust.

The general evolution of the Arabian–Nubian Shield is best viewed as a threefold tectonic–magmatic series of events between 900 and 550 Ma: (1) rifting of a supercontinent (Rodinia?) ~900 Ma; followed by (2) sea-floor spreading, arc and back-arc basin formation, and accretion of juvenile shield crust; ending with (3) E–W shortening (640–550 Ma) as East and West Gondwana collided (Stoeser and Camp 1985, Schandelmeir 1990, 1994, Stern 1993, Stern and Kröner 1993).

## 5.12 INDIAN CRATON

Late Proterozoic sediments are distributed mainly in the northern part of the subcontinent, filling the very large Vindhyan Basin in the north–central part and, to the west, the smaller Trans-Aravalli Vindhyan Basin in Rajasthan. In addition, small pockets (Kurnool) are present in the northern and central parts of Cuddapah Basin and, possibly, Bhima Basin in the south (Fig. 5-17, Table 2-14, columns 1 and 5). For convenience, the Vindhyan Basin is considered here as the late Proterozoic representative despite the presence of mid-Proterozoic strata in the lower stratigraphic parts. Elsewhere in the subcontinent, a scattering of 70–400 Ma metamorphic ages along the east coast may reflect the presence of a NE-trending 'Pan-African' mobile belt, product of the Indian Ocean Orogeny. In the Lesser Himalayas (Krol nappe and eastern equivalents), late Sinian diamictite-bearing successions with Ediacaran fossils are reported locally.

### 5.12.1 VINDHYAN BASIN

The Vindhyan Supergroup occupies a large, southwest-convex, sickle-shaped basin just south of the Great Gangetic Plain, with an outcrop area of 105 000 km² (Fig. 5-17). The maximum stratigraphic thickness of the dominantly sandstone–shale–limestone sequence is 4250 m. The sediments are generally horizontal or nearly so and only slightly metamorphosed, constituting the best preserved Proterozoic sedimentary sequences in India (Pichamuthu 1967, Mathur 1982, Raha and Sastry 1982, McMenamin et al 1983, Prasad 1984, Soni et al 1987).

Lower Vindhyan sediments are dominantly calcareous and probably marine, comprising thick limestones, porcellanite and sandstone with rare mafic dykes and rhyolite–andesite volcanic rocks. The unconformably overlying Upper Vindhyans are mainly arenaceous and largely fluviatile or deltaic in origin, consisting chiefly of fine-grained, hard red sandstones, which, since the days of King Asoka (273–232 BC), have been used in Indian architecture. Widespread ripple marks and persistent red colour suggest shallow-water deposition in mainly land-locked or marginal basins. For the most part, Vindhyan sediments are attributed to deposition in shallow-marine tidal flats upon a tectonically stable platform (Gupta 1977).

On a biostratigraphic basis, Raha and Sastry (1982) consider the lower Vindhyan sediments to be early- to mid-Riphean (1650–1000 Ma) and the upper Vindhyans to be mid- to late-Riphean (younger than 940–910 Ma). Thus the Vindhyan Supergroup appears to straddle the mid–late Proterozoic boundary and may, in fact, extend to the Paleozoic boundary.

### 5.12.2 SRI LANKA

The island of Sri Lanka, a pear-shaped unit some 400 km long by up to 220 km wide, which is 'suspended' from the southeastern tip of India, is underlain, except for some coastal Miocene deposits, by Precambrian crystalline rocks of varying tectono-metamorphic overprints. Of these, three major divisions are recognized on the basis of lithology, structure and age: (1) Highland Complex (formerly Highland Group and Southwestern Group); (2) Wanni Complex (formerly western part of Vijyan Complex); and (3) Vijyan Complex (now restricted to eastern part only of former Vijyan Complex) (Fig. 5-17, inset) (Dahanayake and Jayasema 1983, Perera 1983, Liew et al 1987, Cooray

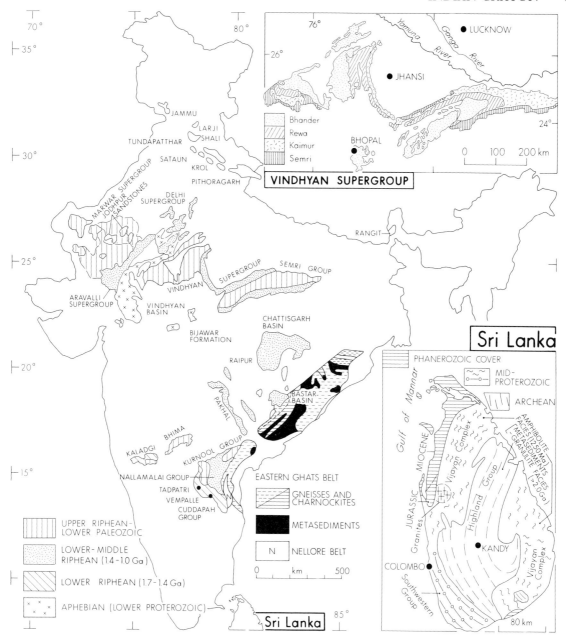

**Fig. 5-17.** Sketch map showing the mid to late Proterozoic basins of India (from Raha and Sastry 1982, Fig. 1). Upper inset map shows the gross stratigraphy of the Vindhyan Basin (from McMenamin et al 1983, Fig. 1). Lower inset map shows the main geologic features of the Sri Lanka (from Dahanayake and Jayasena 1983, Fig. 1). All illustrations reproduced with permission of the authors.

1994, Kröner et al 1994a). (1) The Highland Complex, which is concentrated in the central highlands of Sri Lanka, comprises mainly interbanded ortho- and paragneisses including pelitic gneiss, metaquartzite, marble and charnockitic gneiss. Sequences are commonly about 7 km thick, but there is no preserved primary stratigraphy. Outliers of these rocks to the east represent tectonic klippen.

(2) The Vijyan Complex comprising gneiss, migmatites, granitoids and scattered metasediments, extends inland from the eastern coast towards the foothills of the central highlands.

(3) The Wanni Complex consists of migmatite, gneiss, charnockitic gneiss, minor metasediments and granitoids lying west of the Highland Complex.

According to an earlier view, the 'Highland Series' is younger than the 'Vijyan Series'. However, the Highland Complex is now known to be mainly older than the Vijyan Complex, the boundary between them being a thrust contact of still unresolved position and nature.

Recent geochronologic data indicate: (1) Highland Complex sedimentation and basic igneous activity (lava flows, sills, dykes) occurred at ~2.0 Ga, with some model ages to 3.4 Ga; (2) earliest granitoid intrusions in the Highland Complex took place at 1.94–1.80 Ga; (3) orthogneisses of Wanni Complex were intruded at 1.1–0.67 Ga, with model ages to 2.0 Ga; (4) orthogneisses of Vijyan Complex were intruded at 1.04–1.03 Ga, with model ages to 1.8 Ga; (5) pervasive high-grade metamorphism of Highland, Vijyan and Wanni complexes occurred at 670–455 Ma, with peak metamorphism at ~610 Ma.

Accordingly, Sri Lankan lithologic components were variably developed in early–mid-Proterozoic times and were pervasively overprinted in Eocambrian time (Liew et al 1987, Cooray 1994, Hiroi et al 1994, Holz et al 1994, Kröner et al 1994b, Malisenda et al 1994, Prame and Pohl 1994). The distinct crustal age provinces are separated one-from-the-other by tectonic discontinuities; they represent distinct terranes accreted just prior to the main Pan-African granulite metamorphism (~610 Ma).

## 5.13 AUSTRALIAN CRATON

Late Proterozoic to Paleozoic (1100-450 Ma) platform cover is exposed discontinuously along a broadly NW-trending zone that crosses central Australia with a strike length (Tasmania–Kimberley) of 3300 km and a maximum width (Officer Basin–Georgina Basin) of 1800 km (Fig. 1-5h). The principal depocentres are (1) the Adelaide Geosyncline of South Australia; (2) the Officer, Amadeus, Ngalia and Georgina basins of Central Australia, which contain the thickest sequences and maximum widths; (3) southeastern Kimberley Region, marking the northwestern limit; and (4) additional small zones in Tasmania at the southeastern limit. (Key references: Plumb et al 1981, Preiss and Forbes 1981, Preiss et al 1981, Plumb 1985, Preiss 1987, Walters et al 1995).

### 5.13.1 ADELAIDE GEOSYNCLINE

The Adelaide Geosyncline lies between (1) the Gawler Craton (stabilized by about 1500 Ma) and its cratonic cover, the Stuart shelf, on the west, from which it is separated by the Torrens Hinge Zone; and (2) the Curnamona Craton (Nucleus) and associated basement inliers, including the Willyama (Block) Domain on the east (Fig. 3-23). The 'geosyncline' including the Flinders and Adelaide basins, is viewed as a complex of rift-controlled intracratonic basins, grading to a miogeoclinal-type shelf in the southeast. Although thicknesses of Adelaidean sediments are highly variable, aggregate maximum thicknesses generally increase from west to east across the geosyncline within the 5–15 km range.

The three main late Proterozoic depositional features in the Adelaide sector of South Australia comprise (1) the near-horizontal Stuart Shelf sequence to the west; (2) a major NNW-trending transitional Torrens Hinge Zone of moderately folded rocks; and (3) the thick, late Proterozoic to Cambrian Adelaide Geosyncline sequence to the east, which has been relatively strongly deformed by the early Paleozoic (Cambro–Ordovician) Delamerian Orogeny (~500 Ma), to produce a complex, sinuous and branching system of folds (Preiss et al 1981, Preiss 1987, fig. 5.10, Preiss 1990).

The Adelaide Geosyncline, the stratotype for the Adelaidean period, contains the most complete record of Adelaidean sedimentation in Australia. The late Proterozoic to Cambrian deposits fall into three major sedimentary sequences (supergroups), each of distinctive lithologic–tectonic–paleogeographic pattern, and mutually separated by unconformities or disconformities as illustrated (Table 5-5). Of these, the two lower Warrina and Hysen supergroups encompass all of Adelaidean time including Willouran, Torrensian, Sturtian and Marinoan (Ediacaran) chronostratigraphic divisions. The uppermost Moralana Supergroup is entirely Cambrian. The best available Adelaidean age constraint is a U-Pb zircon date of 802 Ma on the Rook Tuff, a unit that lies stratigraphically well below the Umberatana Group (Christie-Blick et al 1995).

Certain distinctive lithologies characterize particular stratigraphic levels in the Adelaidean succession, although not necessarily uniquely, including (1) Callanna Group mafic volcanic rocks and replaced evaporites; (2) Burra sedimentary magnesite; and (3) Umberatana tillites (Lower and Upper glacials). Tillites and associated rock types are the best markers for Australia-wide stratigraphic correlations (Schmidt et al 1991). Also, the uppermost Pound Subgroup of the Wilpena Group contains the well known Ediacara assemblage of soft-bodied metazoan fossils (Glaessner and Walter 1981, Jenkins 1984, Preiss 1987), precursors to the seemingly

Table 5-5. Summary of chronostratigraphic and lithostratigraphic units in the Adelaide Geosyncline.

| Chronostratigraphic units | | | Major lithostratigraphic units | |
|---|---|---|---|---|
| PALEOZOIC CAMBRIAN | | MORALANA SUPERGROUP | Lake Frome Group | Clastic-filled troughs |
| | | | Kanmantoo Group | Red-beds |
| | | | | Sands-silts-carbonates transgression |
| | | | | Volcanics |
| | | | Normanville Group | Uratanna clastics fill erosional channels in Pound Quartzite |
| | | | | *disconformity* |
| LATE PROTEROZOIC ADELAIDEAN | Ediacaran | HEYSEN SUPERGROUP | Wilpena Group (100–6000 Ma) | Pound Quartzite (subgroup) (metazoan fossils) |
| | | | | Dolomite ('cap'), shale (Brachina), sandstone (ABC) |
| | Marinoan | | Umberatana Group (4000–9000 m) | Marinoan tillites (Upper glacials) siltstone, dolomite, quartzites |
| | | | | Oolitic limestone-dolomite ⎫ |
| | Sturtian | | | siltstone, feldspathic sandstone ⎬ Interglacials |
| | | | | Carbonaceous siltstone (Tapley Hill) ⎭ |
| | | | | Tillites (Sturt, Appila) (Lower glacials—upper phase) (up to 5400 m) |
| | | | | Ironstones (Braemar), dolomites |
| | | | | Tillite (Pualco) (Lower glacials—lower phase) |
| | | | | *disconformity* |
| | Torrensian | WARRINA SUPERGROUP | Burra Group (2230–10 400 m) | Feldspathic sandstone, laminated siltstone (Belair) |
| | | | | Siltstone, sandstone, dolomite |
| | | | | Dolomite, local chert, magnesite (Skillogalee) |
| | | | | Sandstone, feldspathic quartzite |
| | | | | Siltstone, phyllite, dolomite |
| | | | | Dolomite, dolomitic phyllites, sandstone (Blyth) |
| | | | | Sandstone-conglomerate (Rhynie-Aldgate) |
| | | | | *conformable transition?* |
| | Willouran | | Callana Group (430–8400 m) | Cyclic repetitions of siltstones-sandstones-stromatolitic carbonates with evaporites (300–8400 m) |
| | | | | Mafic volcanics (Wooltana, Beda) ⎫ |
| | | | | Carbonates ⎬ (130–2700 m) |
| | | | | Basal clastics ⎭ |
| | | | | *major unconformity* |
| Pre-Adelaidean | | | | Archean to early Proterozoic metamorphic complexes; mid-Proterozoic granitoids, volcanic and sedimentary rocks |

After Preiss et al (1981)

abrupt appearance and radiation of Cambrian skeletogenous organisms (Mount and McDonald 1992). *Stromatolites* of the Adelaide Geosyncline have been studied and described and their biostratigraphic application and problems summarized (Preiss 1977, Preiss and Forbes 1981, Preiss 1987). Most are newly described forms, not previously known in other regions. They apparently represent forms endemic to South Australia, although a close comparison can be made of those from the Umberatana Group with certain of those from the late Riphean and Vendian of the Patom region of Russia.

More than two dozen *microfossil* types have been recognized in Adelaidean black cherts, falling into five broad categories—cellular filaments, non-septate filaments, solitary unicells, colonial unicells and unicells containing organelle-like bodies (Preiss 1987). Black cherts in South Skillogalee dolomite are locally richly fossiliferous, containing solitary eucaryotic cells and multicellular colonies that may represent green algae. The most abundant microfossils closely resemble modern mat-building Cyanobacteria. In addition, there are types similar to modern mound-forming colonial cells, and a possible fungal filament.

The *Ediacara assemblage* of soft-bodied metazoan fossils lies in the lower Rawnsley quartzite of the Ediacara Range where first discovered in 1947. Ediacara-bearing strata lie between the youngest late Proterozoic tillites and the base of the Cambrian, a dated interval from 590 to 540 Ma. Ediacara fauna have now been described from more than 25 localities worldwide, representing every continent except Antarctica (Hofmann et al 1990). In all cases the fossils indicate a shallow marine environment of normal salinity with temperate to tropical oxygenated waters and unrestricted water circulation (Cruse et al 1993). It is noted that metazoan fossils of Ediacaran affinity have also been discovered from the Stirling Range Formation near the SE margin of the Yilgarn Block, West Australia (Cruse and Harris 1994).

### 5.13.2 OFFICER, AMADEUS, NGALIA AND GEORGINA BASINS

The Centralian Superbasin, as defined (Walters et al 1995), encompasses the late Proterozoic to Cambrian fill of the Amadeus, Ngalia, Officer and Savoury (extreme NW) basins. Regional crustal sagging that initiated the superbasin began ~800 Ma, following Musgrave-Arunta stabilization (1200–1100 Ma). The resulting superbasin, with up to 10 km in stratigraphic thickness, was disrupted internally at 600–540 Ma (Petermann Ranges Orogeny) and again at 320 Ma (Alice Springs Orogeny) to form the presently existing structural basins listed above. New results from acritarch biostratigraphy and isotope chemostratigraphy in conjunction with conventional lithostratigraphy and sequence analysis, allow greatly improved stratigraphic resolution across the superbasin. Parts of the sequence along the NE margin of the Amadeus Basin were deposited under the influence of salt movement within the underlying Bitter Springs Formation (Kennedy 1993).

The Adelaidean succession of central Australia, as a whole, accumulated as a southward thickening wedge, thickest at or just south of the present southern edge of the Amadeus Basin. A thinner shelf zone, with local depressions and ridges, extended across the Arunta Inlier from the northern Amadeus Basin to the Ngalia Basin. Much of the Adelaidean in the southern Georgina Basin accumulated in a series of narrow, deep, NW-trending grabens or aulacogens, indented into the southern edge of the North Australian Craton (Plumb et al 1981).

### 5.13.3 KIMBERLEY REGION

Mildly disturbed, late Adelaidean glacial successions are well exposed in three separate areas of the Kimberley and Victoria River regions (Fig. 1-5h). The successions, which unconformably overlie a variety of rocks and are unconformably overlain by early Cambrian Antrim Plateau Volcanics, are, respectively, about 500, 5000 and 2500 m in thickness.

### 5.13.4 TASMANIA

Exposures of Precambrian rocks in Tasmania are confined to the western half of the island, in two main segments—Rocky Cape and Tyenna blocks (Fig. 1-5h). The structural complexity, separation of segments and scarcity of radiometric data make regional stratigraphic interpretations difficult and uncertain (Williams 1978b).

The late Proterozoic is tentatively subdivided into three main sequences, representing distinct depositional basins: (1) the oldest group of siltstone, orthoquartzite, dolomite and subgreywacke; (2) a middle group Burnie of mudstone, quartz–wacke and minor pillow lavas; and (3) an uppermost group of stromatolitic carbonate and chert with thick diamictites (Preiss and Forbes 1981).

## 5.14 ANTARCTIC CRATON

Late Proterozoic supracrustal rocks are reported from Denman Glacier, western Dronning Maud Land, and possibly the Shackleton Range and Prince Charles Mountains (James and Tingey 1983, Goodge et al 1993).

In the Denman Glacier area (~100°E), Cambrian sediments unconformably overlie greenschist facies metabasalts, for which Ravich and Grikurov (1976) cite an age of 610 Ma.

Late Proterozoic low grade metasediments are abundant in Oates Land, in Marie Byrd Land, and widely distributed in the Transantarctic Mountains, forming the younger frame of the East Antarctic Metamorphic Shield (Stump et al 1988).

Yoshida and Kizaki (1983) interpret the tectonic–metamorphic history of the Lutzow–Holm Bay Region as including an event dated at ~560 Ma, involved high-amphibolite facies metamorphism with E-trending upright open folds, followed by a fourth event at ~460 Ma (early Paleozoic). Similar events are reported from almost all the coastal areas of the Indian Ocean sector of east Antarctica, from about 0° to 100°E (Yoshida and Kizaki 1983). In the Transantarctic Mountains there is evidence for the Beardmore Orogeny at about 650 Ma (Laird 1981). Also, high-grade metamorphic tectonites in the central Transantarctic Mountains form a major ductile shear zone that provides dates between ~540 and 520 Ma; this zone records oblique plate convergence along the late Proterozoic–Cambrian Antarctic margin of Gondwanaland (Goodge et al 1993).

In summary, in the Gondwanaland reconstruction of Craddock (1982), the ~500 Ma (i.e. Pan-African) event strongly affected that part of East-Antarctica adjoining Africa (30°W–30°E) but only sparingly in the remaining parts to the east adjoining India and especially Australia, where the ~1000 Ma event is well documented.

# Chapter 6

# Evolution of the Continental Crust

## 6.1 INTRODUCTION

Three interrelated factors have controlled the dynamic evolution of the continental crust: (1) bulk Earth heat production and loss; (2) sialic fractionation and cratonization; and (3) the rise of atmospheric oxygen level. These three factors, together with pervasive gravitational control, operated cumulatively and all first-order crustal features have developed in response to them. These features are conveniently grouped into (a) endogenous; and (b) exogenous processes and products, in order to distinguish phenomena operating respectively mainly within Earth from those operating mainly upon its surface.

Earth's continental crust preserves a record of mantle-derived silicate melts produced during at least 3.96 Ga and possibly 4.2 or more. This continuing, largely irreversible differentiation is closely linked to the evolution of the atmosphere, which, like the crust, is of secondary origin.

## 6.2 ENDOGENOUS PROCESSES AND PRODUCTS

### 6.2.1 ARCHEAN HEAT FLOW AND GEOTHERMAL GRADIENTS

There is general agreement that Earth's heat production in the Archean was two to three times greater than at present and has declined steadily from an estimated initial level of 53 pW/kg (picowatts per kilogram) some 4.6 Ga ago to the present level of 11 pW/kg, as judged from potassium, thorium and uranium (O'Nions et al 1978) (see further below, including Fig. 6-7, upper graph).

Indeed, higher Archean heat flow is supported by the restriction of high-Mg komatiitic lava flows to Archean sequences (Condie 1992). Specifically, komatiitic lavas with liquid compositions more magnesian than 22% MgO are restricted to the Archean, the most magnesian (33% MgO) with an estimated extrusion temperature of about 1650°C. In contrast, post-Archean time is marked by a sharp decrease to 19–22% maximum MgO content in volcanic rocks, implying a maximum extrusion temperature of about 1350°C.

At present, Earth loses about half its heat during plate creation and destruction (Bickle 1984). It follows that during the still hotter Archean, there was some equivalent mechanism of heat loss, and that it operated at a significantly faster rate. The operation of any such mechanism implies the recycling of mantle-derived volcanic material and hence the formation at surface followed by downsinking (subduction or sagduction) of some sort of oceanic crust on its return to the mantle. This is accepted as a direct consequence of higher Archean heat flow. Does this mean that Archean 'plate tectonics' was necessarily the exact analogue or even resembled that of the present? Not at all. In addition to enhanced mantle convection, Archean 'plate tectonics' would involve lithospheric 'plates' with distinct thickness, composition and rheologic properties. Thus the nature of the Archean process involved distinctive ocean crust and, presumably, sialic nuclei (proto-continents). As yet we know little about this process, a fruitful target of future research.

Despite higher Archean heat production, it is known that Archean provinces at present have considerably lower heat flow than their post-Archean counterparts (Taylor and McLennan 1985). It is further known that, of this total heat flow, the crustal component accounts for about 40% and the mantle component for about 60%. This is best explained by lower crustal abundances of the common heat-producing elements, K, Th and U, in the existing Archean provinces. As discussed below, the reason for the lower heat levels in Archean provinces is that Archean crust is dominated by Na-rich TTG rocks with comparatively low K–Th–U levels. This most significant aspect of Archean crust has major relevance to the dynamic evolution of the continental crust.

## 6.2.2 THE NATURE OF ARCHEAN OCEANIC CRUST

Modelling of Archean mantle convection (Campbell and Jarvis 1984) shows that Archean plumes, rising at spreading centres, would attain maximum temperatures about 150°C above that of modern plumes. This would have profound influence on the nature of the resulting oceanic crust. As a consequence of higher temperature, both the amount of melt and the percentage of partial melting would have been greater at Archean spreading centres than at modern equivalents. Archean oceanic crust must therefore have had a higher average MgO content than modern oceanic crust, probably comparable to that of high MgO komatiites. Indeed, Archean oceanic crust may have consisted of predominant komatiites with subordinate accompanying basalts formed by lower degrees of partial melting (Arndt 1983).

The higher density of komatiite compared to basalt may have resulted in the sinking and remelting of all Archean ocean crust, for none has been recognized so far in the geologic record (Bickle et al 1994).

It has been emphasized (Chapter 2) that tholeiitic basalts dominate in the lower levels of many Archean greenstone sequences, including those of early- to mid-Archean age. These volcanic rocks are commonly pillowed, indicating subaqueous accumulation. Were these earlier Archean sequences produced at Archean spreading centres analogous to those of present day? The answer is probably no, for two reasons: (1) the tholeiitic basalts do not have the anticipated high MgO content; and (2) they lack many characteristics of modern ophiolites. Instead, these tholeiitic basalts may represent partial melts of convection cells associated with back-arc-type spreading. If the plumes associated with these cells rose from within the mantle rather than from the postulated core–mantle boundary, they would be appreciably cooler and therefore melt at shallower depths, giving rise to tholeiitic melts (Campbell and Jarvis 1984).

However, as Earth cooled and convection slowed, basalt production became increasingly important at oceanic ridges relative to komatiite, this reflecting a decrease in the degree of melting of the ultramafic sources at ridges (Condie 1980a). As Earth's temperature continued to fall, and basalt became proportionately more abundant in oceanic crust, widespread partial melting of wet basalt (garnet amphibolite) commenced in down-sinking environments to produce tonalitic magmas on a major scale which thereupon rose to intrude the overlying volcanic rocks. Thus the production of tonalites—major builders of the continental crust in quantity—is integral to the cooling Earth scenario (see further below).

Furthermore, it has been estimated (McCulloch 1993) that, as Earth cooled, average oceanic lithosphere in the normal cooling process, would have become negatively buoyant and suitable for subduction when the mantle was only 50°C hotter than at present, which condition was reached as recently as 1.4–0.9 Ga ago. Prior to this time, a different tectonic model would have operated such as subcrustal delamination (Davies 1992). Thereafter plate tectonics in the conventional sense may have generated, in sharp contrast to the earlier process. Furthermore, it has been estimated that the mid-Proterozoic ocean crust was two to three times as thick as at present, the sharp decline occurring at about 1.0 Ga (Moores 1993).

## 6.2.3 GRANITOID ASSOCIATIONS

Granitoid rocks are the predominant component of the continental crust, with the mean composition of the upper continental crust approximately that of granodiorite. It follows that evolution of the continental crust is best represented by such rocks. What evidence exists of an evolutionary trend in Earth's older granitoid rocks? As we have observed, Precambrian terrains of the world contain two main granitoid associations, an older Na-rich TTG (or Na-tonalite) suite and a younger K-rich granodiorite–monzonite–adamellite–granite (or K-granite) suite. In fact, the Na-tonalite suite predominated throughout most of the Archean. However, in latest Archean time and subsequently, the K-granite suite becomes important. What bearing does this have on crustal evolution?

### The production of granitoid rock

It is now broadly accepted that the bulk of granitoid rocks are derived by partial melting processes involving anatexis, rather than by *in situ* processes (granitization) or by fractional crystallization of basalt or other mafic material (Wyllie 1977, 1983).

To reiterate, Archean granitoid rocks define two geochemical trends: (a) the TTG (or Na-tonalite) trend, characterized by Na-enrichment; and (b) the quartz monzonite–adamellite–granite (or K-granite) trend, which is calc–alkalic and characterized by K-enrichment. Most Archean gneissic complexes lie on the Na-tonalite trend, whereas many later Archean massive batholiths and plutons lie on the calc–alkalic or K-rich trend, being also enriched in Ba and Rb relative to Sr (Condie 1981b).

On a global scale, why should there be a paucity of K-granites prior to about 2.6 Ga, with some notable local exceptions? The most direct explanation appears to lie in the corresponding paucity of suitably thickened continental crust until after the major global 2.7–2.6 Ga-old tonalite event marking the Archean–Proterozoic boundary. Accordingly, it appears that the continents thickened significantly at 2.7–2.6 Ga, chiefly by tonalite-trondhjemite magmatism, and that the root zones thereafter began to melt to produce K-granites in quantity.

Condie (1986b) has summarized pertinent data on *Na-tonalite* suite production. Tonalites can be produced by fractional crystallization or batch melting of a mafic or ultramafic source under wet conditions (Barker and Arth 1976, Abbot and Haffman 1984, Campbell and Jarvis 1984, Rapp et al 1991, Luais and Hawkesworth 1994). In brief, the pertinent data demonstrate that the composition of the parent material is all important to tonalite production, with komatiite being out of bounds but tholeiite permissible. Indeed, Archean tonalites are now considered to have formed mainly by batch melting of a mafic source under relatively wet conditions (Fig. 6-1). Stated otherwise, the most successful model for Archean tonalite production involves batch melting of a wet mafic source (Martin 1986).

How might normal ocean ridge basalt become enriched in incompatible elements and thereby provide a suitable tonalite parent? The most likely mechanism is deep-sea alteration of ocean ridge basalts. This includes both deep-sea weathering and hydrothermal alteration, of which the latter may occur to depths of 8 km. Alternative mechanisms include derivation from an enriched mantle or enrichment by metasomatism in the lithosphere above descending slabs (Condie 1986b).

The *K-granite suite* (grandodiorite–monzonite–

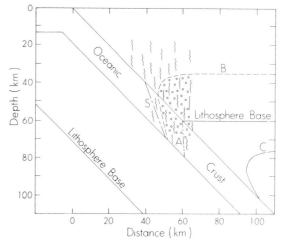

**Fig. 6-1.** Idealized cross-section of descending Archean oceanic crust, showing the major site of tonalite production: S - serpentine dehydration reaction; A - amphibole-eclogite transition; B - hydrous basalt solidus; C - hydrous-herzolite solidus. Modified from Wyllie (1979) for an Archean upper mantle 150°C hotter than at present. Curly lines denote water from dehydration of descending slab; stippled pattern shows major sites of tonalite production. (From Condie 1986b, Fig. 2, and reproduced with permission of the author.)

adamellite–granite) occurs typically as batholiths intruding both granitoid-greenstone and gneiss–migmatite terrains. Although locally the dominant lithology (Annhaeusser and Robb 1981, Condie 1981b, Luais and Haekesworth 1994, Taylor and McLennan 1985). K-granitoid rocks form about 30% on average of large, varied Archean provinces. Rocks of this type first appeared locally in significant quantity about 3.2 Ga ago (e.g. Kaapvaal Craton) but their greatest period of development was in the late Archean, specifically 2.7–2.5 Ga. In general, such rocks follow after development of the Na-tonalite suite and associated greenstone belts, the K-rich suite being intruded into the upper crust either very late in, or following the development of, the greenstone belts. Such a time relationship suggests that the prior existence of such greenstone-bearing, tonalite-rich sialic crust is a necessary condition for K-granite production.

This K-granite magmatism in late Archean time is probably the single most important period of granite generation on record. It played a major role in the evolution of the continental crust, both in terms of effecting widespread cratonization of the crust and as potent repositories in the upper continental crust of the main heat-producing elements, K, Th and U.

K-granitic rocks are also common in Proterozoic terrains. For example, the widespread occurrence of rapakivi granites and associated rhyolites, with or without anorthosites, in mid-Proterozoic terrains, appears to be distinctive and unique. The geochemical and isotopic compositions of such rocks, which range from granodiorite to granite with minor trondhjemite, are consistent with a crustal origin from a mafic–intermediate source with a short crustal history (Taylor and McLennan 1985), and especially to partial melting of hydrous, calc-alkalic to high-K calc-alkalic, mafic to intermediate metamorphic rocks in the crust (Roberts and Clemens 1993).

The net effect of progressive granitoid development in the evolving continental crust was to enhance the chemical contrast of typical Archean and later- to post-Archean granitoid rocks, and thereby provide a major temporal demarcation in continental evolution at the Archean–Proterozoic boundary.

## 6.2.4 GROWTH RATE OF CONTINENTAL CRUST

Numerous growth rates for the continental crust have been proposed (Fig. 6-2) (Hurley 1968, Hurley and Rand 1969, Fyfe 1978, Veizer and Jansen 1979, Armstrong 1981, 1991, Reymer and Schubert 1984, Taylor and McLennan 1985, Condie 1986b). They range from the extremes of early (pre 4.0 Ga), rapid continental growth, followed by continuous recycling of the resulting continental crust through the mantle, to a continuous or quasi-continuous growth throughout geologic time. However, current geologic and isotopic data generally, though not exclusively (Armstrong 1991), support an accelerated growth model lying between these extremes with >50% of the continental crust formed by 2.7 Ga.

As discussed below, the preferred growth model followed in this book allows for rapid late Archean growth to about 60% level by 2.5 Ga, followed by (1) moderate yet significant early Proterozoic growth to a total of 80% by 1.7 Ga; and (2) modest continuous growth of the remaining 20% to the present.

### Comparative island-arc accretion model

Based on recent Nd studies and volumetric calculations, Reymer and Schubert (1986, 1987) and Schubert (1991) provide a preliminary comparison of Precambrian continental growth rates to those based on the popular island-arc accretion model. According to the calculations, the growth rates for four investigated Precambrian (Archean to late Proterozoic) shields and provinces are a factor of 6–10 higher than the average Mesozoic–Cenozoic arc-addition rate of 30 $(km^3/km)/Ma$ (Reymer and Schubert 1984). In the absence of acceptable alternatives, it is concluded that some other crust-forming mechanism, such as underplanting, must have operated in the Precambrian to account for such high growth rates.

## 6.2.5 COMPOSITION OF CONTINENTAL CRUST

Taylor and McLennan (1985) provide a preliminary comparison of the chemical composition of Archean and post-Archean continental crust respectively based on typical clastic sedimentary assemblages. The basic assumption made is that the sediments are representative of a mixture of complex source lithologies that collectively represent the upper continental crust as exposed to subaerial erosion at the time of sedimentation.

Three tentative conclusions are reached: (1) Archean sedimentary rocks have a different composition to post-Archean sedimentary rocks. The difference is directly attributed to the respective compositions of exposed Archean and post-Archean continental crusts. In brief, the Archean crust was comparatively less differentiated and more mafic. (2) Across post-Archean time there are no substantial changes in the composition of sedimentary rocks that can be related to significant changes in composition of the upper continental crust. This constancy in post-Archean upper crustal composition indicates that any post-Archean additions to

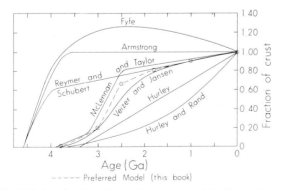

**Fig. 6-2.** A selection of crustal growth models, (from Taylor and McLennan 1985, Fig. 10.1, and reproduced with permission of the authors), to which has been added the preferred crustal growth model as used in this book.

the crust were compositionally indistinguishable from that of the upper crust itself. (3) The single most important change in the bulk composition of the continental crust occurred in late Archean time. As discussed above, this change in composition is considered to be the direct outcome of the widespread addition to the upper continental crust of the K-granite suite derived by intracrustal partial melting in lower levels of the recently thickened late Archean continental crust. The Archean–Proterozoic boundary thereby becomes unique in the dynamic evolution of the continental crust.

## 6.2.6 HIGH GRADE METAMORPHIC TERRAINS

Precambrian crystalline terrains of the world comprise on average 62% amphibolite-facies, 25% granulite-facies (i.e. high-grade), and 13% greenschist and lower-facies metamorphism. Such an average metamorphic exposure corresponds to a bulk removal mainly by erosion of 12–15 km of upper crust to expose the requisite crustal depth zones—a prolific source of clastic–chemical detritus, amongst other considerations. Furthermore, the exposed high-grade metamorphic terrains require exhumation from even greater crustal depth of 17–31 km. Clearly, the continental crust, including the lower part, has been geodynamically active throughout Precambrian time (Moore et al 1986, Harley and Black 1987, Newton 1987, Van Reenan et al 1987, Harley 1989, Clemens 1992). Indeed, understanding the origin and evolution of the deeper crust remains a fundamental challenge (Passchier et al 1990).

Granulite complexes vary in size from the small ($0.5 \times 10^3$ km$^2$) but intensely studied Lewisian Complex (2.8–2.6 Ga) (Park and Tarney 1987), to the very large ($100 \times 10^3$ km$^2$) Ashuanipi Complex (2.6 Ga) (Percival and Girard 1988). They occur in Precambrian terrains of all ages without lithologic distinction by age (Martignole 1992).

Granulite facies rocks are characterized by (1) dryness (pyroxenes); (2) depletion in some but not necessarily all of Rb, Th, U, K and heavy REEs; (3) $CO_2$-rich primary fluid inclusions; (4) high metamorphic temperatures (650–950°C) and pressures (5–11 kbar); (5) common shallow-marine (shelf-type) protoliths (marbles, quartzites, metapelites, para-amphibolites); and (6) intrusive charnockites (pyroxene-bearing quartzofeldspathic rocks) with associated anorthosite massifs. Virtually all granulite complexes show evidence of large-scale horizontal tectonics, e.g. recumbent folding, usually early in the high grade metamorphic cycle. The high pressure range implies some unique transient metamorphic control at crustal depths of 17–30 km; the extreme burial of surficial protoliths implies great crustal thickening prior to granulite-facies metamorphism.

Considered worldwide, possible mechanisms of granulite-facies metamorphism and the return transfer (exhumation) of metasedimentary-bearing granulites to the present surface, include (1) continental-scale underthrusting; (2) accordian-style thickening; (3) nappe-stacking; and (4) hot-spot development, with crustal thinning and underplating over a mantle upwell. Although applicable to one or other granulite complexes, no single mechanism explains them all.

Indeed, the origin of granulite (charnockitic) massifs including mechanisms of incompatible element depletion is a major unresolved problem of petrology which must bear fundamentally on the origin of continents (Newton 1992).

## 6.2.7 MAFIC DYKE SWARMS

Mafic dyke swarms, present in all Precambrian shields of the world, represent repeated expressions of crustal extension, involving the transfer of enormous volumes of basaltic and related magmas from the mantle to the upper continental crust. The majority of continental dyke swarms are either Proterozoic or late Phanerozoic in age (Halls 1987).

In Precambrian shields, dyke swarms of different trend criss-cross one another, often in profusion. Major intrusion episodes are centred at about 2.9, 2.5, 2.1 and 0.1 Ga, with a broad peak at 1.3–1.1 Ga (Halls 1987, Cadman et al 1993). Many swarms extend for hundreds of kilometres and in extreme cases cover areas exceeding $10^6$ km$^2$. For many swarms total crustal extensions of several per cent have been documented. Giant radiating dyke swarms have been genetically linked to megaplumes (Campbell and Griffiths 1991).

Paleomagnetic data have generated Precambrian polar wander paths that are generally dissimilar for each of the major continents. This implies that the continental fragments have moved independently throughout most of Proterozoic time.

Only a few Precambrian dyke swarms have been definitely established as feeders for extrusive sequences, which, if they ever existed, have either been covered, eroded, or degraded (altered or deformed) beyond recognition (Ernst and Baragar

1992). Significantly, continuing studies of some other silicate planets reveal prevalent 'basaltic overplating' involving unusually large volcanic constructs, e.g. Olympic Mons on Mars (Lowman 1989). This raises the presently unresolved question as to whether such major crustal construction occurred on Earth.

## 6.2.8 THE EXPANDING EARTH THEORY

Coincident with Wegener's (1924) first bold and imaginative step in concluding that a single supercontinent, Pangea, had disrupted into separate diverging continents, the expanding Earth theory was propounded.

Carey (1956, 1976), a keen and stimulating exponent of an expanding Earth, marshalls considerable evidence opposed to the now-popular theory of a constant dimension Earth. Recent supporters include Glikson (1983) and Krogotkin and Yefremov (1992), amongst others.

However, the conclusion of modern paleomagnetic analysis (McElhinney et al 1978, Schmidt and Clark 1980) and other considerations (Wilson 1985), is that Earth has never expanded and, in fact, may have slightly contracted. Although the paleomagnetic test itself may be challenged, the lack of supporting evidence for expansion on such planetary bodies as Moon, Mars and Mercury is a telling point against the expanding Earth theory (Taylor and McLennan 1985).

Although the scientific community is almost totally opposed to the theory, more evidence is needed to prove or disprove that Earth's diameter has changed significantly. In the meantime, no matter how unlikely, an expanding Earth cannot be completely excluded, whether or not in conjunction with plate tectonic processes.

## 6.3 EXOGENOUS PROCESSES AND PRODUCTS

Exogenetic evolution encompasses the response of Earth's surficial environment to the enlargement of the continents, with resulting changes in the biosphere–atmosphere–hydrosphere, dominated by increased oxygen content. Whereas photolytic dissociation had doubtless produced some oxygen continuously throughout Archean time (Towe 1983, 1990, 1991), this small production was quickly absorbed in local oxygen 'sinks'. It was not until the advent of widespread photosynthesis in stable platform environments that oxygen was produced in sufficient quantity to satisfy progressively the main oxygen 'sinks', starting with ferrous iron and working through to the global production of sulphates and carbonates (Cloud 1976, 1983a,b). The gradual evolution of a more oxygenating environment, as succinctly summarized by Cloud, is marked by a series of steps in the rock record: first, the widespread occurrence of easily oxidized detrital uraninite and pyrite in continental sediments older than ~2.3 Ga; next, the great bulge in BIF deposition prior to 2.0 Ga; then the massive development of continental redbeds starting at ~2.0 Ga; finally, widespread carbonate and sulphate deposition. The great watershed in this succession of events is interpreted to be marked by the attainment of an oxygen level of ~1% present atmospheric level (PAL), which occurred at ~2.0 Ga and is marked in the rock record by the BIF–redbed changeover mentioned above.

The essential links are (1) the growth of the continents, products of Earth's endogenous processes; (2) the consequent changeover from mantle-dominated to continent-dominated activities in the exogenous environment; and (3) the subsequent rapid biogenic to hydrospheric–atmospheric evolution. The changeover occurred in the late Archean to early Proterozoic interval (2.6–2.0 Ga).

## 6.3.1 SEA WATER COMPOSITION

The composition of sea water, a fundamental product of mantle outgassing (Holland 1984, Wayne 1991), is a sensitive indicator of crustal interactions. The evolution in sea water composition provides an independent check on continental evolution and provides strong support to the interpretation of a dramatic growth in the volume of continental crust in late Archean time.

As reviewed by Veizer (1984), the present day fluxes which control the composition of sea water are (Fig. 6-3) (1) continental river discharge, $F_c$; (2) interaction between sea water and oceanic crust (seafloor basalts) ('mantle' flux), $F_m$; (3) efflux via, and interaction with, sediments, $F_s$; and possibly, (4) subduction of sediments and trapped water in subduction zones, $F_d$.

Specifically, the buffering of ocean chemistry during the Archean, which was dominated by high mantle heat generation and with only small continental presence, should have been dominated by $F_m$, whereas in post-Archean times, characterized by large continents and diminished mantle heat

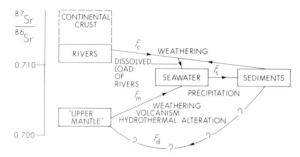

**Fig. 6-3.** Exogenic cycle of strontium. Present day fluxes controlling the composition of seawater are: (1) continental river discharge ($F_C$); (2) seawater–ocean basalt interaction ('mantle flux') ($F_M$); (3) efflux and interaction with sediments ($F_S$); (4) possible subduction of sediments ($F_D$). (From Veizer 1984, Fig. 1, and reproduced with permission of the author.)

flux, the dominant controlling flux would have been the river flux, $F_c$. In fact, the Sr-isotopic trend, based on the signature in marine carbonates, corresponds to the upper mantle growth curve until 2.5 Ga ago, at which point the Sr-isotopic trend shows a sudden increase in the ratio which continues across Proterozoic time (Fig. 6-4). This sudden increase in ratio, accomplished during late Archean to early Proterozoic time (2.5–2.0 Ga), marks the transition from a 'mantle' to a 'river' buffered ocean (Veizer and Compston 1976).

The $^{87}Sr/^{86}Sr$ sea water curve, in brief, documents the sudden changeover in the buffering systematics of the oceans beginning at ~2.5 Ga. The correspondence of this curve with the general growth curve of the continents discussed above, based on completely different parameters, is obvious and thereby provides independent corroboration of the reliability of the continental growth curve.

## 6.3.2 URANIFEROUS CONGLOMERATES

The presence in several Precambrian formations of detrital uraninite ($UO_2$), and sulphides (mostly pyrite, $FeS_2$) is now well established (Roscoe 1973, Grandstaff 1980). Uraniferous (–pyritiferous) conglomerates of economic grade, such as at Blind River in Ontario, Witwatersrand in South Africa and Jacobina in Brazil, do not occur later than mid-Aphebian (early Proterozoic) time (~2.3 Ga), despite the occurrence of some exotic detrital uraninite and sulphides in recent sediments.

Both uraninite and pyrite are unstable under oxidizing conditions. The formation of large uraniferous conglomerates in Archean and early Proterozoic times indicates that the oxygen partial pressure was below $10^{-2}$–$10^{-6}$ PAL to permit the survival of detrital uraninite (Grandstaff 1980). Thus the

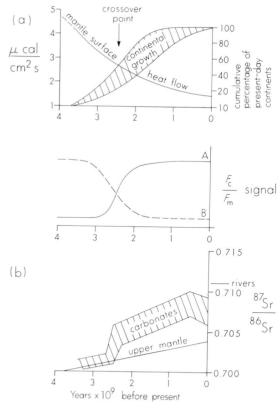

**Fig. 6-4.** Generalized $^{87}Sr/^{86}Sr$ variations in seawater during geologic history. (a) Schematic presentation of mantle surface heat flow and of areal continental growth during geologic history. The continental growth curve was obtained by utilizing 2–$4 \times 10^{-10}$ $y^{-1}$ as the average recycling constants. The associated $F_C/F_M$ signal in sea water composition should be of the type A if the transition results in a gain (i.e. $^{87}Sr/^{86}Sr$, Na?, $^{18}O/^{16}O$) and of type B if in a loss (i.e. Fe, Mn, Ba, Sr?) of an entity. Note that this general shape of the signal will evolve regardless of the details of the two exponential (or other type) curves of opposing slopes. For radiogenic isotopes (i.e. Sr) the flat parts of the curves will be modified by radioactive decay. Modified from Veizer et al (1982). (b) Generalized $^{87}Sr/^{86}Sr$ variations in sea water during geologic history. Modified from Veizer and Compston (1976). Note the departure of 'sea water' curve from mantle values at ~2.5 Ga ago. (From Veizer 1984, Fig. 2, and reproduced with permission of the author.)

relations provide independent evidence of an increasing oxygen content across early Proterozoic time.

## 6.3.3 BANDED IRON FORMATION (BIF)

BIF is a useful description term which characterizes a thinly layered sedimentary rock consisting mainly of silica and iron minerals; it is mainly a chemogenic precipitate modified by diagenesis and metamorphism. Various classifications of BIF have been proposed to accommodate the wide range of types based on lithology, supracrustal association, and tectonic setting (Goodwin 1956, 1982b, Gross 1983, James 1983, Walker et al 1983, Klein and Beukes 1993, Manikyamba et al 1993).

BIF types range widely in the geologic record, though each type has a prefered age range. Thus Algoma-type BIF of cherty iron oxide–sulphide–carbonate varieties, is common in Archean greenstone belts. Superior and Hamersley types, of cherty hematitic–magnetitic varieties, are characteristic of early Proterozoic time. Rapitan type, a cherty hematitic variety associated with diamictites including tillites, is restricted to the late Proterozoic to early Phanerozoic period (0.7–0.4 Ga) (Yeo 1986, Kirschvink 1992, Klein and Beukes 1993).

BIF has been deposited practically throughout known Earth history. Thus Isua metasediments of Greenland, 3.8–3.5 Ga, include large commercially attractive magnetite-bearing BIF. At the other end of the time scale, ironstones of Jurassic (135 Ma) and even Pliocene (1–10 Ma) age exist in northern Asia. Nevertheless, BIF is quantitatively a Precambrian phenomenon, especially in the early Proterozoic (2.5–1.8 Ga), as discussed below.

James (1983) provides a useful summation of the main BIF districts of the world, including estimated tonnages by district. Five very large districts, each with an estimated $10^{14}$ t of BIF dominate Earth's BIF distribution; Hamersley Range, Australia; Transvaal–Griquatown, South Africa; Minas Gerais, Brazil; Labrador Trough and extensions, Canada–USA; and Krivoy Rog–KMA, Russia. Together these five very large BIF repositories contain an overwhelming 90% of Earth's total estimated BIF ($5.76 \times 10^{14}$ t).

Furthermore, James' time assessment of BIF deposition (Fig. 6-5) reveals an extraordinary concentration of preserved BIF, amounting to an estimated 92% of Earth's total BIF, deposited in early Proterozoic time (2.5–1.9 Ga) (see also Klein and Beukes 1993, fig. 2). Thus the early Proterozoic was the major epoch of ferric iron sedimentation in Earth's history, thereby marking a major milestone in the evolution of Earth's continental crust.

Cloud (1983b), amongst others, suggests that an important historical change in oxygen level occurred in the 2.5–2.0 Ga interval (see below). Before this time, oxygen levels remained below 1% PAL. Starting about 2.5 Ga ago, however, the oxygen level began to rise. This gradual increase coincided with the appearance of the first oxygen-producing photosynthesizers (chlorophyl-a) in the form of blue-green algae or their progenitors.

The favoured BIF deposition model is modified

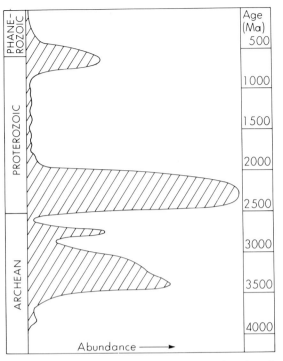

**Fig. 6-5.** Estimated abundance of iron formation deposited through geologic time. Horizontal scale is non-linear, approximately logarithmic: range 0–$10^{15}$ t. (From James 1983, Fig. 12-1, and reproduced with permission of the author.)

after a number of previously published ideas and concepts, notably those of James (1954), Cloud (1973), Holland (1973), Drever (1974), Trendall (1975), Button (1976), Goodwin (1982b), Mel'nik (1982), Ewers (1983), Gross (1983) and Morris (1993). Essential to this model is the timing of great BIF deposition (2.5–1.9 Ga), following the great period, no matter how diachronous, of continental cratonization (2.7–2.5 Ga), with its potential, for the first time in Earth's history, of providing for the development of stable continental platforms and consequent algal bloom on a truly grand scale.

The favoured model, in brief, involves a Precambrian ocean with an upper oxic layer overlying a much larger volume of anoxic water representing an enormous reservoir of diversely derived dissolved iron and silica. Large-scale precipitation of BIF followed the upwelling of the deep, Fe- and $SiO_2$-rich waters to the newly developed shallow continent-margin basins and shelves. However, this deep ocean source of dissolved Fe and $SiO_2$ was replenished repeatedly by contemporaneous hydrothermal discharge accompanying submarine volcanism (Jacobsen and Pimentel-Klose 1988, Kimberley 1989, Christiaens 1994), augmented by pervasive chemical leaching of iron and silica in response to fluid circulation in the fresh oceanic crust, a form of submarine weathering traditionally called halmyrolysis. The role of organic catalyst in the precipitation of silica remains unresolved (Heaney and Veblen 1991), but may have been significant.

It seems clear that the major production of BIF stopped relatively suddenly, about 1.8 Ga ago. This may have been the time when the photosynthetic oxygen source finally exceeded the rate of supply of reduced matter to the combined atmosphere and hydrosphere, whereupon the oceans were largely swept free of reduced compounds, including iron, and a stable aerobic environment became established (Cloud 1973, Walker et al 1983, Christiaens, 1994).

## 6.3.4 REDBEDS AND PALEOSOLS

Redbeds are detrital sedimentary rocks with reddish-brown ferric oxide pigments coating mineral grains, filling pores, or dispersed in a clay matrix (van Houten 1973). They generally form in fluvial or alluvial environments on land, thereby representing terrestrial oxidation products. They are generally regarded as requiring the presence of free oxygen in the atmosphere (Walker et al 1983, Long and Lyons 1992).

Redbeds are rare or absent in the Archean. This may be reasonably attributed to an oxygen-deficient Archean atmosphere (Walker et al 1983). The well known transition from drab-coloured rocks in lower horizons of the ~2.3 Ga old Huronian Supergroup in Canada to redbeds higher in the sequence has been interpreted as indicating the time that free oxygen became available in at least the local environment (Roscoe 1973). From this time on, redbeds become increasingly common in the preserved record on all continents.

Paleosols (Martini and Chesworth 1992) or fossil soil profiles, on the other hand, have been studied in a number of Archean terranes as old as 3.0 Ga (Schau and Henderson 1983, Walker et al 1983, Holland and Zbinden 1988, Gall 1992, Sutton and Maynard 1993, Retallack and Krinsley 1993, Macfarlane et al 1994a,b). They typically show a deficiency of iron and other chemical alterations near the surface that is generally attributed to surface leaching. Some may have been misidentified and, in fact, represent metamorphic alteration zones (Palmer et al 1989, Wiggering and Beukes 1990). Thus these Archean–earliest Proterozoic (to 2.2 Ga) paleosol profiles are consistent with weathering under anaerobic conditions, without, however warranting a firm conclusion regarding the precise level of atmospheric oxygen at that time.

## 6.3.5 GLACIGENE AND EVAPORITE DEPOSITS

Convincing evidence of Precambrian glaciation is restricted to two time periods only: (1) early Proterozoic, between 2.5 and 2.0 Ga, in North America, South Africa and Australia; and (2) late Proterozoic, 1.0–0.57 Ga, in all continents, probably including Antarctica. It is not known whether the early Proterozoic glacigene deposits are products of a single widespread episode or a number of geographically and temporally independent episodes.

Based on paleomagnetic data, the widespread late Proterozoic glaciations with closely associated BIF have been assigned to low paleolatitude (<25°) glaciers (McWilliams and McElhinny 1980, Schmidt et al 1991), suggesting an episode of exclusive equatorial, severe continental glaciation. This has led to a number of theories attempting to explain their occurrence, including that of an ice-charged, highly reflective 'snowball-type' Earth (Kirshvink 1992). However, based on new and re-analysed paleomagnetic data combined with climate models, Meert and Van der Voo (1994) demonstrate that the glaciations may well have occurred at latitudes above 25°. They offer an alternative explanation that is consistent with the waxing and waning of intermediate latitude ice sheets to form the late Proterozoic tillites and other conformable sequences of alternating warm climate–cold climate strata.

Significantly, the early Proterozoic glacigene deposits of the Huronian Supergroup, Canada, occupy a stratigraphic niche coinciding with the apparent transition from anoxic to oxidizing environment, as discussed below.

Sulphate evaporite deposits, preserved in the form of sulphate minerals or pseudomorphs, have formed at a more or less steady state on most of the continents since at least 3.5 Ga ago (Walker et al 1983). Evaporite precipitation unconditionally requires arid conditions, whether at low or high temperatures. On the basis of morphologic and mineralogic evidence, Walker et al (1983) conclude that the record of Precambrian evaporites suggests that the primitive climate on early Earth (since 3.5 Ga at least) was not very different from later Phanerozoic climates.

Taken together, the data on glacigene and evaporite deposits, although sparse, suggest that Precambrian climatic conditions were essentially similar to those prevailing in the Phanerozoic, with the average surface temperatures during most of the Precambrian neither above 60°C nor below a few degrees Centigrade. Within this broad range it is not yet possible to say whether temperatures were generally lower or higher than they are today. However, it is clear that on at least two occasions, i.e. 2.5–2.0 Ga and 1.0–0.57 Ga, temperatures low enough to permit continental glaciation prevailed, the second occurring even near the equator.

### 6.3.6 OXIDATION STATE

The causative link in the changeover from a non-oxidizing environment to one characterized by free oxygen is considered to be the growth of the continents; this triggered both the changeover from mantle-dominated to continent-dominated lithologic and biologic activities, and the consequent enhanced biomass with greatly increased production of photosynthetic oxygen. This changeover is considered to have occurred in late Archean to early Proterozoic time in the interval from ~2.5 to ~1.8 Ga (Eriksson and Cheney 1992) and probably between 2.2 and 1.9 Ga (Holland and Beukes 1990).

The vast oxygen-consuming entities ('sinks') of Archean and earliest Proterozoic times included all the reduced atmospheric components inherited from Earth's initial aggregation, the reduced products of copious volcanic outgassing, and a high rate of flux of reduced matter in solution, both from weathering and volcanic exhalations, i.e. ferrous iron, carbon monoxide, sulphite and various sulphides (Cloud 1983b).

It is stressed by Cloud (1983b) that the role of burial of organic matter has been crucial to the development of an oxidizing environment. For example, it has been calculated (Holland 1978) that the now-buried reduced carbon, sulphides and ferrous iron would require a mere 3 Ma to eliminate all resident oxygen in the present atmosphere. Our present oxidizing atmosphere is thus the result of a kinetic lag between oxygen 'sinks' and oxygen sources. The principal current oxygen 'sinks' are (1) the exposure by erosion or other activities of once buried reduced carbon, sulphides, and ferrous iron; and (2) volcanic gases. These account for about 75% and 25% respectively, of the approximately $400 \times 10^6$ t of oxygen produced each year as a result of photosynthesis (Holland 1978). These same processes must have operated, though not necessarily at the same rate, during Precambrian times.

### 6.3.7 BIOGENESIS

A consequence of major outgassing in primitive Earth and the development of the atmosphere and hydrosphere was chemical evolution, leading in stages through amino acids, polypeptides, RNA and DNA to anaerobic heterotrophs (organisms unable to manufacture their own food) and procaryotes (unicellular organisms lacking a nucleus) (Cloud 1976, 1977) (Fig. 6-6). Highly metamorphosed sedimentary rocks in the 3.8–3.5 Ga Isua assemblage of West Greenland, of a type alluded to by Glover (1992), yield some indications of photoautotrophic biologic activity in the form of carbon isotope ratios. The oldest known microfossils, however, occur in the 3.5 Ga Warrawoona Group of Western Australia, where wavy to planar laminated charts, considered to be stromatolitic, contain well-preserved filamentous microfossils (Awramik et al 1983), not an unchallenged claim, however (Buick 1988). Other detected early Archean (3.4–3.2 Ga) stromatolites with microfossils occur in the Swaziland Supergroup, South Africa. Younger Archean (i.e. 3.0–2.5 Ga) fossiliferous strata are more widespread including stromatolites and cyanobacteria in the 2.75 Ga Fortescue Group, West Australia. Still younger (i.e. Proterozoic or post 2.5 Ga) fossil assemblages include: (1) diverse cyanobacteria (heterocysts), bacteria and stromatolites in the 2.0 Ga Gunflint Formation, Ontario, first independently discovered by Tyler (Tyler and Barghoorn 1954) and later collaboratively expanded (Barghoorn and Tyler 1965); and (2) eucaryotes in 1.3 Ga Beck Spring and 0.9 Ga Bitter Spring strata, progressively culminating in Ediacaran metazoons multicellular organisms) in the 0.6–0.5 Ga (i.e. Eocambrian) Adelaidean System, Australia, including sessile and vagile bethonic and planktonic marine forms. However, the bogenicity of all stromatolites older than 3.2 Ga has been recently challenged (Lowe 1994), who ascribes certain conical-domal structures to evaporate precipitation and/or soft-sedimentary deformation. This author, however, does not question the widespread occurrence of bacterial communities on Earth prior to 3.2 Ga.

With the rapid increase of the atmospheric oxygen level to about 1% PAL (Fig. 6-7) about 2.0 Ga ago, intracellular isolation of anaerobic vital pro-

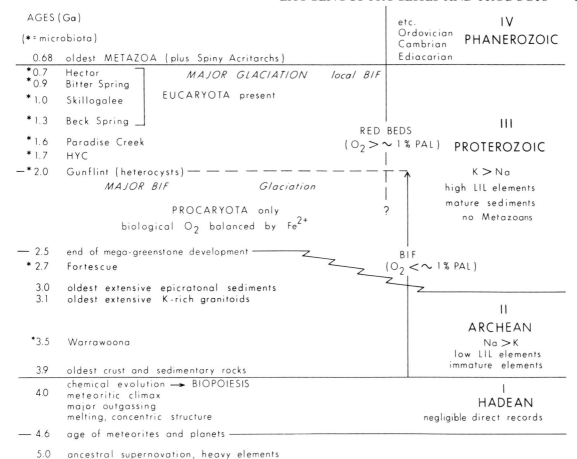

**Fig. 6-6.** Abbreviated outline of Precambrian history with biostratigraphic emphasis. (Slightly modified from Cloud 1983a, Fig. 1, and reproduced with permission of the author.)

cesses became essential. This led to the diversification of the eucaryotic cell about 1.4 Ga ago, exemplified by Beck Spring (1.3 Ga) and Bitter Spring (0.9 Ga) eucaryota. The next big step on the evolutionary path was the origin of sexuality. This occurred about 0.7 Ga ago, the age of the oldest known metazoans. The first metazoans were dependent on simple diffusion for oxygen; exoskeletons appeared later, perhaps 0.6 Ga ago, when increasing oxygen levels favoured emergence of more advanced respiratory systems. Subsequent evolution produced the oldest trilobite 0.57 Ga ago and the oldest hominids 0.004 Ga ago (Cloud 1976, Hofmann 1981, Runnegar 1982, Schopf et al 1983, Saul 1994).

Future research on ancient biosphere will seek new research directions driven by new technology, new concepts, and new requirements for information. The desire to test the Gaia Hypothesis will spur much of this research. This will lead to a fuller understanding of biologic evolution and the relationship to the physical evolution of Earth (Nowlan 1993).

### 6.3.8 SEDIMENTATION

The changing pattern of sedimentary facies across Precambrian time is an important aspect of exogenic evolution. The dominant trend is from immature and chemically primitive Archean greenstone type assemblages to mature and 'continental' Proterozoic assemblages, eventually including common redbeds and sulphate deposits. This unidirectional change, in brief, supports the interpretation of an enlarging stable continental presence.

### Archean sequences

As outlined by Lowe (1980, 1982) *Archean greenstone belts* are characterized by thick, cyclic, mafic–felsic volcanogenic–volcaniclastic accumulations

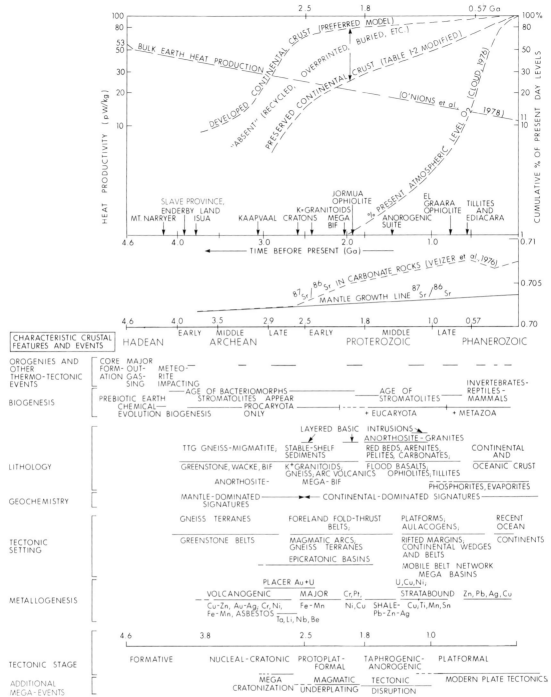

**Fig. 6-7.** Crustal controls and trends in development and preservation of Earth's continental crust, together with characteristic crustal features and events arranged according to fivefold tectonic stages. (Modified after Goodwin 1981b, Fig. 1).

that are frequently capped by comparatively thin units of chemical sediments.

In the lower–middle volcanic-dominated parts, interflow sediments comprise banded chert, silicified volcanic rocks, BIF, and, locally, carbonates, barite, and minor terrigenous clastic rocks (Lowe 1982). This sedimentation apparently occurred in quiet, low-relief, clastic-starved subaqueous environments.

The middle–upper stratigraphic parts are characterized by coarse, poorly stratified felsic volcanic components, including ashflow tuffs, conglomerates, flow rocks and associated high level plutons. The building of these large vent complexes produced broad subaerial volcanic edifices of considerable relief which became important sources of loose sediments and provided a range of sedimentary environments (Lowe 1982).

The uppermost stratigraphic parts, in turn, consist predominantly of sedimentary rocks, commonly terrigenous detritus with negligible volcaniclastic components. The sedimentary composition indicates a provenance composed of uplifted volcanic–sedimentary terrains supplemented by a variety of plutonic and metamorphic rocks.

This transition from lower volcanic-dominated to overlying sedimentary sequences marks a profound change in the tectonic and magmatic histories of the greenstone belts, the overall direction being towards the formation of substantial areas of subaerially exposed, tectonically stable sialic crust. Thus the termination of greenstone belt development in a particular region involved intrusion, deformation and cratonization of the belts themselves (Lowe 1982).

*Archean medium to high grade terrains* also include substantial sedimentary material, now more or less transformed to granitoid gneiss. Indeed, the original nature of some of these now-granulite facies rocks is difficult to interpret. However, shallow-water sedimentary facies, including quartzites, pelites and carbonates (calcsilicates), in addition to turbiditic facies, are common. In general, these little studied, higher grade terrains seem to contain a sedimentary association distinct from those of the adjoining greenstone belts, and to thereby represent separate and distinctive sedimentary environments compared to the volcanic-dominated greenstone belts themselves (Lowe 1982).

The oldest substantial example of *Archean epicratonic sediments* is provided by orthoconglomerate–quartzitic (–arkosic) sandstone–shale–carbonate–BIF assemblages in the 3.0 Ga old Pongola Supergroup in southern Africa which unconformably overlies 3.5–3.2 Ga Swaziland greenstones. Similar relations are found in southern India and West Australia (Rogers 1993). Elsewhere, epicratonic–cratonic assemblages of this type are commonly developed by 2.8–2.4 Ga, in response to diachronous cratonization.

## Proterozoic sequences

During the *early Proterozoic*, the marked increase in the proportions of mature orthconglomerate, quartzite, sandstone and shale at the expense of immature greywacke and siltstone reflects the presence of greatly expanded epicontinental seas provided by preceding cratonization. These and other related sedimentary changes, which are also marked by carbon isotope trends (Des Marais 1994), collectively represent the vital transition from immature and chemically more primitive ('mantle-like') to mature and 'continental', a diachronous change coinciding with the general Archean-to-Proterozoic transition (Eriksson and Donaldson 1986).

The *early to mid-Proterozoic* transition (~1.8 Ga) is marked by a dramatic rise in the quantity of bimodal volcanic–quartzite–arkose assemblages of common redbed character, a manifestation of widespread taphrogenic rifting of thick, stable continental crust operating in an aerobic environment (Condie 1982). The oldest known sulphates of substantial dimension are tidal-flat gypsum–anhydrite deposites of age 1.8 Ga (Eriksson and Truswell 1978). Widespread oxidation of sulphur to sulphates in sea water, for which atmospheric oxygen was a prerequisite, must have predated this.

### 6.3.9 PHOSPHORITES

The oldest known phosphorites are apatite-rich mesobands (a few centimetres thick) in certain 2.6–2.0 Ga old BIF. Thereafter, phosphatic stromatolites are known, as for example in the 1.6–1.2 Ga old Aravalli Group of India. Phosphorites are somewhat more abundant in sedimentary sequences younger than about 900 Ma (Cook and McElhinny 1979). However, the most striking increase in the abundance of phosphorites occurs at the Proterozoic–Cambrian boundary, i.e. ~600 Ma (Sheldon 1981). This dominant bulge may be related to, if not the cause of, increased biological activity, accepting phosphorus as a globally limiting nutrient.

### 6.3.10 MINERAL DEPOSITS

The broad sweep of exogenous evolution is well represented by changing metallogenic patterns across the Precambrian (Anhaeusser 1981, Hutchinson 1981, Meyer 1981, 1988, Folinsbee 1982, Lambert 1983, Rickard 1987, Sawkins 1989, Barley and Groves 1993, Stowe 1994). These are briefly summarized in nine successive metallogenic associations, as follows:

(1) Volcanogenic base metal (Cu–Pb–Zn) sulphide deposits and related precious metals (Au–Ag) associated with greenstone belts mainly of late Archean age (~2.7 Ga). Cr and asbestos hosted in large ultramafic–mafic silt-like complexes in Zimbabwe greenstone belts (3.4 Ga). Ni–Pt in Stillwater Complex (2.6 Ga). Attributed to prevailing primitive (mantle-like), tectonically unstable environments.

(2) Placer Au–U (e.g. Witwatersrand, 2.7 Ga) and U (e.g. Elliot Lake, 2.4 Ga) deposits with pyrite. Attributed to an oxygen-deficient continental environment.

(3) Major BIF deposition, 2.6–2.0 Ga. Attributed to rising hydrospheric oxygen, product of the bulge in photosynthesis stemming from the development of stable shelves at the Archean–Proterozoic boundary.

(4) Ni–Cu–Pt–Cr–Ti–Sn–V deposits in layered mafic–ultramafic intrusions within stable continental shields, notably Bushveld (2.0 Ga) in South Africa and Sudbury (1.9 Ga) in Canada. Cu–Zn massive sulphide deposits, e.g. Outokumpu, Finland (1.9 Ga). Kiruna-type massive magnetite–apatite ores (1.9 Ga). Significantly, the 2000–800 Ma (i.e. 1200 Ma) Proterozoic interval marks an apparent worldwide gap in (Cr deposition; the 1200 interval is also marked by comparatively low (Cr/Fe ratios in layered igneous complexes and rare ophiolites(Stowe 1994), relationships tentatively attributed to the presence of thicker, thereby subduction-resistant oceanic crust (Moores 1993, Stowe 1994). After 800 Ma, ophiolite-hosted podiform chrome deposits predominantly indicating plate tectonic systems similar to those of present time.

(5) Stratiform Pb–Zn sulphide deposits in clastic-hosted environments, e.g. Broken Hill-Mount Isa-McArthur River, Australia (1.69 Ga); Bergslagen, Sweden (1.89 Ga); Aggeneys and Gamsberg, Namaqualand (~2.0 Ga); Sullivan, Canada (1.4 Ga); Balmat, New York (1.1 Ga). Attributed to the development of sea water sulphates which, under the influence of biogenic reduction, provides the co-metal sulphide precipitate. Metal sources are commonly attributed to exhalative and/or metal-leaching processes.

(6) U (Au–Ag–Cu–Ni–Co) deposits related to unconformities in the period 2.0–1.0 Ga (e.g. Athabasca Basin, Canada, 1.3–1.1 Ga; Cahill Basin and Olympic Dam, Australia 1.6 Ga). These large concentrations of metal were probably moved in oxidizing waters of meteoric origin.

(7) Sedimentary Cu deposits, 1400–600 Ma, e.g. Zambia, Africa (1.1 Ga), White Pine, USA (1.1 Ga). Extensive oxidation and redbed conditions in a tectonically inactive region with internal drainage. Under appropriate conditions, streams could transport copper for deposition in neighbouring euxenic basins.

(8) Mississippi Valley type Pb–Zn deposits—late Proterozoic (~6.0 Ga), e.g. Nanisivik and Gayna River, Canada. Attributed to metal deposition from brines in platform carbonates undergoing karst-type erosion.

(9) Large Proterozoic oil pools and giant gas fields from late Proterozoic assemblages are known especially in Oman, Siberia and China. Biomarker techniques now allow convincing demonstrations that the oil in nearby Phanerozoic reservoirs has a Proterozoic source (Walter 1991).

## 6.4 SUMMARY CRUSTAL DEVELOPMENT BY STAGE

Earth's nine Precambrian cratons reveal a remarkably consistent pattern of crustal development one-to-the-other, with due allowance for diachronous cratonization. This unfolding pattern is characterized by progressive Archean—especially late Archean—cratonization, subsequent early Proterozoic craton consolidation with widespread platformal development and craton margin interactions; followed by large, stable, sustained mid-Proterozoic cratons characterized by interior rifting including basalt-redbed-filled anlacogens, impressive auorogenic magmatism and culminating continental collisions and concluding with present-stature late Proterozoic to early Paleozoic continental plates operating under the influence of actualistic plate tectonic processes.

Various schemes of crustal development may be devised depending on the desired focus. The following highly generalized scheme, focuses on stage-by-stage craton growth with attendant endogenic–exogenic evolution. Development of continental crust is divided into five main stages, each with distinctive characteristics (Fig. 6-7). Stage boundaries are highly arbitrary.

### 6.4.1 HADEAN (FORMATIVE) STAGE (4.6–3.9 Ga)

Model Pb ages from Earth and radiometric ages from meteorites and the Moon suggest that the earliest terrestrial crust formed during or shortly after the late stages of planetary accretion, whether homogeneous or inhomogeneous (Kaula 1980), at ~4.5 Ga (Condie 1989, Wetherill 1990). This is supported by ages of 4563 Ma from the oldest dated zircons in the solar system (Ireland and Wlotzka 1992), and 4270 Ma for the oldest dated terrestrial zircon (Compston and Pidgeon 1986).

McCulloch (1994) proposes an accretion interval for planet Earth of ~60–100 Ma, leading to final accretion by ~4480 Ma, with either an early (~4540 Ma) or late (~4480 Ma) giant impact origin for the Moon (Cameron and Benz 1991). Earth accretion was accompanied or followed by core formation (Kaula 1980), with major outgassing. This accretion history allows for an extended interval for cooling of the terrestrial magma ocean, and hence avoids the dilemma of a highly differentiated early Earth (McCulloch 1994). The resulting nascent Earth was subjected to intense meteorite impacting in the period 4.2–3.9 Ga, involving an estimated 2500–3000 terrestrial impact structures, each more than 100 km wide (Grieve 1980). Thus, despite final accretion of Earth at ~4500 Ma, the moderately well-preserved terrestrial record extends back only as far as ~3800 Ma, with a single predecessor date at 3962 Ma. However, this ancient and limited terrestrial evidence is now being significantly augmented by well-dated records of relevant events in the solar system obtained from less active planetary bodies, together with complementary sources of information from astronomical theory and observation (Wetherill 1990, Anderson 1989, Teisseyre et al 1992).

It seems clear that Earth lost almost all of its primordial atmosphere, if such in fact existed at all, at the time of giant impacting and especially during Moon formation, and that the present atmosphere was acquired later. The most probably hypothesis for acquisition of a secondary atmosphere involves the incorporation of volatile materials into the planet as it accreted in various stages, with resulting outgassing of our atmosphere–hydrosphere (Wayne 1991). $H_2O$ condensed to form oceans, with liquid $H_2O$ first existing at temperatures well above 100°C (Holland 1984); the bulk of $CO_2$ eventually formed sedimentary carbonate rocks, leaving outgassed $N_2$ to accumulate and become the most abundant species in the present atmosphere. Apparently unique in the solar system, Earth's atmosphere quickly evolved to produce an environment suitable for the creation/retention and support of life. With life came the eventual growth of our present $O_2$-rich atmosphere (Wayne 1991), the exact composition at any time determined by biological processes acting in concert with physical and chemical changes in accord with the Gaia hypothesis.

### 6.4.2 ARCHEAN STAGE (3.9–2.5 Ga)

The main facets of preserved Archean crust are summarized in Section 2.1.2 and Table 2-2. Salient aspects include: (1) predominant medium–high grade granitoid components; (2) subordinate medium–low grade greenstone belts with basalt–komatiite–turbidite sequences and rare platform associates; and (3) widespread early recumbent and later vertical deformations. Reliable pre-3.8 Ga terrestrial dates are currently restricted to eight widespread sites. The oldest coherent rock assemblages, documented in southern Africa and West Australia, fall in the 3.6–3.3 Ga range. Stabilized cratons able to support platform cover existed by 3.0 Ga locally in four extant southern (Gondwanaland) continents. Elsewhere, craton stability was widely attained by ~2.5 Ga.

The bulk of preserved Archean crust is attributed to fractionation–aggregation processes operating during successive mega-crust-forming, craton-spawning events, notably at 3.8–3.5, 3.1–2.9 and 2.7–2.6 Ga, each involving: (1) major production of juvenile, mantle-derived Na-rich (TTG) granitoids; (2) common medium–high grade, low pressure metamorphism; and (3) pervasive tectonism involving napperian recumbency with overthrusting and compressional isoclinal folding–faulting–shearing. Of these mega-events, the terminal Archean (2.7–2.6 Ga) was a particularly potent craton-building operation.

The origin and assemblage of common Archean granitoid–greenstone belts pose largely unresolved problems. According to one popular view (Taira et al 1992), greenstone belts originated by seafloor-spreading followed by subduction–accretion processes involving island arcs, melanges, ophiolites, terigenous sediments and 'micro-continental' fragments, all transported significant distances (thousands of kilometres) where tectonically accreted to enlarging crustal entities in the form of displaced (allochthonous) exotic terranes as outlined for some Phanerozoic belts by Coney et al (1980). Others, more impressed by the simple uniformity of the granitoid – greenstone components, and denying the presence of oceanic crust in the Archean greenstone belts (Bickle et al 1994), prefer to interpret them as products of a more restricted, essentially *in situ* arc-magmatism process involving limited lateral–vertical accretionary displacements leading to the assemblage of discrete fault-bounded lithotectonic domains, more of an autochthonous platelet–tectonic-style Archean predecessor process, in the general manner outlined by Lowe (1994) for the Barberton greenstone belt (3.5–3.2 Ga), in an Earth lacking large stable cratons.

As stressed by Hamilton (1993), many current Archean interpretations have been based on extrapolations of modern plate-tectonic processes backwards through time. However, when considered forward from the viewpoint of planetology, especially the developing understanding of Venus, Earth's closest counterpart as known, it becomes apparent that much of Archean magmatism–tectonism may be better explained by models of voluminous magmatism, such as expressed by plume-generated phenomena, including great upwellings in the form of lava lakes and magma oceans. Considering that as much as 99% of Vesuvian heat-loss is by conduction (Taylor 1991), Archean Earth may likewise have lost most of its heat by processes of conduction and rising melts rather than through windows opened into the interior by plate spreading (Hamilton 1993).

Furthermore, at the higher Archean mantle temperatures, where pressure-release melting starts deeper and generates thicker basaltic or komatiitic crust compared to the present-day situation, the resulting compositional stability renders mechanisms of conventional plate tectonics ineffective, with the growing Archean 'cratons' essentially stabilized on top of compositionally stratified roots (Vlaar et al 1994).

In brief, a forward extrapolation of this type, emphasizing an early plume-generated, voluminous magmatism scenario, is required, to better serve as a counterbalance to the undoubted appeal of the modern plate- tectonic paradigm, and hence establish a more balanced stage-by-stage view of Earth's crustal development.

## 6.4.3 EARLY PROTEROZOIC STAGE (2.5–1.8 Ga)

The main facets of preserved early Proterozoic crust are summarized in Section 3.1.2 and Table 3-2. The history of this era is characterized by stable shelf sedimentation including mega-BIF and placer Au–U concentrations, in association with other decisive tectonic elements, ranging from intracratonic rifts and grabens, through substantial pericratonic troughs and basins, to major multi-zoned intercratonic mobile belts, collectively products of protocontinental break-up, enlargement and eventual re-aggregation during the terminal era orogeny. Included are: (1) greenstone–turbidite assemblages, products of juvenile magmatic arcs; (2) the oldest well-developed ophiolites, tillites and continental redbeds; (3) a stable enhanced oxygenic environment resulting from enhanced biogenic photosynthesis; and (4) exceptional mafic–ultramafic layered igneous complexes, some of proposed meteorite-impact origin.

Certain lithologic elements closely resemble Archean precursors; others are practically unique to the era; still others appear for the first time to reappear consistently thereafter in younger eras. The early Proterozoic era thereby represents a major watershed in Earth's crustal history, ushering a global scene that has persisted to the present.

Thus the actualistic plate tectonic model can be fitted with reasonable success to early Proterozoic crust (Hoffman 1989a). Accretionary wedges, ophiolites, fore-arc and back-arc basins, magmatic arcs and rifted margin sedimentary wedges make their appearance although different in important ways from modern equivalents (Hamilton 1993). The problem then arises as to the appropriate blend of predecessor Archean and actualistic plate-tectonic processes when interpreting the early Proterozoic and younger rock record.

## 6.4.4 MID-PROTEROZOIC STAGE (1.8–1.0 Ga)

The main facets of preserved mid-Proterozoic crust are summarized in Section 4.1.2 and Table 4-2. Broadly cratonized protocontinental crust provided the setting for remarkable craton-wide taphrogenic systems of rifts, troughs, grabens and other assorted aulacogens, commonly arranged along orthogonal patterns, leading to the eventual break-up of former coherent cratons with incipient coeval development of rift-induced interior subsidence basins, some very large.

Continental rifting took extravagant forms with the development of long, irregular intracratonic rift systems, some with spectacular continental arc magmatism involving voluminous mantle-derived flood basalts and local crustal-derived felsic associates. Widespread anorogenic magmatism commonly took the form of granitoid plutonism–volcanism, including numerous alkaline ring structures. Its most flamboyant manifestation was in the form of the supercontinent-wide anorogenic suite involving mantle-derived anorthosites and associates, and coeval lower continental crust-derived rapakivi granites–rhyolites and associates, all collectively arranged in spectacular alignments. These were the products of long-sustained linear subcrustal thermal regimes that affected exceptionally large, stable supercontinental craton(s), products of early–mid-Proterozoic craton enlargement–aggregation.

In sharp contrast, long, linear, mid-era to end-of-era pericratonic mobile belts, some with major juvenile crustal accretions and others almost exclusively with tectonically reworked older crust, represent major 'geosynclinal' elements involving recognizable offshore arcs, wedges, basins and melanges, products of protocontinental break-up and tectonic re-aggregation.

### 6.4.5 LATE PROTEROZOIC STAGE (1.0–0.57 Ga)

The main facets of preserved late Proterozoic crust are summarized in Section 5.1.2 and Table 5-2. Broadening of the preceding craton-wide aulacogen stage to full-scale platform subsidence led to the development of major interior basins, sites of shallow-marine to fluviatile redbed, stromatolitic carbonate and phosphate accumulations. Concurrent pericratonic rifting with development of long, linear mobile belts effectively defined the future terminal-Precambrian continents.

The last 250 million years of Precambrian history (800–530 Ma), corresponding to the Sinian period, is one of the most extraordinary periods known in Earth history (Brookfield 1994). Life forms changed from simple cells to diverse multicellular differentiated skeletalized organisms. World-wide cooling triggered widespread glaciations which stripped the cratons of much of their sedimentary cover. Great glacio-eustatic sea-level changes induced by glaciations controlled the deposition of sediments in extensive and possibly globally correlatable megacycles. Rifting progressively fractured and split apart a supercontinent (Rodinia?), starting about 725 Ma and terminating at about 600 Ma, with opening of the Pacific Ocean in the west, with eventual assemblage of Gondwanaland in part along the Mozambique belt in east Africa. This set the global scene for oncoming Phanerozoic activities.

## 6.5 PREFERRED MODEL FOR THE EVOLUTION OF THE CONTINENTAL CRUST

### 6.5.1 INTRODUCTION

Overall control of the evolution of Earth's continental crust has been exerted by declining bulk Earth heat production, which has decreased steadily from an initial estimated level of 53 pW/kg to the present level of 11 pW/kg. The bulk Earth heat production profoundly influenced, successively, core-mantle differentiation and development, the fractionation of the mantle to produce oceanic crust with distinctively evolving chemical–rheologic parameters, and partial melting of this mafic crust to produce the sialic fraction, which has progressively aggregated in the form of growing cratons to eventually produce the modern continents. Cratonization, in turn, controlled the availability of stable shelf environments which profoundly influenced the exogenic environment, notably the organically controlled photosynthetic production of oxygen levels, all important to organic evolution (Fig. 6-7).

Available data overwhelmingly support a model of episodic growth of continental crust across geologic time. Periods of accelerated growth occurred at 3.8–3.5, 3.2–3.0, 2.8–2.5, 2.0–1.7, 1.2–1.0 and 0.6–0.0 Ga. Of these, the late Archean (2.8–2.5 Ga) interval was extraordinarily productive of continental crust, with the early Proterozoic interval (2.0–1.7 Ga) of lesser yet still substantial dimensions.

According to the preferred primary growth curve (uppermost curve in Fig. 6-7) the proportion of developed continental crust was 20% by 3.0 Ga, 65% by 2.5 Ga, 80% by 1.8 Ga, 90% by 1.0 Ga, and 92% by 0.6 Ga. This implies that (1) the production of pre 3.0 Ga sialic crust was small and sporadic; (2) the bulk of the continental crust was produced during late Archean time (2.9–2.5 Ga), the period of mega-cratonization; and (3) smaller yet significant accretions of juvenile crust occurred during Proterozoic time, notably the early Proterozoic. The overall result was, in brief, the episodic separation of the crust from its mantle source, with progressive sialic development and cratonization to form the growing continents. It follows that the most profound changeover in mantle–crust relations occurred in late Archean time immediately prior to the early Proterozoic. Thermal mantle plumes were major players in the growth of the continental crust, with magmatic underplating an important process alongside lateral plate collisions and accretions.

The growth curve for developed continental crust across geologic time discussed above diverges sharply from that for preserved (now-measurable) continental crust (Fig. 6-7, middle curve). The latter curve, as presently drawn, is based on the data summarized in Table 1-2, with minor modifications (Goodwin 1991), with due allowance for Poldervaart's (1955) calculation to the effect that Earth's Precambrian crust constitutes 72% of Earth's total (Precambrian + Phanerozoic) continental crust.

The resulting control points on the preserved continental crust curve (Fig. 6-7, middle curve) are: 8% (estimated) at 3.0 Ga, 12% at 2.5 Ga, 30% at 1.8 Ga, 47% at 1.0 Ga and 72% at 0.6 Ga. The difference between the curves for developed and preserved continental crust (upper and middle curves respectively in Fig. 6-7) represents ostensibly missing or unidentified continental crust. The bulk of this ostensibly 'absent' fraction lies in successively younger mobile belts in the form of recycled and overprinted crust, in tectonically buried and otherwise unobserved or unrecognized sites, and in eroded and resedimented components upon or marginal to the Precambrian cratons themselves. This divergence in the two curves attains a maximum in late Archean–early Proterozoic times, converging thereafter to the present.

The salient feature of the preferred growth curve of developed continental crust (i.e. upper curve in Fig. 6-7) is that half of the total crust separated from its mantle source in the comparatively short (500 Ma) interval of late Archean time (3.0–2.5 Ga). The withdrawal of this large sialic-producing fraction dramatically altered the composition of the mantle source, rendering it geochemically less fertile and tectonically more inert, especially the subcratonic parts. Aggregation of the juvenile sialic crust into large, thick, tectonically stable cratons created appropriate environments both for (1) endogenous deep intracrustal melting, with the more highly mobile fraction ascending to form the K-granitoid-rich upper continental crust; and for (2) exogenous accumulation of distinctive platform sediments upon the newly formed stable continental shelves. The profound exogenic impact of the growth of continental shelves is reflected both in the dramatic rise in initial strontium ratios in carbonate rocks (Fig. 6-7), a direct outgrowth of elevated continental as opposed to dominant mantle contributions to sea water, and in an explosion in photosynthetically derived atmosphere–hydrosphere oxygen levels (Fig. 6-7, lowest curve), with profound impact on biologic evolution.

Further to the preferred growth curve of developed continental crust (Fig. 6-7, upper curve), the shoulder in the curve at 2.5 Ga, reflecting the sudden epi-Archean decline in growth rate following extravagant late Archean cratonic growth, is placed at the 65% level of present day crust; this is in response to the twin imperatives of the unique stature of the Archean-to-Proterozoic geologic transition on the one hand and, on the other, the continuing albeit greatly diminished role of juvenile crustal accretion during the Proterozoic, especially early Proterozoic, time. This preferred growth curve is therefore in essential accord with pertinent general geologic and geochemical parameters.

Eleven geologic milestones mark out the route of continental growth: (1) Mount Narryer zircons (West Australia) at 4.2 Ga, oldest recorded terrestrial material and product of nascent crust fractionation; (2) Slave Province and Napier Complex gneisses at ~3.9 Ga, the oldest known terrestrial rock masses; (3) Isua metasediments (Greenland), >3.9 Ga, with sialic detritus; (4) Kaapvaal craton (southern Africa), 3.1 Ga, oldest stable craton supporting epicratonic basins; (5) global plethora of large stable cratons by 2.6 Ga, product of spectacular sialic development–aggregation in late Archean time, viewed as the single most important continental event; (6) the rapid subsequent formation of K-granitoid-rich upper continental crust (2.6–2.5 Ga), product of pervasive intracrustal fractionation and potent agent of craton stability; (7) mega-BIF accumulation (2.5–2.0 Ga), in response to rising oxygen production; (8) Jormua ophiolite (Finland) at 1.9 Ga, amongst the oldest recognized obducted oceanic crust; (9) bimodal anorogenic suite (mantle-derived anorthosite and crust-derived rapakivi granite suites) at 1.5 Ga, product of long-sustained craton stability and linear subcrustal thermal regimes; (10) El Graara ophiolite (Bou Azzer, northwest Africa), 0.8 Ga, one of many late Proterozoic ophiolites; and (11) the twin association of glaciogenic diamictites and soft-bodied metazoan fauna (Ediacara) at 0.6 Ga.

### 6.5.2 CONCLUDING STATEMENT

The above model does not support the concept of early large-scale generation of continental crust (Hargraves 1976, Fyfe 1978, Lowman 1989, 1992) including that of a near steady-state, non-continental-growth Earth (Armstrong 1981, 1991), nor the strongly non-actualistic early sialic generation model of Shaw (1972, 1976, 1980). Neither does it support the early intermediate composition crustal model of Lowman (1984). In terms of episodic continental growth, it is in accord with models developed by Goodwin (1976), Condie (1980a, 1986b, 1989), Moorbath (1984), Windley (1984), Windley et al (1981), Kröner (1984, 1985a) and Fyfe (1978, 1993).

According to the preferred model, the operating tectonic processes changed significantly throughout geologic times, as did the nature of the evolving ocean–continental crust. It is unlikely that actualistic plate tectonic processes *per se* operated, especially in Archean times, under such different mantle–crust thermal–rheologic regimes. On the other hand, some modified form of mantle convection-driven tectonic process presumably operated in Archean time, e.g. lava-lake (–magma ocean) plate tectonics (Duffield 1972, Hamilton 1993) sagduction (Goodwin 1978), delamination (Kröner 1981, 1985a, Davies 1992, Kay and Mahlburg Kay 1993), hot-spot and/or magma underplating (Fyfe 1978, 1993, Hamilton 1993, Passchier 1994, Reymer and Schubert 1984). The exact nature of these predecessor processes, which probably involved distinctive oceanic crust, a higher convection turnover and comparatively small, restricted cratons during much of Archean time (Lowe 1994), is highly conjectural. All that can be reasonably concluded, in the apparent absence of certain critical modern plate tectonic signatures, is that the operating Archean processes were, at least in detail, non-actualistic.

By early Proterozoic time, however, the continental crust was well developed, with the mantle approaching modern day status. The late-era geologic record (by 1.9 Ga) includes at least local evidence of modern plate tectonic operation (Hoffman 1980, 1989, Kontinen 1987) with evidence for long-lived, coupled crust–mantle systems (Goodwin 1985, Groves et al 1987). By this time, then, the door is open for the operation of a range of crustal processes, including those inherited from Archean time successively joined by progressive plate-tectonic innovations including the marginal accretion of orogenic belts by collisional tectonics (Bird and Dewey 1970, Hsu 1979, Hamilton 1989, 1993, Ellis 1992), and the lateral accretion of allochthonous, displaced, exotic or suspect terranes (Coney et al 1980, Howell 1989, Hamilton 1990). In this regard, the early to mid-Proterozoic record includes strong evidence in favour of the significant presence of ensialic fold belts, for which a number of exclusively intracratonic processes have been proposed, including, for example, delamination of the underlying lithosphere (Kröner 1981, 1985a), a view not shared by Ellis (1992) however. Thus a spectrum of tectonic possibilities is available, a flexibility of interpretation that is in accord with the mixed, essentially transitional nature of the early- to mid-Proterozoic record.

On the other hand, by the late Proterozoic there is widespread evidence—in the form of combined ophiolite, blue schist, arc-type volcanic and abundant calc-alkalic igneous associations—of the operation of subduction-style plate tectonics and, at least in the future Atlantic region, of repeated Wilson cycles. Thus, by 1.0 Ga at least, a critical thermal watershed had been reached, past which the operating plate tectonic process was pronounced actualistic.

A major challenge, then, is to sort out the nature of the changing tectonic processes across Precambrian time, a challenge that requires constant attention to, and interpretation of, the preserved Precambrian geologic record. Correctly interpreting this record in the light of modern plate tectonics (Hamilton 1990) requires a judicious blend of application, adaptation, rejection and innovation, the first two warranted in the presence of certain critical signatures, the last two in their absence. The danger in the complacent application of a popular paradigm lies in unwarranted speculation based on invalid models. All models carry with them a particular viewpoint that produces sharp focus on certain aspects but blurred vision elsewhere.

Comparative planetology is likewise relevant to our understanding of the origin and development of Earth's continental crust (Hamilton 1993). It is now known that the silicate planets, including Moon, Mars, Mercury, Venus and the meteorite parent bodies, underwent early 'global' differentiation (i.e. non-plate-tectonic style as opposed to patchy, protocontinental style) by igneous processes during or shortly after their accretion, followed by basin-forming heavy meteorite bombardment to ~3.9 Ga, and long basaltic 'second' differentiation to 2.5 Ga, which marks the terminal event of Moon and Mercury and possibly Mars but not of Venus and Earth (Lowman 1989). Indeed, the surface of Venus shows continental-like plateaux, but the low population of impact craters and the occurrence of undoubted volcanic land forms show them to be relatively young constructs rather than primitive crust (Head and Crumpler 1987). Indeed, the formation of corona, large, complex, generally circular, volcanic-tectonic structures on Venus and Mars from mantle diapirs suggests that similar structures may have formed in Earth's lithosphere (Watters and James 1995). Herrick (1994) proposes that geologic activity on Venus changed abruptly at ~800 Ma, from globally highly mobile and heavily deforming, through a brief period of global volcanic flooding to the present activity concentrated around large volcanic-tectonic centres. This marks a changeover from the cessation of "plate-tectonics" and subsequent shift to "hot-spot tectonics" as mantle cooling made the lithosphere positively buoyant (Herrick 1994). Thus fundamental changeovers in tectonic style within a planet are to be expected. Did Earth experience similar changes in subduction-controlling lithosphere buoyancy? The physiography of Mars, though complicated by extensive volcanism, incipient crustal fragmentation, and the effects of erosion and deposition, shows little evidence of patches of ancient crust that could be construed as continental nuclei. Comparable products appear to be present on Mercury and Mars (Lowman 1989). Did Earth follow a similar evolution with development of an originally global primordial crust, the result of an evolutionary sequence shared by the other silicate planets to some degree, with the early differentiated crust later destroyed by bombardment and other processes? Was there a comparable basaltic 'second' differentiation phase on Earth and did it involve massive plume-generated basaltic underplating and overplating? At what stage in Earth's crustal evolution did actualistic plate tectonics enter the scene and how many predecessor processes and variations were there, leading up to the modern subduction-style process? What is the significance, if any, of sharply competing tectonic theories such as surge tectonics (Meyerhoff et al 1992)?

Regarding current initial explorations of the outer planets and their spectacularly novel observations, the only certainty is that further exploration in space will also reveal the completely unexpected. As with our expanding exploration in outer space, so with our continuing search back across Precambrian time. To restrict the geologic interpretation on Earth at any stage to one or other paradigm alone would be to hobble the search for potentially critical evidence which otherwise might lead, step-by-step, to the eventual understanding of novel predecessor tectonic processes which undoubtedly characterized, stage-by-stage, the dynamic evolution of Earth's continental crust.

# References

Abbot D H and Hoffman S E (1984) Archean plate tectonics revisited. Heat flow, spreading rate, and the age of subducting oceanic lithosphere and their effects on the origin and evolution of continents. Tectonics 3: 429–448

Affaton P, Rahaman M A, Trompette R and Sougy J (1991) The Dahomeyide Orogen: tectonothermal evolution and relationships with the Volta basin. In: The West African Orogens and Circum-Atlantic Correlatives (eds Dallmeyer R D and Lecorche J P) Springer-Verlag, Berlin: 107–122

Affaton P, Sougy J and Trompette R (1980) The tectono-stratigraphic relationship between the upper Precambrian and Lower Volta Basin and the Pan-African Dahomeyide orogenic belt (West Africa). Am J Sci 280: 224–248

Ager D V (1980) The Geology of Europe. McGraw-Hill, London

Ahmad T and Tarney J (1994) Geochemistry and petrogenesis of late Archaean Aravalli volcanics, basement enclaves and granitoids. Prec Res 65: 1–23

Aksenov Y M, Keller B M, Sokolov B S et al (1978) A general scheme for upper Precambrian stratigraphy of the Russian Platform. Int Geo Rev 22: 444–457

Aldiss DT and Carney J N (1992) The geology and regional correlation of the Proterozoic Okwa Inlier, western Botswana. Prec Res 56: 255–274

Allaart J H (1976) Kotilidion mobile belt in South Greenland. In: Geology of Greenland (eds Escher A and Watt WS). The Geological Survey of Greenland, Copenhagen, Denmark, pp 120–151

Allen P, Condie K and Bowing G P (1986) Geochemical characteristics and possible origins of the southern Closepet Batholith, South India. J Geol 94: 283–299

Almeida F F M and Hasui Y (1984) O Pré-Cambriano de Brasil. Edigar Blucher São Paulo

Almeida F F M, Hasui Y, de Brito Neves B B and Fuck R A (1981) Brazilian structural provinces: an introduction. In: The Geology of Brazil (eds Mabesoone J M, de Brito Neves B B and Sial A N). Earth Sci Rev 17: 1–29

Alvi S H and Raza M (1991) Nature and magma type of Jagannathpur Volcanics, Singhbhum, eastern India. J Geol Soc Ind 38: 524–531

Ames L, Titton GR and Zhou G (1993) Timing of collision of the Sino–Korean and Yangtse cratons: U–Pb zircon dating of coesite-bearing eclogites. Geol 21: 339–342

Andersen L S and Unrug R (1984) Geodynamic evolution of the Bangweulu Block, northern Zambia. Prec Res 25: 187–212

Anderson D L (1982) Hotspots, polar wander, Mesozoic convection and the geoid. Nature 297: 391–393

Anderson D L (1989) Comp Earth Sci 243: 367–370

Anderson J G C (1965) The Precambrian of the British Isles. In: The Precambrian Vol. 2 (ed. Rankama K). Interscience Publishers, New York, pp 25–111

Anderson J G C (1978) The Structure of Western Europe. Pergamon Press, Oxford

Anderson J L (1983) Proterozoic anorogenic granite plutonism of North America. In: Proterozoic Geology: Selected Papers from an International Proterozoic Symposium (eds Medaris L G, Byers C W, Mickelson D M and Shanks W C). Geol Soc Am Memoir 161: 133–154

Anderson J L and Bender E E (1989) Nature and origin of Proterozoic A-type granitic magmatism in the southwestern United States of America. Lithos 23: 19–52

Anderson J L, Cullers R L and Van Schmus W R (1980) Anorogenic metaluminous and peraluminous granite plutonism in the mid-Proterozoic of Wisconsin USA. Contrib Mineral Petrol 74: 311–328

Andersson U B (1991) Granitoid episodes and mafic-felsic magma interaction in the Svecofennian of the Fennoscandian Shield, with main emphasis on the ~1.8 Ga plutonics. Prec Res 51: 127–149

Anderton R, Bridges P M, Leader M R (1979) A dynamic stratigraphy of the British Isles. George Allen and Unwin, London, 301 p

Andreoli M A G (1984) Petrochemistry, tectonic evolution and metasomatic mineralizations of Mozambique belt granulites from S. Malawi and Tete (Mozambiquie). Prec Res 25: 161–186

Andreoli M A G, Smith C B, Watkeys M, Moore J M, Ashwal L D and Hart R J (1994) The geology of the Steenkampskraal monazite deposit, South Africa: Implications for REE–Th–Cu mineralization in charnockite-granulite terranes. Econ Geol 89: 994–1016

Andrews-Speed C P (1989) The mid-Proterozoic Mporokoso Basin, northern Zambia: sequence stratigraphy, tectonic setting and potential for gold and uranium mineralisation. Prec Res 44: 1–17

Anhaeusser C R (ed.) (1983) Contributions to the geology of the Barberton Mountain Land. Geol Soc S Africa Spec Public 9

Anhaeusser C R (1984) Structural elements of the Archaean granite-greenstone terranes as exemplified by the Barberton Mountain Land, southern Africa. In: Precambrian Tectonics Illustrated (eds Kröner A and Greiling R). Schweizerbart'sche Verlagsbuchhandlung, Stuttgart, pp 57–78

Anhaeusser C R (1985) Archean layered ultramafic complexes in the Barberton Mountain Land, South Africa. In: Evolution of Archean Supracrustal Sequences (eds Ayres L D, Thurston P C, Card K D and Weber W). Geol Ass Can Spec Pap 28: 281–301

Anhaeusser C R (1992) Structures in granitoid gneisses and associated migmatites close to the granulite boundary of the Limpopo Belt, South Africa. Prec Res 55: 81–92

Anhaeusser C R and Burger A J (1982) An interpretation of U-Pb zircon ages for Archaean tonalite gneisses from the Johannesburg-Pretoria granite dome. Trans Geol Soc S Afr 85: 111–116

Anhaeusser C R and Maske S (eds) (1986) Mineral Deposits of South Africa. Geol Soc S Afr, Spec Public, vols. 1 and 2: 2335 pp

Anhaeusser C R and Robb L J (1981) Magmatic cycles and the evolution of the Archaean granitic crust in the eastern Transvaal and Swaziland. In: Archaean Geology (eds Glover J E and Groves D I). Geol Soc Austral Spec Public 7: 457–467

Annhaeusser C R (1992) Structures in granitoid gneisses and

associated migmatites close to the granulite boundary of the Limpopo Belt, South Africa. Prec Res 55: 81–92

Annovitz L M and Essene E J (1990) Thermobarometry and pressure-temperature paths in the Grenville Province of Ontario. J Petrol 31: 197–241

Ansdell K M and Kyser T K (1991) Plutonism, deformation and metamorphism in the Proterozoic Flin Flon greenstone belt, Canada: Limits on timing provided by the single-zircon Pb-evaporation technique. Geol 19: 518–521

Armstrong R A, Compston W, Retief E A, Williams I S and Weke H J (1991) Zircon ion microprobe studies bearing on the age and evolution of the Witwatersrand triad. Prec Res 53: 243–266

Armstrong R L (1981) Radiogenic isotopes: the case for crustal recycling on a near-steady-state no-continental-growth Earth. Philos Trans R Soc Lond (Ser A) 301: 443–472

Armstrong R L (1991) The persistent myth of crustal growth. Austral J Earth Sci 38: 613–630

Arndt N T (1983) Role of thin, komatiite-rich crust in the Archean plate tectonic process. Geol 11: 372–375

Arndt N T and Goldstein S L (1987) Use and abuse of crust-formation ages. Geol 15: 893–895

Arndt N T, Nelson D R, Compston W, Trendall A F and Thorne A M (1991) The age of the Fortescue Group, Hamersley Basin, Western Australia, from ion microprobe zircon U-Pb results. Austral J Earth Sci 38: 261–281

Arora M and Naqvi S M (1993) Geochemistry of Archaean arenites formed by anoxic exogenic processes—An example from Bababudan Schist belt, India. J Geol Soc Ind, 42: 247–268

Ashton K E (1982) Further geological studies of the Woodburn Lake Group, northwest of Tehek Lake, District of Mackenzie. Geol Surv Can Pap 82-lA: 151–157

Ashwal L D, Wiebe R A, Wooden J L, Whitehouse M J and Snyder D (1992) Pre-Elsonian mafic magmatism in the Nain igneous complex, Labrador: the Bridges layered intrusion. Prec Res 56: 73–87

Aubouin J (1980) Geology of Europe: a synthesis. Episodes 1980, No. 1: 3–8

Auvray B (1979) Genèse et évolution de la croûte continentale dans le nord du Massif Armoricain. University of Rennes, unpublished thesis

Auvray B, Blais S, Jahn B-M and Piquet D (1982) Komatiites and the komatiitic series of the Finnish greenstone belts. In: Komatiites (eds Arndt N T and Nisbet E G). Allen and Unwin, London, pp 131–146

Avramtchev L (1985) Carte géologique du Québec, carte 2000 (échelle 1:1500000). Québec Min Rich Nat DV 84–92

Awramik S M, Schopf J W and Walter M R (1983) Filamentous fossil bacteria from the Archean of Western Australia. Prec Res 20: 357–374

Ayres L D (1978) Metamorphism in the Superior Province of northwestern Ontario and its relationship to crustal development. In: Metamorphism in the Canadian Shield, (eds Fraser A J and Heywood W W) Geol Surv Can Pap 78-10: 25–36

Ayres L D and Thurston P C (1985) Archean supracrustal sequences in the Canadian Shield: an overview. In: Evolution of Archean Supracrustal Sequences (eds Ayres L D, Thurston P C, Card K D and Weber W). Geol Ass Can Spec Pap 28: 343–380

Baadsgaard H (1976) Further U-Pb dates on zircons from the early Precambrian rocks of the Godthåbsfjord area, West Greenland. Earth Planet Sci Lett 33: 261–267

Baadsgaard H and McGregor V R (1981) The U-Th-Pb systematics of zircons from the type Nûk gneisses, Godthåbsfjord, West Greenland. Geochim Cosmochim Acta 45: 1099–1109

Baadsgaard H, Nutman A P, Bridgwater D et al (1984) The zircon geochronology of the Akilia association and Isua supracrustal belt, West Greenland. Earth Planet Sci Lett 68: 221–228

Baker B H (1967). The Precambrian of the Seychelles Archipelago In: The Precambrian, Vol. 3: (ed Rankama K). Interscience Publisher, New York, pp 123–132

Balashov Y A, Bayanova T B and Mitrofanov F P (1993) Isotope data on the age and genesis of layered basic–ultrabasic intrusions in the Kola Peninsula and northern Karelia, northeastern Baltic Shield. Prec Res 64: 197–205

Bally A W, Scotese C R and Ross M I (1989) North America; Plate-tectonic setting and tectonic elements. In: The Geology of North America—An Overview. Geol Soc Am, The Geology of North America, A: 1–15

Banerjee D M and Bhattacharya P (1994) Petrology and geochemistry of greywackes from the Aravalli Supergroup, Rajasthan, India and the tectonic evolution of a Proterozoic sedimentary basin. Prec Res 67: 11–35

Banerjee D M, Deb M and Strauss H (1992) Organic carbon isotopic composition of Proterozoic sedimentary rocks of India: preliminary results. In: Early Organic Evolution: Implications for Mineral and Energy Resources (eds. Schidlowski M, Golubic S, Kimberley M M, McKirdy D M and Trudinger P A). Springer-Verlag, Berlin: 232–240

Baragar W R A and Scoates R F J (1981) The Circum-Superior Belt: a Proterozoic plate margin? In: Precambrian Plate Tectonics (ed Kröner A). Elsevier, Amsterdam, pp 297–330

Baragar W R A and Scoates R F J (1987) Volcanic geochemistry of the northern segments of the Circum-Superior Belt of the Canadian Shield. In: Geochemistry and Mineralization of Proterozoic Volcanic Suites, (eds Pharaoh T C, Beckinsale R D and Rickard D). Geol Soc Spec Public 33: 113–131

Barbey P, Convert J, Moreau B et al (1984) Petrogenesis and evolution of an early Proterozoic collisional orogenic belt: the granulite belt of Lapland and the Belomorides (Fennoscandia). Bull Geol Soc Fin 56: 161–168

Barghoorn E S and Tyler S A (1965) Microorganisms from the Gunflint chert. Sci 147: 563–577

Barker F and Arth J (1976) Generation of trondhjemitic-tonalitic liquids and Archean bimodal trondhjemitic-basalt suites. Geol 4: 596–600

Barley M E (1993) Volcanic, sedimentary and tectonostratigraphic environments of the ~3.46 Ga Warrawoona Megasequence: a review. In: Archean and Early Proterozoic Geology of the Pilbara Region, Western Australia (eds Blake T S and Meakins A L). Prec Res 60: 47–67

Barley M E and Groves D I (1992) Supercontinent cycles and the distribution of metal deposits through time. Geol 20: 291–294

Barley M E, Groves D I and Blake T S (1992) Archaean metal deposits related to tectonics: Evidence from Western Australia. In: The Archaean: Terrains, Processes and Metallogeny (eds. Glover J E and Ho S E). The Univ W Austral. Public, 22: 307–324

Barley M E, McNaughton N J, Williams I S and Compston W (1994) Age of Archaean volcanism and sulphide mineralization in the Whim Creek Belt, west Pilbara. Austral J Earth Sci 41: 175–177

Barnes R G and Willis I L (1989) The stratigraphic setting of the Pb–Zn–Ag mineralization at Broken Hill—a discussion. Econ Geol 84: 188–191

Barovich K M, Patchett P J, Peterman Z E and Sims P K (1989) Nd isotopes and the origin of 1.9–1.7 Ga Penokean continental crust of the Lake Superior region. Geol Soc Am Bull 101: 333–338

Barrett T J, Cattalani S, Hoy L, Riopel J and Lafleur P-J (1992) Massive sulfide deposits of the Noranda area, Quebec. IV. The Mobrun mine. Can J Earth Sci 29: 1349–1374

Barrie C T, Ludden J N and Green T H (1993) Geochemistry of volcanic rocks associated with Cu–Zn and Ni–Cu deposits in the Abitibi subprovince. Econ Geol 88: 1341–1358

Barton E S and Burger A J (1983) Reconnaissance isotopic investigations in the Namaqua mobile belt and implications for

Proterozoic crustal evolution—Upington geotraverse. Spec Public Geol Soc S Afr 10: 173–191

Barton E S, Atermann W, Williams I S and Smith C B (1994) U–Pb zircon age for a tuff in the Campbell Group, Griqualand West Sequence, South Africa: Implications for Early Proterozoic rock accumulation rats. Geol 22: 343–346

Barton J M (1981) The pattern of Archean crustal evolution in southern Africa as deduced from the evolution of the Limpopo Mobile Belt and the Barberton granite-greenstone terrain. In: Archean Geology (eds Glover J E and Groves D I). Geol Soc Austral Spec Publ 7: 21–31

Barton J M and Key R M (1981) The tectonic development of the Limpopo Mobile Belt and the evolution of the Archean craton of southern Africa. In: Precambrian Plate Tectonics, (ed Kröner A). Elsevier, Amsterdam: 185–212

Barton J M and Van Reenen D D (1992) When was the Limpopo Orogeny? Prec Res 55: 7–16

Barton J M, Holzer L, Kamber B and three others (1994) Discrete metamorphic events in the Limpopo belt, southern Africa: Implications for application of P–T paths in complex metamorphic terrains. Geol 22: 1035–1038

Basu A K (1986) Geology of parts of the Bundelkhand Granite Massif, central India. Geol Surv Ind 117: 61–124

Bauer P W and Williams M L (1994) The age of Proterozoic orogenesis in New Mexico, U.S.A. Prec Res 67: 349–356

Bayley R W and James H L (1973) Precambrian iron-formations of the United States. Econ Geol 68: 934–959

Bayly B (1990) The Vredefort structure: estimates of energy for some internal sources and processes. Tectonophys 171: 153–167

Beakhouse G P (1985) The relationship of supracrustal sequences to a basement complex in the western English River Subprovince. In: Evolution of Archean Supracrustal Sequences (eds Ayres L D, Thurston P C, Card K D and Weber W). Geol Ass Can Spec Pap 28: 169–178

Beardsmore T J, Newbery S P and Laing W P (1988) The Maronan Supergroup: an inferred early volcanosedimentary rift sequence in the Mount Isa Inlier, and implications for ensialic rifting in the middle Proterozoic of Northwest Queensland. Prec Res 40/41: 487–507

Beckinsale R D, Drury S A and Holt R W (1980a) 3360 Myr old gneisses from the South Indian craton. Nature 283: 469–470

de Beer J H and Stettler E H (1992) The deep structure of the Limpopo Belt from geophysical studies. Prec Res 55: 173–186

Beeson J, Delor C and Harris L (1988) A structural and metamorphic transverse across the Albany Mobile Belt, Western Australia. Prec Res 40/41: 117–136

Behrendt J C, Green A G, Cannon W F et al (1988) Crustal structure of the Midcontinent rift system: results from GLIMPCE deep seismic reflection profiles. Geol 16: 81–85

Belevtsev Ya N, Belevtsev R Ya and Siroshtan R I (1983) The Krivoy Rog Basin. In: Iron-Formation: Facts and Problems (eds Trendall A F and Morris R C). Dev Prec Geol 6. Elsevier, Amsterdam, pp 211–251

Bell R and Jefferson C W (1987) An hypothesis for an Australian–Canadian connection in the Late Proterozoic and the birth of the Pacific Ocean. PacRim Congress 1987, Parkville, Australia: 39–50

Bell T H (1992) The role of thrusting in the structural development of the Mount Isa Mine and its relevance to exploration in the surrounding region. Econ Geol 86: 1602–1625

Ben Otham D, Polve M and Allègre C J (1984) Neodymium strontium composition of granulites: constraints on the evolution of the lower crust. Nature 307: 510–516

Bennett V C and De Paolo D J (1987) Proterozoic crustal history of the western United States as determined by neodymium isotopic mapping. Geol Soc Am, Bull 99: 674–685

Bennett V C, Nutman A P and McCulloch M T (1993) Nd isotopic evidence for transient, highly depleted mantle reservoirs in the early history of the Earth. Earth Planet Sci Lett 119: 299–317

Bergeron R (1957) Late Precambrian rocks of the north shore of the St Lawrence River and of the Mistasini and Otish Mountain areas, Quebec. In: The Proterozoic of Canada. R Soc Can Spec Public 2: 101–111

Bernard-Griffiths J, Peucat J J, Postaire B et al (1984) Isotopic data (U-Pb, Rb-Sr, Pb-Pb and Sm-Nd) on mafic granulites from Finnish Lapland. Prec Res 23: 325–348

Bernasconi A (1983) The Archaean terranes of central eastern Brazil: a review. Prec Res 23: 107–131

Berrangé J P (1982) The Eastern Bolivia mineral exploration project 'Proyecto Precambrico'. Episodes 1982(4): 3–8

Bertrand J M L and Lasserre M (1976) Pan-African and pre-Pan-African history of the Hoggar (Algerian Sahara) in the light of new geochronological data from the Aleksod area. Prec Res 3: 343–362

Bertrand J M, Michard A, Dautel D and Pillot M (1984) Ages U-Pb éburnéens et pan-africains au Hoggar central (Algérie). Conséquences géodynamiques. C R Acad Sci 298 sér 2: 643–646

Bertrand J M, Michard A, Boullier A M and Dautel D (1986) Structure and U-Pb geochronology of the Central Hoggar (Algeria): a reappraisal of its Pan-African evolution. Tectonics 5: 955–972

Bertrand J M, Roddick J C, Van Kronendonk M J and Ermanovics I (1993) U-Pb geochronology of deformation and metamorphism across a central transect of the Early Proterozoic Torngat Orogen, North River map area, Labrador. Can J Earth Sci 30: 1470–1489

Bertrand-Safarti J and Potin B (1994) Microfossiliferous cherty stromatolites in the 2000 Ma Franceville Group, Gabon. Prec Res 65: 341–356

Besairie H (1967) The Precambrian of Madagascar. In: The Precambrian, Vol. 3 (ed. Rankama K). Interscience Publishers, New York, pp 133–142

Besairie H (1971) Carte géologique au 1/2000000 et notice explicative. Docum Bur Géol Madagascar, No. 184

Bessoles B (1977) Géologie de l'Afrique. Le craton Ouest African. Mem Bur Rech Géol Min Paris 88

Bessoles B and Lasserre M (1978) Le complexe de base du Cameron. Bull Soc Géol Fr XIX: 1083–1090

Bessoles B and Trompette R (1980) Géologie de l'Afrique. La chaîne panafricaine 'zone mobile d'Afrique central (partie sud) et zone mobile soudanaise'. Mem BRGM, Orléans 92

Bettencourt J S et al (1987) The Rondonian tin-bearing anorogenic granites and associated mineralization. Int Symp on Granites and Associated Mineralizations, Excursion Guides, Salvador, Brazil, pp 49–87

Beukes N J (1983) Palaeoenvironmental setting of iron-formations in the depositional basin of the Transvaal Supergroup, South Africa. In: Iron-Formation: Facts and Problems (eds Trendall A F and Morris R C). Dev Prec Geol 6. Elsevier, Amsterdam, pp 131–209

Bevier M L and Gebert J S (1991) U-Pb geochronology of the Hope Bay–Elu Inlet area, Bathurst Block, northeastern Slave Structural Province, Northwest Territories. Can J Earth Sci 28: 1925–1930

Bibikova E V (1984) The most ancient rocks in the USSR territory by U-Pb data on accessory zircons. In: Archean Geochemistry, (eds Kröner A, Hanson G N and Goodwin A M). Springer-Verlag, Berlin, 235–250

Bibikova E V and Krylov I N (1983) Isotopic age of Archean acid volcanic rocks. DAN USSR 268: 1231–1234 (in Russian)

Bibikova E V, Grinenko V A, Kiselevsky M A and Shokolyukov V A (1982b) Geochronological and oxygen-isotope study of the granulites of the USSR. Geochem 12: 1718–1782 (in Russian)

Bibikova E V, Shul-diner V I, Gracheva T V et al (1984) Isotopic age of granulites on the West of Stanovoy region. Daklady Academii Nauk SSSR 276(6): 1471–1474

Bibikova E V, Drugova G M, Duk V L and Leysky L (1986) Geochronology of the Aldan-Vitim Shield. In: Isotopic Methods in Geology and Geochronological Scale. (ed. Yu A Shukolyukov). Izd 135–159, Nauka, Moscow (in Russian)

Bibikova E V, Belov A N, Gracheva T M et al (1987) On the age of the granulite metamorphism of the Anabar Shield. In Isotopic Dating of Metamorphic and Metasomatic Processes. Transactions of the 23rd Session on Isotope Geochron. Nauka, Moscow, pp 71–85 (in Russian)

Bickford M E (1984) U-Pb zircon chronology of early and middle Proterozoic igneous events in the Gunnison, Salida, and Wet Mountains areas, Colorado. Geol Soc Am (Abs with Programs) 16: 215

Bickford M E (1988) The formation of continental crust. Geol Soc Am Bull 100: 1375–1391

Bickford M E, Van Schmus W R and Zietz I (1986) Proterozoic history of the midcontinent region of North America. Geol 14: 492–496

Bickford M E, Collerson K D and Lewry J F (1994) Crustal history of the Rae and Hearne provinces, southwestern Canadian Shield, Saskatchewan: constraints from geochronological and isotopic data. Prec Res 68: 1–21

Bickle M J (1984) Variation in tectonic style with time: Alpine and Archean systems. In: Patterns of Change in Earth History, (eds Holland H D and Trendall A F). Dahlem Konferenzen, 1984. Springer-Verlag, Berlin, pp 357–370

Bickle M J and Eriksson K A (1982) Evolution and subsidence of early Precambrian sedimentary basin. Trans R Soc Lond 305: 225–247

Bickle M J and Nisbet E G (eds) (1993) The geology of the Belingwe greenstone belt, Zimbabwe. Geol Soc Zimbabwe, Spec Publicic 2: 239 pp

Bickle M J, Morant P, Bettenay L F et al (1985) Archean tectonics of the Shaw Batholith, Pilbara Block, Western Australia: structural and metamorphic tests of the batholith concept. In: Evolution of Archean Supracrustal Sequences (eds Ayres L D, Thurston P C, Card K D and Weber W). Geol Ass Can Spec Pap 28: 325–341

Bickle M J, Bettenay L F, Campbell I et al (1986) Formation of the continental crust—the first billion years of the Pilbara Archean Terra Cognita 6: 125

Bickle M J, Bettenay L F, Chapman H J, Groves D I, McNaughton N J, Campbell I H and de Laeter J R (1993) Origin of the 3500–3300 Ma calc-alkaline rocks in the Pilbara Archaean: isotopic and geochemical constraints from the Shaw Batholith. In: Archean and Early Proterozoic Geology of the Pilbara Region, Western Australia (eds. Blake T S and Meakins A L). Prec Res 60: 117–149

Bickle M J, Nisbet E G and Martin A (1994) Archean greenstone belts are not oceanic crust. J Geol 102: 121–138

Bird J M and Dewey J F (1970) Lithosphere plate-continental margin tectonics and the evolution of the Appalachian Orogen. Geol Soc Am Bull 81: 1031–1060

Bischoff A A (1988) The history and origin of the Vredefort Dome. S Afr J Sci 84: 413–417

Bishop A C, Road R A and Adams C J D (1975) Precambrian rocks within the Hercynides. Geol Soc Spec Rep 6: 102–107

Bjørnerud M, Craddock C and Wills C J (1990) A major late Proterozoic tectonic event in southwestern Spitsbergen. Prec Res 48: 157–165

Black L P (1988) Isotopic resetting of U-Pb zircon and Rb-Sr and Sm-Nd whole-rock systems in Enderby Land, Antarctica: implications for the interpretation of isotopic data from polymetamorphic and multiply deformed terrains. Prec Res 38: 355–365

Black L P and James P R (1983) Geological history of the Archaean Napier Complex of Enderby Land. In: Antarctic Earth Science (eds Oliver R L, James P R and Jago J B). Cambridge University Press: 11–15

Black L P, Moorbath S, Pankhurst R J and Windley B R (1973) $^{207}Pb/^{206}Pb$ whole-rock age of the Archean granulite facies metamorphic event in West Greenland. Nature 244: 50–53

Black L P, Williams I S and Compston W (1986a) Four zircon ages from one rock: the complex history of a 3930 Ma old granulite from Mount Sones, Enderby Land, Antarctica. Contrib Mineral Petrol 94: 427–437

Black L P, Sheraton J W and James P R (1986b) Late Archean granites of the Napier Complex, Enderby Land, Antarctica: a comparison of Rb-Sr, Sm-Nd and U-Pb isotopic systematics in a complex terrain. Prec Res 32: 343–368

Black L P, Harley S L, Sun S S and McCulloch M T (1987) The Rayner Complex of East Antartica: complex isotopic systematics within a Proterozoic mobile belt. J Metam Geol 5: 1–26

Black L P, Kinny P D, Sheraton J W and Delor C P (1991) Rapid production and evolution of late Archaean felsic crust in the Vestfold Block of East Antarctica. Prec Res 50: 283–310

Black R (1980) Precambrian of West Africa. Episodes 4: 3–8

Black R and Liégeois J P (1993) Cratons, mobile belts, alkaline rocks and the continental lithospheric mantle: the Pan-African testimony. J Geol Soc Lond 150: 89–98

Black R, Ba H, Ball E et al (1979a) Outline of the Pan-African geology of Adrar des Iforas (Republic of Mali). Geol Rdsch 68: 543–564

Black R, Caby R, Moussine-Pouchkine A, Bayer R et al (1979b) Evidence for late Precambrian plate tectonics in West Africa. Nature 278: 223–227

Black R, Latouche L, Liégeois J P, Caby R and Bertrand J M (1994) Pan-African displaced terranes in the Tuareg shield (central Sahara). Geol 22: 641–644

Blackburn C E, Bond W D, Breaks F W et al (1985) Evolution of Archean volcanic-sedimentary sequences of the western Wabigoon Subprovince and its margins: a review. In: Evolution of Archean Supracrustal Sequences, (eds Ayres L D, Thurston P C, Card K D and Weber W). Geol Ass Can Spec Pap 28: 89–116

Blackburn W H and Srivastava D C (1994) Geochemistry and tectonic significance of the Ongarbira metavolcanic rocks, Singhbum District, India. Prec Res 67: 181–206

Blais S (1989) Les ceintures de roches vertes archéennes de Finlande orientale: géologie, pétrologie, géochimie et évolution géodynamique. Mem Docum Centre Armoricain d'Etude Structural des Socles 22

Blake D H (1980) Volcanic rocks of the Paleohelikian Dubawnt Group in the Baker Lake-Angikuni Lake area, District of Keewatin, NWT Geol Surv Can Bull 309

Blake D H (1987) Geology of the Mount Isa Inlier and environs, Queensland and Northern Territory. Austral Bur Miner Resour Bull 225

Blake D H and Groves D I (1987) Continental rifting and the Archean-Proterozoic transition. Geol 15: 229–232

Blake D H and McNaughton N J (1984) A geochronological framework for the Pilbara region. Univ Western Australia, Geology Department and University Extension, Public 9: 1–22

Blake D H and Page R W (1988) The Proterozoic Davenport Province, Central Australia; Regional geology and geochronology. Prec Res 40/41: 329–340

Blenkinsop T G, Fedo C M, Bickle M J, Eriksson K A, Martin A, Nisbet E G and Wilson J F (1993) Ensialic origin of the Ngezi Group, Belingwe greenstone belt, Zimbabwe. Geol 21: 1135–1138

Blignault H J, van Aswegen G, Der Merwe S W and Colliston W P (1983) The Namaqualand Geotraverse and environs: part of the Proterozoic Namaqua Mobile Belt. Spec Public Geol Soc S Africa 10: 1–29

Boak J L and Dymek R F (1982) Metamorphism of the ca. 3800 Ma supracrustal rocks at Isua, West Greenland: implications for the early Archaean crustal evolution. Earth Planet Sci Lett 59: 155–176

Boardman S J and Condie K C (1986) Early Proterozoic bimodal volcanic rocks in central Colorado, USA, Part 2: geochemistry, petrogenesis and tectonic setting. Prec Res 34: 37–68

Boer R H (1991) Bibliography on the geology of the Okiep copper district, Namaqualand, South Africa, 1685–1990. Econ Geol Res Unit, Univ Witwatersrand, Johannesburg, Info Circ 241, 15 pp

Boerboom T J and Zartman R E (1993) Geology, geochemistry, and geochronology of the central Giants Range batholith, northeastern Minnesota. Can J Earth Sci 30: 2510–2522

Bogdanova S V and Bibikova E V (1993) The "Saamian" of the Belomorian Mobile Belt: new geochronological constraints. Prec Res 64: 131–152

Bogdanova S V, Bibikova E V and Gorbatschev R (1994) Palaeoproterozoic U–Pb zircon ages from Belorussia: new geodynamic implications for the East European Craton. Prec Res 68: 231–240

Bonavia F F and Chorowicz J (1993) Neoprotozoic structures in the Mozambique Orogenic belt of southern Ethiopia. Prec Res 62: 307–322

Bond G C, Nickerson P A and Kominz M A (1984) Breakup of a supercontinent between 625 Ma and 555 Ma: New evidence and implications for continental histories. Earth Planet Sci Lett 70: 325–345

Bonhomme M G, Gauthier-Lafaye F and Weber F (1982) An example of lower Proterozoic sediments: the Francevillian in Gabon. Prec Res 18: 87–102

Borg G (1988) The Koras-Sinclair-Ghanzi Rift in southern African. Volcanism, sedimentation, age relationships and geophysical signature of a late middle Proterozoic rift system. Prec Res 38: 75–90

Borg S G and DePaolo D J (1994) Laurentia, Australia, and Antarctica as a Late Proterozoic supercontinent: Constraints from isotopic mapping. Geol 22: 307–310

Bouillier A M (1991) The Pan-African Trans-Saharan belt in the Hoggar shield (Algeria, Mali, Niger): A review. In: R D Dallmeyer and J P Lécorché (eds). The West African Orogens and Circum-Atlantic Correlatives. Springer-Verlag, Berlin

Bowen T B, Marsh J S, Bowen M P and Eales H V (1986) Volcanic rocks of the Witwatersrand Triad, South Africa. I: description, classification and geochemical stratigraphy. Prec Res 31: 297–324

Bowes D R (1980a) Structural sequence in the gneissose complex of eastern Finland as a basis for correlation in the Presvecokarelides. Acta Geol Pol 30: 15–26

Bowes D R, Halden N M, Koistinen T J and Park A F (1984) Structural features of basement and cover rocks in the eastern Svecokarelides, Finland. In: Precambrian Tectonics Illustrated (eds Kröner A and Greiling R). Schweizerbart'sche Verlagsbuchhandlung, Stuttgart, pp 147–171

Bowring S A and Karlstrom K E (1990) Growth, stabilization and reactivation of Proterozoic lithosphere in the southeastern United States. Geol 18: 1203–1206

Bowring S A, Williams I S and Compston W (1989b) 3.96 Ga gneisses from the Slave Province, Northwest Territories, Canada. Geol 17: 971–975

Box S E, Moll-Stalcup E J, Wooden J L and Bradshaw J Y (1990) Kilbuck terrane: Oldest known rocks in Alaska. Geol 18: 1219–1222

Boyd N K and Smithson S B (1993) Moho in the Archean Minnesota gneiss terrane: Fossil, alteration, or layered intrusion? Geol 21: 1131–1134

van Breemen O, Davis W J and King J E (1992) Temporal distribution of granitoid plutonic rocks in the Archean Slave Province, northwest Canadian shield. Can J Earth Sci 29: 2186–2199

van Breemen O and Higgins M D (1993) U–Pb zircon age of the southwest lobe of the Havre-Saint-Pierre Anorthosite Complex, Grenville Province, Canada. Can J Earth Sci 30: 1453–1457

van Breemen O, Aftalion M and Allaart J H (1974) Isotopic and geochronologic studies on granites from the Ketilidian mobile belt of South Greenland. Bull Geol Soc Am 85: 403–412

Briden J C, Piper J D A, Henthorn D I and Rex D C (1971a) New paleomagnetic results from Africa and related potassium-argon age determinations. Ann Rep Res Inst Afr Geol, University of Leeds 15: 46–50

Bridgwater D, Escher A, Nash D F and Watterson J (1973) Investigations on the Nagssugtoqidian boundary between Holsteinsborg and Kangamiut, central West Greenland. Rapp Grønlands Geol Unders 55: 22–25

Bridgwater D, Collerson K D, Hurst R W and Jesseau C W (1975) Field characteristics of the Early Precambrian rocks from Saglek, Coast of Labrador. Geol Surv Can Pap 75-1A: 287–296

Bridgwater D, Keto L, McGregor V R and Myers J S (1976) Archaean gneiss complex of Greenland. In: Geology of Greenland (eds Escher A and Watt W S). Geol Surv Greenland, Copenhagen, pp 18–75

Bridgwater D, Collerson K D and Myers J S (1978) The development of the Archaean Gneiss complex of the North Atlantic region. In: Evolution of the Earth's Crust (ed Tarling D H). Academic Press, London, pp 19–69

Brito Neves B B (1986) Tectonic regimes in the Proterozoic of Brazil. XII Simposio de Geologia do Nordeste, pp 235–251

Brito Neves B B and Cordani U G (1991) Tectonic evolution of South America during the Late Proterozoic. Prec Res 53: 23–40

Brito Neves B B, Beurlen H and dos Santos F (1982) Characteristics and mineralizations of the Archean and Early Proterozoic of the Barborema Province, Brazil. Rev Brasil Geosciencias 12 (1–3): 234–239

Bronner G, Roussel J, Trompette R (1980) Genesis and geodynamic evolution of the Taoudeni cratonic basin (Upper Precambrian and Paleozoic), Western Africa. In: Dynamics of Plate Interiors, (eds Bally A W, Bender P L, McGetchin T R, and Walcott R I). Geophys Geol Soc Am, pp 81–90

Brookfield M E (1993) Neoproterozoic Laurentia—Australia fit. Geol 21: 683–686

Brookfield M E (1994) Problems in applying preservation, facies and sequence models to Sinian (Neoproterozoic) glacial sequences in Australia and Asia. Prec Res 70: 113–143

Bros R, Stille P, Gauthier-Lafaye F, Weber F and Clauer N (1992) Sm–Nd isotope dating of Proterozoic clay materials: An example from the Francevillian sedimentary series, Gabon. Earth Planet Sci Lett 113: 207–218

Brown M and D'Lemos R S (1991) The Cadomian granites of Mancellia, northeast Armorican Massif of France: relationship to the St. Malo migmatite belt, petrogenesis and tectonic setting. Prec Res 51: 393–427

Brown M, Burwell A D, Friend C R L and McGregor V R (1981) The late Archaean Qôrqut granite complex of southern West Greenland. J Geophys Res 86: 10617–10632

Bruguier O, Dada S S and Lancelot J R (1994) Early Archaean component (> 3.5 Ga) within a 3.05 Ga orthogneiss from northern Nigeria: U–Pb zircon evidence. Earth Planet Sci Lett 125: 89–103

Bruneton P (1993) Geological environment of the Cigar Lake uranium deposit. Can J Earth Sci 30: 653–673

Buchan K L, Mortensen J K and Card K D (1994) Integrated paleomagnetic and U–Pb geochronologic studies of mafic intrusions in the southern Canadian Shield: implications for the Early Proterozoic polar wander path. Prec Res 69: 1–10

Buhl D, Granert B and Raith M (1983) U–Pb zircon dating of Archean rocks from the South Indian Craton: results from the amphibolite to granulite facies transition zone at Kabbal Quarry, Southern Karnataka. Fort Mineral 61: 43–45

Bühn B, Stanistreet I G and Okrusch M (1992) Late Proterozoic outer shelf manganese and iron deposits at Otjosondu (Namibia) related to the Damaran ocean opening. Econ Geol 87: 1393–1411

Buick R (1988) Carbonaceous filaments from North Pole, Western Australia: are they fossil bacteria in Archaean stromatolites? A reply (to discussion). Prec Res 39: 311–317

Bullard E, Everett J E and Smith A G (1965) The fit of the continents around the Atlantic. Philos Trans R Soc Lond (Ser A) 258: 41–51

Bunting J A (1986) Geology of the eastern part of the Nabberu Basin, Western Australia. Geol Surv W Austral Bull 131

Burger A J, Clifford T N and McG Miller R (1976) Zircon U-Pb ages of the Franzfontein granitic suite, northern South West Africa. Prec Res 3: 415–431

Burke K C A and Dewey J F (1973) Plume-generated triple junctions: key indicators in applying plate tectonics to old rocks. J Geol 81: 406–433

Burke K and Sengor A M (1986) Tectonic escape in the evolution of the continental crust. Am Geophys Union, Geodynamics Ser 14: 41–53

Burke K, Kidd W S F and Kusky T (1985) Is the Ventersdorp rift system of southern Africa related to a continental collision between the Kaapvaal and Zimbabwe cratons at 2.64 Ga ago? Tectonophys 115: 1–24

Butler R F (1992) Paleomagnetism: Magnetic Domains to Geologic Terrains. Boston, Blackwell Scientific Publications: 319 pp

Button A (1976) Transvaal and Hamersley basins—review of basin development and mineral deposits. Min Sci Eng 8: 262–293

Button A (1981a) The Transvaal Supergroup. In: Precambrian of the Southern Hemisphere (ed Hunter D R). Elsevier, Amsterdam: 527–536

Button A (1981b) The Ventersdorp Supergroup. In: Precambrian of the Southern Hemisphere (ed D R Hunter). Elsevier, Amsterdam: 520–527

Caby R (1987) The pan-African belt of West Africa from the Sahara desert to the Gulf of Benin. In: Anatomy of Mountain Ranges (eds Schaer J P and Rogers J). Princeton University Press, Princeton: 129–170

Caby R, Bertrand J N L and Black R (1981) Pan-African ocean closure and collision in the Hogar-Iforas segment, central Sahara. In: Precambrian Plate Tectonics (ed Kröner A). Elsevier, Holland: 407–434

Caby R, Andreopoulos-Renaud U and Pin C (1989) Late Proterozoic arc–continental and continent–continent collision in the pan-African trans-Saharan belt of Mali. Can J Earth Sci 26: 1136–1146

Caddeg S W, Bachman R L, Campbell T J, Reid R R and Otto R P (1991) The Homestake Gold Mine, an early Proterozoic iron-formation-hosted gold deposit. U.S. Geol Surv Bull 1857-J: 67 pp

Cadman A C, Heaman L, Tarney J, Wardle R and Krogh T E (1993) U–Pb geochronology and geochemical variation within two Proterozoic mafic dyke swarms, Labrador. Can J Earth Sci 30: 1490–1504

Cadman A C, Tarney J, Baragar W R A and Wardle R J (1994) Relationship between Proterozoic dykes and associated volcanic sequences: evidence from the Harp Swarm and Seal Lake Group, Labrador, Canada. Prec Res 68: 357–374

Cahen L (1982) Geochronological correlation of the Late Precambrian sequences on and around the stable zones of Equatorial Africa. Prec Res 18: 73–86

Cahen L, Kröner A and Ledent D (1979) The age of the Vista Alegre Pluton and its bearing on the reinterpretation of the Precambrian geology of northern Angola. Ann Soc Geol Belg T.102: 265–275

Cahen L, Delahl J and Lavreau J (1976) The Archaean of Equatorial Africa: a review. In: The Early History of the Earth (ed Windley B F). John Wiley, London pp 489–498

Cahen L, Snelling N J, Delhal J and Vail J R (1984) The Geochronology and Evolution of Africa. Clarendon Press, Oxford

Cambray F W (1978) Plate tectonics as a model for the environment of deposition and deformation of the early Proterozoic (Precambrian X) of northern Michigan, (Abs). Geol Soc Am (Abs. with Programs) 7: 376

Cameron A G W and Benz W (1991) The origin of the Moon and the single impact hypothesis. Icarus 92: 204–216

Campbell D S, Treloar P J and Bowes D R (1979) Metamorphic history of staurolite-bearing schist from the Svecokarelides, near Heinävaara, eastern Finland, Geol Fören Fürhand 101: 105–118

Campbell F H A (1978) Geology of the Helikian rocks of the Bathurst Inlet area, Coronation Gulf, Northwest Territories. Geol Surv Can Pap 78-1A: 97–106

Campbell F H A and Cecile M P (1981) Evolution of the Early Proterozoic Kilohigoki Basin, Bathurst Inlet-Victoria Island, Northwest Territories. In: Proterozoic Basins of Canada (ed Campbell F H A). Geol Surv Can Pap 81-10: 103–131

Campbell I H (1987) The distribution of orthocumulate textures in the Jimerlana Intrusion. J Geol 95: 35–54

Campbell I H and Griffiths R W (1990) Implications of mantle plume structure for the evolution of flood basalts. Earth Planet Sci Lett 99: 79–93

Campbell I H and Griffiths R W (1991) Megaplumes and giant radiating dyke swarms. Geol Ass Can, Joint Ann Meeting, Program with Abs: A19

Campbell I H and Griffiths R W (1992) The changing nature of mantle hotspots through time: Implications for the chemical evolution of the mantle. Earth Planet Sci Lett (in Press).

Campbell I H and Hill R I (1988) A two-stage model for the formation of the granite-greenstone terrains of the Kalgoorlie–Norseman area, W Australia. Earth Planet Sci Lett 90: 11–25

Campbell I H and Jarvis G T (1984) Mantle convection and early crustal evolution. Prec Res 26: 15–56

Campbell I H and Taylor S R (1983) No water, no granites— no oceans, no continents. Geophys Res Lett 10: 1061–1064

Cannon W F (1994) Closing of the Midcontinent rift—A far-field effect of Grenvillian compression. Geol 22: 155–158

Cannon W F, Green A G, Hutchinson D R et al (1989) The North American Midcontinent rift beneath Lake Superior from GLIMPCE seismic reflection profiling. Tectonics 8: 305–332

Cannon W F, Lee M W, Hinze W J, Schulz K J and Green A G (1991) Deep crustal structure of the Precambrian basement beneath northern Lake Michigan midcontinent North America. Geol 19: 207–210

Cannon W F, Peterman Z E and Sims P K (1993) Crustal-scale thrusting and origin of the Montreal River monocline—a 35-km-thick cross section of the Midcontinent Rift in northern Michigan and Wisconsin. Tectonics 12: 728–744

Card K D (1982) Progress report on regional geological synthesis, central Superior Province. Geol Surv Can Pap 82-1A: 23–28

Card K D (1990) A review of the Superior Province of the Canadian Shield: a product of Archean accretion. Prec Res 48: 99–156

Card K D and King J E (1992) The tectonic evolution of the Superior and Slave provinces of the Canadian Shield: introduction. Can J Earth Sci 29: 2059–2065

Card K D, Gupta V K, McGarth P H and Grant F S (1984) The Sudbury structure: its regional geological and geophysical setting. In: The Geology and Ore Deposits of the Sudbury Structure (eds Pye E G, Naldrett A J and Giblin P E). Ont Geol Surv Spec Vol 1: 25–43

Carey S W (1956) The Expanding Earth. Elsevier, Amsterdam

Carey S W (1976) The Expanding Earth—an essay review. Earth Sci Rev 11: 105–143

Carl C, von Pechmann E, Höhndorf A and Ruhrmann G (1992)

Mineralogy and U/Pb, Pb/Pb, and Sm/Nd geochronology of the Key Lake uranium deposit, Athabasca basin, Saskatchewan, Canada. Can J Earth Sci 29: 879–895

Carpena J, Kienast J R, Ouzegane K and Jehanno C (1988) Evidence of the contrasted fission-track clock behaviour of the apatites from In Ouzzal carbonatites (northwest Hoggar): the low-temperature thermal history of an Archean basement. Bull Geol Soc Am 100: 1237–1248

Carvalho H (1972) Noticia explicative da cart géologica de Angola, folha 377, Vila de Almoster, esc 1:100000. Bol Serv Geol Minas Angola

Carvalho H (1983) Geologia de Angola. Geol map 1:1000000. Lab Nac Invest Cient Tropical, Lisbon

Cassidy K F, Barley M E, Groves D I, Perring C S and Hallberg J A (1991) An overview of the nature, distribution and inferred tectonic setting of granitoids in the late-Archean Norseman–Wiluna belt. Prec Res 51: 51–83

Castaing C, Triboulet C, Feybesse J L and Chèvremont P (1993) Tectonometamorphic evolution of Ghana, Togo and Benin in the light of the Pan-African/Brasiliano orogeny. Tectonophys 218: 323–342

Castaing C, Feybesse J L, Thiéblemont D, Triboulet C and Chèvrement P (1994) Palaeogeographical reconstructions of the Pan-African/Brasiliano orogen: closure of an oceanic domain or intracontinental convergence between major blocks? Prec Res 69: 327–344

Cavell P A, Wijbrans J R and Baadsgaard H (1992) Archean magmatism in the Kaminak Lake area, District of Keewatin, Northwest Territories: ages of the carbonatite-bearing alkaline complex and some host granitoid rocks. Can J Earth Sci, 29: 896

Cawthorn R G and Meyer F M (1993) Petrochemistry of the Okiep copper district basic intrusive bodies, northwestern Cape Province, South Africa. Econ Geol 88: 590–605

Chadwick B (1986) Malene stratigraphy and late Archaean structure: new data from Ivisârtoq, inner Godthåbsfjord, southern West Greenland. Rapp Grønlands Geol Unders 130: 74–85

Chadwick B (1990) The stratigraphy of a sheet of supracrustal rocks within high-grade orthogneisses and its bearing on Late Archaean structure in southern West Greenland. J Geol Soc, Lond 147: 639–652

Chadwick B and Coe K (1983) Descriptive text to accompany 1:1000000 sheet, Buksefjorden 63 v.1 Nord. Rapp Grønlands Geol Unders

Chadwick B and Crewe M A (1986) Chromite in the early Archean Akilia association (ca. 3800 m.y.) Ivisartoq region, inner Gothabsfjord, southern Greenland. Econ Geol 81: 184–191

Chadwick B, Ramakrishnan M and Viswanatha M N (1981a) Structural and metamorphic relations between Sargur and Dharwar supracrustal rocks and Peninsular gneiss in central Karnataka. J Geol Soc Ind 22: 557–569

Chadwick B, Ramakrishnan M and Viswanatha M N (1981b) The stratigraphy and structure of the Chitradurga region: an illustration of cover-basement interaction in the late Archean evolution of the Karnataka craton, southern India. Prec Res 16: 31–54

Chadwick B, Ramakrishnan M and Viswanatha M N (1985a) A comparative study of tectonic fabrics and deformation mechanisms in Dharwar grits and phyllites and Sargur quartzites on the west of the Chitradurga supracrustal belt, Karnataka. J Geol Soc Ind 26: 526–546

Chadwick B, Ramakrishnan M, Viswanatha M N (1985b) Bababudan—a late Archean intracratonic volcanosedimentary basin, Karnataka, south India, Parts I and II. J Geol Soc Ind 26: 769–821

Chadwick B, Ramakrishanan M and Viswanatha M N (1986) Detrital chromite and zircon in Archaean Sargur metasedimentary rocks, Karnataka. Geol Surv Ind Spec Public 12: 87–96

Chadwick B, Vasuder V N and Jayaram S (1988) Stratigraphy and structure of late Archaean, Dharwar volcanic and sedimentary rocks and their basement in a part of the Shimoga Basin, east of Bhadrayathi, Karnataka. J Geol Soc Ind 32: 1–19

Chadwick B, Ramakrishnan M, Vasudev V N and Viswanatha M N (1989) Facies distributions and structure of a Dharwar volcanosedimentary basin: evidence for late Archaean transpression in Southern India? J Geol Soc Lond 146: 825–834

Chadwick B, Vasudev V N, Krishna R B and Hegde G V (1991) The stratigraphy and structure of the Dharwar Supergroup adjacent to the Honnali Dome: Implications for late Archaean basin development and regional structure in the western part of Karnataka. J Geol Soc Ind 38: 457–484

Chai G and Naldrett A J (1992) Characteristics of Ni–Cu–PGE mineralization and genesis of the Jinchuan deposit, Northwest China. Econ Geol 87: 1475–1495

Chandler F W (1982) The structure of the Richmond Gulf graben and the geological environments of lead-zinc mineralization and of iron-manganese formation in the Nastapoka Group, Richmond Gulf area, New Quebec—Northwest Territories. In: Current Research, Part A; Geol Surv Can Pap 82-1A: 1–10

Chappell B W and White A J R (1974) Two contrasting granite types. Pac Geol 8: 173–174

Chaudhuri A and Howard J D (1985) Ramgundam sandstone: a middle Proterozoic shoal-bar sequence. J Sediment Petrol 55: 392–397

Chauvel C, Arndt N T, Kielinzchuk S and Thom A (1987) Formation of Canadian 1.9 Ga old continental crust. 1. Nd isotopic data. Can J Earth Sci 24: 396–406

Chauvet A, Faure M, Dossin I and Charvet J (1994) A three-stage structural evolution of the Quadrilátero Ferrifero: consequences for the Neoproterozoic age and the formation of gold concentrations of the Ouro Preto area, Minas Gerais, Brazil. Prec Res 68: 139–167

Chemale F, Rosière C A and Endo I (1994) The tectonic evolution of the Quadrilátero Ferrifero, Minas Gerais, Brazil. Prec Res 65: 25–54

Chen J, Zhang H, Xing Y and Ma G (1981) On the upper Precambrian (Sinian Suberathem) in China. Prec Res 15: 207–228

Chen Tingyu, Niu Baogui, Liu Zhigan, Fu Yunlian and Ren Jishun (1992) Geochronology of Yanshanian magmatism and metamorphism in the hinterland of the Dabie Mountains and their geologic implications. Acta Geologica Sinica 5: 155–163

Cheng Y, Bai J and Sun D (1982) The lower Precambrian of China. Rev Brasil Geociencias 12(1-3): 65–73

Cheng Y, Sun D and Wu J (1984) Evolutionary mega-cycles of the Early Precambrian proto-North China Platform. J Geodyn 1: 251–277

Chevé S R and Machado N (1988) Reinvestigation of the Castignon Lake carbonatite complex, Labrador Trough, New Quebec Geol Assoc Can—Mineral Ass Can, Program with Abs, 13: A20

Chikhaoui M, Dupuy C and Dostal J (1980) Geochemistry and petrogenesis of late Proterozoic volcanic rocks from northwestern Africa. Contr Min Petrol 73: 375–388

Chown E H and Caty J L (1973) Stratigraphy, petrography and paleocurrent analysis of the Aphebian clastic formation of the Mistassini-Otish Basin. In: Huronian Stratigraphy and Sedimentation (ed Young G M). Geol Ass Can Spec Pap 12: 49–71

Chown E H, Daigneault R, Mueller W and Mortensen J K (1992) Tectonic evolution of the Northern Volcanic Zone, Abitibi Belt, Quebec. Can J Earth Sci 29: 2211–2225

Christiaens, L J (1992) Explaining the origin of Precambrian banded iron-formations: A matter of coming to terms with equilibrium thermodynamics. Prec Res. in press

Chumakov N M and Semikhatov M A (1981) Riphean and Vendian of the USSR. Prec Res 15: 229–253

Claesson S, Huhma H, Kinny P D and Williams I S (1993) Svecofennian detrital zircon ages—implications for the Precambrian evolution of the Baltic Shield. Prec Res 64: 109–130

Clauer R, Caby R, Daniel J and Trompette R (1982) Geochronology of sedimentary and metasedimentary Precambrian rocks of the West African Craton. Prec Res 18: 53–71

Clemens J D (1992) Partial melting and granulite genesis: a partisan overview. Prec Res 55: 297–301

Cliff R A and Rickard D (1990) Isotope systematics of the Kiruna magnetite ores, Sweden: Part I. Age of the ore. Econ Geol 85: 1770–1776

Cliff R A and Rickard D (1992) Isotope systematics of the Kiruna magnetite ores, Sweden: Part 2. Evidence for a secondary event 400 m.y. after ore formation. Econ Geol 87: 1121–1129

Clifford T N (1970) The structural framework of Africa. In: African Magmatism and Tectonics (eds Clifford T N and Gass I G). Oliver and Boyd, Edinburgh: 1–26

Clifford T N, Stumpfl E F, Burger A J et al (1981) Mineralchemical and isotopic studies of Namaqualand granulites, South Africa: a Grenville analogue. Contr Min Petrol 77: 225–250

Cloud P (1973) Paleoecological significance of the banded iron-formation. Econ Geol 68: 1135–1143

Cloud P (1976) Beginnings of biosphere evolution and their biogeochemical consequences. Paleobiol 2: 351–387

Cloud P (1983a) Aspects of Proterozoic biogeology. Geol Soc Am Mem 161: 245–251

Cloud P (1983b) Early biogeologic history: the emergence of a paradigm. In: Origin and Evolution of Earth's Earliest Biosphere: An interdisciplinary Study (ed Schopf J W). Princeton University Press

Cloud P (1988) Oasis in Space: Earth History from the Beginning. W W Norton and Co, New York

Cloud P E (1987) Trends, transitions, and events in Cryptozoic history and their calibration: apropos recommendations by the Subcommission on Precambrian Stratigraphy. Prec Res 37: 256–264

Clowes R M, Cook F A, Green A G, Keen C E, Ludden J N, Percival J A, Quinlan G M and West G F (1992) Lithoprobe: a new perspective on crustal evolution. Can J Earth Sci 29: 1813–1864

Coats C J, Quirke T T, Bell C K et al (1972) Geology and mineral deposits of the Flin Flon, Lynn Lake and Thompson area, Manitoba and the Churchill-Superior front of the western Precambrian Shield. 24th Int Geol Congr, Montreal 1972, Guidebook Excursions A31 and C31

Coe K (1980) Nûk gneisses of the Buksefjorden region, southern West Greenland, and their enclaves. Prec Res 11: 357–371

Coertze F J, Burger A J, Walraven F et al (1978) Field relations and age determinations in the Bushveld Complex. Trans Geol Soc S Afr 81: 1–11

Cogley J G (1984) Continental margins and the extent and number of the continents. Rev Geophys Space Phys 22: 101–122

Collerson K D and McCulloch M T (1983) Field and Sr-Nd isotopic constraints on Archean crust and mantle evolution in the East Pilbara Block, Western Australia. Geol Soc Austral Abs. 9: 167–168

Collerson K D, Jesseau C W and Bridgwater D (1976) Crustal development of the Archean gneiss complex: eastern Labrador. In: The Early History of the Earth, (ed Windley B F). John Wiley, London, pp 237–253

Collerson K D, McCulloch M T and Nutman A P (1989) Sr and Nd isotope systematics of polymetamorphic Archaean gneisses from southern West Greenland and northern Labrador. Can J Earth Sci 26: 446–466

Colley H and Westra L (1987) The volcano-tectonic setting and mineralization of the early Proterozoic Kemiö-Orijärvi-Lohja belt, SW Finland. In: Geochemistry and Mineralization of Proterozoic Volcanic Suites (eds Pharaoh T C, Beckinsale R D and Rickard D). Geol Soc Spec Public 33: 95–107

Collins W J, Beams S D, White A J R and Chappell B W (1982) Nature and origin of A-type granites with particular reference to southeastern Australia. Contr Min Petrol 80: 189–200

Colliston W P and Reimold W U (1990) Structural studies in the Vredefort Dome: Preliminary interpretations of results on the southern portion of the structure. Econ Geol Res Unit, Univ Witwatersrand, Johannesburg, Info Circ 229, 31 pp

Colliston W P, Praekelt H E and Schoch A E (1991) A progressive ductile shear model for the Proterozoic Aggeneys Terrane, Namaqua mobile belt, South Africa. Prec Res 49: 205–215

Compston W and Kröner A (1988) Multiple zircon growth within early Archean tonalitic gneiss from the Ancient Gneiss Complex, Swaziland. Earth Planet Sci Lett 87: 13–28

Compston W and Pidgeon R T (1986) Jack Hills, evidence of more very old detrital zircons in Western Australia. Nature 321: 766–769

Compston W and Williams I S (1982) Protolith ages from inherited zircon cores measured by a high mass resolution ion microprobe. Fifth International Conference on Geochronology, Cosmochronology and Isotope Geology, Nikko, Japan. Abs. 63–64

Compston W, Williams I S, McCulloch M T et al (1981) A revised age for the Hamersley Group. Fifth Australian Geological Convention, Perth Geol Soc Austral Abs. 3: 40

Compston W, Williams I S and Myer C (1984) Uranium-lead geochronology of zircons from lunar breccia 73217 using a sensitive high mass-resolution ion microprobe. Proc Lunar Planet. Sci Conf 14, J Geophys Res 89 (suppl.): B525–534

Compston W, Froude D O, Ireland T R et al (1985) The age of (a tiny part of) the Australian continent. Nature 317: 559–560

Compton P (1978) Rare earth evidence for the origin of the Nûk gneisses, Buksefjorden region, southern West Greenland. Contr Min Petrol 66: 283–293

Condie K C (1976a) Plate Tectonics and Crustal Evolution. Pergamon Press, New York

Condie K C (1976b) The Wyoming Archean Province in the western United States. In: The Early History of the Earth (ed Windley B F). John Wiley, London, pp 499–510

Condie K C (1980a) Origin and early development of the Earth's crust. Prec Res 11: 183–197

Condie K C (1980b) Proterozoic tectonic setting in New Mexico. New Mex Geol 2: 27–29

Condie K C (1981a) Archean Greenstone Belts. Elsevier 434 p

Condie K C (1981b) Geochemical and isotopic constraints on the origin and source of Archean granites. In: Archean Geology (eds Glover J E and Groves D I). Geol Soc Austral Spec Public 7: 753–756

Condie K C (1982a) Plate-tectonics model for Proterozoic continental accretion in the southwestern United States. Geol 10: 37–42

Condie K C (1986a) Geochemistry and tectonic setting of Early Proterozoic supracrustal rocks in southwestern United States. J Geol 94: 845–864

Condie K C (1986b) Origin and early growth rate of continents. Prec Res 32: 261–278

Condie K C (1987) Early Proterozoic volcanic regimes in southwestern North America. In: Geochemistry and Mineralization of Proterozoic Volcanic Suites (eds Pharaoh T C, Beckinsale R D and Rickard D). Geol Soc Spec Public 33: 211–218

Condie K C (1989) Origin of the Earth's crust. Palaeogeogr Palaeoclimatology, Palaeoecol (Global Planet Change Sect) 75: 57–81

Condie K C (1992) Evolutionary changes at the Archaean–Proterozoic boundary. In: The Archaean: Terrains, Processes and

Metallogeny (eds. Glover J E and Ho S E). The Univ W Austral. Public. 22: 177–189

Condie K C (ed.) (1993) Proterozoic Crustal Evolution. Elsevier, Amsterdam, 537 pp

Condie K C and Allen P (1984) Origin of Archaean charnockites from southern India. In: Archean Geochemistry (eds Kröner A, Hanson G N and Goodwin A M). Springer-Verlag, Berlin 182–203

Condie K C and Rosen O M (1994) Laurentia–Siberia connection revisited. Geol 22: 168–170

Coney P J, Jones D L and Monger J W H (1980) Cordilleran suspect terranes. Nature 288: 329–333

Connelly J N and Heaman L M (1993) U–Pb geochronological constraints on the tectonic evolution of the Grenville Province, western Labrador. Prec Res 63: 123–142

Conrad J E and McKee E H (1992) $^{40}Ar/^{39}Ar$ dating of vein amphibole from the Bayan Obo iron-rare earth element-niobium deposit, Inner Mongolia, China: constraints on mineralization and deposition of the Bayan Obo Group. Econ Geol 87: 185–188

Constanzo-Alvarez V and Dunlop P J (1993) Paleomagnetism of the Red Lake greenstone belt, northwestern Ontario: Possible evidence for the timing of gold mineralization. Earth Planet Sci Lett 119: 599–615

Cook F A (1992) Racklan Orogen. Can J Earth Sci 29: 2490–2496

Cook F A and Van der Velden A J (1993) Proterozoic crustal transition beneath the Western Canada sedimentary basin. Geol 21: 785–788

Cook F A, Dredge M and Clark E A (1992) The Proterozoic Fort Simpson structural trend in northwestern Canada. Geol Soc Am Bull 104: 1121–1137

Cook N D J and Ashley P M (1992) Meta-evaporite sequence, exhalative chemical sediments and associated rocks in the Proterozoic Willyama Supergroup, South Australia: implications for metallogenesis. Prec Res 56: 211–226

Cook P J and McElhinny M W (1979) A re-evaluation of the spatial and temporal distribution of sedimentary phosphate deposits in the light of plate tectonics. Econ Geol 74: 315–330

Cooper M R (1990) Tectonic cycles in southern Africa. Earth Sci Rev 28: 321–364

Cooray P G (1994) The Precambrian of Sri Lanka: a historical review. Prec Res 66: 3–18

Cordani U G and de Brito Neves B B (1982) The geologic evolution of South America during the Archaean and Early Proterozoic. Rev Basil Geociencias, 12(1–3): 78–88

Cordani U G, Tassinari C C G and Kawashita K (1984) A Serra dos Carajas como região limítrofe entre provincias tectónicas. Cliencias Terra 9: 6–11

Cordani U G, Sato K and Marinho M M (1985) The geological evolution of the ancient granite-greenstone terrane of central-southern Bahia, Brazil. Prec Res 27: 187–213

Corfu F (1993) The evolution of the southern Abitibi greenstone belt in the light of precise U–Pb geochronology. Econ Geol 88: 1323–1340

Corfu F and Andrews A J (1986) A U–Pb age for mineralized Nipissing Diabase, Gowganda. Can J Earth Sci 23: 107–109

Corfu F and Stott G M (1993a) Age and petrogenesis of two late Archean magmatic suites, Northwestern Superior Province, Canada: zircon U–Pb and Lu–Hf isotopic relations. J Petrol 34: 817–838

Corfu F and Stott G M (1993b). U–Pb geochronology of the central Uchi Subprovince, Superior Province. Can J Earth Sci 30: 1179–1196

Corrigau D, Culshaw N G and Mortensen J K (1994) Pre-Grenvillian evolution and Grenvillian overprinting of the Parautochthonous Belt in Key Habour, Ontario: U–Pb and field constraints. Can J Earth Sci. 31: 583–596

Cotterill P (1979) The Selukwe Schist belt and its chromite deposits. Geol Soc S Afr Spec Public 5: 229–245

Coward M P (1983) A tectonic history of the Damaran belt. In: Evolution of the Damara Orogen, South West Africa/Namibia (ed Miller R M). Geol Soc S Afr Spec Public 11: 409–421

Coward M P and Daly M C (1984) Crustal lineaments and shear zones in Africa: their relationship to plate movements. Prec Res 24: 27–45

Coward M P, James P R and Wright L (1976) Northern margin of the Limpopo Mobile Belt. Bull Geol Soc Am 87: 601–611

Cowie J W and Bassett M G (compilers) (1989) Global stratigraphic chart. Episodes, 12: no. 2, supplement.

Craddock C (1982) Antarctica and Gondwanaland (Review Paper). In: Antarctic Geoscience (ed Craddock C). University of Wisconsin Press, Madison, pp 3–13

Craig G Y (Editor) (1991) Geology of Scotland. 3rd edn. The Geological Society, Bath: 612 pp

Crawford A R and Compston W (1970) The age of the Vindhyan System of Peninsular India. Q J Geol Soc Lond 125: 351–371

Creaser R A and Cooper J A (1993) U–Pb geochronology of middle Proterozoic felsic magmatism surrounding the Olympic Dam Cu–U–Ag–Ag and Moonta Cu–Au–Ag deposits. South Australia. Econ Geol 88: 186–197

Creaser R A and Fanning C M (1993) A U–Pb zircon study of the Mesoproterozoic Charleston Granite, Gawler Craton, South Australia. Austral J Earth Sci 40: 519–526

Crocker C H, Collerson K D, Lewry J F and Bickford M E (1993) Sm–Nd, U–Pb and Rb–Sr geochronology and lithostructural relationships in the southwestern Rae Province: constraints on crustal assembly in the western Canadian Shield. Prec Res 61: 27–50

Crook K A W (1989) Why the Precambrian time-scale should be chronostratigraphic; A response to recommendations by the Subcommission on Precambrian Stratigraphy. Prec Res 41: 143–150

Cruse T and Harris L B (1994) Ediacaran fossils from the Stirling Range Formation, Western Australia. Prec Res 67: 1–10

Cruse T, Harris L B and Rasmussen B (1993) Geological Note. The discovery of Ediacaran trace and body fossils in the Stirling Range Formation, Western Australia: Implications for sedimentation and deformation during the "Pan African" orogenic cycle. Austral J Earth Sci 40: 293–296

Culotta R C, Pratt T and Oliver J (1990) A tale of two sutures: COCORP's deep seismic surveys of the Grenville province in the eastern U.S. midcontinent. Geol 18: 646–649

Cumming G L and Krstic D (1992) The age of unconformity-related uranium mineralization in the Athabasca Basin, northern Saskatchewan. Can J Earth Sci 29: 1623–1639

Cumming G L, Krstic D and Wilson J A (1987) Age of the Athabasca Group, northern Alberta. Geol. Assoc. Canada—Mineral Assoc Canada, Joint Annual Meeting, Program with Abstracts 12: 35

Dahanayake K (1982) Structural and petrological studies on the Precambrian Vijayan Complex of Sri Lanka. Revista Brasileira de Geociências 12: 88–93

Dahanayake K and Jayasena H A H (1983) General geology and petrology of some Precambrian crystalline rocks from the Vijayan Complex of Sri Lanka. Prec Res 19: 301–316

Dahlberg H H (1974) Granulites of sedimentary origin associated with rocks of the charnockite suite in the Bakhuys Mountains, NW Surinam. Ninth Inter-Guyana Geological Conference, Spec Public No. 6. Ministerior de Mines et Hidrocarburos, Venezuela, pp 415–423

Daigneault R and Allard G O (1994) Transformation of Archean structural inheritance at the Grenvillian Foreland Parautochthon Transition Zone, Chibougamau, Quebec. Can J Earth Sci 31: 470–488

Dallmeyer R D and Garcia E M (editors) (1990) Pre-Mesozoic Geology of Iberia. Springer-Verlag, New York: 416 pp

Dalrymple G B and Langphere M A (1969) Potassium-argon

dating. Principles techniques and applications to geochronology. W H Freeman, San Francisco

Daly M C (1986a) The intracratonic Irumide Belt of Zambia and its bearing on collision orogeny during the Proterozoic of Africa. In: Collision Tectonics, (editors Coward M P and Ries A). Geol Soc Spec Public 19: 321–328

Daly M C (1986b) Crustal shear zones and thrust belts: their geometry and continuity in Central Africa. Philos Trans R Soc Lond (Ser A) 317: 111–128

Daly M C and Unrug R (1982) The Muva Supergroup of northern Zambia: A craton to mobile belt sedimentary sequence. Trans Geol Soc S Afr 85: 155–165

Daly J S, Mitrofanov F P and Morozova L N (1993) Late Archaean Sm–Nd model ages from the Voche-Lambina area: implications for the age distribution of Archaean crust in the Kola Peninsula, Russia. Prec Res 64: 189–195

Dalziel I W D (1991) Pacific margins of Laurentia and East Antarctica—Australia as a conjugate rift pair: Evidence and implications for an Eocambrian supercontinent. Geol 19: 598–601

Dalziel I W D (1992a) Antarctica; a tale of two supercontinents? Ann Rev Earth Planet Sci 20: 501–526

Dalziel I W D (1992b) On the organization of American plates in the Neoproterozoic and the breakout of Laurentia. GSA Today, 2: 237, 240–241

Dalziel I W D (1994) Precambrian Scotland as a Laurentia–Gondwana link: origin and significance of cratonic promontories. Geol 22: 589–592

Dalziel I W D, Salda L H D and Gahagan L M (1994) Paleozoic Laurentia–Gondwana interaction and the origin of the Appalachian–Andean mountain system. Geol Soc Am Bull 106: 243–252

Dann J (1991) A Proterozoic ophiolite in central Arizona; Implications for models of crustal growth. Geol 19: 590–593

Danni J C M, Fuck R A and Leonardos J R (1982) Archean and lower Proterozoic units in Central Brazil. Geol Rdsch 71(1): 291–317

Dardenne M A and Walde D H G (1979) A estratigrafia dos Gropos Bambuí e Macaúbas no Brasil Central. Atas do I simposio de Geologia de Minas Gerais, Diamantina

Dasgupta S, Roy S and Fukuoka M (1992) Depositional models for manganese oxide and carbonate deposits of the Precambrian Sausar Group, India. Econ Geol 87: 1412–1418

Dasgupta S, Sanyal S, Sengupta P and Fukuoka M (1994) Petrology of granulites from Anakapalle—Evidence for Proterozoic decompression in the Eastern Ghats, India. J Petrol 35: 433–459

DaSilva Filho A F, Guimarães I P and Thompson R N (1993) Shoshonitic and ultrapotassic Proterozoic intrusive suite in the Cachoeirinha–Salgueiro belt, NE Brazil: a transition from collisional to post-collisional magmatism. Prec Res 62: 323–342

Davidson A (1985) Tectonic framework of the Grenville Province in Ontario and Western Quebec. In: Deep Proterozoic crust in North Atlantic Provinces, (eds Tobi A C and Touret J L R) D Reidel Publicising Co, Dordrecht: 133–149

Davidson A (1986) New Interpretations in the southwestern Grenville Province. In: The Grenville Province (eds Moore J M, Davidson A and Baer A J). Geol Ass Can Spec Pap 31: 61–74

Davidson A and Van Breemen O (1987) Northeastern extension of the Proterozoic Igneous terranes of Mid-contental North America. Institute on Lake Superior Geology, Proc Abs. 33rd Ann Mg 33: 22

Davies G F (1992) On the emergence of plate tectonics. Geol 20: 963–966

Davies C J and Coward M P (1982) The structural evolution of the Gariep Arc in southern Namibia (South-west Africa). Prec Res 17: 173–198

Davis D W and Sutcliffe R H (1984) U-Pb ages from the Nipigon Plate. Geol Ass Can (Abs. with Program) 9: 57

Davis D W and Jackson M C (1988) Geochronology of the Lumby Lake greenstone belt: A 3 Ga complex within the Wabigoon subprovince, northwest Ontario. Geol Soc Am Bull 100: 818–824

Davis D W, Hirdes W, Schaltegger U and Nunoo E A (1994) U–Pb age constraints on deposition and provenance of Birimian and gold-bearing Tarkwaian sediments in Ghana, West Africa. Prec Res 67: 89–107

Davison I and Santos R A (1989) Tectonic evolution of the Sergipano Fold Belt, NE Brazil, during the Brasíliano orogeny. Prec Res 44: 319–342

Davison I, Teixeira J B G, Silva M G et al (1988) The Rio Itapicuru greenstone belt, Bahia Brazil: structure and stratigraphic outline. Prec Res 42: 1–17

Davis W F, Fryer B J and King J E (1994) Geochemistry and evolution of Late Archean plutonism and its significance to the tectonic development of the Slave craton. Prec Res 67: 207–241

Dawes P R (1976) Precambrian to Tertiary of northern Greenland. In: Geology of Greenlands (eds Escher A and Watt W S). The Geological Survey of Greenland, Copenhagen, pp 248–303

Dazhong S and Songnian L (1985) A subdivision of the Precambrian of China. Prec Res 28: 137–162

De Beer J H and Meyer R (1983) Geoelectrical and gravitational characteristics of the Namaqua-Natal mobile belt and its boundaries. Geol Soc S Afr Spec Publicic 10: 91–100

De Beer J H and Stettler E H (1988) Geophysical characteristics of the southern African continental crust. J Petrol Special Volume: Oceanic and Continental Lithosphere: 163–184

De Laeter J R, Libby W G and Trendall A F (1981) The older Precambrian geochronology of Western Australia. Geol Soc Austral Spec Public 7: 145–157

De Matties T A (1994) Early Proterozoic volcanogenic massive sulfide deposits in Wisconsin: An overview. Econ Geol 89: 1122–1151

De Paolo D J and Wasserburg G J (1979) Sm-Nd age of the Stillwater complex and the mantle evolution curve for neodymium. Geochim Cosmochim Acta 43: 999–1008

De Wit M J (1982) Gliding and overthrust nappe tectonics in the Barberton greenstone belt. J Struc Geol 4: 117–136

De Wit J J and Stern C R (1980) A 3500 Ma ophiolite complex from the Barberton greenstone belt, South Africa: Archean oceanic crust and its geotectonic implications. 2nd Int Archean Symp, Perth. Extended Abs Geol Soc Austral: 85–87

De Wit M J, Fripp R E P and Stanistreet I G (1983) Tectonic and stratigraphic implications of new field observations along the southern part of the Barberton greenstone belt. Geol Soc S Afr Spec Public 9: 21–29

De Wit M J, Hart R A and Hart R J (1987) The Jamestown ophiolite complex, Barberton mountain belt: a section through 3.5 Ga oceanic crust. J Afr Earth Sci 5: 681–730

De Wit M J, Jones M G and Buchanan D L (1992) The geology and tectonic evolution of the Pietersburg Greenstone Belt, South Africa. Prec Res 55: 123–153

Dearnley R (1965) Orogenic fold-belts and continental drift. Nature 206: 1083–1087

Deb M and Sarkar S C (1990) Proterozoic tectonic evolution and metallogenesis in the Aravalli-Delhi orogenic complex, northwestern India. Prec Res 46: 115–137

Delaney G D (1981) The Mid-Proterozoic Wernecke Supergroup, Wernecke Mountains, Yukon Territory. In: Proterozoic Basins of Canada (ed Campbell F H A). Geol Surv Can Pap 81-10: 1–23

Delian F, Dasgupta S, Bolton B R, Harlya Y, Momoi H, Miura H, Jiaju L and Roy S (1992) Mineralogy and geochemistry of the Proterozoic Wafangzi ferromanganese deposit, China. Econ Geol 87: 1430–1440

Denis E and Dabard M P (1988) Sandstone petrography and geochemistry of late Proterozoic sediments of the Armorican Massif (France)—a key to basin development during the Cadomian Orogeny. Prec Res 42: 189–206

Derry L A, Kaufman A J and Jacobsen S B (1992) Sedimentary cycling and environmental change in the Late Proterozoic: Evidence from stable and radiogenic isotopes. Geochim et Cosmochim Acta 56: 1317–1329

Des Marais D J (1994) Tectonic control of the crustal organic carbon reservoir during the Precambrian. Chem Geol 114: 303–314

Dickin A P and Higgins M D (1992) Sm/Nd evidence for a major 1.5 Ga crust forming event in the central Grenville province. Geol 20: 137–140

Dickinson W R (1993) Making composite continents. Nature 364: 284–285

Dietz R S (1961) Vredefort ring structure: meteorite impact scar? J Geol 69: 499–516

DiMarco M J and Lowe D R (1989) Stratigraphy and sedimentology of an early Archean felsic volcanic sequence, eastern Pilbara Block, Western Australia, with special reference to the Duffer Formation and implications for crustal evolution. Prec Res 44: 147–169

Dirks P H G M, Hoek J D, Wilson C J L and Sims J (1994) The Proterozoic deformation of the Vestfold Hills Block, East Antarctica: implications for tectonic development of adjacent granulite belts. Prec Res 65: 277–295

Doe B R (1970) Lead Isotopes. Springer-Verlag, Berlin

Doig R (1983) Rb-Sr isotope study of Archean gneisses north of the Cape Smith Fold Belt, Ungava, Quebec. Can J Earth Sci 20: 821–829

Doig R (1991) U–Pb zircon dates of Morin anorthosite suite rocks, Grenville Province, Quebec. J Geol 99: 729–738

Donn W L, Donn B G and Valentine W G (1965) On the early history of the earth. Geol Soc Am Bull 76: 287–306

Dorr J V N (1969) Physiographic, stratigraphic and structural development of the Quadrilatero Ferrifero, Minas Gerais, Brazil. US Geol Surv Prof Pap 641A

Dostal J, Dupuy C and Caby R (1994) Geochemistry of the Neoproterozoic Tilemsi belt of Iforas (Mali, Sahara): a crustal section of an oceanic island arc. Prec Res 65: 55–69

Dostal J, Keppie J D, Cousens B L and Murphy J B (1994) 550–580 Ma magmatism in Cape Breton Island (Nova Scotia, Canada): the product of NW-dipping subduction during the final stage of amalgamation of Gondwana. Prec Res, submitted

Dott R H (1983) The Proterozoic red quartzite enigma in the north-central United States: resolved by plate collision? In: Early Proterozoic Geology of the Great Lakes Region (ed Medaris L G). Geol Soc Am Mem 160: 129–141

Douglas R J W, (ed.) (1970) Geology and Economic Minerals of Canada. Geol Surv Can, Econ Geol Rep 1

Dressler B O (1984) General geology of the Sudbury area. In: The Geology and Ore Deposits of the Sudbury Structure (eds Pye E G, Naldrett A J and Giblin P E). Ont Geol Surv Spec Vol 1: 57–82

Drever J I (1974) Geochemical model for the origin of Precambrian banded iron-formation. Bull Geol Soc Am 85: 1099–1106

Drew L J, Meng Qingrun and Sun Weijun (1990) The Bayan Obo iron- rare earth-niobium deposit, Inner Mongolia, China. Lithos 26: 43–65

Drexel J F, Preiss W V and Parker A J (Editors) (1993) The Geology of South Australia. Volume I. The Precambrian. Geol. Surv. South Australia, State Print, South Australia. Bull 54: 242 pp

Drury S A and Holt R W (1983) The tectonic framework of the South Indian Craton: a reconnaissance involving Landsat imagery. In: Structure and Tectonics of Precambrian Rocks in India (ed Sinha-Roy S). Hindustan Publicishing Corp Press, New Delhi, pp 178–185

Drury S A, Holt R W and van Clasteren P C (1983) Sm-Nd and Rb-Sr ages for Archaean rocks in western Karnataka, South India. J Geol Soc Ind 25: 454–459

Drysdall A R, Johnson R L, Moore T A and Thieme J G (1972) Outline of the geology of Zambia. Geol Mijnbouw 51: 265–276

Du Toit A L (1954) The Geology of South Africa. Oliver and Boyd, Edinburgh

Du Toit M C, Van Reenen D D and Roering C (1983) Some aspects of the geology, structure and metamorphism of the southern marginal zone of the Limpopo metamorphic complex. In: The Limpopo Belt (eds Van Biljon W J and Legg J H). Geol Soc S Afr Spec Public 8: 121–142

Duffield W A (1972) A naturally occurring model of global plate tectonics. J Geophys Res 77: 2543–2555

Duncan C C and Turcotte D L (1994) On the breakup and coalescence of continents. Geol 22: 103–106

Dunlop D J (1981) Palaeomagnetic evidence for Proterozoic continental development. Philos Trans R Soc Lond (Ser A) 301: 265–277

Dunlop J S R and Buick R (1981) Archaean epiclastic sediments derived from mafic volcanics, North Pole, Pilbara Block, western Australia. In: Archaean Geology, (eds Glover J E and Groves D I). Geol Soc Austral Spec Public 7: 225–233

Dupret L (1984) The Proterozoic of northeastern Armorican Massif. In: Precambrian in Younger Fold Belts. (ed Zoubeck V). John Wiley, New York

Dupuy C, Michard A, Dostal J, Dautel D and Baragar W R A (1992) Proterozoic flood basalts from the Copper River area, Northwest Territories: isotope and trace element geochemistry. Can J Earth Sci 29: 1937–1943

Durrheim R J, Barker W H and Green R W E (1992) Seismic studies in the Limpopo Belt. Prec Res 55: 187–200

Durrheim R J and Mooney W D (1991) Archean and Proterozoic crustal evolution: Evidence from crustal seismology. Geol 19: 606–609

Dymek R F (1984) Supracrustal rocks, polymetamorphism and evolution of SW Greenland Archean Gneiss Complex. In: Patterns of Change in Earth Evolution, (eds Holland H D and Trendall A F). Dahlem Konferenzen, 1984. Springer-Verlag, Berlin, pp 313–343

Dymek R F and Klein C (1988) Chemistry, petrology and origin of banded iron-formation lithologies from the 3800 Ma Isua Supracrustal Belt, West Greenland. Prec Res 39: 247–302

Eade K E (1966) Fort George River and Kaniapiskau River (west half) map-areas, New Quebec. Geol Surv Can Mem 339

Eales H V, de Klerk W J and Teigler B (1990) Evidence for magma mixing processes within the Critical and Lower zones of the northwestern Bushveld Complex. Chem Geol 88: 261–278

Easton R M (1983) Crustal structure of rifted continental margins: geological constraints from the Proterozoic rocks of the Canadian Shield. Tectonophys, 94: 371–390

Easton R M (1985) The nature and significance of pre-Yellowknife Supergroup rocks in the Point Lake area, Slave Structural Province, Canada. In: Evolution of Supracrustal Sequences (eds Ayres L D, Thurston P C, Card K D and Weber W). Geol Ass Can Spec Pap 28: 153–167

Easton R M (1986) Geochronology of the Grenville Province; Part 1: Compilation of data; Part II: Synthesis and interpretation. In: the Grenville Province (eds Moore J M, Davidson A and Baer A J). Geol Ass Can Spec Pap 31: 127–173

Edwards G E (1992) Mantle decarbonation and Archean high-Mg magmas. Geol 20: 899–902

Ehlers C, Lindroos A and Selonen O (1993) The late Svecofennian granite–migmatite zone of southern Finland—a belt of transpressive deformation and granite emplacement. Prec Res 64: 295–309

Eide E A, McWilliams M O and Liou J G (1994) $^{40}Ar/^{39}Ar$ geochronology and exhumation of high-pressure to ultra high-pressure metamorphic rocks in east-central China. Geol 22: 601–604

Eliasson T and Schöberg H (1991) U–Pb dating of the post-kinematic Sveconorwegian (Grenvillian) Bohus granite, SW Sweden: evidence of restitic zircons. Prec Res 51: 337–350

Ellenberger F and Tamain A L G (1980) Hercynian Europe. Episodes 1980: 22–27

Ellis D J (1992) Precambrian tectonics and the physicochemical evolution of the continental crust. II. Lithosphere delamination and ensialic orogeny. Prec Res 55: 507–524

Elming, S-Å, Pesonen L J, Leino M A H, Khramov A N, Mikhailova N P, Krosnova A F, Mertanen S, Bylund G and Terho M (1993) The drift of the Fennoscandian and Ukrainian shields during the Precambrian: a palaeomagnetic analysis. Tectonophys 223: 177–198

Elston D P (1979) Late Precambrian Sixtymile Formation and orogeny at the top of the Grand Canyon Supergroup. US Geol Surv Prof Pap 1092

Emeleus C H and Upton B G J (1976) The Gardar period in southern Greenland. In: Geology of Greenland (eds Escher A and Watt W S). The Geological Survey of Greenland, Copenhagen, pp 152–181

Emslie R F (1978) Anorthosite massifs, rapakivi granites and late Proterozoic rifting of North America. Prec Res 7: 61–98

Emslie R F (1985) Proterozoic anorthosite massifs. In: The Deep Proterozoic Crust in the North-Atlantic Provinces (eds Tobi A C and Touret J L R). D Reidel, Dordrecht, pp 39–60

Emslie R F (1991) Granitoids of rapikivi granite-anorthosite and related associations. Prec Res 51: 173–192

Emslie R F and Hunt P A (1990) Ages and petrogenetic significance of igneous mangerite-charnockite suites associated with massif anorthosites, Grenville Province. J Geol 98: 213–231

Emslie R F and Loveridge W D (1992) Fluorite-bearing early and middle Proterozoic granites, Okak Bay area, Labrador: Geochronology, geochemistry and petrogenesis. Lithos 28: 87–109

Engel A E J, Dixon T H and Stern R J (1980) Late Precambrian evolution of Afro-Arabian crust from ocean to craton. Bull Geol Soc Am 91: 699–706

Eriksson K A (1981) Archaean platform-to-trough sedimentation, east Pilbara Block, Australia. In: Archean Geology (eds Glover J E and Groves D I). Geol Soc Austral Spec Public 7: 235–244

Eriksson K A and Donaldson J A (1986) Basinal and shelf sedimentation in relation to the Archean-Proterozoic boundary. Prec Res 33: 103–121

Eriksson K A and Truswell J F (1978) Geological Processes and Atmospheric Evolution. In: Evolution of the Earth's Crust (ed Tarling D E). Academic Press, London, pp 219–238

Eriksson P G and Cheney E S (1992) Evidence for the transition to an oxygen-rich atmosphere during the evolution of red beds in the Lower Proterozoic sequences of southern Africa. Prec Res 54: 257–269

Ernst R E and Baragar W R A (1992) Evidence from magnetic fabric for the flow pattern in the Mackenzie giant radiating dyke swarm. Nature 356: 511–513

Escher A and Burri M (1967) Stratigraphy and structural development of the Precambrian rocks in the area northeast of Disko Bugt, West Greenland. Rapp Gronlands Geol Unders 13

Escher A, and Pulvertaft T C R (1968) The Precambrian rocks of the Upernavik-Kraulshavn area (72°–74° 15′N), West Greenland. Rapp Grønlands Geol Unders 15: 11–14

Escher A and Pulvertaft T C R (1976) Rinkian mobile belt of West Greenland. In: Geology of Greenland (eds Escher A and Watt W S). The Geological Survey of Greenland, Copenhagen, pp 104–119

Escher A and Watt W S (1976) Summary of the Geology of Greenland. In: Geology of Greenland (eds Escher A and Watt W S). The Geological Survey of Greenland, Copenhagen, pp 10–16

Escher A, Sorensen K and Zeck H P (1976) Nagssugtoqidian mobile belt in West Greenland. In: Geology of Greenland (eds Escher A and Watt W S). The Geological Survey of Greenland, Copenhagen, pp 76–95

Etheridge M A, Rutland R W R and Wyborn L A I (1987) Orogenesis and tectonic process in the early to middle Proterozoic of Northern Australia. In: Proterozoic Lithospheric Evolution (ed Kröner A). Am Geophys Union (Geodyn Ser) 17: 131–147

Ewers W E (1983) Chemical factors in the deposition and diagenesis of banded iron-formation. In: Iron-Formation Facts and Problems (eds Trendall A F and Morris R C). Elsevier, Amsterdam: 491–512

Fahrig W F (1987) The tectonic settings of continental mafic dyke swarms: failed arms and early passive margin. In: Mafic Dyke Swarms (eds Halls H C and Fahrig W G). Geol Ass Can Spec Pap 34: 331–348

Fanning C M, Flint R B, Parker A J, Ludwig K R and Blissett A H (1988) Refined Proterozoic evolution of the Gawler Craton, South Australia, through U-Pb zircon geochronology. Prec Res 40-41: 363–386

Faure G (1977) Principles of Isotope Geology. John Wiley, New York

Faure G (1986) Principles of Isotope Geology, 2nd edn. John Wiley, New York

Faure G and Powell J L (1972) Strontium Isotope Geology. Springer-Verlag, Berlin

Feng R, Kerrich R, McBride S and Farrar E (1992) $^{40}Ar/^{39}Ar$ age constraints on the thermal history of the Archean Abitibi greenstone belt and the Pontiac Subprovince: implications for terrane collision, differential uplift, and overprinting of gold deposits. Can J Earth Sci 29: 1389–1411

Ferreira-Filho C F, Kamo S L, Fuck R A, Krogh T E and Naldrett A J (1994) Zircon and rutile U–Pb geochronology of the Niquelândia layered mafic and ultramafic intrusion, Brazil: constraints for the timing of magmatism and high grade metamorphism. Prec Res 68: 241–255

Feybesse J-L and Milési J-P (1994) The Archaean/Proterozoic contact zone in West Africa: a mountain belt of décollement thrusting and folding on a continental margin related to the 2.1 Ga convergence of Archaean cratons? Prec Res 69: 199–227

Fitches W R (1971) Sedimentation and tectonics at the northeast end of the Irumide Orogenic Belt, N Malawi and Zambia. Geol Rdsch 60: 589–619

Fitches W R, Ajibade A C, Egbuniwe I G et al (1985) Late Proterozoic schist belts and plutonism in NW Nigeria. J Geol Soc Lond 142: 319–337

Fleck R J, Greenwood W R, Hadley D G et al (1980a) Rubidium-strontium geochronology and plate tectonic evolution of the southern part of the Arabian Shield. US Geol Surv Prof Pap 1131

Fleck R J, Greenwood W R, Hadley D G et al (1980b) Age and evolution of the southern part of the Arabian Shield. In: Evolution and Mineralization of the Arabian-Nubian Shield, Institute of Applied Geology, King Abdulaziz University, Jeddah; 3: 1–17 Pergamon Press, Oxford

Fletcher I R, Rosman K J R, Williams I R et al (1984) Sm-Nd geochronology of greenstone belts in the Yilgarn Block, Western Australia. Prec Res 26: 333–361

Flöttmann T and Oliver R (1994) Review of Precambrian–Palaeozoic relationships at the craton margins of southeastern Australia and adjacent Antarctica. Prec Res 69: 293–306

Foden J D, Buick I S and Mortimer G E (1988) The petrology and geochemistry of granitic gneisses from the East Arunta Inlier, central Australia: implications for Proterozoic crustal development. Prec Res 40/41: 233–259

Folinsbee R E (1982) Variations in the distribution of mineral deposits with time. In: Mineral Deposits and the Evolution of the Biosphere (eds Holland H D and Schidlowski M). Dahlem Workshop. Springer-Verlag, Berlin: 219–236

Foose M P and Weiblen P W (1986) The physical and chemical setting and textural and compositional characteristics of sulfide ores from the south Kawishiwi intrusion, Duluth Complex, Minnesota, U.S.A. In: Geology and Metallogeny of Copper Deposits (eds. Friedrich, G H, Gentin A D, Naldrett A J, Ridge J D, Stillitoe R H and Vokes F M). Berlin–Heidelberg, Springer-Verlag: 8–24

Francis D, Ludden J and Hynes A (1983) Magma evolution in a Proterozoic rifting environment. J Petrol 24: 556–582

Franke D (1993) The southern border of Baltica—a review of the present state of knowledge. Prec Res 64: 419–430

Franssen L and André L (1988) The Zadinian Group (Late Proterozoic, Zaire) and its bearing on the origin of the West Congo Orogenic Belt. Prec Res 38: 215–234

Frarey M J (1977) Geology of the Huronian belt between Sault Ste Marie and Blind River, Ontario. Geol Surv Can Mem 383

French B M (1990) Absence of shock-metamorphic effects in the Bushveld Complex, South Africa: result of an intensive search. Tectonophys 171: 287–301

French B M and Nielson R L (1990) Vredfort Bronzite Granophyre: chemical evidence for origin as a meteorite impact melt. Tectonophys 171: 119–138

Frey H (1980) Crustal evolution of the early Earth: the role of major impacts. Prec Res 10: 195–216

Friend C R L (1984) The origins of the Closepet Granite and the implications for the crustal evolution of southern Karnataka. J Geol Soc Ind 25: 73–84

Friend C R L (1985) Evidence for fluid pathways through Archaean crust and the generation of the Closepet granite, Karnataka, South India. Prec Res 27: 239–250

Friend C R L and Nutman A P (1991) SHRIMP U–Pb geochronology of the Closepet Granite and Peninsular Gneiss, Karnataka, south India. J Geol Soc Ind 38: 357–368

Friend C R L and Nutman A P (1994) Two Archaean granulite–facies metamorphic events in the Nuuk–Maniitsoq region, southern West Greenland: correlation with the Saglek block, Labrador. J Geol Soc 151: 421–424

Friend C R L, Nutman A P and McGregor V R (1987) Late-Archaean tectonics in the Faeringehavn-Tre Brødre area, south of Buksefjorden, southern West Greenland. J Geol Soc Lond 144: 369–376

Friend C R L, Nutman A P and McGregor V R (1988) Late Archean terrane accretion in the Godthåb region, southern West Greenland. Nature (Lond) 335: 535–538

Frisch T (1982) Precambrian geology of the Prince Albert Hills, Western Melville Peninsula, District of Franklin. Geol Surv Can Bull 346

Frost C D (1993) Nd isotopic evidence for the antiquity of the Wyoming province. Geol 21: 351–354

Froude D O, Ireland T R, Kinny P D et al (1983) Ion Microprobe identification of 4100–4200 Ma old terrestrial zircons. Nature 304: 616–618

Fyfe W S (1978) Evolution of the earth's crust: modern plate tectonics to ancient hot spot tectonics? Chem Geol 23: 89–114

Fyfe W S (1993) Hot spots, magma underplating, and modification of continental crust. Can J Earth Sci 30: 908–912

Gaál G and Gorbatschev R (1987) An outline of the Precambrian evolution of the Baltic Shield. Prec Res 35: 15–52

Gabert G (1984) Structural-lithological units of Proterozoic rocks in East Africa, their base, cover and mineralization. In: African Geology (eds Klerx J and Michot J). Volume in honour of L Cahen, 12th colloq Afr Geol 6–8 April 1983. Brussels: 11–21

Gall Q (1992) Precambrian paleosols in Canada. Can J Earth Sci 29: 2530–2536

Gao Z, Pen C, Li Y et al (1980) The Sinian System and its glacial deposits in Quruqtag, Xinjian. In: Research on Precambrian Geology; Sinian Suberathem in China. Tianjin Science and Technology Press, pp 186–213

Garris M A and Postnikov D V (1970) Precambrian geochronology of the east of the Russian Platform and the miogeosynclinal regions of the Urals. In: Prcambrian/Geochronology, Izdvo, Nauko, Moscow, pp 74–96

Gastil G (1960) The distribution of mineral dates in time and space. Am J Sci 258: 1–35

Gauthier-Lafaye F and Weber F (1989) The Francevillian (Lower Proterozoic) uranium ore deposit of Gabon. Econ Geol 84: 2267–2285

Gauthier-Lafaye F, Weber F and Ohmoto H (1989) Natural fission reactors of Oklo. Econ Geol 84: 2286–2295

Gebauer D (1986) The development of the continental crust of the European Hercynides since the Archean based on radiometric data. In: Proceedings of the Third workshop on the European Geotraverse (EGT) Project. The Central Segment (eds Freeman R, Mueller S and Giese P). European Science Foundation, pp 15–23

Gee R D (1979) Structure and tectonic style of the western Australian Shield. Tectonophys 58: 327–369

Gee R D, Baxter J L, Wilde S A and Williams J R (1981) Crustal development in the Archaean Yilgarn Block, western Australia. In: Archaean Geology (eds Glover J E and Groves D I). Geol Soc Austral Spec Public 7: 43–56

Gee R D, Myers J S and Trendall A F (1986) Relations between Archean high-grade gneiss and granite-greenstone terrains in Western Australia. Prec Res 33: 87–102

Gélinas L, Brooks C, Perrault G et al (1977) Chemostratigraphic divisions within the Abitibi Volcanic Belt, Rouyn-Noranda District, Quebec. In: Volcanic Regimes in Canada (eds Baragar W R A, Coleman L C and Hall J M). Geol Ass Can Spec Pap 16: 265–295

Gibb R A (1983) Model for suturing of Superior and Churchill plates: an example of double indentation tectonics. Geol 11: 413–417

Gibb R A and Walcott R I (1971) A Precambrian suture in the Canadian Shield. Earth Planet Sci Lett 10: 417–422

Gibbs A K and Barron C N (1983) The Guiana Shield reviewed. Episodes 1983: 7–14

Gibbs A K and Olszewski W J (1982) Zircon U-Pb ages of Guyana greenstone-gneiss terranes. Prec Res 17: 199–214

Giles C W (1988) Petrogenesis of the Proterozoic Gawler Range Volcanics, South Australia. Prec Res 40/41: 407–427

Gill J E (1949) Natural divisions of the Canadian Sheild. R Soc Can v.43, Sect 4, Ser 3: 61–69

Glaessner M G and Walter M R (1981) Australian Precambrian paleobiology. In: Precambrian of the Southern Hemisphere (ed Hunter D R). Elsevier, Amsterdam, pp 361–396

Glikson A Y (1983) Geochemical, isotopic and paleomagnetic tests of early sial-sima patterns: The Precambrian crustal enigma revisited. Geol Soc Am Mem 161: 95–117

Glikson A Y (1993) Asteroids and early Precambrian crustal evolution. Earth Sci Rev 35: 285–319

Glover J E (1992) Sediments of Early Archaean coastal plains: a possible environment for the origin of life. Prec Res 56: 159–166

Glover J E and Ho S E (editors) (1992) The Archaean: Terrains, Processes and Metallogeny. Geol Department and University Extension. The Univ Western Australia. Public. No. 22. 3rd Intl. Archaean Symposium (3IAS)

Goldich S S and Wooden J L (1980) Origin of the Morton Gneiss, southwestern Minnesota. Part 3: Geochronology. Geol Soc Am Spec Pap 182: 77–94

Goldich S S, Hedge C E, Stern T W et al (1980) Archean rocks of the Granite Falls area, southwestern Minnesota. Geol Soc Am Spec Pap 182: 19–43

Goode A D T (1981) Proterozoic geology of Western Australia. In: Precambrian of the Southern Hemisphere (ed Hunter D R). Elsevier, Amsterdam: 105–203

Goode A D T, Hall W D M and Bunting J A (1983) The Nabberu Basin of Western Australia. In: Iron-Formation: Facts and Problems (eds Trendall A F and Morris R C) Elsevier, Amsterdam: 295–323

Goodge J W, Walker N W and Hensen V L (1993) Neoproterozoic–Cambrian basement-involved orogenesis within the Antarctic margin of Gondwana. Geol 21: 37–40

Goodwin A M (1956) Facies relations in the Gunflint iron-formation. Econ Geol 51: 565–595

Goodwin A M (1968) Evolution of the Canadian Shield: Presidential Address. Proc Geol Ass Can 19: 1–14

Goodwin A M (1974) Precambrian belts, plumes and shield development. Am J Sci 274: 987–1028

Goodwin A M (1976) Giant impacting and the development of continental crust. In: The Early History of the Earth (ed Windley B J). John Wiley, New York: 77–95

Goodwin A M (1977) Archean volcanism in Superior Province. In: Volcanic Regimes in Canada (eds Baragar W R A, Coleman L C and Hall J M). Geol Ass Can Spec Pap 16: 205–241

Goodwin A M (1978) The nature of Archean crust in the Canadian Shield. In: Evolution of the Earth's Crust (ed Tarling D H). Academic Press, London, pp 175–218

Goodwin A M (1981a) Archean plates and greenstone belts. In: Precambrian Plate Tectonics (ed Kröner A). Elsevier, Amsterdam: 105–135

Goodwin A M (1981b) Precambrian perspectives. Sci 213: 55–61

Goodwin A M (1982a) Archean volcanoes in southwestern Abitibi Belt, Ontario and Quebec: form, composition and development. Can J Earth Sci 19: 1140–1155

Goodwin A M (1982b) Distribution and origin of Precambrian banded iron formation. Revista Brasileira Geosciencias 12(1–3): 457–462

Goodwin A M (1985) Rooted Precambrian ring-shields: growth, alignment and oscillation. Am J Sci 285: 481–531

Goodwin A M (1991) Precambrian Geology. Academic Press, London

Goodwin A M and Ridler R H (1970) The Abitibi orogenic belt. In: Symposium on Basins and Geosynclines of the Canadian Shield (ed Baer A J). Geol Surv Can Pap 70-40: 1–30

Gopalan K, Macdougall J D, Roy A B and Murali A V (1990) Sm-Nd evidence for 3.3 Ga old rocks in Rajasthan, northwestern India. Prec Res 48: 287–297

Gorbatschev R (1980) The Precambrian development of southern Sweden. Geol Fören Förhand 102(2): 129–140

Gorbatschev R (editor) (1993) The Baltic Shield. Special Volume. Prec Res 64: 1–430

Gorbatschev R and Bogdanova S (1993) Frontiers in the Baltic Shield. Prec Res 64: 3–21

Gorbunov G I, Zagorodny V G and Robonen E I (1985) Main features of the geological history of the Baltic Shield and the epochs of ore formation. Bull Geol Surv Fin 333: 3-41

Gordon M B and Hempton M R (1986) Collision-induced rifting; the Grenville Orogeny and the Keweenawan rift of North America. Tectonophys, 127: 1–25

Gordon T M, Hunt P A, Bailes A H and Syme E C (1990) U–Pb ages from the Flin Flon and Kisseynew belts, Manitoba: Chronology of crust formation at an early Proterozoic accretionary margin. Geol Ass Can Spec Pap 37: 177–200

Gow P A, Wall V J, Oliver N H S and Valenta R K (1994) Proterozoic iron oxide (Cu–U–Au–REE) deposits: Further evidence of hydrothermal origins. Geol 22: 633–636

Gower C F and Tucker R D (1994) Distribution of pre-1400 Ma crust in the Grenville province: implications for rifting in Laurentia–Baltica during geon 14. Geol 22: 827–830

Gower C F, Heaman L M, Loveridge W D, Schärer U and Tucker R D (1991) Grenvillian magmatism in the eastern Grenville Province, Canada. Prec Res 51: 315–336

Gower C F, Schärer U and Heaman L M (1992) The Labradorian orogeny in the Grenville Province, eastern Labrador, Canada. Can J Earth Sci 29: 1944–1957

Grandstaff D E (1980) Origin of uraniferous conglomerates at Elliot Lake, Canada and Witwatersrand, South Africa: implications for oxygen in the Precambrian atmosphere. Prec Res 13: 1–26

Grant N K (1973) Orogeny and reactivation to the west and southeast of the West African Craton. In: The Ocean Basins and Margins (eds Nairn A E M and Stehli F G). Plenum Press, New York, pp 447–492

Grant N K (1978) Structural distinction between a metasedimentary cover and an underlying basement in the 600 m.y. old Pan-African domain of northwestern Nigeria, West Africa. Bull Geol Soc Am 89: 50–58

Green A G, Milkereit B, Davidson A et al (1988) Crustal structure of the Grenville front and adjacent terranes. Geol 16: 788–792

Green J C (1977) Keweenawan Plateau volcanism in the Lake Superior region. In: Volcanic Regimes in Canada. Geol Ass Can Spec Pap 16: 407–422

Green J C (1982) Geology of Keweenawan extrusive rocks. In: Geology and Tectonics of the Lake Superior Basin (eds Wold R J and Hinze W J). Geol Soc Am Mem 156: 47–55

Greenberg J K and Brown B A (1983) Lower Proterozoic volcanic rocks and their setting in the southern Lake Superior district. In: Early Proterozoic Geology of the Great Lakes Region (ed Medaris L G). Geol Soc Am Mem 160: 67–84

Greenwood W R, Anderson R E, Fleck R J and Roberts R J (1980) Precambrian geologic history and plate tectonic evolution of the Arabian Shield. Bull Saudi Arabian Dir Gen Min Res 24: 35pp

Greenwood W R, Stoeser D B, Fleck R J and Stacey J S (1982) Late Proterozoic island-arc complexes and tectonic belts in the southern part of the Arabian Shield, Kingdom of Saudi Arabia. Saudi Arabian Deputy Ministry for Mineral Resources Open File Report USGS-OF-02-8

Gresse P G and Germs G J B (1993) The Nama foreland basin: sedimentation, major unconformity bounded sequences and multisided active margin advance. Prec Res 63: 247–272

Grew E S (1978) Precambrian basement at Molodezhnay Station, East Antarctica. Bull Geol Soc Am 89: 801–813

Grew E S, Manta W I and James P R (1988) U–Pb data in granulite facies rocks from Fold Island, Kemp Coast, East Antarctica. Prec Res 42: 63–75

Grieve R A F (1980) Impact bombardment and its role in protocontinental growth on the early Earth. Prec Res 10: 217–247

Grieve R A F and Head J W (1981) Impact cratering: A geological process on the planets. Episodes 1981-2: 3–9

Griffin W L, Taylor P N, Hakkinen J W et al (1978) Archean and Proterozoic crustal evolution in Lofoten-Vesteralen, N. Norway. J Geol Soc Lond 135: 629–647

Gross G A (1965) Iron-formation of Snake River area, Yukon and Northwest Territories. Geol Surv Can, Pap 65-1

Gross G A (1968) Geology of iron deposits in Canada. Vol 3—Iron Ranges of the Labrador Geosyncline. Geol Surv Can, Econ Geol Rep 22

Gross G A (1983) Tectonic systems and the deposition of iron-formation. In: Developments and Interactions of the Precambrian Atmosphere, Lithosphere and Biosphere (eds Nagy B, Weber R, Guerrero J C and Schidlowski M). Elsevier, Amsterdam, pp 63–79

Gross G A and Zajac I S (1983) Iron-formation in fold belts marginal to the Ungava Craton. In: Iron-Formation: Facts and Problems (eds Trendall A F and Morris R C). Elsevier, Amsterdam: 253–294

Groves D I (1982) The Archaean and earliest Proterozoic evolution and metallogeny of Australia. Rev Brasil Geociencias 12(1–3): 135–148

Groves D I and Batt W D (1984) Spatial and temporal variations of Archaean metallogenic associations in terms of evolution of granitoid-greenstone terrains with particular emphasis on the Western Australian Shield. In: Archaean Geochemistry (eds Kröner A, Hanson G N and Goodwin A M). Springer-Verlag, Berlin: 73–98

Groves D I, Ho S E, Rock N M S et al (1987) Archean cratons, diamond and platinum: Evidence for coupled long-lived crust-mantle systems. Geol 15: 801–805

Gruau G, Martin H, Leveque B and Capdevila R (1985) Rb-Sr and Sm-Nd geochronology of lower Proterozoic granite-greenstone terrains in French Guiana, South America. Prec Res 30: 63–80

Gulson B L (1972) The Precambrian geochronology of granitic rocks from Northern Sweden. Geol Fören Förhand 94: 229–244

Gulson B L and Krogh T E (1972) U/Pb zircon studies on the age and origin of post-tectonic intrusions from South Greenland. Rapp Grønlands Geol Unders 45: 48–53

Gulson B L, Perkins W G and Mizon K J (1983) Lead isotope studies bearing on the genesis of copper orebodies at Mount Isa, Queensland. Econ Geol 78: 1466–1504

Gupta J N, Pandey B K, Chabria T et al (1984) Rb-Sr geochronological studies on the granites of Vinukonda and Kanagiri, Prakasam district, Andhra Pradesh, India. Prec Res 26: 105–109

Gupta V J (1977) Indian Precambrian Stratigraphy. Hindustan Publishing Corp Press New Delhi

Gurnis M and Torsvik T H (1994) Rapid drift of large continents during the late Precambrian and Paleozoic: Paleomagnetic constraints and dynamic models. Geol 22: 1023–1026

Haddoum H, Choukroune P and Peucat J J (1994) Evolution of the Precambrian In-Ouzzal block (Central Sahara, Algeria). Prec Res 65: 155–166

Hagemann S G, Groves D I, Ridley J R and Vearncombe J R (1992) The Archean lode gold deposits at Wiluna, Western Australia: high-level brittle-style mineralization in a strike-slip regime. Econ Geol 87: 1022–1053.

Halden N M and Bowes D R (1984) Metamorphic development of cordierite-bearing layered schist and mica schist in the vicinity of Savonranta, eastern Finland. Bull Geol Soc Fin 56: 3–23

Halden N M, Bowes D R and Dash B (1982) Structural evolution of migmatites in granulite facies terrane: Precambrian crystalline complex of Angit, Orissa, India. Trans R Soc Edinburgh (Earth Sci) 73: 109–118

Hall W D M and Goode A D T (1978) The early Proterozoic Nabberu Basin and associated iron formations of Western Australia. Prec Res 7: 129–184

Hallberg J A (1986) Archaean basin development and crustal extension in the northeastern Yilgarn Block, Western Australia. Prec Res 31: 133–156

Hallberg J A and Glikson A Y (1981) Archaean granite-greenstone terranes of Western Australia. In: Precambrian of the Southern Hemisphere (ed Hunter D R). Elsevier, Amsterdam: 33–103

Halls H C (1978) The Late Precambrian central North American rift system—survey of recent geological and geophysical investigations. In: Tectonics and Geophysics of Continental Rifts (eds Newman E R and Rambergs I B). NATO Advanced Study Institute, Series C 37. D Reidel, Boston: 111–123

Halls H C (1987) Introduction; Concluding remarks. In: Mafic Dyke Swarms (eds Halls H C and Fahrig W F). Geol Ass Can Spec Pap 34: 1–3, 483–492

Halm J K E (1935) An astronomical aspect of the evolution of the earth. Presidential Address. Astronom Soc S Afr IV: 1–28

Hamilton J M, Bishop D T, Morris H C and Owens O E (1982) Geology of the Sullivan orebody, Kimberley, BC, Canada. In: Precambrian Sulphide Deposits, (eds Hutchinson R W, Spence C D and Franklin J M). Geol Ass Can Spec Pap 25: 597–665

Hamilton P J (1977) Sr isotope and trace element studies of the Great Dyke and Bushveld mafic phase and their relation to Early Proterozoic magma genesis in southern Africa. J Petrol 18: 24–52

Hamilton P J, O'Nions R K, Bridgwater D and Nutman A (1983) Sm-Nd studies of Archean metasediments and metavolcanics from West Greenland and their implications for the earth's early history. Earth Planet Sci Lett 62: 263–272

Hamilton W (1970) Bushveld Complex—product of impacts? Geol Soc S Afr Spec Public 1: 367–379

Hamilton W B (1989) Crustal geologic processes of the United States. In: Geophysical framework of the continental United States (eds Pakiser, L C and Mooney W D). Boulder, Colorado, Geol Soc of Am Mem 172: 743–781

Hamilton W B (1990) On terrane analysis. Phil Trans R Soc Lond A 331: 511–522

Hamilton W B (1993) Evolution of Archean mantle and crust. In: Precambrian: Conterminous (eds Reed J C et al.). Geol Soc Am, The Geology of North America, C-2: 597–614

Hancock S L and Rutland R W R (1984) Tectonics of an early Proterozoic geosuture: the Halls Creek orogenic subprovince, northern Australia. J Geodyn 1: 387–432

Hanmer S and McEachern S (1992a) Kinematical and rheological evolution of a crustal-scale ductile thrust zone, Central MetaSymposium; G A Gross and J A Donaldson (eds); Geol Surv Can, Open Fill 2163: 36–56

Hanmer S and McEachern S (1992b) Kinematical and rheological evolution of a crustal-scale ductile thrust zone, Central Metasedimentary Belt, Grenville orogeny, Ontario. Can J Earth Sci 29: 1779–1790

Hansen B T, Oberli F and Steiger R H (1973) The geochronology of the Scoresby Sûnd area. 4: Rb/Sr whole-rock and mineral ages. Rapp Grønlands Geol Unders 58: 55–58

Hansen B T, Oberli F and Steiger R H (1974) The geochronology of the Scoresby Sûnd area, central East Greenland. 6: Rb/Sr whole-rock and U-Pb ages. Rapp Grønlands Geol Unders 66: 32–38

Hansen E C, Newton R C and Janardhan A S (1984) Pressures, temperatures and metamorphic fluids across an unbroken amphibolite facies to granulite facies transition in southern Karnataka, India. In: Archaean Geochemistry (eds Kröner A, Hanson G N and Goodwin A M). Springer-Verlag, Berlin, pp 161–203

Hansen V L, Goodge J W, Keep M and Oliver D H (1993) Asymmetric rift interpretation of the western North American margin. Geol 21: 1067–1070

Hanson G N (1980) Rare earth elements in petrogenetic studies of igneous systems. Ann Rev Earth Planet Sci 8: 371–406

Hanson G N, Krogstad E J and Rajamani V (1988) Tectonic setting of the Kolar Schist Belt, Karnataka, India (abstract). J Geol Soc Ind 31: 40–42

Hanson R E, Wilson T J and Wardlaw M S (1988) Deformed batholiths in the Pan-African Zambezi belt, Zambia: age and implications for regional Proterozoic tectonics. Geol 16: 1134–1137

Hanson R E, Wardlaw M S, WIlson T J and Mwale G (1993) U-Pb zircon ages from the Hook granite massif and Mwembeshi dislocation: constraints on Pan-African deformation, plutonism, and transcurrent shearing in central Zambia. Prec Res 63: 189–209

Hargraves R B (1976) Precambrian geologic history. Sci 193: 363–371

Hargraves R B (1986) Faster spreading or greater ridge length in the Archean? Geol 14: 750–752

Harlan S S (1993) Paleomagnetism of Middle Proterozoic diabase sheets from central Arizona. Can J Earth Sci 30: 1415–1426

Harland W G (1983) The Proterozoic glacial record. In: Proterozoic Geology: Selected Papers from an International Proterozoic Symposium (eds Medaris L G, Mickelson D M, Byers C W and Shanks W C). Geol Soc Am Mem 11: 279–288

Harley S L (1989) The origin of granulites: a metamorphic perspective. Geol Mag 126: 215–247

Harper C T (ed.) (1973) Geochronology: Radiometric Dating of Rocks and Minerals. Dowden, Hutchinson and Ross, Stroudsburg, Pennsylvania

Harris A L, Baldwin C T, Bradbury H J et al (1978) Ensialic basin sedimentation: the Dalradian Supergroup. In: Crustal Evolution in northwestern Britain and adjacent regions (eds Bowes D R and Leake B E). Geol J Spec Issues 10: 115–138

Harris N B W, Santosh M and Taylor P N (1994) Crustal evolution in South India: Constraints from Nd isotopes. J Geol 102: 139–150

Harrison J E (1972) Precambrian Belt Basin in NW United States: its geometry, sedimentation and copper occurrences. Bull Geol Soc Am 83: 1215–1240

Hartnady C J H (1991) About turn for supercontinents. Nature 352: 476–478

Hartnady C J H (1993) Supercontinents and geotectonic megacycles. Am J Sci in press

Hartnady C J H, Newton A R and Theron J N (1974) The stratigraphy and structure of the Malmesbury Group in the southwestern Cape. Bull Precambrian Res Unit, University of Cape Town 15: 193–213

Hartnady C, Joubert P and Stowe C (1985) Proterozoic crustal evolution in southwestern Africa. Episodes 8: 236–244

Haslam H W, Darbshire D P F and Davies A E (1983) Irumide and post-Mozambique plutonism in Malawi. Geol Mag 120: 254–269

Hasui Y and Almeida F F M (1985) The Central Brazil Shield Reviewed. Episodes 8(1): 29–37

Hawkesworth C J, Menzies M A and van Calsteren P (1986) Geochemical and tectonic evolution of the Damara Belt, Namibia. In: Collision Tectonics (eds Coward M P and Ries A C). Geol Soc Spec Public 19: 305–319

Head J W and Crumpler L S (1987) Evidence for divergent plate-boundary characteristics and crustal spreading on Venus. Sci 238: 1380–1385

Heaman L M (1989) U–Pb dating of mafic dyke swarms: what are the options?

Heaman L M and Grotzinger J P (1992) 1.08 Ga diabase sills in the Pahrump Group, California: Implications for development of the Cordilleran miogeocline. Geol 20: 637–640

Heaman L M and LeCheminant A N (1988) U–Pb baddeleyite ages of the Muskox Intrusion and Mackenzie Dyke Swarm, NWT, Canada (Abs.). Geol Ass Canada, Joint Ann Mtg, St John's Newfoundland. (Program with Abs.) 13: A53

Heaman L M, LeCheminant A N and Rainbird R H (1992) Nature and timing of Franklin igneous events, Canada: Implications for a Late Proterozoic mantle plume and the break-up of Laurentia. Earth Planet Sci Lett 109: 117–131

Heaney P J and Veblen D R (1991) An examination of spherulitic dubiomicrofossils in Precambrian banded iron formations using the transmission electron microscope. Prec Res 49: 355–372

Hedge C E, Marvin R F and Naeser C W (1975) Age provinces in the basement rocks of Liberia. J Res US Geol Surv 3: 425–429

Hedge C E, Houston R S, Tweto O L et al (1986) The Precambrian of the Rocky Mountain region. US Geol Surv Prof Pap 1241D

Hegner V E, Kroner A and Hofmann A W (1984) Age and isotope geochemistry of the Archean Pongola and Usushwana suites, Swaziland, southern Africa: a case for crustal contamination of the mantle derived magma. Earth Planet Sci Lett 70: 267–279

Henderson J B (1981) Archaean basin evolution in the Slave Province, Canada; (ed Kröner A E). In: Precambrian Plate Tectonics. Elsevier, Amsterdam: 213–235

Henderson J R (1984) Description of a virgation in the Foxe Fold Belt, Melville Peninsula, Canada. In: Precambrian Tectonics Illustrated (eds Kröner A and Greiling R). Schweizerbart'sche Verlagsburchhandlung, Stuttgart, pp 251–261

Henriksen N and Higgins A K (1976) East Greenland Caledonian fold belt. In: Geology of Greenland (eds Escher A and Watt W S). The Geological Survey of Greenland, Copenhagen, pp 182–246

Herz N (1969) Anorthosite belts, continental drift, and the anorthosite event. Sci 164: 944–947

Herz N, Hasui Y, Costa J B S and Malta M A Da S (1989) The Araguaia Belt, Brazil: a reactivated Brasíliano–Pan-African cycle (550 Ma) suture. Prec Res 42: 371–386

Heubeck C and Lowe D R (1994) Depositional and tectonic setting of the Archean Moodies Group, Barberton Greenstone Belt, South Africa. Prec Res 68: 257–290

Hickman A H (1983) Geology of the Pilbara Block and its environs. Western Australia, Geol Surv Bull 127

Higgins M D and van Breemen O (1992) The age of the Lac-Saint-Jean Anorthosite Complex and associated mafic rocks, Grenville Province, Canada. Can J Earth Sci 29: 1412–1423

Hill R I and Campbell I H (1993) Age of granite emplacement in the Norseman region of Western Australia. Austral J Earth Sci 40: 559–574

Hill R I, Campbell I H and Compston W (1989) Age and origin of granitic rocks in the Kalgoorlie–Norseman region of W. Australia: implications for the origin of Archean crust Geochim Cosmochim Acta 53: 1259–1275

Hill R I, Campbell I H and Chappell B W (1992) Crustal growth, crustal reworking and granite genesis in the southeastern Yilgarn Block, Western Australia. In: The Archaean: Terrains, Processes and Metallogeny (ed. Glover J E). Univ West Austral Public 22: 203–213

Hinze W J and Wold R J (1982) Lake Superior geology and tectonics—overview and major unresolved problems. In: Geology and Tectonics of the Lake Superior Basin (eds Wold R J and Hinze W J). Geol Soc Am Mem 156: 273–280

Hirdes W, Davis D W and Eisenlohr B N (1992a) Reassessment of Proterozoic granitoid ages in Ghana on the basis of U/Pb zircon and monazite dating. Prec Res 56: 89–96

Hirdes W, Davis D W and Eisenlohr B N (1992b) Reassessment of Proterozoic granitoid ages on the basis of U–Pb zircon and monazite dating. Prec Res 56: 89–96.

Hiroi Y, Ogō Y and Namba K (1994) Evidence for prograde metamorphic evolution of Sri Lankan pelitic granulites, and implications for the development of continental crust. Prec Res 66: 245–263

Hirt A M, Lowrie W, Cleudenen W S and Kigfield R (1993) Correlation of strain and the anisotropy of magnetic susceptibility in the Onaping Formation: evidence for a near-circular origin of the Sudbury Basin. Tectonophys 225: 231–254

Hoal B G (1993) The Proterozoic Sinclair Sequence in southern Namibia: intracratonic rift or active continental margin setting? Prec Res 63: 143–162

Hobbs B E, Archibald N J, Etheridge M A and Wall V J (1984). Tectonic history of the Broken Hill Block, Australia. In: Precambrian Tectonics Illustrated (eds Kröner A and Greiling R). Schweizerbart'sche Verlagsbuchhandlung, Stuttgart, pp 353–368

Hoffman P F (1973) Evolution of an early Proterozoic continental margin: the Coronation geosyncline and associated aulacogens of the northwestern Canadian Shield. Philos Trans R Soc Lond (Ser A) 273: 547–581

Hoffman P F (1980) Wopmany Orogen: a Wilson cycle of Early Proterozoic age in the northwest of the Canadian Shield, In: The Continental Crust and its Mineral Deposits, (ed. D W Strangway). Geol Ass Can Spec Pap 20: 523–549

Hoffman P F (1981) Autopsy of Athopuscow Aulacogen: a failed arm affected by three collisions. In: Proterozoic Basins of Canada (ed. Campbell F H A). Geol Surv Can Pap 81-10: 97–102

Hoffman P F (1985) Is the Cape Smith Belt (northern Quebec) a klippe? Can J Earth Sci 22: 1361–1369.

Hoffman P F (1988). United plates of America, the birth of a craton. Ann Rev Earth Planet Sci 16: 543–603.

Hoffman P F (1989a) Precambrian geology and tectonic history of North America. In: The Geology of North America—An Overview (eds. Bally A W and Palmer A R). Boulder, Colorado, The Geol Soc Am. The Geology of North America, vol. A: 447–512.

Hoffman P F (1989b) Precambrian geology and tectonic history of North America. In: The Geology of North America—An overview (eds Bally A W and Palmer A R). Geol Soc Am, The Geology of North America A: 447–512

Hoffman P F (1990a) Dynamics of the tectonic assembly of northeast Laurentia in geon 18 (1.9–1.8 Ga). Geosci Can 17: 222–226

Hoffman P F (1990b) Subdivision of Churchill Province and extent of the Trans-Hudson Orogen. In: The Early Proterozoic Trans-Hudson Orogen of North America (eds Lewry J F and Stauffer M R. Geol Ass Can, Spec Pap 37: 15–40

Hoffman P F (1991) Did the breakout of Laurentia turn Gondwanaland inside out? Sci 252: 1409–1412

Hoffman P F, Bell I R, Hildebrand R S, and Thorstad L (1977) Geology of the Athapuscow Aulacogen, East Arm of Great Slave Lake, District of Mackenzie. In: Report of Activities, Part A. Geol Surv Can Pap 77-1A: 117–129.

Hoffman P F and St Onge M R (1981) Contemporaneous thrusting and conjugate transcurrent faulting during the second collision in Wopmay Orogen: implications for the subsurface structure of post-orogenic outliers. In Current Research, Part A. Geol Surv Can Pap 81-1A: 251–257

Hoffman P F and Bowring S A (1984) Short-lived 1.9 Ga continental margin and its destruction, Wopmay Orogen, northwest Canada. Geol, 12: 68–72

Hoffman P F, Card K D and Davidson A (1982). The Precambrian: Canada and Greenland (ed. Palmer A R). Geol Soc Am D-NAG Spec Public 1: 3–6

Hofmann H (1990) Precambrian time units and nomenclature—The geon concept. Geol 18: 340–341

Hofmann H (1991) Precambrian time units—Geon or geologic unit? Geol 19: 958–959

Hofmann H J (1981) Precambrian fossils in Canada—The 1970s in retrospect. In: Proterozoic Basins of Canada, (ed. F H A Campbell). Geol Surv Can Pap 81-10: 419–443

Hofmann H J, and Jackson G D (1969) Precambrian (Aphebian) microfossils from Belcher Islands, Hudson Bay. Can J Earth Sci 6: 1137–1144

Hofmann H J and Masson M (1994) Archean stromatolites from Abitibi greenstone belt, Quebec, Can Geol Soc Am Bull 106: 424–429

Hofmann H J, Narbonne G M and Aitken J D (1990) Ediacaran remains from intertillite beds in northwest Canada. Geol 18: 1199–1202

Holland H D (1973) The oceans: a possible source of iron in iron-formations. Econ Geol 68: 1169–1172

Holland H D (1984) The Chemical Evolution of the Atmosphere and Oceans. Princeton University Press, Princeton, New Jersey, U.S.A.: 582 pp

Holland H D and Beukes N J (1990) A paleoweathering profile from Griqualand West, South Africa: evidence for a dramatic rise in atmospheric oxygen between 2.2 and 1.9 BYBP. Am J Sci 290A: 1–34

Holland H D and Zbinden E A (1988) Paleosols and the evolution of the atmosphere, Part I. In: Physical and Chemical Weathering in Geochemical Cycles (eds Lerman A and Meybeck M). Reidel, Dordrecht, pp 61–82

Holm D K, Holst T B and Lux D R (1993) Postcollisional cooling of the Penokean orogen in east–central Minnesota. Can J Earth Sci 30: 913–917

Holmes A (1951) The sequence of pre-Cambrian orogenic belts in south and central Africa. 18th Int Geol Congr, Lond (1948) 14: 254–269

Holmes A and Lawson R W (1927) Factors involved in the calculation of the ages of radioactive minerals. Am J Sci 13: 327–344

Hölzl S, Hofmann A W, Todt W and Köhler H (1994) V–Pb geochronology of the Sri Lankan basement. Prec Res 66: 123–149

Horodyski R J (1983) Sedimentary geology and stromatolites of the Middle Proterozoic Belt Supergroup, Glacier National Park, Montana. Prec Res 20: 391–425

Horwitz R C (1990) Paleogeographic and tectonic evolution of the Pilbara Craton, Northwestern Australia. Prec Res 48: 327–340

Houston R S, Erslev E A, Frost C D et al. (1993) The Wyoming Province. In: Precambrian:Conterminous U.S. (eds. Reed J C, Bickford M E, Houston R S et al.). Geol Soc Am. The Geology of North America, C-2: 121–170

van Houten F B (1973) Origin of red beds. A review. Ann Rev Earth Planet Sci 11: 39–61Howell D G (1989) Tectonics of suspect terranes. London: Chapman and Hall (232 pp)

Hsü K J (1979) Thin skinned plate tectonics during neo-alping orogenesis. Am J Sci 279: 353–366

Hunter D R (ed.) (1981) Precambrian of the Southern Hemisphere. Elsevier, Amsterdam

Hunter D R and Hamilton P J (1978) The Bushveld Complex. In: Evolution of the Earth's crust, (ed. D H Tarling) Academic Press, London: 107–173

Hunter D R and Pretorius D A (1981) Structural framework (southern Africa) In: Precambrian of the southern Hemisphere, (ed. D R Hunter) Elsevier, Amsterdam: 397–422

Hunter D R and Wilson A H (1988) A continuous record of Archean evolution from 3.5 Ga to 2.6 Ga in Swaziland and northern Natal. S Afr J Geol 91(1): 57–74

Hunter D R, Barker F and Millard H T (1984) Geochemical investigation of Archaen bimodal and Dwalile metamorphic suites, Ancient Gneiss Complex, Swaziland. Prec Res 24: 131–155

Hurley P M (1968) Absolute abundance and distribution of Rb, K and Sr in the earth. Geochim Cosmochim Acta 32: 273–283

Hurley P M and Rand J R (1969) Pre-drift continental nuclei. Sci 164: 1229–1242

Hurley P M, Fairbairn H W and Gaudette H E (1976) Progress report on Early Archean rocks in Liberia, Sierra Leone and Guyana, and their general stratigraphic setting. In: The Early History of the Earth, (ed. Windley B F) John Wiley, New York, pp 511–521

Hurley P M, Leo G W, White R W and Fairbairn H W (1971) Liberian age Province (about 2700 m.y.) and adjacent provinces in Liberia and Sierra Leone. Bull Geol Soc Am 82: 3483–3490

Hurst G S W (1990) Geology and Mineral Resources of Western Australia (Mem. 3) with maps. Director Geol Surv W Austral, Mem 3: 826 pp

Hutchinson R W (1981) Mineral deposits as guides to supracrustal evolution. In: O'Connell R J and Fyfe W S, eds., Evolution of the earth. Geodyn Ser 5: 120–140

Hynes A and Francis D M (1982) A transect of the early Proterozoic Cape Smith fold belt, New Quebec. Tectonophys, 88: 23–59.

Ilyin A V (1990) Proterozoic supercontinent, its latest Precambrian rifting, breakup, dispersal into smaller continents, and subsidence of its margins: Evidence from Asia. Geol 18: 1231–1234

Indares A (1993) Eclogitized gabbros from the eastern Grenville Province: textures, metamorphic context, and implications. Can J Earth Sci 30: 159–173

Ireland T R and Wlotzka F (1992) The oldest zircons in the solar system. Earth Planet Sci Lett 109: 1–10

Irvine T N (1970) Crystallization sequences in the Muskox Intrusion and other layered intrusions. I, Olivine-pyroxene-plagioclase relations. Geol Soc S Afr Spec Public 1: 441–476

Irvine T N (1980) Magmatic infiltration metasomatism, double-diffusive fractional crystallization, and adcumulus growth in the Muskox Intrusion and other layered intrusions. In: Physics of Magmatic Processes, (ed. Hargraves R B) Princeton University Press, pp 325–383

Irvine T N, Keith D W and Todd S G (1983) The J M platinum-palladium reef of the Stillwater Complex, Montana: II. Origin by double-diffusive convection magma mixing and implications for the Bushveld Complex. Econ Geol 78: 1287–1334

Irving E and McGlynn J C (1981) On the coherence, rotation and paleolatitude of Laurentia in the Proterozoic. In: Precambrian Plate Tectonics, (ed. Kröner A) Elsevier, Amsterdam: 561–598

Isachsen C E and Bowring S A (1994) Evolution of the Slave craton. Geol 22: 917–920

Jackson E D (1970) The cyclic unit in layered intrusions—a comparison of repetitive stratigraphy in the ultramafic rocks of the Stillwater, Great Dyke and Bushveld Complexes. Geol Soc S Afr Spec Public 1: 391–424

Jackson G D and Iannelli T R (1981) Rift-related cyclic sedimentation in the Neohelikian Borden Basin, northern Baffin Island. In: Proterozoic Basins of Canada, (ed. Campbell F H A). Geol Surv Can Pap 81-10: 269–302.

Jackson M C (1991) The Dominion Group—A review of the late Archaean volcano-sedimentary sequence and implications for the tectonic setting of the Witwatersrand Supergroup, South Africa. Econ Geol Res Unit, Univ of Witwatersrand, Info Circ 240: 40 pp

Jackson M J, Muir M D and Plumb K A (1987) Geology of the southern McArthur Basin, Northern Territory. Bull Bur Min Res Austral 220

Jackson S L, Fyon J A and Corfu F (1994) Review of Archean supracrustal assemblages of the southern Abitibi greenstone belt in Ontario, Canada: product of microplate interaction within a large-scale plate-tectonic setting. Prec Res 65: 183–205

Jackson S L, Sutcliffe R H, Ludden J H, Hubert C, Green A G, Milkrereit B, Mayrand L, West GF and Verpaelst P (1990) Southern Abitibi greenstone belt: Archean crustal structure from seismic-reflection profiles. Geol 18: 1086–1090

Jacobsen S B and Pimentel-Klose M R (1988) A Nd isotopic study of the Hamersley and Michipicoten banded iron formations: the source of REE and Fe in Archean oceans. Earth Planet Sci Lett 87: 29–44

Jahn B-M (1990) Early Precambrian basic rocks of China. In: Early Precambrian Basic Magmatism, Editors R P Hall and D J Hughs. Blackie, Bishopbriggs, Glasgow: 294–316

Jahn B-M and Ernst W G (1990) Late Archean Sm-Nd isochron age for mafic-ultramafic supracrustal amphibolite from the northeastern Sino-Korean Craton, China. Prec Res 46: 295–306

Jahn B-M and Schrank A (1983) REE geochemistry of komatiites and associated rocks from Piumhi, southeastern Brazil. Prec Res 21: 1–20

Jahn B-M and Zhang Z Q (1984a) Archean granulite gneisses from eastern Hebei Province, China: rare earth geochemistry and tectonic implications. Cont Mineral Petrol 85: 225–243

Jahn B-M and Zhang Z Q (1984b) Radiometric ages (Rb-Sr, Sm-Nd, U-Pb) and REE geochemistry of Archean granulite gneisses from eastern Hebei Province, China. In: Archean Geochemistry: (eds Kröner A, Hanson G N and Goodwin A M) Springer-Verlag, Berlin: 204–234

Jahn B-M, Vidal P and Kröner A (1984) Multi-chronometric ages and origin of Archean tonalitic gneisses in Finnish Lapland: a case for long crustal residence time. Cont Mineral Petrol 86: 398–408

Jahn B-M, Auvray B, Cornichet J et al (1987) 3.5 Ga old amphibolites from eastern Hebei Province, China: field occurrence, petrography, Sm-Nd isochron age and REE geochemistry. Prec Res 34: 311–346

Jahn B-M, Bertrand-Safarti J, Morin N and Mace J (1990) Direct dating of stromatolitic carbonates from the Schmidtsdrif Formation (Transvaal Dolomite), South Africa, with implications on the age of the Ventersdorp Supergroup. Geol 18: 1211–1214

James H L (1954) Sedimentary facies of iron-formation. Econ Geol 49: 235–293

James H L (1955) Zones of regional metamorphism in the Precambrian of northern Michigan Bull Geol Soc Am 66: 1455–1487

James H L (1983) Distribution of banded iron-formation in space and time. In: Iron-Formations: Facts and Problems (eds Trendall A F and Morris R C). Elsevier, Amsterdam; pp 471–490

James P R and Black L P (1981) A review of the structural evolution and geochronology of the Archaean Napier Complex of Enderby Land, Australian Antarctic Territory. In: Archaen Geology, (eds Glover J E and Groves D I). Geol Soc Austral Spec Public 7: 71–83

James P R and Tingey R J (1983) The Precambrian geological evolution of the East Antarctic Metamorphic Shield- a review. In: Antarctic Earth Science, (eds Oliver R L, James P R and Jago J B). Cambridge University Press, pp 5–10

Janardhan A S, Srinkantappa C, Ramachandra H M (1978) The Sargur schist complex—an Archaean high-grade terrain in southern India. In: Developments in Precambrian Geology I. Archaean Geochemistry (eds Windley B F and Naqvi S M). Elsevier, Amsterdam; pp 127–150

Janardhan A S, Jayananda M and Shankara M A (1994) Formation and tectonic evolution of granulites from the Biligiri Rangan and Niligiri Hills, S. India: Geochemical and isotopic constraints. J Geol Soc Ind, 44: 27–40

Janasi V de A and Ulbrich H H G J (1991) Late Proterozoic granitoid magmatism in the state of São Paulo, southeastern Brasil. Prec Res 51: 351–374

Jansen H (1981) The Waterberg and Soutpansberg Groups. In: Precambrian of the Southern Hemisphere: (ed. Hunter D R). Elsevier, Amsterdam: 536–556

Jayananda M and Mahabaleswar B (1991) The generation and emplacement of Closepet granite during late Archaean granulite metamorphism in southeastern Karnataka. J Geol Soc Ind 38: 418–426

Jenkins R J F (1984) Ediacaran events: boundary relationships and correlation of key sections, especially in 'Armorica'. Geol Mag 121: 635–643

Jensen L S (1985) Stratigraphy and petrogenesis of Archean metavolcanic sequences, southwestern Abitibi Subprovince, Ontario. In: Evolution of Archean Supracrustal Sequences, (eds Ayres L D, Thurston P C, Card K D and Weber W). Geol Ass Can Spec Pap 28: 65–87

Johansson L and Kullerud L (1993) Late Sveconorwegian metamorphism and deformation in southwestern Sweden. Prec Res 64: 347–360

Johansson Å. Meier M, Oberli F and Wilkman H (1993) The early evolution of Southwest Swedish Gneiss Province: geochronological and isotopic evidence from southernmost Sweden. Prec Res 64: 361–388

Johnson M R, Cornell D H and Walraven F (1989) A revised Precambrian time scale for South Africa. Geological Survey, South African Committee for Stratigraphy, Chronostratigraphic Series, no. 1: 6 pp.

Jolly W T and Smith R E (1972) Degradation and metamorphic differentiation of the Keweenawan Tholeiitic Lavas of Northern Michigan, USA. J Petrol 13: 273–309

Jost H and Oliveira A M de (1991) Stratigraphy of the greenstone belts, Goiás, central Brazil. J S Am Earth Sci 4: 201–214

Joubert P (1981) The Namaqualand Metamorphic Complex. In:

Precambrian of the Southern Hemisphere: (ed Hunter D R). Elsevier, Amsterdam pp 671–705

Jourde, G (1983) La chaîne de Lurio (Mozambique): un témoin de l'existence de chaînes kibariennes en Afrique orientale. Abs. 12th Coll Afr Geol Brussels; p 50

Kale V S (1991) Constraints on the evolution of the Purana basins of Peninsular India. J Geol Soc Ind 38: 231–252

Kalliokoski J (1965) Geology of north-central Guayana Shield, Venezuela. Bull Geol Soc Am 76: 1027–1049

Kalsbeek F and Myers J S (1973) The geology of the Fiskenaesset region. Rapp Grønlands Geol Unders 51: 5–18

Kalsbeek F and Taylor P N (1985) Pb-isotopic studies of Proterozoic igneous rocks, West Greenland, with implications on the evolution of the Greenland Shield. In: The Deep Proterozoic Crust in the North Atlantic Provinces (eds Tobi A C and Touret J L R). D Reidel, Dordrecht; pp 237–245

Kalsbeek F, Taylor P N and Pidgeon R T (1988) Unreworked Archaean basement and Proterozoic supracrustal rocks from northeastern Distro Bugt, West Greenland: implications for the nature of Proterozoic mobile belts in Greenland. Can J Earth Sci 25: 773–782

Kalsbeek F, Australrheim H, Bridgwater D, Hansen B T, Pederson S and Taylor P N (1993a) Geochronology of Archaean and Proterozoic events in the Ammassalik area, South-East Greenland, and comparisons with the Lewisian of Scotland and the Nagssugtoqidian of West Greenland. Prec Res 62: 239–270

Kalsbeek F, Nutman A P and Taylor P N (1993b) Palaeoproterozoic basement province in the Caledonian fold belt of North-East Greenland. Prec Res 63: 163–178

Kamo S L and Davis D W (1994) Reassessment of Archean crustal development in the Barberton Mountain Land, South Africa, based on U–Pb dating. Tectonics 13: 167–192

Karlstrom K E and Bowring S A (1993) Proterozoic orogenic history of Arizona. In: Precambrian Conterminous U.S. (eds. Reed J C Jr et al). Geol Soc Am, Geology of North America, C-2: 188–211

Katz M B (1971) The Precambrian metamorphic rocks of Ceylon. Geol Rundsch 60: 1523–1549

Kaufman A J, Knoll A H and Awramik S M (1992) Biostratigraphic and chemostratigraphic correlation of Neoproterozoic sedimentary successions: Upper Tindir Group, northwestern Canada, as a test case. Geol 20: 181–185

Kaula W M (1980) The beginning of the Earth's thermal evolution. In: The Continental Crust and its Mineral Deposits, (ed. Strangway D W). Geol Ass Can, Spec Pap 20: 25–34

Kay R W and Mahlburg Kay S (1993) Delamination and delamination magmatism. In: A G Green, A Kröner, H-J Götze and N. Pavlenkove (Editors), Plate Tectonic Signatures in the Continental Lithosphere. Tectonophys 219: 177–189

Kazansky V I and Moralev V M (1981) Archean geology and metallogeny of the Aldan Shield, USSR In: Archean Geology (eds Glover J E and Groves D I). Geol Soc Austral Spec Public 7: 111–120

Kearns C, Ross G M, Donaldson J A and Geldsetzer H J (1981) The Helikian Hornby Bay and Dismal Lakes Groups, District of Mackenzie. In: Proterozoic Basins in Canada (ed. Campbell F H A). Geol Surv Can Pap 81-10: 157–182

Kellet R L, Barnes A E and Rive M (1994) The deep structure of the Grenville Front: a new perspective from western Quebec. Can J Earth Sci. 31: 282–292

Kennedy M (1993) The Undoolya Sequence: Late Proterozoic salt influenced deposition, Amadeus Basin, central Australia. Austral J Earth Sci 40: 217–228

Kerr A, Krogh T E, Corfu F, Schärer U, Gandhi S S and Kwok Y Y (1992) Episodic Early Proterozoic granitoid plutonism in the Makkovik Province, Labrador: U–Pb geochronological data and geological implications. Can J Earth Sci 29: 1166–1179

Kerr J W (1980) A plate tectonic contest in Arctic Canada. In: The Continental Crust and its Mineral Deposits (ed. Strangway D W). Geol Ass Can Spec Pap 20: 457–486

Ketchum J W F, Jamieson R A, Heaman L M, Culshaw N G and Krogh T E (1994) 1.45 Ga granulites in the southwestern Grenville province: Geologic setting, P–T conditions, and U–Pb geochronology. Geol 22: 215–218

Key R M, Charlesley T J, Hackman B D et al (1989) Superimposed upper Proterozoic collision-controlled orogenies in the Mozambique orogenic belt of Kenya. Prec Res 44: 197–225

Khain V E (1985) Geology of the USSR. First part. Old Cratons and Paleozoic Fold Belts. Gebruder Born Traeger, Berlin

Khan R M K, Govil P K and Naqvi S M (1992) Geochemistry and genesis of banded iron formation from Kudremukh schist belt, Karnataka Nucleus, India. J Geol Soc Ind, 40: 311–328

Kimberley M M (1989) Exhalative origins of iron formations. Ore Geol Rev 5: 13–145

Kinloch E D and Peyerl W (1990) Platinum-group minerals in various rock types of the Merensky Reef: genetic implications. Econ Geol 85: 537–555

Kinny D D (1986) 3820 Ma zircons from a tonalitic Amîtsoq gneiss in the Godthåb district of southern West Greenland. Earth Planet Sci Lett 79: 337–347

Kinny P D, Williams I S, Froude D O, Ireland T R and Compston W (1988) Early Archaean zircon ages from orthogneisses and anorthosites at Mount Narryer, Western Australia. Prec Res 38: 325–341

Kirby G A (1979) The Lizard complex as an ophiolite. Nature 282: 58–61

Kirschvink J L (1992) Late Proterozoic low-latitude global glaciation: The snowball Earth. In: The Proterozoic Biosphere: A Multidisciplinary Study (eds. Schopf, J W and Klein C). Cambridge University Press: 51–53

Klasner J S, Cannon W F and Van Schmus W R (1982) The pre-Keweenawan tectonic history of southern Canadian Shield and its influence on formation of the Midcontinent Rift. Geol Soc Am Mem 156: 27–46

Klein C and Beukes N J (1990) Geochemistry and sedimentology of a facies transition from limestone to iron-formation deposition in the Early Proterozoic Transvaal Supergroup, South Africa. Econ Geol 84: 1733–1774

Klein C and Beukes N J (1993) Sedimentology and geochemistry of the glaciogenic late Proterozoic Rapitan iron-formation in Can. Econ Geol 88: 542–565

Klemperer S L (1992) Introduction: deep crustal probing. Prec Res 55: 169–172

Klerkx J (1971) Caractéres metamorphiques et structuraux du socle pré-cambrien de la région d'Uweinat (Libie). CR Acad Sci Paris D272: 3246–3248

Klerkx J, Liegeois J P, Lavreau J and Claessens W (1987) Crustal evolution of the northern Kibaran Belt, eastern and central Africa. In: Proterozoic Lithospheric Evolution (ed. Kröner A). Am Geophys Union (Geodyn Ser) 17: 217–233

Knight I and Morgan W C (1981) The Aphebian Ramah Group, Northern Labrador. In: Proterozoic Basins of Canada (ed. Campbell F H A). Geol Surv Can Pap 81-10: 313–330

Knoll A H and Butterfield N J (1989) New window on Proterozoic life. Nature 337: 602–603

Knutson J, Donnelly T H, Eadington P J and Tonkin D G (1992) Hydrothermal alteration of Middle Proterozoic Basalts, Stuart Shelf, South Australia—A possible source for Cu mineralization. Econ Geol 87: 1054–1077.

de Kock G S (1992) Forearc basin evolution in the Pan-African Damara Belt, central Namibia: the Hureb Formation of the Khomas Zone. Prec Res 57: 167–194

Koeberl C, Reimold W U and Boer R H (1993) Geochemistry and mineralogy of Early Archean spherule beds, Barberton Mountain Land, South Africa: evidence for origin by impact doubtful. Earth Planet Sci Lett 119: 441–452

Koeppel V (1968) Age and history of the uranium mineralization of the Beaverlodge area, Saskatchewan. Geol Surv Can, Pap 67-31

Koistinen T J (1981) Structural evolution of an early Proterozoic strata-bound Cu-Co-Zn deposit, Outokumpu, Finland. Trans R Soc Edinburgh (Earth Sci) 72: 115–158

Kontinen A (1987) An early Proterozoic ophiolite—the Jormua mafic-ultramafic complex, northeastern Finland. Prec Res 35: 313–341

Kralik M (1982) Rb-Sr age determinations on Precambrian carbonate rocks of the Carpentarian McArthur Basin, Northern Territory, Australia. Prec Res 18: 157–170

Krapez B (1993) Sequences stratigraphy of the Archean supracrustal belts of the Pilbara Block, Western Australia. In: Archean and Early Proterozoic Geology of the Pilbara Region, Western Australia (eds. Blake T S and Meakins A L). Prec Res 60: 1–45

Krogh T E (1973) A low-contamination method for hydrothermal decomposition of zircon and extraction of U and Pb for isotopic age determinations. Geochim Cosmochim Acta 46: 637–649

Krogh T E (1982) Improved accuracy of U-Pb zircon ages by the creation of more concordant systems using an air abrasion technique. Geochim Cosmochim Acta 46: 637–649

Krogh T E and Davis G L (1975) The production and preparation of $^{205}$Pb for use as a tracer for Isotope Dilution Analyses. Carnegie Inst Washington, Yearbook 74: 416–417

Krogh T E, McNutt R H and Davis G L (1982) Two high precision U–Pb Zircon ages for the Sudbury Nickel Irruptive. Can J Earth Sci 19: 723–728

Krogh T E, Davis D W and Corfu F (1984) Implications of precise U-Pb dating for the geological evolution of the Superior Province. Geol Ass Can (Abs with Program) 9: 79

Krogstad E J, Hanson G N and Rajamani V (1991) U–Pb ages for zircon and sphene for two gneiss terranes adjacent to the Kolar schist belt, south India: Evidence for separate crustal evolution histories. J Geol 99: 801–816

Kröner A (1974) Late Precambrian formations in the western Richtersveld, northern Cape Province. Trans R Soc S Afr 41: 375–433

Kröner A (1978) The Namaqua Mobile Belt within the framework of Precambrian crustal evolution in southern Africa. Geol Soc S Afr Spec Public 4: 181–187

Kröner A (1980) Chronologic evolution of the Pan African Damara belt in Namibia, South West Africa. Prec Res 5: 311–357

Kröner A (1981) Precambrian plate tectonics. In: Precambrian Plate Tectonics (ed Kröner A). Elsevier Amsterdam: 57–90

Kröner A (1984) Evolution, growth and stabilization of the Precambrian lithosphere. Phys Chem Earth 15: 69–106

Kröner A (1985a) Evolution of the Archean continental crust. Ann Rev Earth Planet Sci 13: 49–74

Kröner A (1985b) Ophiolites and the evolution of tectonic boundaries in the Late Proterozoic Arabian-Nubian Shield of northeast Africa and Arabia. Prec Res 27: 277–300

Kröner A and Blignault H J (1976) Towards a definition of some tectonic and igneous provinces in western South Africa and southern South West Africa. Trans Geol Soc S Afr 79: 232–238

Kröner A and Correia H (1980) Continuation of the Pan-African Damara belt into Angola: a proposed correlation of the Chela group in southern Angola with the Nosib group in northern Namibia (SWA). Trans Geol S Afr 83: 5–16

Kröner A and Tegtmeyer A (1994) Gneiss-greenstone relationships in the Ancient Gneiss Complex of southwestern Swaziland, southern Africa, and implications for early crustal evolution. Prec Res 67: 109–139

Kröner A and Todt W (1988) Single zircon dating constraining the maximum age of the Barberton Greenstone Belt, Southern Africa: J Geophy Res 93: 15329–15337

Kröner A, Stern R J, Dawoud A S, Compston W and Reischmann T (1987a) The Pan-African continental margin in northeastern Africa: evidence from a geochronological study of granulites at Sabaloka, Sudan. Earth Planet Sci Lett 85: 91–104

Kröner A, Compston W, Tegtmeyer A and Milisenda C (1987b) Growth of early Archaean crust in the Ancient Gneiss Complex of Swaziland and adjacent Barberton Greenstone Belt, southern Africa (Abs.) Workshop on the Growth of Continental Crust. Lunar and Planetary Institute, Houston

Kröner A, Compston W and Williams I S (1989) Growth of early Archaean crust in the Ancient Gneiss Complex of Swaziland as revealed by single zircon dating. Tectonophys 161: 271–298

Kröner A, Linnebacher P, Stern R J, Reischmann T, Manton W and Hussein I M (1991) Evolution of the Pan-African island arc assemblages in the southern Red Sea Hills, Sudan, and in southwestern Arabia as exemplified by geochemistry and geochronology. Prec Res 53: 99–118

Kröner A, Pallister J S and Fleck R J (1992a) Age of initial oceanic magmatism in the Late Proterozoic Arabian Shield. Geol 20: 803–806

Kröner A, Todt W, Hussein I M, Mansour M and Rashwan A A (1992b) Dating of late Proterozoic ophiolites in Egypt and the Sudan using the single grain zircon evaporation technique. Prec Res 59: 15–32

Kröner A, Kehelpannala K V W and Kriegsman L M (1994a) Origin of compositional layering and mechanism of crustal thickening in the high-grade gneiss terrain of Sri Lanka. Prec Res 66: 21–37

Kröner A, Jaeckel P and Williams I S (1994b) Pb-loss patterns in zircons from a high-grade metamorphic terrain as revealed by different dating methods: U–Pb and Pb–Pb ages for igneous and metamorphic zircons from northern Sri Lanka. Prec Res 66: 151–181

Kropotkin P N and Yefremov V N (1992) Changes in the earth's radius in the geological past. Geotect 26: 263–271

Krupp R, Oberthür T and Hirdes W (1994) The early Precambrian atmosphere and hydrosphere: Thermodynamic constraints from mineral deposits. Econ Geol 89: 1581–1598

Kumar G R R and Raghavan V (1992) The incipient charnockites of Transition Zone, Granulite Zone and Khondalite Zone of South India: Contrasting mechanisms and controlling factors. J Geol Soc Ind, 39: 293–302

Kumarapeli P S and Saul V A (1966) The St Lawrence valley system: a North American equivalent of the East African rift valley system. Can J Earth Sci 3: 639–658

Kuyumjian R M and Dardenne M A (1982) Geochemical characteristics of the Crixas greenstone belt, Goias Brazil. Rev Brasil Geociencias 12(103): 324–330

Labotka T C, Albee A L, Lanphere M A and McDowell S D (1980) Stratigraphy, structure and metamorphism in the central Panamint Mountains (Telescope Peak quadrangle), Death Valley area, California: Summary. Bull Geol Soc Am 91: 125–129

Ladeira E A and Noce C M (1990) New U-Pb ages for Precambrian rocks of the Quadrilatero Ferrifero, Brazil: Soc Econ Geol, Newslett 1: 9

Laflèche M R, Dupuy C and Dostal J (1992) Tholeiitic volcanic rocks of the late Archean Blake River Group, southern Abitibi greenstone belt: origin and geodynamic considerations. Can J Earth Sci 29: 1448–1458

Laird M G (1981) Lower Palaeozoic rocks of Antarctica. In: Lower Palaeozoic of the Middle East, Eastern and Southern Africa, and Antarctica. (ed. Holland C H) John Wiley, New York, pp 257–314

Lamb S H (1984) Structures on the eastern margin of the Archaean Barberton greenstone belt, northwest Swaziland. In: Precambrian Tectonics Illustrated (eds Kröner A and Greiling R) E Schwerzerbartsche Verlagsbuchhandlong, Stuttgart: 19–39

Lambert I B (1983) The major stratiform lead-zinc deposits of the Proterozoic. Geol Soc Am Mem 161: 209–226

Lambert M B (1976) The Back River Volcanic Complex, District of Mackenzie Geol Surv Can Pap 76-1A: 363–367

Lambert M B (1978) The Back River Volcanic Complex—a cauldron subsidence structure of Archean Age. Geol Surv Can Pap 78-1A: 153–157

Lancelot J R, Vitrac A and Allègre C J (1976) Uranium and lead isotope dating with grain-by-grain zircon analysis: a study of complex geological history with a single rock. Earth Planet Sci Lett 29: 357–366

Larson E E, Patterson P E and Mutschler F E (1994) Lithology, chemistry, age and origin of the Proterozoic Cardenas Basalt, Grand Canyon, Arizona. Prec Res 65: 255–276

Lasserre M (1978) Mise au point les granitoides dits 'Ultimes' du Cameroon: gisement, pétrographie et géochronologie. Bull Bur Réch Géol Min Paris, 2$^{eme}$ Série, Sect 4, No 2-1978: 143–159

Latouche L (1978) Etude pétrographique et structurale du Précambrien de la région des Gour Oumelalen (Nord-Est de l'Ahaggar, Alberie). Thèse Doct Etat Univ Paris

Latouche L and Vidal Ph (1974) Géochronologie du Précambrien de la région des Gour Oumelalen (Nord-Est de l'Ahaggar, Algérie). Un exemple de mobilisation du strontium radiogénique. Bull Soc Géol Fr 16(7): 195–203

Lauerma R (1982) On the ages of some granitoid and schist complexes in northern Finland. Bull Geol Soc Fin 54: 85–100

Lavreau J (1982b) The Archean and Lower Proterozoic of Central Africa. Rev Brasil Geociencias 12(1-3): 187–192

Law J D M, Bailey A C, Cadle A B, Phillips G N and Stanistreet J G (1990) Reconstructive approach to the classification of Witwatersrand "quartzites". S Afr J Geol 93 (1): 83–92

Lawson A C (1885) Report on the geology of the Lake-of-the-Woods region with special reference to the Keewatin (Huronian?) belt of the Archean rocks. Geol Surv Canada Ann Rep (New Series), Vol. 1, Rep CC

Lawson A C (1934) The Eparchean peneplain: an exploitation of the doctrine of isostasy. Bull Geol Soc Am 45: 1059–1072

Layer P W, Kröner A and York D (1992) Pre-3000 Ma thermal history of the Archean Kaap Valley pluton, South Africa. Geol 20: 717–720

Leblanc M (1975) Ophiolites précambriennes et gites arséniés de cobalt: Bou-Azzer (Maroc) Thèse Univ Paris (VI)

Leblanc M (1976) Proterozoic oceanic crust at Bou Azzer. Nature 261: 34–35

Leblanc M (1981) The Late Proterozoic ophiolites of Bou Azzer (Morocco): evidence for Pan-African plate tectonics, In: Precambrian Plate Tectonics, (ed. Kröner A). Elsevier, Amsterdam: 435–451

Leblanc M and Moussine-Pouchkine A (1994) Sedimentary and volcanic evolution of a Neoproterozoic continental margin (Bleida, Anti-Atlas, Morocco). Prec Res 70: 25–44

LeCheminant A N, Miller A R and LeCheminant G M (1987) Early Proterozoic alkaline igneous rocks, District of Keewantin, Canada: petrogenesis and mineralization. In: Geochemistry and Mineralization of Proterozoic Volcanic Suites, (eds Pharaoh T C, Beckinsale R D and Rickard D). Geol Soc Spec Publ 33: 219–240

Ledru P, N'dong J E, Johan V et al (1989) Structural and metamorphic evolution of the Gabon Orogenic Belt: collision tectonics in the lower Proterozoic. Prec Res 44: 227–240

Ledru P, Pons J, Milési J P, Feybesse J L and Johan V (1991) Transurrent tectonics and polycyclic evolution in the Lower Proterozoic of Senegal–Mali. Prec Res 50: 337–354

Ledru P, Johan V, Milési J P and Tegyey M (1994) Markers of the last stages of the Palaeoproterozoic collision: evidence for a 2 Ga continent involving circum South Atlantic provinces. Prec Res 69: 169–191

LeGallais C J and Savoie S (1982) Basin evolution of the Lower Proterozoic Kaniapiskan Supergroup, central Labrador Miogeosyncline (Trough), Quebec. Bull of Can Petrol Geol 30: 150–166

Legault F, Francis D, Hynes A and Budkewitsch P (1994) Proterozoic continental volcanism in the Belcher Islands: implications for the evolution of the Circum Ungava Fold Belt. Can J Earth Sci 31: 1536–1549

Lehtonen M, Manninen T, Rastas P et al (1985) Keski-Lapin geologisen Kartan selitys. Summary and discussion: explanation to the geological map of Central Lapland. Geol Surv Fin Rep Invest 71: 1–35

Leube A, Hirdes W, Mauer R and Kesse G O (1990) The early Proterozoic Birimian Supergroup of Ghana and some aspects of its associated gold mineralization. Prec Res 46: 139–165

Lewry J F and Sibbald T I I (1980) Thermotectonic evolution of the Churchill Province in northern Saskatchewan. Tectonophys 68: 45–82

Lewry J F, Sibbald T I I and Rees C J (1978) Metamorphic patterns and their relation to tectonism and plutonism in the Churchill Province in northern Saskatchewan. In: Metamorphism in the Canadian Shield (eds Fraser J A and Heywood W W) Geol Surv Can Pap 78-10: 139–154

Lewry J F, Stauffer M R and Fumerton S (1981) A Cordilleran-type batholithic belt in the Churchill Province in northern Saskatchewan. Prec Res 14: 277–313

Lewry J F, Sibbald T I I and Schledewitz D C P (1985) Variations in character of Archean rocks in the western Churchill Province and its significance. In: Evolution of Archean Supracrustal Sequences (eds Ayres L D, Thurston P C, Card K D and Weber W). Geol Ass Can Spec Pap 28: 239–261

Lewry J F, Macdonald R, Livesey C et al (1987) U-Pb geochronology of accreted terranes in the Trans-Hudson Orogen, northern Saskatchewan, Canada. In: Geochemistry and Mineralization of Proterozoic Volcanic Suites, (eds Pharaoh T C, Beckinsale R D and Rickard D) Geol Soc Spec Pub 33: 147–166

Lewry J F, Hajnal Z, Green A, Lucas S B, White D, Stauffer M R, Ashton K E, Weber W and Clowes R (1994) Structure of a Paleoproterozoic continent–continent collision zone: a LITHOPROBE seismic reflection profile across the Trans-Hudson Orogen, Canada. Tectonophys 232: 143–160

Libby W G, de Laeter J R and Myers J S (1986) Geochronology of the Gascoyne Province. Geol Surv W Austral, Rep 20

Liégeois J P, Claessens W, Camara D and Klerkx J (1991) Short-lived Eburnean orogeny in southern Mali. Geol, tectonics, U–Pb and Rb–Sr geochronology. Prec Res 50: 111–136

Liégeois J P, Black R, Navez J and Latouche L (1994) Early and late Pan-African orogenies in the Aïr assembly of terranes (Tuareg shield, Niger). Prec Res 67: 59–88

Liew T C, Kröner A, Milisenda C and Hofmann A W (1987) Crust formation and recycling signatures in high-grade metamorphic rocks of Sri Lanka: U-Pb zircon and Sm-Nd isotopic evidence. Terra cognita 7: 332–333

Lightfoot P C, De Souza H and Doherty W (1993) Differentiation and source of the Nipissing Diabase intrusions, Ontario, Canada. Can J Earth Sci 30: 1123–1140

Link P K et al. (1993) Middle and Late Proterozoic stratified rocks of the western U.S. Cordillera, Colorado Plateau, and Basin and Range province. In: Precambrian: Conterminous. (eds Reed J C et al.). Geol Soc Am, The Geology of North America, C-2: 463–595

Liou J G, Graham S A, Marayama S et al (1989) Proterozoic blueschist belt in western China: Best documented Precambrian blueschists in the world. Geol 17: 1127–1131

Lipin B R (1994) Pressure increases, the formation of chromite seams, and the development of the ultramafic series in the Stillwater Complex, Montana J Petrol 34: 955–976

Litherland M, Klinck B A, O'Connor E A and Pitfield P E J (1985) Andean-trending mobile belts in the Brazilian Shield. Nature 314: 345–348

Litherland M, Annells R N, Appleton J D et al (1986) The

geology and mineral resources of the Bolivian Precambrian Shield. Br Geol Surv Overseas Mem 9

Litherland M, Annells R N, Hawkins M P et al (1989) The Proterozoic of eastern Bolivia and its relation to the Andean mobile belt. Prec Res 43: 157–174

Liu D Y, Page R W, Compston W and Wu J (1985) U-Pb zircon geochronology of late Archean metamorphic rocks in the Taihangshan-Wutaishan area, North China. Prec Res. 27: 85–109

Liu D Y, Nutman A P, Compston W, Wu J S and Shen Q H (1992) Remnants of $\geq$ 3800 Ma crust in the Chinese part of the Sino–Korean craton. Geol 20: 339–342

Lobach-Zhuchenko S B and Vrevsky A B (1984) Granite-Greenstone Terrains. In: The Oldest Rock of the USSR. Nauka, Leningrad

Lobach-Zhuchenko S B, Levchenkov O A, Chekulaev V P and Krylov I N (1986) Geological evolution of the Karelian granite-greenstone terrain. Prec Res 33: 45–65

Lobach-Zhuchenko S B, Chekulayev V P, Sergeev S A, Levchenkov O A and Krylov I N (1993) Archaean rocks from southeastern Karelia (Karelian granite-greenstone terrain). Prec Res 62: 375–397

Loosveld R J H (1989) The intra-cratonic evolution of the central eastern Mount Isa Inlier, northwest Queensland, Australia. Prec Res 44: 243–276

López-Martinez M, York D and Hanes J A (1992) $^{40}$Ar/$^{39}$Ar geochronological study of komatiites and komatiitic basalts from the Lower Onverwacht Volcanics: Barberton Mountain Land, South Africa. Prec Res 57: 91–119

Lovelock J E (1988) The ages of Gaia: A biography of our living Earth. New York, W. W. Norton

Lovelock J E (1994) The Gaia hypothesis. In: Environmental Evolution: Effects of the origin and evolution of life on planet Earth. Cambridge, Massachusetts, MIT Press

Lowe D R (1980) Archean sedimentation. Ann Rev Earth Planet Sci 8: 145–167

Lowe D R (1982) Comparative sedimentology of the principal volcanic sequence of Archean greenstone belts in South Africa, Western Australia and Canada: implications for crustal evolution. Prec Res 17: 1–29

Lowe D R (1994a) A biological origin of described stromatolites older than 3.2 Ga. Geol 22: 387–390

Lowe D R (1994b) Accretionary history of the Archean Barberton Greenstone Belt (3.55–3.22 Ga), southern Africa. Geol 22: 1099–1102

Lowe D R and Knauth L P (1977) Sedimentology of the Onverwacht Group (3.4 billion years), Transvaal, South Africa, and its bearing on the characteristics and evolution of the early Earth. J Geol 85: 699–723

Lowe D R, Byerly G R, Ransom B L and Nocita B W (1985) Stratigraphy and sedimentological evidence bearing on structural repetition in early Archean rocks of the Barberton greenstone belt, South Africa. Prec Res 27: 165–186

Lowe D R, Byerly G R, Asaro F and Kyte F J (1989) Geological and geochemical record of 3400-million-year-old terrestrial meteorite impacts. Sci 245: 959–962

Lowell G R (1991) The Butler Hill Caldera: a mid-Proterozoic ignimbrite-granite complex. Prec Res 51: 245–263

Lowman P D (1984) Formation of the earliest continental crust: inferences from the Scourian Complex of Northwest Scotland and geophysical models of the lower continental crust. Prec Res 24: 199–215

Lowman P D (1989) Comparative planetology and the origin of continental crust. Prec Res 44: 171–195

Lowman P D (1992) Geophysics from orbit: the unexpected surprise. Endeavour, New Series, vol. 16, no. 2: 50–58

Luais B and Hawkesworth C J (1944) The generation of continental crust: An integrated study of crust-forming processes in the Archaean of Zimbabwe. J Petrol 35: 43–93

Lucas S B and St-Onge M R (1992) Terrane accretion in the internal zone of the Ungava orogen, northern Quebec. Part 2: Structural and metamorphic history. Can J Earth Sci, 29: 765–782

Ludden J (1994) The Abitibi–Grenville Lithoprobe transect seismic reflection results. Part I: the western Grenville Province and Pontiac Subprovince. Can J Earth Sci. 31: 227–228

Lundquist Th (1979) The Precambrian of Sweden. Sveriges Geol Unders Serie C SR 768

Luukkonen E J (1985a) Structural and U-Pb isotopic study of late Archaen migmatitic gneisses of the Presvecokarelides, Lylyvaara, eastern Finland. Trans R Soc Edinburgh (Earth Sci) 76: 401–410

Luukkonen E J (1985b) The structure and stratigraphy of the northern part of the late Archaen Kuhmo greenstone belt, eastern Finland. Geol Surv Fin Spec Pap 4: 71–96

Maas R and McCulloch M T (1991) The provenance of Archean clastic metasediments in the Narryer Gneiss Complex, western Australia: Trace element geochemistry, Nd isotopes, and U-Pb ages for detrital zircons. Geochim Cosmochim Acta, 55: 1915–1932

Macfarlane A, Crow M J, Arthurs J W and Wilkinson A F (1981) The geology and mineral resources of northern Sierra Leone. Overseas Mem Inst Geol Sci Lond No. 7

Macfarlane A W, Danielson A and Holland H D (1994a) Geology and major and trace element chemistry of the late Archean weathering profiles in the Fortescue Group, Western Australia: implications for atmospheric $PO_2$. Prec Res 65: 297–317

Macfarlane A W, Danielson A, Holland H D and Jacobsen S B (1994b) REE chemistry and Sm–Nd systematics of late Archean weathering profiles in the Fortescue Group, Western Australia. Geochim Cosmochim Acta 58: 1777–1794

Macgregor A M (1951) Some milestones in the Precambrian of Southern Rhodesia. Proc Geol Soc S Afr 55: 27–71

Machado N, Lindenmayer Z, Krogh T E and Lindenmayer D (1991) U–Pb geochronology of Archean magmatism and basement reactivation in the Carajás area, Amazon shield, Brazil. Prec Res 49: 329–354

Machado N, Noe C M, Ladeira E A and De Oliveira O B (1992) U–Pb geochronology of Archean magmatism and Proterozoic metamorphism in the Quadrilátero Ferrífero, southern São Francisco craton, Brazil. Geol Soc Am Bull 104: 1221–1227

Machado N, David J, Scott D J, Lamothe D, Philippe S and Gariépy C (1993) U–Pb geochronology of the western Cape Smith Belt, Canada: new insights on the age of initial rifting and arc magmatism. Prec Res 63: 211–223

Mahalik N K (1994) Geology of the contact between the Eastern Ghats Belt and North Orissa Craton, India. J Geol Soc Ind, 44: 41–51

Malisa E and Muhongo S (1990) Tectonic setting of gemstone mineralization in the Proterozoic metamorphic terrane of the Mozambique Belt in Tanzania. Prec Res 46: 167–176

Manikyamba C, Balaram V and Naqvi S M (1993) Geochemical signatures of polygenetic origin of a banded iron formation (BIF) of the Archean Sandur greenstone belt (schist belt) Karnataka nucleus, India. Prec Res 61: 137–164

Margulis L and West O (1993) Gaia and the colonization of Mars. GSA Today, 3: 277–280, 291

Mariano J and Hinze W J (1994) Structural interpretation of the Midcontinent Rift in eastern Lake Superior from seismic reflection and potential field studies. Can J Earth Sci 31: 619–628

Marmo J S and Ojakangas R W (1984) Lower Proterozoic glaciogenic deposits, eastern Finland. Geol Soc Am Bull 95: 1055–1062

Marsh J S, Bowen M P, Rogers N W and Bowen T B (1989) Volcanic rocks of the Witwatersrand Triad, South Africa, II. Petrogenesis of mafic and felsic rocks of the Dominion Group. Prec Res 44: 39–65

Martignole J (1992) Exhumation of high-grade terranes—a review. Can J Earth Sci 29: 737–745

Martignole J, Machado N and Indares A (1994) The Wakeham Terrane: a Mesoproterozoic terrestrial rift in the eastern part of the Grenville Province. Prec Res 68: 291–306

Martin H (1986) Effect of steeper Archean geothermal gradient on geochemistry of subduction-zone magmas. Geol 14: 753–756

Martin H and Porada H (1977) The intracratonic branch of the Damara orogen in South West Africa. Part I. Discussion of geodynamic models. Prec Res 5: 311–357

Martin H, Auvray B, Blais S et al (1984a) Origin and geodynamic evolution of the Archean crust of eastern Finland. Bull Geol Soc Fin 56: 135–160

Martin H, Chauvel C and Jahn B.-M et al (1984b) Isotopic data (U-Pb, Rb-Sr, Pb-Pb and Sm-Nd) on mafic granulites from Finnish Lapland. Prec Res 23: 325–348

Martin M W and Walker J D (1992) Extending the western North American Proterozoic and Paleozoic continental crust through the Mojare Desert. Geol 20: 753–756

Martini I P and Chesworth A T (eds) (1992) Weathering, Soils and Paleosols. Developments in Earth Surface Processes, 2. Elsevier. 618 pp.

Martini J E J (1991) The nature, distribution and genesis of the coesite and stishovite associated with the pseudotachylite of the Vredefort Dome, South Africa. Earth Planet Sci Lett 103: 285–300

Martini J E J (1994) A Late Archaean-Palaeoproterozoic (2.6 Ga) palaeosol on ultramafics in the Eastern Transvaal, South Africa. Prec Res 67: 159–180

Martyn J E (1987) Evidence for structural repetition of the Kalgoorlie district, western Australia. Prec Res 37: 1–18

Mascarenhas J F and da Silva Sa J H (1982) Geological and metallogenic patterns in the Archean and Early Proterozoic of Bahia State, eastern Brazil. Rev Brasil Geociencias 12(1-3): 193–214

Mathur S M (1982) Precambrian sedimentary sequences of India: their geochronology and correlation. Prec Res 18: 139–144

Matthews P E (1972) Possible Precambrian obduction and plate-tectonics in southeastern Africa. Nature 240: 37–39

Matthews P E (1981) Eastern or Natal sector of the Namaqua-Natal mobile belt in Southern Africa. In: Precambrian of the southern Hemisphere, (ed. Hunter D R). Elsevier, Amsterdam, pp 705–715

McCarthy T S (Editor) (1990) Geological studies related to the origin and evolution of the Witwatersrand Basin and its mineralization. S Afr J Geol 93 (1): 309 pp

McCarthy T S, Stanistreet I G and Robb L J (1990) Geological studies related to the origin of the Witwatersrand Basin and its mineralization—an introduction and a strategy for research and exploration. S Afr J Geol 93 (1): 1–4

McConnell R B (1951) Rift and shield structures in East Africa. 18th Int Geol Congr Lond (1948) 14: 199–207

McCourt S and van Reenen D (1992) Structural geology and tectonic setting of the Sutherland Greenstone Belt, Kaapvaal Craton, South Africa. Prec Res 55: 93–110

McCourt S and Vearncombe J R (1987) Shear zones bounding the central zone of the Limpopo mobile belt, southern Africa. J Struct Geol 9: 127–137.

McCulloch M T (1993) The role of subducted slabs in an evolving earth. Earth Planet Sci Lett 115: 89–100

McCulloch M T (1994) Primitive $^{87}Sr/^{86}Sr$ from an Archean barite and conjecture on the Earth's age and origin. Earth Planet Sci Lett 126: 1–13

McCulloch M T and Wasserburg G J (1980) Early Archean Sm-Nd model ages from a tonalitic gneiss, northern Michigan. Geol Soc Am Spec Pap 182: 135–138

McCulloch M T, Collerson K D and Compston W (1983) Growth of Archaean crust within the Western Gneiss Terrain, Yilgarn Block, Western Australia. J Geol Soc Austral 30: 149–153

McEachern S J and van Breemen O (1993) Age of deformation within the Central Metasedimentary Belt boundary thrust zone, southwest Grenville Orogen: constraints on the collision of the Mid-Proterozoic Elzevir terrane. Can J Earth Sci 30: 1155–1165

McElhinny M W and McWilliams M O (1977) Precambrian geodynamics—a paleomagnetic view. Tectonophys 40: 137–159

McElhinny M W, Taylor S R and Stevenson D J (1978) Limits to the expansion of the Earth, Moon, Mars, and Mercury and to changes in the gravitational constant. Nature 271: 316–321

McGregor V R (1968) Field evidence of very old Precambrian rocks in the Godthâb area, West Greenland. Rapp Grønlands Geol Unders 15: 31–35

McGregor V R (1973) The early Precambrian gneisses of the Godthâb district, West Greenland. Philos Trans R Soc (Ser A) 273: 343–358

McGregor V R (1979) Archean grey gneisses and the origin of the continental crust, In: Trondhjemites, Dacites and Related Rocks, (ed. Barker F). Elsevier, Amsterdam: 169–204

McLelland J M and Chiarenzelli J (1990) Constraints on emplacement age of anorthositic rocks, Adirondack Mts., New York. J Geol 98: 19–43

McLennan S M and Taylor S R (1982) Geochemical constraints on the growth of the continental crust. J Geol 90: 347–361

McMechan M E (1981) The Middle Proterozoic Purcell Supergroup in the southwestern and southeastern Mountains, British Columbia and the initiation of the Cordilleran miogeocline, southern Canada and adjacent United States. Bull Can Petrol Geol 29: 583–621

McMenamin M S and McMenamin D L S (1990) The emergence of animals: The Cambrian breakthrough. New York, Columbia University Press: 217 pp

McMenamin D W, Kumar S and Awramik S M (1983) Microbial fossils from the Kheinjua Formation, Middle Proterozoic Semri Group (Lower Vindyan), Son Valley area, central India. Prec Res 21: 247–271

McWilliams M O (1981) Palaeomagnetism and Precambrian tectonic evolution. In Precambrian Plate Tectonics, (ed Kröner A) Elsevier, Amsterdam: 649–687

McWilliams M O and McElhinny M W (1980) Late Precambrian paleomagnetism of Australia: the Adelaide Geosyncline. J Geol 88: 1–26

Meen J K, Rogers J J W and Fullagar P D (1993) Lead isotopic compositions of the Western Dharwar Craton, southern India: Evidence for distinct Middle Archean terranes in a Late Archean craton. Geochim Cosmochim Acta, 56: 2455–2470

Meert J G and Van der Voo R (1994) The Neoproterozoic (1000–540 Ma) glacial interials: No more snowball earth? Earth Planet Sci Lett 123: 1–13

Meert J G, Van der Voo R and Patel J (1994) Paleomagnetism of the Late Archean Nyanzian System, western Kenya. Prec Res 69: 113–131

Meijerink A M J, Rao D P and Rupke J (1984) Stratigraphic and structural development of the Precambrian Cuddapah Basin, SE India. Prec Res 26: 57–104

Mel'nik Y P (1982) Precambrian Banded Iron-Formations. Elsevier, Amsterdam

Mendelsohn F (1981) Precambrian geology of Zaire and Zambia. In: Precambrian of the Southern Hemisphere, (ed. Hunter D R). Elsevier, Amsterdam: 721–739

Mernagh T P, Heinrich C A, Leckie J F et al. (1994) Chemistry of low temperature hydrothermal gold, platinum and palladium (I uranium) mineralization at Coronation Hill, Northern Territory, Australia. Econ Geol 89: 1053–1073

Meyer C (1988) Ore deposits as guides to geologic history of the Earth. Ann Rev of Earth Planet Sci 16: 147–171

Meyer M T, Bickford M E and Lewry J F (1992) The Wathaman batholith: An Early Proterozoic continental arc in the Trans-Hudson orogenic belt, Canada. Geol Soc Am Bull 104: 1073–1085

Meyerhoff A A, Taner I, Morris A E L, Martin B D, Agocs W B and Meyerhoff H A (1992) Surge tectonics: a new hypothesis of Earth dynamics. In Chatterjee S and N Hotton (eds.) New Concepts in Global Tectonics. Texas Tech University Press, Lubbock, USA: 309–409

Michard-Vitrac A, Lancelot J, Allègre C J and Moorbath S (1977) U-Pb ages on single zircons from the early Precambrian rocks of West Greenland and the Minnesota River Valley. Earth Planet Sci Lett 35: 138–139

Mikkola A (1980) The metallogeny of Finland. Bull Geol Surv Fin 305

Milisenda C C, Liew T C, Hofmann A W and Köhler H (1994) Nd isotopic mapping of the Sri Lanka basement: update and additional constraints from Sr isotopes. Prec Res 66: 95–110

Milkereit B, Forsyth D A, Greene A G, Davidson A, Hanmer S, Hutchinson D R, Hinze W J and Mereu RF (1992a) Seismic images of a Grenvillian terrane boundary. Geol 20: 1027–1030

Milkereit B, Green A and Sudbury Working Group (24 individuals) (1992b) Deep geometry of the Sudbury structure from seismic reflection profiling. Geol 20: 807–811

Miller A R, Cumming G L and Krstic D (1989) U–Pb, Pb–Pb, and K–Ar isotopic study and petrography of uraniferous phosphate-bearing rocks in the Thelon Formation, Dabawnt Group, Northwest Territories. Can J Earth Sci 26: 867–880

Miller K C and Hargraves R B (1994) Paleomagnetism of some Indian kimberlites and lamproites. Prec Res 69: 259–267

Miller R M (1983) The Pan-African Damara Orogen of South West Africa/Namibia. Geol Soc S Afr Spec Public 11: 431–515

Minter W E L, Feather C E and Glatthaar C W (1988) Sedimentological and mineralogical aspects of the newly discovered Witwatersrand placer deposit that reflect Proterozoic weathering, Welkom Gold Field, South Africa. Econ Geol 83: 481–491

Minter W E L, Goedhart M, Knight J and Frimmel H E (1993) Morphology of Witwatersrand gold grains from the Basal Reef: evidence for their detrital origin. Econ Geol 88: 237–265

Mitchelmore M D and Cook F A (1994) Inversion of the Proterozoic Wernecke basin during tectonic development of the Racklan Orogen, northwest Canada. Can J Earth Sci 31: 447–457

Mogk D W, Mueller P A and Wooden J L (1992) The nature of Archean terrane boundaries: an example from the northern Wyoming Province. Prec Res 55: 155–168

Montalvao R M G, Hildred P D, Bezerra P E L et al (1982) Petrographic and chemical aspects of the mafic-ultramafic rocks of the Crixas, Guarinos, Pilar de Goias-Hidrolina and Goias greenstone belts, central Brazil. Rev Brasil Geociencias 12(1-3): 331–347

Montgomery C W (1979) Uranium-lead geochronology of the Archean Imataca Series, Venezuelan Guyana Shield. Cont Mineral Petrol 69: 167–176

Mooney W D and Braile L W (1989) The seismic structure of the continental crust and upper mantle of North America. In: The Geology of North America—An overview (eds A W Bally and A R Palmer), The Geology of North America, vol. A. Geol Soc Am

Moorbath S (1976) Age and isotope constraints for the evolution of Archean crust. In: The Early History of the Earth, (ed. Windley B F). John Wiley, New York: 351–360

Moorbath S (1977) The oldest rocks and the growth of continents. Sci Am 236: 92–104

Moorbath S (1984) Patterns and geological significance of age derminations in continental blocks. In: Patterns of Change in Earth Evolution, (eds Holland H D and Trendall A F). Dahlem Konferenzen, 1984. Springer-Verlag, Berlin: 207–219

Moorbath S (1986) The most ancient rocks revisited. Nature 321: 725

Moorbath S and Taylor P N (1988) Early Precambrian crustal evolution in Eastern India: the ages of the Singhbhum Granite and included remnants of older gneiss. Geol Soc Ind 31: 82–84

Moorbath S, O'Nions R K and Pankhurst R J (1973) Early Archaean age for the Isua iron-formation, West Greenland. Nature 245: 138–139

Moorbath S, Taylor P N and Goodwin R (1981) Origin of granitic magma by crustal mobilization: Rb-Sr and Pb/Pb geochronology and geochemistry of the late Archaean Qôrqut Granite Complex of southern West Greenland. Geochim Cosmochim Acta, 45: 1051–1060

Moore J M (1986) Introduction: the 'Grenville Problem' then and now. In: The Grenville Province (eds Moore J M, Davidson A and Baer A J). Geol Ass Can Spec Pap 31: 1–11

Moore J M, Davidson A and Baer A J (eds) (1986) The Grenville Province. Geol Ass Can Pap 31

Moore J M, Watkeys M K and Reid D L (1990) The regional setting of the Aggeneys/Gamsberg base metal deposits, Namaqualand, South Africa. In: Regional Metamorphism of Ore Deposits (eds. Spry P G and Brynzia T). Utrecht, V.S.P: 77–95

Moore M, Davis D W, Robb L J, Jackson M C and Grobler D F (1993) Archean rapakivi granite-anorthosite-rhyolite complex in the Witwatersrand basin hinterland, southern Africa. Geol 21: 1031–1034

Moores E M (1991) Southwest U.S.–East Antarctic (SWEAT) connection: A hypothesis. Geol 19: 425–428

Moores E M (1993) Neoproterozoic oceanic crustal thinning, emergence of continents, and origin of the Phanerozoic ecosystem: A model. Geol 21: 5–8

Morey G B (1983a) Animikie Basin, Lake Superior region, USA. In: Iron-Formation: Facts and Problems (eds Trendall A F and Morris R C). Elsevier, Amsterdam, pp 13–67

Morey G B (1983b) Lower Proterozoic stratified rocks and the Penokean orogeny in east-central Minnesota. In: Early Proterozoic Geology of the Great Lakes Region (ed. Medaris L G). Geol Soc Am Mem 160: 97–112

Morgan W C (1975) Geology of the Precambrian Ramah Group and basement rocks in the Nachvak Fiord-Saglek Fiord area, North Labrador. Geol Surv Can Pap 75-54, 42 pp

Morris R C (1993) Genetic modelling for banded iron-formation of the Hamersley Group, Pilbara Craton, Western Australia. In: Archean and Early Proterozoic Geology of the Pilbara Region, Western Australia (eds Blake T S and Meakins A L). Prec Res 60: 243–286

Morse S A (1982) A partisan review of Proterozoic anorthosite. Am Mineral 67: 1807–1100

Mortensen J K (1993a) U–Pb geochronology of the eastern Abitibi subprovince. Part I: Chibougamau–Matagami–Joutel region. Can J Earth Sci 30: 11–28

Mortensen J K (1993b) U–Pb geochronology of the eastern Abitibi subprovince. Part 2. Noranda–Kirkland Lake area. Can J Earth Sci 30: 29–41

Mortensen J K and Card K D (1993) U–Pb age constraints for the magmatic and tectonic evolution of the Pontiac Subprovince, Quebec. Can J Earth Sci 30: 1970–1980

Mortensen J K, Thorpe R I, Padgham W A, King J E and Davis W J (1988) U–Pb zircon cages for felsic volcanism in Slave Province, NWT. Radiogenic Age and Isotopic Studies. Report 2. Geol Surv Can Pap 88-2: 85–95

Mortimer G E, Cooper J A, Paterson H L et al (1988) Zircon U-Pb dating in the vicinity of the Olympic Dam Cu-U-Au-deposit, Roxby Downs, South Australia. Econ Geol 83: 694–709

Mosley P N (1993) Geological evolution of the late Proterozoic "Mozambique Belt" of Kenya. Tectonophys 221: 223–250

Mount J F and McDonald C (1992) Influences of changes in climate, sea level, and depositional systems in the fossil record of the Neoproterozoic—Early Cambrian metazoan radiation, Australia. Geol 20: 1031–1034

Moussine-Pouchkine A and Bertrand-Safarti J (1978) Le Gourma: un aulacogene du Précambrien superieur? Bull Soc Géol Fr, 7, 6: 851–857

Mueller P A, Wooden J L, Henry D J and Bowes D R (1985) Archean crustal evolution of the eastern Beartooth Mountains, Montana and Wyoming. Montana Bur Mines Geol Spec Public 92: 9–20

Mueller P A, Wooden J L and Nutman A P (1992) 3.96 Ga zircons from an Archean quartzite, Beartooth Mountains, Montana. Geol 20: 327–330

Mueller P A, Shuster R D, Wooden J L, Erslev E A and Bowes D R (1993) Age and composition of Archean crystalline rocks from the southern Madison Range, Montana: Implications for crustal evolution in the Wyoming craton. Bull Geol Soc Am 105: 437–446

Mueller P A, Heatherington A L, Wooden J L, Shuster R D, Nutman A P and Williams I S (1994) Prec zircons from the Florida basement: A Gondwanan connection. Geol 22: 119–122

Mueller W and Donaldson J A (1992) Development of sedimentary basins in the Archean Abitibi belt, Canada: An overview. Can J Earth Sci 29: 2249–2265

Mueller W, Donaldson J A and Doucet P (1994) Volcanic and tectono-plutonic influences on sedimentation in the Archaean Kirkland Basin, Abitibi greenstone belt, Canada. Prec Res 68: 201–230

Muhling J R (1988) The nature of Proterozoic reworking of early Archaean gneisses, Mukalo Creek Area, Southern Gascoyne Province, Western Australia. Prec Res 40-41: 341–362

Muhling P C and Brakel A T (1985) Geology of the Bangemall Group: the evolution of an intracratonic Proterozoic Basin. Geol Surv W Austral, Bull 128

Muir R J, Fitches W R and Maltman A J (1992) Rhinns complex: A missing link in the Proterozoic basement of the North Atlantic region. Geol 20: 1043–1046

Muir T L and Peredery W V (1984) The Onaping Formation. In: The Geology and Ore Deposits of the Sudbury Structure, (eds Pye E G, Naldrett A J and Giblin P E). Ont Geol Surv Spec Vol 1: 139–199

Myers J S (1981) The Finkenaesset anorthosite complex—a stratigraphic key to the tectonic evolution of the West Greenland Gneiss Complex 3000–2800 MY ago. In: Archean Geology (eds Glover J E and Groves D I). Geol Soc Austral Spec Public 7: 351–360

Myers J S (1984) Archean tectonics in the Fiskenaesset region of southwest Greenland. In: Precambrian Tectonics Illustrated (ed. Kröner A and Greiling R). Schweizerbart'sche Verlagsbuchhandlung, Stuttgart, pp 95–112

Myers J S (1988a) Early Archaean Narryer Gneiss Complex, Yilgarn Craton, Western Australia. Prec Res 38: 297–307

Myers J S (1988b) Oldest known terrestrial anorthosite at Mt Narryer, Western Australia. Prec Res 38: 309–323

Myers J S (1990) Precambrian tectonic evolution of part of Gondwana, southwestern Australia. Geol 18: 537–540

Myers J S (1993) Precambrian history of the West Australian craton and adjacent orogens. Ann Rev Earth Planet Sci 1993: 453–485

Myers J S and Williams I R (1985) Early Precambrian crustal evolution at Mount Narryer, Western Australia. Prec Res 27: 153–163

Myers R E, McCarthy T S, Bunyard M, Cawthorn R G, Falatsa T M, Hewitt T, Linton P, Myers J M, Palmer K J and Spencer R (1990a) Geochemical stratigraphy of the Klipriviersberg Group volcanic rocks. S Afr J Geol 93 (1): 224–238

Myers R E, McCarthy T S and Stanistreet I G (1990b) A tectono-sedimentary reconstruction of the development and evolution of the Witwatersrand Basin with particular emphasis on the Central Rand Group. S Afr J Geol 93 (1): 180–201

Nadeau L and Hammer S (1992) Deep crustal break-back stacking and slow cooling of the continental footwall beneath a thrusted marginal basin, Grenville orogen, Canada. Tectonophysics. In press

Nagaraja Rao B K, Rajurkar S T, Ramalingaswamy G and Ravindra Babu B (1987) Stratigraphy, structure and evolution of the Cuddapah basin. In: Purana Basins of Peninsular India. Geol Soc Ind Mem 6: 33–86

Naha K (1983) Structural-stratigraphic relations of the pre-Delhi rocks of south-central Rajasthan: a summary. Recent Res Geol 10: 40–71

Naha K, Srinivasan R and Jayaram S (1991) Sedimentational, structural and migmatitic history of the Archaean Dharwar tectonic province, southern India. Proc Indian Acad Sci (Earth Planet Sci) 100: 413–433

Naidoo D D, Bloomer S H, Saquaque A and Hefferan K (1991) Geochemistry and significance of metavolcanic rocks from the BouxAzzer-El Graara ophiolite (Morocco). Prec Res 53: 79–97

Nakai S, Masuda A, Shimizu H and Qi Lu (1989) La-Ba dating and Nd and Sr isotope studies on the Baiyun Obo rare earth element ore deposits, Inner Mongolia, China. Econ Geol 84: 2296–2299

Naldrett A J (1984) Summary, Discussion, and Synthesis. In: The Geology and Ore Deposits of the Sudbury Structure, (eds Pyre E G, Naldrett A J and Giblin P E). Ont Geol Surv Spec Vol 1: 533–569

Naldrett A J (1989) Magmatic Sulfide Deposits. Oxford Monographs on Geology and Geophysics No 14. Oxford University Press, Oxford

Nalivkin D V (1970) The geology of the USSR—a short outline. Pergamon Press, New York

Nance R D and Murphy J B (1994) Contrasting basement isotopic signatures and the palinspastic restoration of peripheral orogens: Example from the Neoproterozoic Avalonian–Cadomian belt. Geol 22: 617–620

Nance R D, Murphy J B, Strachan R A, D'Lemos R S and Taylor G K (1991) Late Proterozoic tectonostratigraphic evolution of the Avalonian and Cadomian terranes. Prec Res 53: 41–78

Nanyaro J R, Basu N K, Muhongo S M et al (1983) Structural evolution of the Ubendian belt: preliminary results of a traverse between Kerma and Mpanda (Tanzania). Mus R Afr Centr Tervuren Belg., Dept Geol Min 81/82: 147–152

Naqvi S M (1981) The oldest supracrustals of the Dharwar Craton, India. J Geol Soc Ind 22: 458–469

Naqvi S M and Rogers J J W (1987) Precambrian Geology of India. Clarendon Press, New York

Narain H and Subrahmanyam C (1986) Precambrian tectonics of the South Indian Shield inferred from geophysical data. J Geol 94: 187–198

Narbonne G M, Kaufman A J and Knoll A H (1994) Integrated chemostratigraphy and biostratigraphy of the Windermere Supergroup, northwestern Canada: Implications for Neoproterozoic correlations and the early evolution of animals. Bull Geol Soc Am, 106: 1281–1292

Natarajan M, Rao B B, Parthasarathy R, Kumar A and Gopalan K (1994) 2.0 Ga old pyroxenite-carbonatite complex of Hogenakal, Tamil Nadu, South India. Prec Res 65: 167–181

National Geographic Society Base-Map (1981) The World. National Geographic Society, Washington, 1981

Nedelec A, Nsifa E N and Martin H (1990) Major and trace element geochemistry of the Archaean Ntem plutonic complex (south Cameroon): petrogenesis and crustal evolution. Prec Res 47: 35–50

Needham R S, Stuart-Smith P G and Page R W (1988) Tectonic evolution of the Pine Creek Inlier, Northern Territory. Prec Res 40/41: 543–564

Nelson B K and De Paolo D J (1984) 1700-Myr greenstone volcanic successions in southwestern North America and isotopic evolution of Proterozoic mantle. Nature 33: 143–146

Nelson D R, Trendall A F, de Laeter J R, Grobler N J and Fletcher I R (1992) A comparative study of the geochemical and isotopic systematics of late Archaean flood basalts from the Pilbara and Kaapvaal Cratons. Prec Res 54: 231–256

Nemchin A A, Pidgeon R T and Wilde S A (1994) Timing of Late Archaean granulite facies metamorphism in the southwestern Yilgarn Craton of Western Australia: evidence from U–Pb ages of zircons from mafic granulites. Prec Res 68: 307–321

Newton R C (1987) Petrologic aspects of Precambrian granulite facies terrains bearing on their origins. In: Proterozoic Lithospheric Evolution, (ed. Kröner A). Am Geophys Union (Geodyn Ser). 17: 11–26

Newton R C (1992) An overview of charnockite. Prec Res 55: 399–405

Neymark L A, Kovach V P, Nemchin A A, Morozova I M, Kotov A B, Vinogradov D P, Gorokhovsky B M, Ovchinnikova G V, Bogomolova L M and Smelov A P (1993) Late Archaen intrusive complexes in the Olekma granite-greenstone terrain (eastern Siberia): geochemical and isotopic study. Prec Res 62: 453–472

Nicolaysen L O (1990) The Vredefort structure: an introduction and guide to recent literature. Tectonophys 171: 1–6

Nicolaysen L O and Ferguson J (1981) Diapirs driven by high pore fluid pressure. J Struct Geol 3: 89–95

Nicolaysen L O and Reimold W U (editors) (1990) Cryptoexplosions and catastrophes in the geological record, with a special focus on the Vredefort structure. Tectonophysics 171 (Special Issue. Papers from a workshop held in Parys, South Africa, in July 1987)

Nilson A A, Santos M M and Cuba E A (1982) The nickel-copper sulfide deposit in the Americano do Brasil mafic-ultramafic complex, Goias, Brazil. Rev Brasil Geociencias 12(1-3): 487–497

Nisbet E G (1987) The Young Earth: An Introduction to Archaean Geology. Allen and Unwin, London: 402 p

Nisbet E G (1991) Of clocks and rocks. The four aeons of Earth. Episodes 14: 327–330

Nisbet E G, Bickle M J and Martin A (1977) The mafic and ultramafic lavas of the Belingwe greenstone belt, Rhodesia. Petrol 18: 521–566

Nisbet E G, Wilson J F and Bickle M J (1981) The evolution of the Rhodesian Craton and adjacent Archean terrain: tectonic models. In: Precambrian Plate Tectonics (ed. Kröner A). Elsevier, Amsterdam: 161–183

Nisbet & Walker (1982) Arch komatiite derived from melt layer by corrosion of the (buoyant peridotite) cap was Arch. Earth's principal means of dissipating excess heat data and geological implications. Can J Earth Sci 29

Noble S R and Lightfoot P C (1992) U–Pb baddeleyite ages for the Kerns and Triangle Mountain intrusions. Nippissing Diabase, Ontario. Can J Earth Sci 29: 1424–1429

Nocita B W and Lowe D R (1990) Fan-delta sequence in the Archean Fig Tree Group, Barberton Greenstone Belt, South Africa. Prec Res 48: 375–393

Notman A P, Chernyshev I V, Baadsgaard H and Smelov A P (1992) The Aldan Shield of Siberia, USSR: the age of its Archaean components and evidence for its widespread reworking in the mid-Proterozoic. Prec Res 54: 195–210

Nowlan G S (1993) The ancient biosphere. Geosci Can 20: 113–122

Nutman A and Bridgwater D (1983) Deposition of Malene supracrustal rocks on the Amîtsoq gneiss basement in outer Ameralik, Southern West Grønlands. Geol Unders 112: 43–51

Nutman A P (1986) The early Archaean to Proterozoic history of the Isukasia area, southern West Greenland. Bull Geol Surv Greenland 154

Nutman A P and Bridgewater D (1986) Early Archaean Amitsoq tonalites and granites of the Isukasia area, southern West Greenland: development of the oldest known sial. Contr Mineral Petrol 94: 137–148

Nutman A P and Cordani U G (1993) SHRIMP U–Pb zircon geochronology of Archaean granitoids from the Contendas-Mirante area of the São Francisco Craton, Bahia, Brazil. Prec Res 63: 179–188

Nutman A P, Allaart J H, Bridgwater D et al (1984) Stratigraphic and geochemical evidence for the depositional environment of the early Archean Isu supracrustal belt, southern West Greenland. Prec Res. 25: 365–396

Nutman A P, Fryer B J and Bridgwater D (1989) The early Archaean Nulliak (supracrustal) assemblage, northern Labrador. Can J Earth Sci 26: 2159–2168

Nutman A P, Chadwick B, Ramakrishnan M and Viswanatha M N (1992) SHRIMP U–Pb ages of detrital zircon in Sargar supracrustal rocks in western Karnataka, southern India. J Geol Soc Ind 39: 367–374

Nutman A P, Friend C R L, Kinny P D and McGregor V R (1993) Anatomy of an Early Archean gneiss complex: 3900 to 3600 Ma crustal evolution in southern West Greenland. Geol 21: 415–418

Nyman M W, Karlstrom K E, Kirby E and Graubard C M (1994) Mesoproterozoic contractional orogeny in western North America: Evidence from ca. 1.4 Ga plutons. Geol 22: 901–904

Oehler D Z (1978) Microflora of the middle Proterozoic Balbirini dolomite (McArthur Group) of Australia. Alcheringa 2: 269–309

Ogasawara M (1988) Geochemistry of the early Proterozoic granitoids in the Halls Creek Orogenic subprovince, Northern Australia. Prec Res 40/41: 469–486

Öhlander B, Skiöld T, Elming S-A et al (1993) Delineation and character of the Archaean–Proterozoic boundary in northern Sweden. Prec Res 64: 67–84

Ojakangas R W (1985) Review of Archean clastic sedimentation, Canadian Shield: major felsic volcanic contributions to turbidite and alluvial fan-fluvial facies associations. In: Evolution of Archean Supracrustal Sequences, (eds Ayres L D, Thurston P C, Card K D and Weber W). Geol Ass Can Spec Pap 28: 23–47

Ojakangas R W and Morey G B (1982a) Keweenawan pre-volcanic quartz sandstones and related rocks of the Lake Superior region. Geol Soc Am Memoir 156: 85–96

Ojakangas R W and Morey G B (1982b) Keweenawan sedimentary rocks of the Lake Superior region: a summary. In: Geology and Tectonics of the Lake Superior Basin, (eds Wold R J and Hinze W J). Geol Soc Am Mem 156: 157–164

Oliveira E P, Lima M I C, Carmo U F and Wernick E (1982) The Archean granulite terrain from East Bahia, Brazil. Rev Brasil Geociencias 12(1-3): 356–368

Oliver J, Cook F and Brown L (1983) COCORP and the continental crust. Geophys Res 88: 3329–3347

Oliver R L, Lawrence R W, Goscombe B D et al (1988) Metamorphism and crustal considerations in the Harts Range and neighbouring regions, Arunta Inlier, Central Australia. Prec Res 40/41: 277–295

O'Nions R K, Carter S R, Evensen N M and Hamilton P J (1979) Geochemical and cosmochemical applications of Nd isotope analysis. Ann Rev Earth Planet Sci 7: 11–38

Onstott T C (Ed.) (1994) Preface to Proterozoic paleomagnetism and paleogeography. Special Volume, Prec Res 69: vii–x

Onstott T C, Hargraves R B, York D and Hall C (1984). Constraints on the motions of South America and African Shields during the Proterozoic: I. $^{40}Ar/^{39}Ar$ and paleomagnetic correlations between Venezuela and Liberia. Bull Geol Soc Am 95: 1045–1054

Oreskes N and Einaudi M T (1992) Origin of hydrothermal

fluids at Olympic Dam: Preliminary results from fluid inclusions and stable isotopes. Econ Geol 87: 64–90
Oreskes N O and Einaudi M T (1990) Origin of rare earth element-enriched hematite breccias at the Olympic Dam Cu-U-Au-Ag deposit, Roxby Downs, South Australia. Econ Geol 85: 1–28
Padgham W A (1985) Observations and speculations on supracrustal successions in the Slave Structural Province. In: Evolution of Archean Supracrustal Sequences (eds Ayres L D, Thurston P C, Card K D and Weber W). Geol Ass Can Spec Pap 28: 133–151
Padgham WA (1992) Mineral deposits in the Archean Slave Structural Province; Lithological and tectonic setting. In: Prec-ambrian Metallogeny Related to Plate Tectonics (eds. Poulson K H, Card K D and Franklin J M). Prec Res 58: 1–24
Padgham W A and Fyson W K (1992) The Slave Province: a distinct Archean craton. Can J Earth Sci 29: 2072–2086
Page N J (1977) Stillwater Complex, Montana: rock succession, metamorphism and structure of the complex and adjacent rocks. U S Geol Surv Prof Pap 999
Page R W (1979) Mount Isa Project. In: Geological Branch Summary of Activities, 1978. Bur Min Res Austral Rep 211: 181–182
Page R W (1983) Timing of superposed volcanism in the Proterozoic Mount Isa Inlier, Australia. Prec Res 21: 223–246
Page R W (1988) Geochronology of early to middle Proterozoic fold belts in northern Australia: a review. Prec Res 40-41: 1–19
Page R W and Bell T H (1986) Isotopic and structural responses of granite to successive deformation and metamorphism. J Geol 94: 365–379
Page R W and Hancock S L (1988) Geochronology of a rapid 1.85–1.86 Ga tectonic transition: Halls Creek Orogen, Northern Australia. Prec Res 40/41: 447–467
Page R W and Laing W P (1992) Felsic metavolcanic rocks related to the Broken Hill Pb–Zn–Ag orebody, Australia: Geology, depositional age, and timing of high-grade metamorphism. Econ Geol 87: 2138–2168
Page R W and William I S (1988) Age of the Barramundi Orogeny in Northern Australia by means of ion microprobe and conventional U-Pb zircon studies. Prec Res 40/41: 21–36
Page R W, McCulloch M T and Black L P (1984) Isotopic record of Major Precambrian events in Australia. Prec Geol Proc 27th Int Geol Congr Moscow 1984 5: 25–72
Pallister J S, Stacey J S, Fischer L B and Premo W R (1987) Arabian Shield ophiolites and late Proterozoic microplate accretion. Geol 15: 320–323
Pallister J S, Stacey J S, Fischer L B and Premo W R (1988) Precambrian ophiolites of Arabia: geologic settings, U-Pb geochronology, Pb-isotope characteristics, and implications for continental accretion. Prec Res 38: 1–54
Palmer H C, Tazaki K, Fyfe W S and Zhou Z (1988) Precambrian glass. Prec Res 16: 221–224
Palmer J A, Phillips G N and McCarthy T S (1989) Paleosols and their relevance to Precambrian atmospheric composition. J Geol 97: 77–92
Pankhurst R J, Moorbath S and McGregor V R (1973) Late event in the geological evolution of the Godthaab district, West Greenland. Nature 243: 24–26
Pan Y, Fleet M E and Williams H R (1994) Granulite-facies metamorphism in the Quetico Subprovince, north of Manitouwadge, Ontario. Can J Earth Sci 31: 1427–1439
Park A F (1983) Sequential development of metamorphic fabric and structural elements in polyphase deformed serpentinites in the Svecokarelides of eastern Finland. Trans R Soc Edinburgh (Earth Sci) 74: 33–60
Park A F (1984) Nature, affinities and significance of metavolcanic rocks in the Outokumpu assemblage, eastern Finland. Bull Geol Soc Fin 56: 25–52
Park A F (1985) Accretion tectonism in the Proterozoic Svecokarelides of the Baltic shield. Geol 13: 725–729
Park A F and Bowes D R (1983) Basement-cover relationships during polyphase deformation in the Svecokarelides of the Kaavi district, eastern Finland. Trans R Soc Edinburgh (Earth Sci) 74: 95–118
Park A F and Dash B (1984) Charnockite and related neosome development in the Eastern Ghats, Orissa, India: petrographic evidence. Trans R Soc Edinburgh (Earth Sci) 75: 341–352
Park A F, Bowes D R, Halden N M and Koistinen T J (1984) Tectonic evolution at an early Proterozoic continental margin: the Svecokarelides of eastern Finland. J Geodynam 1: 359–386
Park J K (1994) Palaeomagnetic constraints on the position of Laurentia from middle Neoproterozoic to Early Cambrian times. Prec Res 69: 95–112
Park R G (1992) Plate kinematic history of Baltica during the Middle to Late Proterozoic: A model. Geol 20: 725–728
Park RG (1994) Early Proterozoic tectonic overview of the northern British Isles and neighbouring terrains in Laurentia and Battica. Prec Res 68: 65–79
Park R G and Tarney J (1987) The Lewisian complex: a typical Precambrian high-grade terrain? In: Evolution of the Lewisian and Comparable High Grade Terrains (eds Park R G and Tarney J). Geol Soc Spec Public 27: 13–25
Park R G, Cliff R A, Fettes D G and Stewart A D (1994) Lewisian and Torridonian. In: Precambrian Correlation Report (eds. Gibbons F C and Harris A L). Geol Soc London, in press
Parker A J, Fanning C M, Flint R B, Martin A R and Rankin L R (1988) Archean–Early Proterozoic granitoids, metasediments and mylonites of southern Eyre Peninsula, South Australia. Field Guide Series, Geol Soc, Austral 2: 1–90
Parr J M (1994) The geology of the Broken Hill-type Pinnacles Pb–Zn–Ag deposit, western New South Wales, Australia. Econ Geol 89: 778–790
Partington G A (1990) Environment and structural controls on the intrusion of the giant rare metal Greenbushes pegmatite, Western Australia. Econ Geol 85: 437–456
Passchier C W (1994) Structural geology across a proposed Archaean terrane boundary in the eastern Yilgarn craton, Western Australia. Prec Res 68: 43–64
Passchier C W, Myers J S and Kröner A (1990) Field Geology of High-Grade Gneiss Terrains. Springer-Verlag, 150 pp
Pasteels P and Michot J (1968) Uranium-lead radioactive dating and lead isotope study on sphene and K-feldspar in the Sr Rondane Mountains, Dronning Maud Land, Antarctica. Ecol Geol Helv 63: 239–254
Patchett J and Kouvo O (1986) Origin of continental crust of 1.9–1.7 Ga age: Nd isotopes and U-Pb zircon ages in the Svecokarelian terrain of South Finland. Cont Mineral Petrol 92: 1–12
Patchett P J and Arndt N T (1986) Nd isotopes and tectonics of 1.9–1.7 Ga crustal genesis. Earth Planet Sci Lett 78: 329–338
Patchett P J, Kouvo O, Hedge C E and Tatsumoto M (1981) Evolution of continental crust and mantle heterogeneity: evidence from Hf isotopes. Cont Mineral Petrol 78: 279–297
Patterson J G (1986) The Amer Belt: remnant of an Aphebian foreland fold and thrust belt. Can J Earth Sci 23: 2012–2023
Pearton T N and Viljoen M J (1986) Antimony mineralization in the Murchison greenstone belt–an overview. In: Mineral Deposits of Southern Africa (eds Anhaeusser C R and Maske S). Geol Soc S Afr: 293–320
Pell J and Simony P S (1987) New correlations of Hadrynian strata, southcentral British Columbia. Can J Earth Sci 24: 302–313
Percival J A (1994) The Kapuskasing transect of Lithoprobe: Introduction. Can J Earth Sci 31: 1013–1015
Percival J A and Card K D (1983) Archean crust as revealed in the Kapuskasing uplift, Superior Province, Canada. Geol 11: 323–326

Percival J A and Card K D (1985) Structure and evolution of Archean crust in Central Superior Province, Canada. In: Evolution of Archean Supracrustal sequences (eds Ayres L D, Thurston P C, Card K D and Weber W). Geol Ass Can Spec Pap 28: 179–192

Percival J A and Girard R (1988) Structural character and history of the Ashuanipi complex in the Schefferville area, Quebec-Labrador. Geol Surv Can, Pap 88-1C: 51–60

Percival J A and McGrath P H (1986) Deep crustal structure and tectonic history of the northern Kapuskasing uplift of Ontario: an integrated petrological-geophysical study. Tectonics 5: 553–572

Percival J A and West G F (1994) The Kapuskasing uplift: a geological and geophysical synthesis. Can J Earth Sci 31: 1256–1286

Percival J A, Mortensen J K, Stern R A, Card K D and Bégin N J (1992) Giant granulite terranes of northeastern Superior Province: the Ashuanipi complex and Minto block. Can J Earth Sci 29: 2287–2308

Percival J A, Stern R A, Skulski T and three others (1994) Minto block, Superior province: Missing link in deciphering assembly of the craton at 2.7 Ga. Geol 22: 839–842

Perdahl J-A and Frietsch R (1993) Petrochemical and petrological characteristics of 1.9 Ga old volcanics in northern Sweden. Prec Res 64: 239–252

Peredery W V and Geological Staff (1982) Geology and nickel sulphide deposits of the Thompson Belt, Manitoba. Geol Ass Can Spec Pap 25: 165–209

Perera L R K (1983) The origin of the pink granites of Sri Lanka, another view. Prec Res 20: 17–37

Peterman Z E (1981) Dating of Archean basement in northeastern Wyoming and southern Montana. Geol Soc Am Bull 92: 139–146

Peterman Z E (1982) Geochronology of the southern Wyoming age province—a summary. In: Guide to the 1982 Archean Geochemistry Field Conference, Pt I Field Trips (ed. Goldich S S). Colorado School of Mines

Peterman Z E, Zartman R E and Sims P K (1980) Tonalitic gneiss of early Archean age from northern Michigan. Geol Soc Am Spec Pap 182: 125–134

Philippe S, Lancelot J R, Clauer N and Pacquet A (1993a) Formation and evolution of the Cigar Lake uranium deposit based on U–Pb and K–Ar isotope systematics. Can J Earth Sci 30: 720–730

Philippe S, Wardle R J and Schärer U (1993b) Labradorian and Grenvillian crustal evolution of the Goose Bay region, Labrador: new U–Pb geological constraints. Can J Earth Sci 30: 2315–2327

Piasecki M A J, van Breeman O and Wright A E (1981) Late Precambrian geology of Scotland, England and Wales. In: Geology of the North Atlantic Borderlands (eds Kerr J W and Ferguson J A) Can Ass Petrol Geol, Mem 7: 57–94

Pichamuthu C S (1967) The Precambrian of India. In: The Precambrian, Vol. 3 (ed. Rankama K). Wiley-Interscience, New York, pp 1–96

Pidgeon R T and Horwitz R C (1991) The origin of olistoliths in Proterozoic rocks of the Ashburton Trough, western Australia, using zircon U–Pb isotopic characteristics. Austral J Earth Sci 38: 55–63

Pinna P, Jourde G, Calvez J Y, Mroz J P and Marques J M (1993) The Mozambique Belt in northern Mozambique: Neoproterozoic (1100–850 Ma) crustal growth and tectogenesis, and superimposed Pan-African (800–550 Ma) tectonism. Prec Res 62: 1–59

Piper J D A (1983) Dynamics of the continental crust in Proterozoic times. Geol Soc Am Mem 161: 11–34

Piper J D A (1991) The quasi-rigid premise in Precambrian tectonics. Earth Planet Sci Lett 107: 559–569

Piper J D A (1992) Palaeomagnetic properties of a Precambrian metamorphic terrane: the Lewisian complex of the Outer Hebrides, NW Scotland. Tectonophys, 201: 17–48

Plimer I R (1994) Strata-bound scheelite in meta-evaporites, Broken Hill, Australia. Econ Geol 89: 423–437

Plumb K A (1979) The tectonic evolution of Australia. Earth Sci Rev 14: 205–249

Plumb K A (1985) Subdivision and correlation of Late Precambrian sequences in Australia. Prec Res 29: 303–329

Plumb K A (1988) The Australian Platform. In: Tectonics of Continents and Oceans: Explanatory Note to the International Tectonic Map of the World, Scale 1:15 000 000 (eds Khain V, Leonov Y and Dottin O). CGMW, Subcommission for Tectonic Maps, Moscow: 128–137

Plumb K A (1990) Halls Creek Province and the Granites-Tanami Inlier–Regional geology and minor mineralization. In: Geology of the Mineral Deposits of Australia and Papua-New Guinea (ed. Hughes F). Australian Institute of Mining and Metallurgy, Monograph, in press

Plumb K A and James H L (1986) Subdivision of Precambrian time: recommendations and suggestions by the Subcommission on Prec Stratigraphy. Prec Res 32: 65–92

Plumb K A, Derrick G M, Needham R S and Shaw R D (1981) The Proterozoic of Northern Australia. In: Precambrian of the Southern Hemisphere, (Ed. Hunter D R) Elsevier, Amsterdam: 205–307

Plumb K A, Allen R and Hancock S L (1985) Proterozoic evolution of the Halls Creek Province, Western Australia. Excursion Guide, Conference on Tectonics and Geochemistry of Early to Middle Proterozoic Fold Belts. Austral Bur Miner Res Rec 1985/25

Plumb K A, Ahmed M and Wygralak A S (1990) Regional geology of the Mid-Proterozoic basins of the North Australian Craton. In: Geology of the Mineral Deposits of Australia and Papua New Guinea (ed. Hughes F E). The Australian Institute of Mining and Metallurgy, Melbourne

Podmore F (1970) The shape of the Great Dyke as revealed by gravity surveying. Geol Soc S Afr Spec Public 1: 610–620

Poidevin J-L (1994) Boninite-like rocks from the Palaeoproterozoic greenstone belt of Bogoin, Central African Republic: geochemistry and petrogenesis. Prec Res 68: 97–113

Poldervaart A (1955) Chemistry of the Earth's crust. Geol Soc Am Spec Pap 62: 119–144

Porada H (1989) Pan-African rifting and orogenesis in southern to equatorial Africa and eastern Brazil. Prec Res 44: 103–136

Poulsen K H, Card K D and Franklin J M (1992) Archean tectonic and metallogenic evolution of the Superior Province of the Canadian Shield. In: Prec Metallogeny Related to Plate Tectonics (eds Poulsen K H, Card K D and Franklin J M). Prec Res 58: 25–54

Powell C M, Li Z X, McElhinny M W, Meert J G and Park J K (1993) Paleomagnetic constraints on timing of the Neoproterozoic breakup of Rodinia and the Cambrian formation of Gondwana. Geol 21: 889–892

Prame W K B N and Pohl J (1994) Geochemistry of pelitic and psammopelitic Precambrian metasediments from southwestern Sri Lanka: implications for two contrasting source-terrains and tectonic settings. Prec Res 66: 223–244

Preiss W V (1977) The biostratigraphic potential of Precambrian stromatolites. Prec Res 5: 207–219

Preiss W V (compiler) (1987) The Adelaide Geosyncline. Late Proterozoic stratigraphy, sedimentation, paleontology and tectonics. Bull Geol Surv Austral 53

Preiss W V (1990) A stratigraphic and tectonic overview of the Adelaide Geosyncline, South Australia. In: The Evolution of a Late Precambrian–Early Palaeozoic Rift Complex: The Adelaide Geosyncline. (eds. Jago J. B. and Moore P S). Special Publication 16, Geol Soc Austral: 1–33

Preiss W V and Forbes B G (1981) Stratigraphy, correlation and sedimentary history of Adelaidean (late Proterozoic) basins in Australia. Prec Res 15: 255–304

Preiss W V, Rutland R W R and Murrell B (1981) The Stuart Shelf and Adelaide Geosyncline. In: Precambrian of the Southern Hemisphere (ed. Hunter D R) Elsevier, Amsterdam 327–360

Prendergast M D (1987) The chromite ore field of the Great Dyke, Zimbabwe. In: Evolution of Chromium Ore Fields (ed. Stowe C W). Van Nostrand-Reinhold, New York: 89–108

Pretorius D A (1976) The nature of the Witwatersrand gold-uranium deposits. In: Handbook of Strata-bound and Stratiform Ore Deposits. Vol 7 (ed. Wolf K H). Elsevier, Amsterdam, pp 29–88

Pretorius D A (1981) The Witswatersrand Supergroup. In: Precambrian of the Southern Hemisphere, (ed. Hunter D R). Elsevier, Amsterdam, 511–520

Pretorius D A, Brink W C J and Fouche J (1986) Geological map of the Witwatersrand basin. In: Mineral Deposits of Southern Africa (eds Anhaeusser C R and Mastre S). Geol Soc S Afr, Pretoria

Price R A and Douglas R J W (1972) Variations in tectonic styles in Canada. Geol Ass Can Spec Paper No. 11

Puura V and Huhma H (1993) Palaeoproterozoic age of the East Baltic granulitic crust. Prec Res 64: 289–294

Pye E G, Naldrett A J and Giblin P E (eds) (1984) The geology and ore deposits of the Sudbury structure. Ont Geol Surv Spec Vol 1

Quick J E (1990) Geology and origin of the late Proterozoic Darb Zubaydah ophiolite, Kingdom of Saudi Arabia. Geol Soc Am Bull 102: 1007–1020

Radhakrishna B P (1984) Crustal evolution and metallogeny: evidence from the Indian Shield, a review. J Geol Soc Ind 25: 617–628

Radhakrishna B P (1989) Suspect tectono-stratigraphic terrane elements in the Indian sub-continent. J Geol Soc Ind, 34: 1–24

Radhakrishna B P and Naqvi S M (1986) Precambrian continental crust of India and its evolution. J Geol 94: 145–166

Radhakrishna T, Gopakumar K, Murali A V and Mitchell J G (1991) Geochemistry and petrogenesis of Proterozoic mafic dykes in north Kerala, southwestern India—preliminary results. Prec Res 49: 235–244

Radhakrishna B P, Ramakrishnan M and Mahabaleswar B (eds) (1992) Granulites of South Ind. Geol Soc Ind Mem 17: 502 pp

Raha P K and Sastry M V A (1982) Stromatolites and Precambrian stratigraphy in India. Prec Res 18: 293–318

Rajamani V and Ahmad T (1991) Geochemistry and petrogenesis of the basal Aravalli volcanics near Nathdwara, Rajasthan, India. Prec Res 49: 185–204

Ramaekers P (1981) Hudsonian and Helikian basins of the Athabasca region, northern Saskatchewan. In: Proterozoic Basins in Canada (ed. Campbell F H A). Geol Surv Can Pap 81-10: 219–233

Ramaekers P and Dunn C E (1977) Geology and geochemistry of the eastern margin of the Athabasca Basin. In: Uranium in Saskatchewan. Saskatchewan Geol Soc Spec Public 3: 297–322

Ramakrishnan M, Moorbath S, Taylor P N et al (1984) Rb-Sr and Pb-Pb whole rock isochron ages of basement gneisses in Karnataka Craton. J Geol Soc Ind 25(i): 20–34

Rämö O T (1991) Petrogenesis of the Proterozoic rapakivi granites and related basic rocks of southeastern Fennoscandia: Nd and Pb isotopic and general geochemical constraints. Geol Surv Fin, Bull 355: 161 pp

Rankin D W, Stern T W, McLelland J et al (1983) Correlation chart for Precambrian rocks of the eastern United States. US Geol Surv Prof Pap 1241–E

Rankin D W, Chiarenzelli J R, Drake A A et al. (1993) Proterozoic rocks east and southeast of the Grenville front. In: Precambrian:Conterminous U.S. (eds Reed J C, Bickford M E, Houston R S et al.). Geol Soc Am, The Geology of North America, C-2: 335–461

Rapp R P, Watson E B and Miller C F (1991) Partial melting of amphibolite/eclogite and the origin of Archean trondhjemites and tonalites. Prec Res 51: 1–25

Ravich M G and Grikurov G E (eds) (1976) Explanatory notes to the geological map of Antarctica, scale 1:5 000 000. Research Institute of the Geology of the Arctic, Ministry of Geology of the USSR, Leningrad

Ravich M G and Kamenev E N (1975) Crystalline Basement of the Antarctic Platform. John Wiley, New York

Ray G E (1974) The structural and metamorphic geology of northern Malawi. Geol Soc Lond 130; 427–440

Ray S and Bose M K (1975) Tectonic and petrologic evolution of the Eastern Ghats Precambrian belt. Chayan Geol 1: 1–13

Reed J C, Bickford M E, Premo W R et al (1987) Evolution of the early Proterozoic Colorado province: constraints from U-Pb geochronology. Geol 15: 861–865

Reed J C, Bickford M E, Houston R S et al. (1993a) Prec: Conterminous U.S. Geol Soc Am, The Geology of North America, C-2

Reed J C, Ball T J, Farmer G L and Hamilton W B (1993b) A broader view. In: Precambrian: Conterminous (eds. Reed J C et al.). Geol Soc Am, The Geology of North America, C-2: 597–636

Reid D L and Barton E S (1983) Geochemical characterization of granitoids in the Namaqualand geotraverse. Geol Soc S Afr Spec Public 10: 67–82

Reimold W U and Levin G (1991) The Vredefort Structure, South Africa: A bibliography relating to its geology and evolution. Econ Geol Res Unit, Univ Witwatersrand, Johannesburg, Info Circ 242, 24 pp

Reimold W U, Colliston W P and Wallmach T (1992) Comment on "The nature, distribution and genesis of the coesite and stishovite associated with the pseudotachylite of the Vredefort Dome, South Africa" by J. E. J. Martini. Earth Planet Sci Lett 112: 213–217

Reischmann Th, Bachtadse V, Kröner A and Layer P (1992) Geochronology and paleomagnetism of a late Proterozoic island arc terrane from the Red Sea Hills, northeast Sudan. Earth Planet Sci Lett 114: 1–15

Renner R, Nisbet E G, Cheadle M J, Arndt N T, Bickle M J and Cameron W E (1994) Komatiite flows from the Reliance Formation, Belingwe Belt, Zimbabwe: I. Petrography and Mineralogy. J Petrol 35: 361–400

Retallack G J and Krinsley D H (1993) Metamorphic alteration of a Prec (2.2 Ga) paleosol from South Africa revealed by backscattered electron imaging. Prec Res 63: 27–41

Reymer A P S and Schubert G (1984) Phanerozoic addition rates to the continental crust and crustal growth. Tectonics 3: 63–77

Reymer A P S and Schubert G (1986) Rapid growth of major segments of continental crust. Geol 14: 299–302

Reymer A P S and Schubert G (1987) Phanerozoic and Precambrian crustal growth. In: Proterozoic Lithosphere Evolution, (ed. Kröner A). Am Geophys Union (Geodyn Ser) 17: 1–9

Richter F M (1985) Models for the Archean thermal regimes. Earth Planet Sci Lett 73: 350–360

Rickard D T (1979) Scandinavian metallogenesis. Geojournal 3: 234–252

Ricketts B D and Donaldson J A (1981) Sedimentary history of the Belcher Group of Hudson Bay. In: Proterozoic basins in Canada, (ed. Campbell F H A). Geol Surv Can, Pap 81-10: 235–254

Ridley J (1992) On the origins and tectonic significance of the charnockite suite of the Archaean Limpopo Belt, Northern Marginal Zone, Zimbabwe. Prec Res 55: 407–427

Ring U (1993) Aspects of the kinematic history and mechanisms of superposition of the Proterozoic mobile belts of eastern Central Africa (northern Malawi and southern Tanzania). Prec Res 62: 207–226

Rivalenti G, Girardi V A V, Rossi A and Siena F (1982) The Niquelandia mafic-ultramafic complex of central Goias, Brazil: petrological considerations. Rev Brasil Geociencias 12(1-3): 380–391

Rivers T, Martignole J, Gower C F and Davidson A (1989) New tectonic divisions of the Grenville Province, southeast Canadian Shield. Tectonics, 8: 63–84

Rivers T, van Gool J A M and Connelly J N (1993) Contrasting tectonic styles in the northern Grenville province: Implications for the dynamics of orogenic fronts. Geol 21: 1127–1130

Robb L J and Meyer F M (1990) The nature of the Witwatersrand hinterland: conjectures on the source area problem. Econ Geol 85: 511–536

Robb L J, Davis D W and Kamo S (1989) U-Pb ages on single detrital zircon grains from the Witwatersrand Basin: constraints on the age of sedimentation and on the evolution of granites adjacent to the depository. Econ Geol Res Unit, Univ of Witwatersrand, Info Circ No 208, 16 pp

Robb L J, Meyer F M, Ferraz M F and Drennan G R (1990) The distribution of radioelements in Archaean granites of the Kaapvaal Craton, with implications for the source of uranium in the Witwatersrand Basin. S Afr J Geol 93 (1): 5–40

Robb L J, Davis D W, Samo S L and Meyer F M (1992) Ages of hydrothermally altered granites adjacent to the Witwatersrand Basin: Implications for the origin of Au and U Econ Geol Res Unit, Univ Witwatersrand, Johannesburg, Info Circ 251, 11 pp

Robert F, Sheahan P A and Green S B (1991) Greenstone Gold and Crustal Evolution. NUNA Conference Volume. Geol Ass Can: 252 pp

Roberts M P and Clemens J D (1993) Origin of high-potassium, calc-alkaline, I-type granitoids. Geol 21: 825–828

Robertson I D M and du Toit M C (1981) The Limpopo Belt. In: Precambrian of the Southern Hemisphere (ed. Hunter D R) Elsevier, Amsterdam: 641–671

Robertson I D M and Van Breemen O (1970) The southern satellite dykes of the Great Dyke, Rhodesia. Spec Public Geol Soc S Afr 1: 621–644

Rocci G, Bronner G and Deschamps M (1991) Crystalline basement of the West African craton. In: The West African Orogens and Circum-Atlantic Correlatives (eds, Dallmeyer R D and Lecorche J P). Springer-Verlag, Berlin: 31–61

Rodgers J (1975) Appalachian salients and recesses (Abs.) Geol Soc Am (Abs. with Programs) 7: 111–112

Roering C, Barton J M and Winter H de la R (1990) The Vredefort structure: A perspective with regard to new tectonic data from adjoining terranes. Tectonophys 171: 7–22

Roering C, Van Reenen D D, de Wit M J, Smit C A, de Beer J H and van Schalkwyk J F (1992a) Structural, geological and metamorphic significance of the Kaapvaal Craton—Limpopo Belt contact. Prec Res 55: 69–80

Roering C, van Reenen D D, Smit C A, Barton J M, de Beer J H, de Wit M J, Stettler E H, van Schalkwyk J F, Stevens G and Pretorius S (1992b) Tectonic model for the evolution of the Limpopo Belt. Prec Res 55: 539–552

Rogers J J W (1986) The Dharwar Craton and the assembly of Peninsular India. J Geol 94: 129–143

Rogers J J W (1993) Ind and Ur. J Geol Soc Ind 42: 217–222

Rohon M-L, Vialette Y, Clark T, Roger G, Ohnenstetter D and Vidal Ph, 1993. Aphebian mafic-ultramafic magmatism in the Labrador Trough (New Quebec): its age and the nature of its mantle source. Can J Earth Sci 30: 1582–1593

Rollinson H R (1978) Zonation of supracrustal relics in the Archaean of Sierra Leon, Liberia, Guinea and Ivory Coast. Nature 272: 440–442

Romer R L and Smeds S-A (1994) Implications of U–Pb ages of columbite-tantalites from granitic pegmatites for the Palaeoproterozoic accretion of 1.90–1.85 Ga magmatic arcs to the Baltic Shield. Prec Res 67: 141–158

Romer R L, Martinsson O and Perdahl J-A (1994) Geochronology of the Kiruna iron ores and hydrothermal alterations. Econ Geol 89: 1249–1261

Ronley P R, Ford A B, Williams P and Pride D E (1983) Metallogenic Provinces of Antarctica. In: Antarctic Earth Science, (eds Oliver R L, James P R and Jago J B). Cambridge University Press, Cambridge: 414–419

de Ronde C E J, de Wit M J and Spooner E T C (1994) Early Archean (73.2 Ga) Fe-oxide-rich hydrothermal discharge vents in the Barberton greenstone belt, South Africa. Geol Soc Am Bull 106: 86–104

Roscoe S M (1973) The Huronian Supergroup, a Paleoaphebian succession showing evidence of atmospheric evolution. Geol Ass Can Spec Pap 12: 31–48

Roscoe S M and Card K D (1992) Early Proterozoic tectonics and metallogeny of the Lake Huron region of the Canadian Shield. In: Prec Metallogeny Related to Plate Tectonics (eds. Pousen K H, Card K D and Franklin J M). Prec Res 58: 25–54

Roscoe S M and Card K D (1993) The reappearance of the Huronian in Wyoming: rifting and drifting of ancient continents. Can J Earth Sci 30: 2475–2480

Rosen O M (1989) Two geochemically different types of Precambrian crust in the Anabar Shield, North Siberia. Prec Res 45: 129–142

Ross G M, Parrish R R and Winston D (1992) Provenance and U–Pb geochronology of the Mesoproterozoic Belt Supergroup (northwestern United States): implications for age of deposition and pre-Panthalassa plate reconstructions. Earth Planet Sci Lett 113: 57–76

Rousell D H (1984) Onwatin and Chelmsford Formations. In: The Geology and Ore Deposits of the Sudbury Structure, (eds Pye E G, Naldrett A J and Giblin P E). Ont Geol Surv Spec Vol. 1: 211–218

Roy A B (1988) Stratigraphic and tectonic framework of the Aravalli Mountain Range. Geol Soc Ind Mem 7: 3–31

Roy D W, Woussen G, Dimroth E and Chown E H (1986) The central Grenville Province: a zone of protracted overlap between crustal and mantle processes. In: The Grenville Province (eds Moore J M, Davidson A and Baer A J). Geol Ass Can Spec Pap 31: 51–60

Roy S (1992) Environments and processes of manganese deposition. Econ Geol 87: 1218–1236

Rundle C C and Snelling N J (1977) The geochronology of uraniferous minerals in the Witwatersrand Triad; an interpretation of new and existing U-Pb age data on rocks and minerals from the Dominion Reef, Witwatersrand and Ventersdorp Supergroups. Philos Trans R Soc Lond (Ser A) 286: 657–683

Rundqvist D V and Mitrofanov F P (Editors) (1992) Precambrian Geology of the USSR. Elsevier, Amsterdam

Runnegar B (1982) The Cambrian explosion: animals or fossils? J Geol Soc Austral 29: 395–411

Rutland R W R (1981) Structural framework of the Australian Precambrian. In: Precambrian of the Southern Hemisphere, (ed. Hunter D R). Elsevier, Amsterdam: 1–32

Rutland R W R, Parker A J, Pitt G M et al (1981) The Precambrian of South Australia. In: Precambrian of the Southern Hemisphere (ed. Hunter D R). Elsevier, Amsterdam: 309–360

Ryan B (1981) Volcanisn, sedimentation, plutonism and Grenvillian deformation in the Helikian Basins of Central Labrador. In: Proterozoic Basins of Canada, (ed. Campbell F H A). Geol Surv Can Pap 81-10: 361–378

Ryan B (1991) Makhavinek Lake pluton, Labrador, Canada: geological setting, subdivisions, mode of emplacement, and a comparison with Finnish rapakivi granites. Prec Res 51: 193–225

Sabaté P and Marinho M M (1982) Chemical affinities of low-grade metamorphic formations of the Contendas-Mirante Proterozoic complex (Bahia, Brazil). Rev Brasil Geociencias 12(1-3): 392–402

Sacchi R, Marques J, Costa M and Casati C (1984) Kibaran events in the southernmost Mozambique belt. Prec Res 25: 141–159

Sage R P (1990) Michipicoten Iron-Formation in the Wawa area, Ontario. Field Trip Guidebook prepared for the 8th International Association on the Genesis of Ore Deposits Symposium; G A Gross and J A Donaldson (eds.); Geol Surv Can, Open Fill 2163: 36–56

Sager-Kinsman E A and Parrish R R (1993) Geochronology of detrital zircons from the Elzevir and Frontenac terranes, Central Metasedimentary Belt, Grenville Province, Ontario. Can J Earth Sci 30: 465–473

Saha A K and Ray S L (1984) The structural and geochemical evolution of the Singhbhum granite batholithic complex, India. Tectonophys, 105: 163–176

Saha A K, Ghosh S, Dasgupta D and Mukhopadhay K (1984) Studies on crustal evolution of the Singhbhum-Orissa Iron-Ore Craton. Monogr Crustal Evolution, Ind J Earth Sci pp 1–74

Sahoo K C and Mathur A K (1991) On the occurrence of Sargur-type banded iron-formation in Banded Gneissic Complex of south Rajasthan. J Geol Soc Ind 38: 299–302

Salop L J (1977) Precambrian of the Northern Hemisphere. Elsevier, Amsterdam

Salop L J (1983) Geological Evolution of the Earth During the Precambrian. Springer-Verlag, Berlin

Samson C and West G F (1994) Detailed basin structure and tectonic evolution of the Midcontinent Rift System in eastern Lake Superior from reprocessing of GLIMPCE deep reflection seismic data. Can J Earth Sci 31: 629–639

Santos E J and Brito Neves B B (1984) Provincia Barborema. In: O Précambriano do Brasil (eds Almeida F F M and Hasui Y). Edgard Blucher, São Paulo, pp 123–186

Saquaque A, Admou H, Karson J et al (1989) Precambrian accretionary tectonics in the Bou Azzer – El Graara Region, Anti-Atlas, Morocco. Geol 17: 1107–1110

Sarkar A N (1982) Precambrian tectonic evolution of eastern India: a model of converging microplates. Tectonophys 86: 363–397

Sarkar A N, Bhanumathi L and Balasubrahmanyan M N (1981) Petrology, geochemistry and geochronology of the Chilka Lake igneous complex, Orissa state, India. Lithos 14: 93–111

Sarkar G, Corfu F, Paul D K, McNaughton N J, Gupta S N and Bishui P K (1993) Early Archean crust in Bastar Craton, Central India—a geochemical and isotopic study. Prec Res 62: 127–137

Sarkar S C (1988) Genesis and evolution of the ore deposits in the early Precambrian greenstone belts and adjacent high grade metamorphic terrains of Peninsular India—a study in similarity and contrast. Prec Res 39: 107–130

Sarkar S N and Saha A K (1977) The present status of the Precambrian stratigraphy, tectonics and geochronology of Singhbhum-Keonjhar-Mayurbhanj region, eastern India. Ind J Earth Sci S. Roy Volume: 37–65

Sarkar S N and Saha A K (1983) Structure and tectonics of the Singhbhum-Orissa Iron Ore craton, eastern India. In: Structures and Tectonics of the Precambrian Rocks in India (ed. Sinha-Roy S). Hindustan Publishing Corp Press, New Delhi; Recent Res Geol 10: 1–25

Saul J M (1994) Cancer and autoimmune disease: A Cambrian couple? Geol 22: 5

Saverikko M, Koljonen T and Hoffren V (1985) Paleogeography and paleovolcanism of the Kummitsoiva komatiite complex in northern Finland. Bull Geol Surv Fin 331: 143–158

Sawkins F J (1989) Anorogenic felsic magmatism, rift sedimentation, and giant Proterozoic Pb-Zn deposits. Geol 17: 657–660

Schandelmeier H, Utke A, Harms U and Küster D (1990) A review of the Pan-African evolution of NE Africa: towards a new dynamic concept for continental NE Africa. Berl Geowiss Abh A120: 1–14

Schandelmeier H, Wipfler E, Küster D, Sutton M, Becker R, Stern R and Abdelsalam M G (1994) Atmur–Delgo suture: A Neoproterozoic oceanic basin extending into the interior of northeast Africa. Geol 22: 563–566

Schau M (1975) Volcanogenic rocks of the Prince Albert Group, Melville Peninsula (47 A-d), District of Franklin. Geol Surv Can, Pap 75-1A: 359–361

Schau M (1978) Metamorphism of the Prince Albert Group, District of Keewatin; In: Metamorphism in the Canadian Shield, (eds Fraser A J and Heywood W W). Geol Surv Can, Pap 78-10: 203–214

Schau M (1980) Zircon ages from a granulite-anorthosite complex and a layered gneiss complex, northeast of Baker Lake, District of Keewatin. Geol Surv Can, Pap 80-16: 237–238

Schau M and Henderson J B (1983) Archean chemical weathering at three localities on the Canadian Shield. In: Developments and Interactions of the Precambrian Atmosphere, Lithosphere and Biosphere, (eds Nagy B, Wever R, Guerrero J C and Schidlowski M). Elsevier, Amsterdam: 189–224

Schermerhorn L J G (1981) The West-Congo orogen: a key to Pan-African thermotectonism. Geol Rdsch 70: 850–867

Schiøtte L, Compston W and Bridgwater D (1989a) Ion probe U–Th–Pb Zircon dating polymetamorphic orthogneisses from northern Labrador, Canada. Can J Earth Sci 26: 1533–1556

Schiøtte L, Compston W and Bridgwater D (1989b) U-Th-Pb ages of single zircons in Archaean supracrustals in Nain Province, Labrador, Canada. Can J Earth Sci 26: 2636–2644

Schiøtte L, Noble S and Bridgwater D (1990) U–Pb mineral ages from northern Labrador: Possible evidence for interlayering of Early and Middle Archean tectonic slices. Geosci Can 17: 227–231

Schiøtte L, Nutman A P and Bridgwater D (1992) U–Pb ages of single zircons within "Upernavik" metasedimentary rocks and regional implications for the tectonic evolution of the Archaean Nain Province, Labrador. Can J Earth Sci 29: 260–276

Schiøtte L, Hansen B T, Shirey S B and Bridgwater D (1993) Petrological and whole rock isotopic characteristics of tectonically juxtaposed Archaean gneisses in the Okak area of the Nain Province, Labrador: relevance for terrane models. Prec Res 63: 293–323

Schmidt P W and Clark D A (1980) The response of pealomagnetic data to the expanding earth. Geophys J 61: 95–100

Schmidt P W and Clark D A (1994) Palaeomagnetism and magnetic anisotropy of Proterozoic banded-iron formations and iron ores of the Hamersley Basin, Western Australia. Prec Res 69: 133–155

Schmidt P W, Williams G E and Embleton B J J (1991) Low paleolatitude of Late Proterozoic glaciation: early timing of remanence in haematite of the Elatina Formation, South Australia. Earth Planet Sci Lett 105: 355–367

Schopf, J W (1977) Biostratigraphic usefulness of stromatolitic Precambrian microbiotas: A preliminary analysis. Prec Res 5: 143–173

Schopf J W, Hayes J M and Walter M R (1983) Evolution of Earth's earliest ecosystems: recent progress and unresolved problems. In: Earth's Earliest Biosphere: Its Origin and Evolution (ed. Schopf J W). Princeton University Press, pp 361–384

Schorscher H D, Sautana F C, Polonia J C and Moreira J M P (1982) Quadrilatero Ferrifero-Minas Gerais State: Rio das Velhas greenstone belt and Proterozoic rocks. In: Excursions Annex, International Symposium on Archean and Early Proterozoic Geologic Evolution and Metallogenesis (ISAP); coordenação da Produção mineral da Secretaria das Minas e Energia do Estado da Bahia, Salvador, Bahia, 44 pp

Schubert G (1991) The lost continents. Nature, 354: 358–359

Schürmann H M E (1974) The Pre-Cambrian in North Africa E J Brill, Leiden

Schweitzer J and Kröner A (1985) Geochemistry and petrogenesis of early Proterozoic intracratonic volcanic rocks of the Ventersdorp Supergroup, South Africa. Chem Geol 51: 265–288

Scoates R F J (1981) Volcanic rocks of the Fox River Belt, northeastern Manitoba. Manitoba Dept Energy Mines, Geol Pap GR81-1

Scoon R N and Teigler B (1994) Platinum-Group Element mineralization in the Critical Zone of the western Bushveld Complex: I. Sulfide poor–chromitites below the UG-2. Econ Geol 89: 1094–1121

Scott D J, Helmstaedt H and Bickle M J (1992) Portuniq ophiolite, Cape Smith belt, northern Quebec, Canada: a reconstructed section of Early Proterozoic oceanic crust. Geol 20: 173–176

Scott D J, Machado N, Hanmer S and Gariépy C (1993) Dating ductile deformation using U–Pb geochronology: examples from the Gilbert River Belt, Grenville Province, Labrador, Canada. Can J Earth Sci 30: 1458–1469

Sears J W and Price R A (1978) The Siberian connection: A case for Precambrian separation of the North American and Siberian cratons. Geol 6: 267–270

Sedova I S, Krylov D P and Glebovitsky V A (1993) Ultrametamorphism within the amphibolite-granulite transition zone, upper Aldan river, Siberia. Prec Res 62: 431–451

Seifert K E, Peterman Z E and Thieben S E (1992) Possible crustal contamination of Midcontinent Rift igneous rocks: examples from the Mineral Lake intrusions, Wisconsin. Can. J Earth Sci 29: 1140–1153

Semenenko N P (1972) Comparison between the Ukrainian and Baltic Shield Precambrian. Geotektonika 5: 93–98

Semikhatov M A and Serebryakov S N (1978) Lower Riphean of the Siberian Platform. Tr Geol Inst Akad Nauk SSSR 312: 43–66 (in Russian)

Sengör A M C, Natal'in B A and Burtman V S (1993) Evolution of the Altaid tectonic collage and Palaeozoic growth in Eurasia. Nature, 364: 299–307

Sengupta S, Paul D K, de Laeter J R, McNaughton N J, Bandopadhyay P K and de Smeth J B (1991) Mid-Archaean evolution of the Eastern Indian Craton: geochemical and isotopic evidence from the Bonai pluton. Prec Res 49: 23–37

Shackleton R M (1986) Precambrian collision tectonics in Africa. In: Collision Tectonics (eds Coward M P and Ries A C). Geol Soc Lond Spec Public 19: 329–349

Shackleton R M, Ries A M, Fitches W R and Graham R H (1981) Late Proterozoic tectonics of NE Africa. Abs. 11th Colloq Afr Geol, Open Univ, Milton Keynes, p 13

Sharma R P (1983) Structure and tectonics of the Bundelkhand complex, central India. Recent Res Geol 10: 198–210

Sharma R S (1983) Basement-cover rocks relations in north-central Aravalli Range: a tectonic and metamorphic synthesis. Recent Res Geol 10: 53–71

Shaw D M (1972) Development of the early continental crust. I. Use of trace element distribution coefficient models for the protoarchean crust. Can J Earth Sci 9: 1577–1595

Shaw D M (1976) Development of the early continental crust, Part 2: Pre-Archean, Proto-Archean and later eras. In: The Early History of the Earth, (ed. Windley B F). John Wiley, New York: 33–53

Shaw D M (1980) Evolutionary tectonics of the earth in the light of early crustal structure. In: the Continental Crust and its Mineral Deposits, (ed. Strangway D W). Geol Ass Can Spec Pap 20: 65–73

Shay J and Tréhu A (1993) Crustal structure of the central graben of the Midcontinent Rift beneath Lake Superior. Tectonophys 225: 301–335

Shcherbak N P, Barntnistky E N, Bibikovo E V and Boiko V L (1984) Age and evolution of the Early Precambrian continental crust of the Ukranian Shield. In: Archean Geochemistry; (eds Kröner A, Hanson G N and Goodwin A M). Springer-Verlag, Berlin: 251–261

Sheldon R P (1981) Ancient marine phosphorites. Ann Rev Earth Planet Sci 9: 251–284

Sheraton J W and Black L P (1983) Geochemistry of Precambrian gneisses: relevance for the evolution of the East Antarctic Shield. Lithos 16: 273–296

Sheraton J W, Offe L A, Tingey R J and Ellis D J (1980) Enderby Land, Antarctica—an unusual Precambrian high-grade metamorphic terrain. J Geol Soc Austral 27: 1–18

Shufflebotham M M (1989) Geochemistry and geotectonic interpretation of the Penthievre crystalline massif, northern Brittany, France. Prec Res 45: 247–261

Sibbald T I I, Quirt D H and Grace A J (1990) Uranium deposits of the Athabasca Basin, Saskatchewan. 8th IAGOD Symposium, Field Trip Guidebook, Trip 1, Geol Surv Can, Open File 2166

Siddaiah N S, Hanson G N and Rajamani V (1994) Rare earth element evidence for syngenetic origin of an Archean stratiform gold sulfide deposit, Kolar Schist Belt, south India. Econ Geol 89: 1552–1566

Siedlecka A, Pickering K T and Edwards M B (1989) Upper Proterozoic margin deltaic complex, Finnmark, N. Norway. In: Deltas: Sites and Traps for Fossil Fuels (eds. Whateley M K G and Pickering K T). Geol Soc Lond. Spec Public 41: 205–219

Silva L C (1972) O Macigo gabbro-anorthositico do SW de Angola. Rev Faculdade de Ciènias, Universidad de Lisboa, 17: 253–277

Silvennoinen A, Honkamo M, Juopperi H et al (1980) Main features of the stratigraphy of north Finland. In: Jatulian Geology in the Eastern part of the Baltic Shield (ed. Silvennoinen A). Finnish-Soviet Symp Rovaniemi, Finland, 21–26 August 1979, pp 153–162

Silver L T (1984) Observations on the Precambrian evolution of northern New Mexico and adjacent regions. Geol Soc Am (Abs. with Programs) 16: 256

Simonen A (1980) The Precambrian in Finland. Bull Geol Surv Fin 304

Simonson B M (1992) Geological evidence for a strewn field of impact spherules in the early Precambrian Hamersley Basin of Western Australia. Geol Soc Am Bull 104: 829–839

Sims P K (1980) Boundary between Archean greenstone and gneiss terrains in northern Wisconsin and Michigan. Geol Soc Am Spec Pap 182: 113–124

Sims P K and Peterman Z E (1983) Evolution of Penokean fold belt, Lake Superior region, and its tectonic environment. In: Early Proterozoic Geology of the Great Lakes Region (ed. Medaris L G). Geol Soc Am Mem 160: 3–14

Sims P K and Peterman Z E (1986) Early Proterozoic Central Plains orogen—a major buried structure in the north-central United States. Geol 14: 488–491

Sims P K, Card K D and Lumbers S B (1981) Evolution of early Proterozoic basins of the Great Lakes region. In: Proterozoic Basins of Canada, (ed. Campbell F H A). Geol Surv Can Pap 81-10: 379–397

Sims P K, Van Schmus W R, Schulz K J and Peterman Z E (1989) Tectono-stratigraphic evolution of the Early Proterozoic Wisconsin magmatic terranes of the Penokean Orogen. Can J Earth Sci 26: 2145–2158

Sims P K, Anderson J L, Bauer R L et al (1993) The Lake Superior region and Trans-Hudson orogen. In: Precambrian: Conterminous U.S. (eds. Reed J C, Bickford M E, Houston R S et al.). Geol Soc Am, The Geology of North America, C-2: 11–120

Sinha-Roy S and Mohanty M (1988) Blueschist facies metamorphism in the ophiolitic mélange of the late Proterozoic Delhi fold belt, Rajasthan, India. Prec Res 42: 97–105

Siroschtan R I, Tcherbakov J B and Usenko I S (1978) Ukrainian

Shield. Explanatory text; metamorphic map of Europe. UNESCO, Paris: 39–40

Sivell W J and McCulloch M T (1991) Nd isotope evidence for ultra-depleted mantle in the early Proterozoic. Nature 354: 384–387

Sivoronov A.a., Bobrov A B and Malyuk B I (1984) Morphological types of the Early Precambrian greenstone belts in Middle Dnieper region, Ukrainian Shield. Geotectonika 5: 22–37

Skiöld T (1976) The interpretation of the Rb-Sr and K-Ar ages of late Precambrian rocks in southwestern Sweden. Geol Fören Förhand 98: 3–29

Skiöld T (1988) Implications of new U-Pb zircon chronology to early Proterozoic crustal accretion in northern Sweden. Prec Res 38: 147–164

Skiöld T, Öhlander B, Markkula H, Widenfalk L and Claesson L-A (1993) Chronology of Proterozoic orogenic processes at the Archaean continental margin in northern Sweden. Prec Res 64: 225–238

Skulski T, Hynes A and Francis D (1984) Stratigraphic and lithogeochemical characterization of cyclic volcanism in the LG-3 area, La Grande River greenstone belt, Quebec. In: Chibougamau—stratigraphy and mineralization, (eds Guha J and Chown E H). Can Inst Mining Metall 34: 57–72

Slack J F, Palmer M R, Stevens B P J and Barnes R G (1993) Origin and significance of Tournaline-rich rocks in the Broken Hill district, Australia. Econ Geol 88: 505–541

Sleep N H and Windley B F (1982) Archaean plate tectonics: constraints and inferences. J Geol 90: 363–379

Smith P E, York D, Easton R M, Özdemir Ö and Layer P W (1994) A laser $^{40}$Ar–$^{39}$Ar study of minerals across the Grenville Front: investigation of reproducible excess Ar patterns. Can J Earth Sci. 31: 808–817

Snowden P A (1984) Non-diapiric batholiths in the north of the Zimbabwe Shield. In: Precambrian Tectonics Illustrated (eds Kröner A and Greiling R). Schweizerbart'sche Verlagsbuchhandlung, Stuttgart; pp 135–145

Sobotovich E V, Kamenev E N, Komaristyy A A and Rudnik V A (1976) The oldest rocks in Antarctica (Enderby Land). Int Geol Rev 18: 371–368

Soegaard K and Callahan D M (1994) Late Middle Proterozoic Hazel Fm near Van Horn, Trans-Peros, Texas: Evidence for transpressive deformation in Grenvillian basement. Geol Soc Am Bull 106: 413–423

Söhnge A P G (1986) Mineral provinces of southern Africa. In: Mineral Deposits of South Africa (eds. Anhaeusser C R and Maske S). Geol Soc S Afr Spec Public.: 1–23

Sokolov B S and Fedonkin M A (Editors) (1990) The Vendian system. English translation of (1985) Russian edition by T I Vasiljeva, R. J. Sorkina and R V Fursenko. Springer-Verlag, Berlin

Soni M R, Chakraborty S and Jain V K (1987) Vindhyan Supergroup—a review. In: Purana Basins of Peninsular India. Geol Soc Ind Mem 6: 87–138

Southwick D L, Morey G B and Mossler J H (1986) Fluvial origin of the lower Proterozoic Sioux Quartzite, southwestern Minnesota. Bull Geol Soc Am 97: 1432–1441

Spencer A M (1978) The Late Pre-Cambrian glaciation in Scotland. Geol Soc Lond Mem 6

Spooner E T C and Barrie C T (1993) A special issue devoted to Abitibi Ore Deposits in a modern context: Preface. Econ Geol 88: 1307–1322

Sreenivasa Rao T (1987) The Pakhal Basin—a perspective. In: Purana Basins of Peninsular India. Geol Soc Ind Mem 6: 161–187

Srikantappa C, Raith M and Spiering B (1985) Progressive charnockitization of a leptynite-khondalite suite in southern Kerala, India—evidence for formation of charnockites through decrease in fluid pressure? J Geol Soc Ind 26: 849–872

Stacey J S and Hedge C E (1984) Geochronologic and isotopic evidence for early Proterozoic crust in the eastern Arabian Shield. Geol 12: 310–313

Stanistreet I G and McCarthy T S (1991) Changing tectono-sedimentary scenarios relevant to the development of the Late Archean Witwatersrand Basin. Econ Geol Res Unit, Univ Witwatersrand, Johannesburg, Info Circ 233, 28 pp.

Stern R A, Percival J A and Mortensen J K (1994) Geochemical evolution of the Minto block: a 2.7 Ga continental magmatic arc built on the superior proto-craton. Prec Res 65: 115–153

Stern R J (1981) Petrogenesis and tectonic setting of late Precambrian ensimatic volcanic rocks, Central Eastern desert of Egypt. Prec Res 16: 195–230

Stern R J (1994) Neoproterozoic (900–550 M) arc assembly and continental collision in the East African orogen. Ann Rev Earth Planet Sci (in press)

Stern R J and Hedge C E (1985) Geochronologic and isotopic constraints on late Precambrian crustal evolution in the Eastern Desert of Egypt. Am J Sci 285: 97–127

Stern R J and Kröner A (1993) Late Precambrian crustal evolution in NE Sudan: Isotopic and geochronologic constraints. J Geol 101: 555–574

Stevens B P J, Barnes R G, Brown R E et al (1988) The Willyama Supergroup in the Broken Hill and Euriowie Blocks, New South Wales. Prec Res 40/41: 297–327

Stevens G and van Reenen D (1992) Partial melting and origin of metapelitic granulites in the Southern Marginal Zone of the Limpopo Belt, South Africa. Prec Res 55: 303–319

Stevenson I M (1968) A geological reconnaisance of Leaf River map-area, New Quebec and Northwest Territories. Geol Surv Can Mem 356

Stevenson J S (1990) The volcanic origin of the Onaping Formation, Sudbury, Canada. Tectonophys 171: 249–257

Stewart J H (1972) Initial deposits in the Cordilleran Geosyncline: evidence of a late Precambrian (<850 m.y.) continental separation. Bull Geol Soc Am 83: 1345–1360

Stewart J H (1976) Late Precambrian evolution of North America: plate tectonics implication. Geol 4: 14–16

St Onge M R and Lucas S B (1988) Thermal history models and P-T determinations (Abs.). Geol Ass Can, Joint Ann Mtg, St John's Newfoundland (Program with Abs.) 13: A120

St Onge M R, Lucas S B, Scott D J and Bégin N J (1989) Evidence for the development of oceanic crust and for continental rifting in the tectonostratigraphy of the early Proterozoic Cape Smith Belt. Geosci Can 16: 119–122

St Onge M R, Lucas S B and Parrish R R (1992) Terrane accretion in the internal zone of the Ungava orogen, northern Quebec. Part I: Tectonostratigraphic assemblages and their tectonic implications. Can J Earth Sci 29: 746–764

Stockwell C H (1961) Structural provinces, orogenies and time classification of rocks of the Canadian Shield. Geol Surv Can, Pap 61-17: 108–118

Stockwell C H (1982) Proposals for time classification and correlation of Precambrian rocks and events in Canada and adjacent areas of the Canadian Shield. Part 1: A time classification of Precambrian rocks and events. Geol Surv Can Pap 80-19

Stowe C W (1984) The early Archaean Selukwe nappe, Zimbabwe. In: Precambrian Tectonics Illustrated, (eds Kröner A and Greiling R). Schweizerbart'sche Verlagsbuchhandlung, Stuttgart: 41–56

Stowe C W (1994) Compositions and tectonic settings of chromite deposites through time. Econ Geol 89: 528–546

Strand K O and Laajoki K (1993) Palaeoproterozoic glaciomarine sedimentation in an extensional tectonic setting: the Honkala Formation, Finland. Prec Res 64: 253–271

Streckeisen A (1976) To each rock its proper name. Earth Sci Rev 12: 1–33

Stump E, Miller J M G, Korsch R J and Edgerton D G (1988) Diamictite from Nimrod Glacier area, Antarctica : possible

Proterozoic glaciation on the seventh continent. Prec Res 16: 225–228

Sultan M, Arvidson R E, Sturchio N C and Guinness E A (1987) Lithologic mapping in arid regions with Landsat thematic mapper data: Meatiq dome, Egypt Geol Soc Am Bull 99: 748–762

Sun D Z and Lu S N (1985) A subdivision of the Precambrian of China. Prec Res 28: 134–162

Sun D Z, Bai J, Jin W S et al (1984) The early Precambriana geology of the eastern Hebei. Tianjin Science Technology Press, Tianjin (in Chinese with 20 pages of English summary)

Sun M, Armstrong R L and Lambert S J (1992) Petrochemistry and Sr, Pb and Nd isotopic geochemistry of early Precambrian rocks, Wutaishan and Taihangshan areas, China. Prec Res 56: 1–31

Sundblad K, Ahl M and Schöberg H (1993) Age and geochemistry of granites associated with Mo-mineralizations in western Bergslagen, Sweden. Prec Res 64: 319–335

Sutcliffe R H, Barrie C T, Burrows D R and Beakhouse G P (1993) Plutonism in the southern Abitibi subprovince: a tectonic and petrogenetic framework. Econ Geol 88: 1359–1375

Sutton S J and Maynard J B (1993a) Petrology, mineralogy and geochemistry of sandstones of the lower Huronian Matineuda Formation: resemblance to underlying basement rocks. Can J Earth Sci 30: 1209–1223

Sutton S J and Maynard J B (1993b) Sediment- and basalt-hosted regoliths in the Huronian Supergroup: role of parent lithology in middle Precambrian weathering profiles. Can J Earth Sci 30: 60–76

Suwa K, Tokieda K and Hoshino M (1994) Palaeomagnetic and petrological reconstruction of the Seychelles. Prec Res 69: 281–292

Svoboda J (1966) Regional Geology of Czechoslovakia. Part 1, The Bohemian Massif. Geol Surv Czechoslovakia

Swager C P, Witt W K, Griffin T J, Ahmat A L, Hunter W M, McGoldrick P J and Wyche S (1992) Late Archaean granite-greenstones of the Kalgoorlie Terrane, Yilgarn Craton, Western Australia. In: The Archaean: Terrains, Processes and Metallogeny (eds. Glover J E and Ho S E). The Univ W Austral. Public 22: 107–122

Swami Nath J and Ramakrishnan M (1981) Early Precambrian supracrustals of southern Karnataka: southern India greenstone belts. Geol Surv Ind, Mem 112

Swami Nath J, Ramakrishnan M and Viswanatha M N (1976) Dharwar stratigraphic model and Karnataka craton evolution. Geol Surv Ind 107: 149–175

Sweeney M A, Binda P L and Vaughan D J (1991) Genesis of the ores of the Zambian Copperbelt. Ore Geol Rev 6: 51–76

Syme E C and Bailes A H (1993) Stratigraphic and tectonic setting of Early Proterozoic volcanogenic massive sulfide deposits, Flin Flon, Manitoba. Econ Geol 88: 566–589

Symons D T A (1991) Paleomagnetism of the Proterozoic Wathaman Batholith and the suturing of the Trans-Hudson Orogen in Saskatchewan. Can J Earth Sci 28: 1931–1938

Tack L (1983) Extension du Mayumbien au Bas-Zaïre: le problème de sa délimitation cartographique et implication sur les concepts du cadre géologique général du Précambrien du Bas-Zaïre. Rapp Ann Mus R Afr Centr Tervuren: 127–133

Tack L, De Paepe P, Liégois J P, Nimpagaritse G, Ntungicimpaye A and Midende G (1990) Late Kibaran magmatism in Burundi J Afr Earth Sci 10: 733–738

Tack L, Theunissen K, Lavreau J, Deblond A and Duchesne J C (1992) Ensialic Middle and Upper Proterozoic evolution of the East African subcontinent between Lake Victoria and Lake Tanganyika (abs.). Eos (Transactions, American Geophysical Union) 73: 366

Tack L, Liégeois J P, Deblond A and Duchesne J C (1994) Kibaran A-type granitoids and mafic rocks generated by two mantle sources in a late orogenic setting (Burundi). Prec Res 68: 323–356

Taira A, Pickering K T, Windley B F and Soh W (1992) Accretion of Japanese island arcs and implications for the origin of Archean greenstone belts. Tectonics, 11: 1224–1244

Tanaka H and Idnurm M (1994) Palaeomagnetism of Proterozoic mafic intrusions and host rocks of the Mount Isa Inlier, Australia: revisited. Prec Res 69: 241–258

Tankard A J, Jackson M P A, Eriksson K A et al (1982) Crustal Evolution of Southern Africa. Springer-Verlag, New York

Tanner P W G (1973) Orogenic cycles in East Africa. Bull Geol Soc Am 84: 2839–2850

Tarling D H (1985) Problems in Palaeozoic paleomagnetism. J Geodynam 3: 87–103

Tassinari C C G (1988) Precambrian continental crust evolution of southeastern São Paulo State Brazil, based on isotopic evidences. In: International Conference "Geochemical Evolution of the Continental Crust,- Poços de Caldas. Abs.: 141–147

Tassinari C C G, Siga O and Teixeira W (1984) Épocas metalogenéticas relacionadas à granitogenese do Cráton Amazônico. In: Congr Bras Geol, 33 Rio de Janeiro, (1984) Anais - Rio de Janeiro, SBG 6: 2963–2977

Taylor P N (1975) An early Precambrian age for migmatitic gneisses from Vikan I B O, Vesteralen, north Norway. Earth Planet Sci Lett 27: 35–42

Taylor P N, Chadwick B, Moorbath S et al (1984) Petrography, chemistry and isotopic ages of Peninsular Gneiss, Dharwar acid volcanic rocks and the Chitradurga Granite with special reference to the Late Archean evolution of the Karnataka Craton, Southern India. Prec Res 23: 349–375

Taylor P N, Moorbath S, Leube A and Hirdes W (1988) Geochronology and crustal evolution of early Proterozoic granite-greenstone terrains in Ghana/West Africa. Abstr International Conference on the Geology of Ghana with Special Emphasis on Gold. Comm 75th Anniversary of Ghana Geol Surv Dept, Accra: 43–45

Taylor P N, Moorbath S, Leube A and Hirdes W (1992) Early Proterozoic crustal evolution in the Birimian of Ghana: constraints from geochronology and isotope geochemistry. Prec Res 56: 97–111

Taylor S R and McLennan S M (1985) The Continental Crust: Its Composition and Evolution. Blackwell Scientific Publications, Melbourne

Tegtmeyer A and Kröner A (1985) U-Pb zircon ages for granitoid gneisses in northern Namibia and their significance for Proterozoic crustal evolution of southwestern Africa. Prec Res 28: 311–326

Teisseyre R, Leliwa-Kopystynski J and Lang B (Editors) (1992) Evolution of the Earth and Other Planetary Bodies. Elsevier Science B.V., Amsterdam, The Netherlands

Teixeira J B G, Ohmoto H and Eggles D H (1994) Elemental and oxygen isotope variations in Archean mafic rocks associated with the banded iron-formation at the N4 iron deposit, Carajás, Brazil. Prec Res 3: 584–96

Teixeira W (1982) Geochronology of the southern part of the São Francisco Craton. Rev Brasil Geociencias 12(1-3): 240–250

Teixeira W (1985) A evolução geotectônica da porção meridional do craton de S Francisco. University of São Paulo, unpublished thesis

Teixeira W and Figueiredo M C H (1991) An outline of Early Proterozoic crustal evolution in the São Francisco craton, Brazil: a review. Prec Res 53: 1–22

Teixeira W, Tassinari C C G, Cordani C G and Kawashita K (1989) A review of the geochronology of the Amazonian Craton: tectonic implications. Prec Res 42: 213–227

Thomas D J and Heaman L M (1994) Geologic setting of the Jolu gold mine, Saskatchewan: U–Pb age constraints on plutonism, deformation, mineralization, and metamorphism. Econ. Geol 89: 1017–1029

Thomas R J, Agenbacht A L D, Cornell D H and Moore J M

(1994) The Kibaran of southern Africa: Tectonic evolution and metallogeny. Ore Geol Rev 9: 131–160

Thorne A M and Seymour D B (1986) The sedimentology of a tide-influenced fan delta system in the Early Proterozoic Wyloo Group on the southern margin of the Pilbara Craton, Western Australia. Geol Surv W Austral Rep 19: 70–82

Thorpe R I, Hickman A H, Davis D W, Mortensen J K and Trendall A F (1992) U–Pb zircon geochronology of Archaean felsic units in the Marble Bar region, Pilbara Craton, Western Australia. Prec Res 56: 169–189

Thurston P C and Chivers K M (1990) Secular variation in greenstone sequence development emphasizing Superior Province, Canada. Prec Res 46: 21–58

Tingey R J (1982) The geologic evolution of the Prince Charles Mountains—an Antarctic Archaean cratonic block. In: Antarctic Geoscience, (ed. Craddock C). University of Wisconsin Press, Madison: 455–464

Tingey R J (Editor) (1991a) Geology of Antarctica. Oxford University Press: 710 pp

Tingey R J (1991b) The regional geology of Archaean and Proterozoic rocks in Antarctica. In Tingey R J (Editor)—Geology of Antarctica. Oxford University Press: 1–73

Tingey R J (Compiler) (1991c) Commentary on schematic geological map of Antarctica Scale 1 : 10 000 000. BMR Austral Bull 238: 26 pp

Titton T A (1988) 4.56 Ga-time of 1st condensation solid matter within solar system.

Tobisch O T, Collerson K D, Bhattacharyya T and Mukkhopadhyay D (1944) Structural relationships and Sr–Nd isotope systematics of polymetamorphic granitic gneisses and granitic rocks from central Rajasthan, India: implications for the evolution of the Aravalli craton. Prec Res 65: 319–339

Toteu S F, Michard A, Bertrand J M and Rocci G (1987) U–Pb dating of Precambrian rocks from northern Cameroon, orogenic evolution and chronology of the Pan-African belt of central Africa. Prec Res 37: 71–87

Toteu S F, Van Schmus W R, Penaye J and Nyobé J B (1994) U–Pb and Sm–Nd evidence for Eburnean and Pan-African high-grade metamorphism in cratonic rocks of southern Cameroon. Prec Res 67: 321–347

Toulkeridis T, Goldstein S L, Clauer N, Kröner A and Lowe D R (1994) Sm–Nd dating of Fig Tree clay minerals of the Barberton greenstone belt, South Africa. Geol 22: 199–202

Towe K M (1983) Precambrian atmospheric oxygen and banded iron formations: a delayed ocean model. In: Developments and Interactions of the Precambrian Atmosphere, Lithosphere and Biosphere, (eds Nagy B, Webers R, Guerreraro J C and Schidlowski M). Elsevier, Amsterdam. pp 53–62

Towe K M (1990) Aerobic respiration in the Archaean? Nature 348: 54–56

Towe K M (1991) Aerobic carbon cycling and cerium oxidation: significance for Archean oxygen levels and banded iron-formation deposition. Palaeogeography, Palaeoclimatology, Palaeoecol, 97: 113–123

Treloar P J (1988) The geological evolution of the Magondi Mobile Belt, Zimbabwe. Prec Res 38: 55–73

Treloar P J and Kramers J D (1989) Metamorphism and geochronology of granulites and migmatitic granulites from the Magondi Mobile Belt, Zimbabwe. Prec Res 45: 277–289

Trendall A F (1975) Hamersley Basin. Geol Surv W Austral Mem 2: 119–143

Trendall A F (1976) Geology of the Hamersley Basin. 25th Int Geol Congr, Excursion Guide 43A

Trendall A F (1983) The Hamersley Basin. In: Iron-Formation: Facts and Problems, (eds Trendall A F and Morris R C). Elsevier, Amsterdam: 69–129

Trendall A F and Blockley J G (1968) Stratigraphy of the Dales Gorge Member of the Brockman Iron-Formation, in the Precambrian Hamersley Group of Western Australia. Ann Rep Geol Surv W Austral (1967): 48–53

Trendall A F and Blockley J G (1970) The iron formations of the Precambrian Hamersley Group, Western Australia. Bull Geol Surv W Austral Bull 119

Trendall A F (Compston W, Williams I S, Armstrong R A, Arndt N T, McNaughton N J, Nelson D R, Barley M E, Beukes N J, De Laeter J R, Retief E A and Thorne A M) (1990) Precise zircon U–Pb chronological comparison of the volcano-sedimentary sequences of the Kaapvaal and Pilbara cratons between about 3.1 and 2.4 Ga. Extended Abstracts, 3rd Int Archaean Symposium, Perth: 81–83

Trettin H P (1972) The Innuitian Province. In: Variations in Tectonic Styles in Canada (eds Price R A and Douglas R J W). Geol Ass Can Spec Pap 11: 83–179

Trompette R (1973) Le Précambrien supérieur et le Paleozoique inférieur de l'Adrar de Mauritanie (bordure occidentale du bassin de Taoudeni, Afrique de l'Ouest). Un éxemple de sédimentation de craton. Etude stratigraphique et sédimentologique. Trav Lab Sci Terre, Univ Marseille St Jerome (Serie B) No. 7

Truswell J F (1990) The Transvaal and Griqualand West sequences: Some current issues. Econ Geol Res Unit, Univ Witwatersrand, Johannesburg, Info Circ 232, 69 pp

Tsomondo J M, Wilson J F and Blenkinsop T G (1992) Reassessment of the structure and stratigraphy of the early Archaean Selukwe nappe, Zimbabwe. In: The Archaean: Terrains, Processes and Metallogeny (eds Glover J E and Ho S E). The Univ Western Austral. Public 22: 123–135

Tsunogae T, Miyano T and Ridley J (1992) Metamorphic P–T profiles from the Zimbabwe Craton to the Limpopo Belt, Zimbabwe. Prec Res 55: 259–277

Turchenko S I (1992) Precambrian metallogeny related to tectonics in the eastern part of the Baltic Shield. In: Precambrian Metallogeny Related to Plate Tectonics (eds. Poulsen K H, Card K D and Franklin J M). Prec Res 58: 121–141

Turek A, Sage R P and Van Schmus W R (1992) Advances in the U–Pb zircon geochronology of the Michipicoten greenstone belt, Superior Province, Ontario. Can J Earth Sci 29: 1154–1165

Turekian K K and Clark S P (1969) Inhomogeneous accumulation of the earth from the primitive solar nebula. Earth Planet Sci Lett 6: 346–348

Tyler I M, Fletcher I R, de Laeter J R, Williams I R and Libby W G (1992) Isotope and rare earth element evidence for a late Archaean terrane boundary in the southeast Pilbara Craton, Western Australia. Prec Res 54: 211–229

Tyler S A and Barghoorn E S (1954) Occurrence of structurally preserved plants in Precambrian rocks of the Canadian Shield. Sci 119: 606–608

Uhlein A, Trompette R and da Silva M E (1986) Estruturação tectônica do Supergrupo Espinhaço na região de diamantina (MG). Rev Brasil Geociencias 16(2): 212–216

Unrug R (1988) Mineralization controls and source of metals in the Lufilian fold belt, Shaba (Zaire), Zambia, and Angola. Econ Geol 83: 1247–1258

Unrug R and Unrug S (1990) Paleontological evidence of Paleozoic age for the Walden Creek Group, Ocoee Supergroup, Tennessee. Geol 18: 1041–1045

Upton B G J and Emelens C H (1987) Mid-Proterozoic alkaline magmatism in southern Greenland: the Gardar province. In: Alkaline Igneous Rocks (eds. Fitton J G and Upton B G J). Geol Soc Spec Public 30: 449–471

Urban H, Stribrny B and Lippolt H J (1992) Iron and manganese deposits of the Urucum district, Mato Grosso do Sul, Brazil. Econ Geol 87: 1375–1392

Vaasjoki M, Rämö O T and Sakko M (1991) New U–Pb ages from the Wiborg rapakivi area: constraints on the temporal evolution of the rapakivi granite-anorthosite-diabase dyke association of southeastern Finland. Prec Res 51: 227–243

Vachette M (1979a) Radiochronologie du précambrien de Madagascar. 10$^C$ Colloq Géol Afr Montpellier, Résumés, pp 20–21

Vail J R (1976) Outline of the geochronology and tectonic units of the basement complex of northeast Africa. Proc R Soc Lond (Ser A) 350: 97–114

Vail J R (1978) Outline of the geology and mineral deposits of the Democratic Republic of Sudan and adjacent areas. Overseas Geol Min Res Lond No. 49

Vail J R, Snelling N J and Rex D C (1968) Pre-Katangan geochronology of Zambia and adjacent parts of Central Africa. Can J Earth Sci 5: 621–628

Valenta R (1994) Syntectonic discordant copper mineralization in the Hilton Mine, Mount Isa. Econ Geol 89: 1031–1052

Van Aswegen G, Strydom D, Colliston W P, Praekelt H E, Schoch A E, Blignault H J, Botha B J V and Van der Merwe S W (1987) The structural–stratigraphic development of part of the Namaqua Metamorphic Complex—an example of Proterozoic major thrust tectonics. In: A Kröner (Editor), Proterozoic Lithospheric Evolution (Goedyn. Ser., 17) Am Geophys Union, Washington DC: 207–216

Van der Voo R and Meert J G (1991) Late Proterozoic paleomagnetism and tectonic models: a critical appraisal. Prec Res 53: 149–163

Van Kranendonk M J, St-Onge M R and Henderson J R (1993) Paleoproterozoic tectonic assembly of Northeast Laurentia through multiple indentations. Prec Res 63: 325–347

Van Reenen D D, Barton J M, Roering C et al (1987) Deep crustal response to continental collision: the Limpopo Belt of southern Africa. Geol 15: 11–14

Van Reenen D D, Roering C, Ashwal L D and de Wit M J (1992a) The Archaean Limpopo Granulite Belt: tectonics and deep crustal processes. Prec Res 55: 1–587

Van Reenen D D, Roering C, Ashwal L D and De Wit M J (1992b) Regional geologic setting of the Limpopo Belt. Prec Res 55: 1–5

Van Schalkwyk J F, DeWit M J, Roering C and Van Reenen D D (1993) Tectono-metamorphic evolution of the simatic basement of the Pietersburg greenstone belt relative to the Limpopo Orogeny: evidence from serpentinite. Prec Res 61: 67–88

Van Schmus W R (1976) Early and middle Proterozoic history of the Great Lakes area, North America. Philos Trans R Soc Lond (Ser A) 280: 605–628

Van Schmus W R, Bickford M E, Anderson J L et al. (1993) Transcontinental Proterozoic provinces. In: Precambrian: Conterminous U.S. (eds Reed J C, Bickford M E, Houston R S et al.). Geol Soc Am, The Geology of North America, C-2: 171–334

Van Schmus W R, Bickford M E, Lewry J F and Macdonald R (1987a) U-Pb geochronology in the Trans-Hudson Orogen, northern Saskatchewan, Canada. Can Earth Sci 24: 407–424

Van Schmus W R, Bickford M E and Zietz I (1987b) Early and middle Proterozoic Lithospheric Evolution, (ed. Kröner A). Am Geophys Union (Geodyn Series) 17: 43–68

Vearncombe J R (1983) A dismembered ophiolite from the Mozambique Belt, West Pokot, Kenya. J Afr Earth Sci 1: 133–143

Veizer J (1984) The evolving earth: water tales. Prec Res 25: 5–12

Veizer J and Compston W (1976) $^{87}$Sr/$^{86}$Sr in Precambrian carbonates as an index of crustal evolution. Geochim Cosmochim Acta 40: 905–915

Veizer J and Jansen S L (1979) Basement and sedimentary recycling and continental evolution. J Geol 87: 341–370

Vennemann T W, Kesler S E and O'Neil J R (1992) Stable isotope compositions of quartz pebbles and their fluid inclusions as tracers of sediment provenance: Implications for gold- and uranium-bearing quartz pebble conglomerates. Geol 20: 837–840

Vermaak C F (1981) Kunene anorthosite complex. In: Precambrian of the Southern Hemisphere (ed. Hunter D R). Elsevier, Amsterdam: 578–599

Vermaak C F and Von Gruenewaldt G (1986) Introduction to the Bushveld Complex. In: Mineral Deposits of South Africa (eds. Anhaeusser C R and Maske S). Geol Soc S Afr, Spec Public: 1021–1029

Vidal M and Alric G (1994) The Palaeoproterozoic (Birimian) of Haute-Comoé in the West African craton, Ivory Coast: a transtensional back-arc basin. Prec Res 65: 207–229

Vidal P, Blais S, Jahn B-M and Capdevila R (1980) U-Pb and Rb-Sr systematics of the Suomussalmi Archean greenstone belt (Eastern Finland). Geochim Cosmochim Acta 44: 2033–2044

Vidal Ph, Auvray B, Charlot R and Cogné J (1981) Precambrian relicts in the Armorican Massif. Their age and role in the evolution of the Western and Central European Cadomian-Hercynian belt. Prec Res 14: 1–20

Viljoen M J and Viljoen R P (1970) Archaean volcanicity and continental evolution in the Barberton region, Transvaal. In: African Magmatism and Tectonics, (eds Clifford T N and Gass I G). Oliver and Boyd, Edinburgh: 27–49

Villeneuve M (1977) Précambrien du sud du Lac Kivu (Région du Kivu, République du Zaire). Etude stratigraphique, pétrographique et tectonique. Université d'Aix, Marseille III, thèse inédite

Villeneuve M and Cornée J J (1994) Structure, evolution and palaeogeography of the West African craton and bordering belts during the Neoproterozoic. Prec Res 69: 307–326

Visser J N J and Grobler J N (1985) Syndepositional volcanism in the Rietgat and arenaceous Bothabille formation, Ventersdorp Supergroup (late Archean–early Proterozoic), in South Africa. Prec Res 30: 153–174

Viswanathiah M N (1977) Lithostratigraphy of the Kaladgi and Badami Groups, Karnataka. In: Proceedings of Seminar of Kaladgi, Badami, Bhima and Cuddapah Sediments, (ed. Janardhan A S). Ind Mineral 18: 122–132

Vivallo W and Claesson L A (1987) Intra-arc rifting and massive sulphide mineralization in an early Proterozoic volcanic arc, Skellefte district, northern Sweden. In: Geochemistry and Mineralization of Proterozoic Volcanic Suites, (eds Pharaoh T C, Beckinsale R D and Rickard D). Geol Soc Spec Public 33: 69–79

Vivallo W and Rickard D (1984) Early Proterozoic ensialic spreading-subsidence: evidence from the Garpenberg Enclave, central Sweden. Prec Res 26: 203–221

Vlaar N J, van Keken P A and van den Berg A P (1994) Cooling of the Earth in the Archaean: consequences of pressure-release melting in a hotter mantle. Earth Planet Sci Lett 121: 1–18

Volpe A M and Macdougall J D (1990) Geochemistry and isotopic characteristics of mafic (Phulad ophiolite) and related rocks, Delhi supergroup, Rajasthan, India: Implications for rifting in the Proterozoic. Prec Res 48: 167–191

Walker J C G, Klein C, Schidlowski M, Schopf J W, Stevenson D J and Walter M R (1983) Environmental evolution of the Archean-Early Proterozoic Earth. In: Earth's Earliest Biosphere, (ed. Schopf J W). Princeton University Press, pp 260–290

Wallace H (1981) Keweenawan geology of the Lake Superior Basin. In: Proterozoic Basins of Canada, (ed. Campbell F H A). Geol Surv Can Pap 81-10: 399–417

Walraven F, Armstrong R A and Kruger F J (1990) A chronostratigraphic framework for the north-central Kaapvaal Craton, the Bushveld Complex and the Vredefort structure. Tectonophys 171: 23–48

Walter M R (Editor) (1991) Proterozoic Petroleum. Prec Res 54: 108 pp

Wang K Y, Yan Y H, Yang R Y and Chen K F (1985) REE geochemistry of early Precambrian charnockites and tonalitic-granodioritic gneisses of the Qianan region, east Hebei, north China. Prec Res 27: 63–87

Wang X, Liou J G and Mao H K (1989) Coesite-bearing eclogite from the Dabie Mountains in central China. Geol 17: 1085–1088

Ward P (1987) Early Proterozoic deposition and deformation at the Karelian craton margin in southeastern Finland. Prec Res 35: 71–93

Wardle R J and Bailey D G (1981) Early Proterozoic sequences in Labrador. In: Proterozoic Basins of Canada (ed. Campbell F H A). Geol Surv Can Pap 81-10: 331–358

Wardle R J, Rivers T, Gower C F et al (1986). The northeastern Grenville Province: new insight. In: The Grenville Province, (eds Moore J M, Davidson A and Baer A J). Geol Ass Can Spec Pap 31: 13–29

Wardle R J, Ryan B and Ermanovics I (1990) The eastern Churchill Province, Torngat and New Québec orogens: An overview. Geosci Can, 17 (4): 217–222

Watchorn M A (1980) Fluvial and tidal sedimentation in the 3000 Ma Mozaan basin, South Africa. Prec Res 13: 27–42

Watkeys M K (1983) Brief explanatory notes on the provisional geological map of the Limpopo Belt and environs. Geol Soc S Afr Spec Public 8: 5–8

Watkeys M K, Light M P R and Broderick T J (1983) A retrospective view of the central zone of the Limpopo Belt, Zimbabwe. In: The Limpopo Belt, (eds Van Biljon W J and Legg J H). Geol Soc S Afr Spec Public 8: 65–80

Watkins K P and Hickman A H (1990) Geological evolution and mineralization of the Murchison Province, Western Australia. Geol Surv West Austral Bull, 137

Watson J (1975a) The Lewisian Complex. In: A correlation of the Precambrian rocks in the British Isles (eds Harris A L, Shackleton R M, Watson J et al). Geol Soc Lond Spec Rep 8: 15–29

Watson J (1975b) The Precambrian rocks of the British Isles —a preliminary review. Geol Soc Spec Rep 6: 1–10

Watters B R (1976) Possible late Precambrian subduction zone in South West Africa. Nature 259: 471–473

Wayne R P (1991) Chemistry of Atmosphere. 2nd ed. Clarendon Press, Oxford. 447 pp

Webb A W, Thomson B P, Blissett A H, Daly S J, Flint R B and Parker A J (1986) Geochronology of the Gawler Craton, South Australia. Austral J Earth Sci 33: 119–143

Wegener A (1924) The Origin of Continents and Oceans. (English translation by W A Skerl, London)

Weilers B F (1990) A review of the Pongola Supergroup and its setting on the Kaapvaal Craton. Info Circ 228, Econ Geol Res Unit, Univ Witwatersrand, Johannesburg, 69 pp

Wendt I, Kröner A, Liew T C and Todt W (1987) U-Pb zircon ages and Nd-systematics for Moldanubian granulites of the Bohemian, Massif, Czechoslovakia. Terra Cognita 7: 334–335

Wetherill G W (1990) Formation of the Earth. Ann Rev Earth and Planet Sci 18: 205–256

Wheeler J O, Aitken J D, Berry M J et al (1972) The Cordilleran Structural Province. In: Variations in Tectonic Styles in Canada, (eds Price R A and Douglas R J W). Geol Ass Can Spec Pap 11: 1–81

White D J et al. (1994a) Seismic images of the Grenville Orogen in Ontario. Can J Earth Sci 31: 293–307

White D J et al. (1994b) Paleo-Proterozoic thick-skinned tectonics: Lithoprobe seismic reflection results from the eastern Trans-Hudson Orogen. Can J Earth Sci 31: 458–469

White R W and Clarke G L (1993) Timing of Proterozoic deformation and magmatism in a tectonically reworked orogen, Rayner Complex, Colbeck Archipelago, east Antarctica. Prec Res 63: 1–26

White R W and Leo G W (1969) Geologic reconnaissance in western Liberia. Geol Surv Liberia Spec Pap No. 1

Wiebe R A (1980) Anorthositic magmas and the origin of anorthosite massifs. Nature 286: 564–567

Wiedenbeck M and Watkins K P (1993) A time scale for granitoid emplacement in the Archean Murchison Province, Western Australia, by single zircon geochronology. Prec Res. 61: 1–26

Wiggering H and Beukes N J (1990) Petrography and geochemistry of a 2000–2200 Ma old hematitic paleo-alteration profile on Ongeluk basalt of the Transvaal Supergroup, Griqualand West, South Africa. Prec Res 46: 241–258

Wikström A (1991) Structural features of some granitoids in central Sweden and implications for the tectonic subdivision of granitoids. Prec Res 51: 151–159

Wilkin R T and Bornhorst T J (1992) Geology and geochemistry of granitoid rocks in the Archean Northern complex, Michigan, U.S.A. Can J Earth Sci, 29: 1674–1685

Wilks M E (1988) The Himalayas—a modern analogue for Archaean crustal evolution. Earth Planet Sci Lett 87: 127–136

William-Jones A E and Sawiuk M J (1985) The Karpinka Lake uranium prospect, Saskatchewan: a possible metamorphosed Middle Precambrian sandstone-type uranium deposit. Econ Geol 80: 1927–1941

Williams D A C and Furnell R G (1979) Reassessment of part of the Barberton type area, South Africa. Prec Res 9: 325–347

Williams H (1979) Appalachian Orogen in Canada. Can J Earth Sci 16: 792–807

Williams H R (1978a) The Archaean geology of Sierra Leone. Prec Res 6: 251–268

Williams H R (1978b) Tasman fold belt system in Tasmania. Tectonophys 48: 159–205

Williams H R (1988) The Archaean Kasila Group of western Sierra Leone : geology and relations with adjacent granite-greenstone terrane. Prec Res 38: 201–213

Williams H R and Culver S J (1988) Structural terranes and their relationships in Sierra Leone, J Afr Earth Sci 7: 473–477

Williams I S and Collins W J (1990) Granite-greenstone terranes in the Pilbara Block, Australia, as coeval volcanic-plutonic complexes; evidence from U-Pb zircon dating of the Mount Edgar Batholith. Earth Planet Sci Lett 97 (1–2): 41–53

Williams I S, Page R W, Froude D et al (1983) Early crustal components in the Western Australian Archaean: zircon U-Pb ages by ion microprobe analysis from the Shaw Batholith and Narryer metamorphic belt. Geol Soc Austral (Abs.) 9: 169–170

Williams N (1980) Precambrian mineralization in the McArthur-Cloncurry region, with special reference to stratiform lead-zinc deposits. In: The Geology and Geophysics of Northeastern Australia (eds Henderson R A and Stephenson P J). Geol Soc Austral (Queensland Division), pp 89–107

Williams S J (1986) Geology of the Gascoyne Province, Western Australia. Geol Surv W Austral Rep 15

Wilson A H and Prendergast M D (1989) The Great Dyke of Zimbabwe. I. Tectonic setting, stratigraphy, petrology, structure, emplacement and crystallization. In: Magmatic sulphides–the Zimbabwe volume (eds Prendergast M D and Jones M J). London Inst Mining Metall: 1–20

Wilson A H and Tredoux M (1990) Lateral and vertical distribution of Platinum-Group Elements and petrogenetic controls on the sulfide mineralization in the P1 Pyroxenite Layer of the Darwendale subchamber of the Great Dyke, Zimbabwe. Econ Geol 85: 556–584

Wilson A H and Versfeld J A (1994a) The early Archean Nondweni greenstone belt, southern Kaapvaal Craton, South Africa, Part I. Stratigraphy, sedimentology, mineralization and depositional environment. Prec Res 67: 243–276

Wilson A H and Versfeld J A (1994b) The early Archean Nondweni greenstone belt, southern Kaapvaal Craton, South Africa, Part II. Characteristics of the volcanic rocks and constraints on magma genesis. Prec Res 67: 277–320

Wilson J F (1981) Zimbabwe; Anhaeusser C R and Wilson J F.

The granite gneiss-greenstone shield. In: Precambrian of the Southern Hemisphere, (ed. Hunter D R). Elsevier, Amsterdam: 454–499

Wilson J F, Bickle M J, Hawkesworth C J et al (1978) Granite-greenstone terrains of the Rhodesian Archean craton. Nature, 271: 23–27

Wilson J T (1949) The origin of continents and Precambrian history. Trans R Soc Can 43: 157–182

Wilson J T (1968) Comaprison of the Hudson Bay Arc with some other features. In: Science, History and Hudson Bay, (eds Beals C S and Shenstone D A). Can Dept Energy Min Resources 1015–1033

Wilson M R (1982) Magma types and the evolution of the Swedish Proterozoic. Geol Rdsch 71: 170–184

Windley B F (1993) Proterozoic anorogenic magmatism and its orogenic connection. Geol Soc Lond J 150: 39–50

Windley B F (1984) The Evolving Continents, 2nd edn. John Wiley, New York

Windley B F (1991) Early Proterozoic collision tectonics, and rapakivi granites as intrusions in an extensional thrust-thickened crust: The Ketilidian orogen. Tectonophys 195: 1–10

Windley B F, Herd R K and Bowden A A (1973) The Fiskenaesset Complex, West Greenland, part 1, a preliminary study of the stratigraphy, petrology, and whole rock chemistry from Qeqertarssuatsiaq. Bull Gronlands Geol Unders 106

Windley B F, Bishop F C, Smith J V (1981) Metamorphosed layered igneous complexes in Archean granulite-gneiss belts. Ann Rev Earth Planet Sci 9: 175–198

Windom K E, Van Schmus W R, Seifert K E, Wallin E T and Anderson R R (1993) Archean and Proterozoic tectono-magmatic activity along the southern margin of the Superior Province in northwestern Iowa, United States. Can J Earth Sci 30: 1275–1285

Winsnes T S (1965) The Precambrian of Spitsbergen and Bjornoya. In the Precambrian, Vol. 2 (ed. Rankama K). Interscience Publishers, New York, pp 1–24

Winter H R (1986) A cratonic-foreland model for Witwatersrand basin-development in a continental, back-arc, plate-tectonic setting. University of Witwatersrand, Econ Geol Res Unit, Info Circ 139

Wirth K R, Gibbs A K and Olszewski W J (1985) U-Pb zircon ages of the Grao Para Group and Serra das Carajas Granite, Para, Brasil. Atas II Symp Geol Amazonian Region

Withnall I W, Bain J H C, Draper J J et al (1988) Proterozoic stratigraphy and tectonic history of the Georgetown Inlier, Northeastern Queensland. Prec Res 40/41: 429–446

Worst B G (1960) The Great Dyke of Southern Rhodesia. Geol Surv S Rhodesia. Bull 47: 234 p, also sheets 1–9

Wright J V, Haydon R C and McConachy G W (1987) Sedimentary model for the giant Broken Hill Pb-Zn deposit, Australia. Geol 15: 598–602

Wust H-J, Todt W and Kröner A (1987) Conventional and single grain zircon ages for metasediments and granite clasts from the Eastern Desert of Egypt: evidence for active continental margin evolution in Pan-African times. Terra Cognita, 7: 333–334 (Abs.)

Wyborn L A I (1988) Petrology, geochemistry and origin of a major Australian 1880-1840 Ma felsic volcanic-plutonic suite: a model for intracontinental felsic magma generation. Prec Res 40/41: 37–60

Wyborn L A I, Page R W and Parker A J (1987) Geochemical and geochronological signatures in Australian Proterozoic igneous rocks. In: Geochemistry and Mineralization of Proterozoic Volcanic Suites, (eds Pharaoh T C, Beckinsale R D and Rickard D). Geol Soc Spec Public 33: 377–394

Wynne-Edwards H R (1972) The Grenville Province, In: Variations in Tectonic Style in Canada, (eds Price R A and Douglas R J W). Geol Ass Can Spec Pap 11: 263–334

Wynne-Edwards H R (1976) Proterozoic ensialic orogenesis: millipede model of ductile plate tectonics. Am J Sci 276: 927–953

Xu Shutong, Jiang Laili, Liu Yican and Zhang Yong (1992) Tectonic framework and evolution of the Dabie Mountains in Anhui, Eastern China. Acta Geologica Sinica 5: 221–238

Yang Z, Cheng Y and Wang H (1986) The Geology of China. Clarendon Press, Oxford

Yeo G M (1981) The Late Proterozoic Rapitan glaciation in the northern Cordillera. In: Proterozoic Basins of Canada, (ed. Campbell F H A). Geol Surv Can Paper 81-10: 25–46

Yeo G M (1986) Iron-formations in the Late Proterozoic Rapitan Group, Yukon and Northwest Territories. Can Inst Min Metal Spec Vol. 37: 142–153

York D and Farquhar R M (1972) The Earth's Age and Geochronology. Pergamon Press, New York

Yoshida M and Kizaki K (1983) Tectonic situation of Lützow-Holm Bay in East Antarctica and its significance in Gondwanaland. In: Antarctic Earth Science, (eds Oliver R L, James P R and Jago J B). Cambridge University Press, Cambridge: 36–39

Yoshida M, Kunaki M and Vitanage P W (1992) Proterozoic to Mesozoic East Gondwana: The juxtaposition of India, Sri Lanka, and Antarctica. Tectonics 11: 381–391

Young G M (1981a) Upper Proterozoic supracrustal rocks of North America: a brief review. Prec Res 15: 305–330

Young G M (1981b) The Amundsen Embayment, Northwest Territories; relevance to the Upper Proterozoic evolution of North America. In: Proterozoic Basins of Canada, (ed. Campbell F H A). Geol Surv Can Pap 81-10: 203–218

Young G M (1983) Tectono-sedimentary history of early Proterozoic rocks of the northern Great Lakes region. In: Early Proterozoic Geology of the Great Lakes Region, (ed. Medaris L G). Geol Soc Am Mem 160: 15–32

Yu Y and Morse S A (1993) $^{40}Ar/^{39}Ar$ chronology of the Nain anorthosites, Canada. Can J Earth Sci 30: 1166–1178

Zang W and Walter M R (1992) Late Proterozoic and Early Cambrian microfossils and biostratigraphy, northern Anhui and Jiangsu, central-eastern China. Prec Res 57: 243–323

Zhai M, Yang R, Lu W and Zhou J (1985) Geochemistry and evolution of the Qingyuan Archean granite-greenstone terrain, NE China. Prec Res 27: 37–62

Zhang G W, Bai Y B, Sun Y et al (1985) Composition and evolution of the Archean crust in Central Henan, China. Prec Res 27: 7–35

Zhang J, Dirks P H G M and Passchier C W (1994) Extensional collapse and uplift in a polymetamorphic granulite terrain in the Archaean and Palaeoproterozoic of north China. Prec Res 67: 37–57

Zhang Zh.M, Liou J G and Coleman R G (1984) An outline of the plate tectonics of China. Bull Geol Soc Am 95: 295–312

Zhao Z, Xing Y, Ma G et al (1980) The Sinian System of eastern Yangtze Gorges, Hubei. In: Research on Precambrian Geology; Sinian Suberathem in China. Tianjin Science and Techonology Press, pp 31–55

Zientek M L and Ripley E M (1990) Sulfur isotope studies of the Stillwater Complex and associated rocks, Montana. Econ Geol 85: 376–391

# Index

Page references in **bold typeface** indicate tables and page references in *italics* indicate figures.

Abitibi Belt, 70
Abitibi-Wawa Subprovince, 70
Acasta gneiss, 72, **74**
Adelaide Geosyncline, 48, 256–8, **257**
Adirondack Massif, 194, 233
Africa, **242**
African Craton, 3, *4*, *5*, 15–18, *17*, *18*, 38–44, *39*, *40*, *43*, 104–6, 153–62, 165, 236
   Central Africa, 17–18, 103–4, 209–12, *236*, 240–1
   Northeast Africa, 44, *165*, 165, 251–4
   Northern Africa, 17–18, 212
   Northwest Africa, 41–4, **97**, 163–4, 248–51
   Southern Africa, 15–17, 88–103, *89*, 153–9, 208–11, *236*, 236
Aguapa (Aquajei) Belt, *37*, 84, 205
Ahnet Purple Group, **97**, **250**
Akilia association, 32, 63
Akitkan Group, 10, *11*, 182–3
Alaska, 231–2
Albany-Fraser Province (Belt), *20*, 20, 47, 48, 217
Alberta, 233
Aldan Shield (Domain), 24, 57–8, 128, *129*, 185
Aldan Supergroup, 57
Aldonian Orogeny, 10, *10*
Aleksod group, *18*
Algeman orogeny, 13, *15*
Alligator River Uranium Field, 114, 172
Allochthonous Boundary Thrust, 199
Amadeus-Albany-Fraser Tectonic Zone, 20, 47, 217
Amadeus Basin, 20, 47, 48, 217, 258
Amadeus Transverse Zone, 217
Amalia greenstone belt, 88
Amapa group, *16*
Amazon Basin, *37*
Amazonian Craton, 15, 36, 151, 204–5
Amer Group, 78
Amer Lake Zone, 146
Ameralik Dykes, *12*, 64
Amîtsoq Gneiss, *12*, *14*, 64
Amsaga Gneiss (basement), *18*, *39*, 106
Amundsen Embayment, 230, *231*
Anabar Complex (Gneiss), *11*, 58
Anabar Shield, 25, 58, 128, 223
Ancient Gneiss Complex, 16, 90–191
Angolan Craton, 103, 162
Animikie Basin (Group), 141, *142*, *143*
Anorogenic complexes (intrusions), 191, 193–194
Anorogenic magmatism, 15, 214
Anorthosite massifs, 15, 192, 194, 197
Anorthosite-mangerite-charnockite-rapakivi granite (AMCG) magmatic suites, 15, 73, 193, 197
Antarctic Craton, *4*, *5*, *21*, 21, 49, 49–50, 176, 218, 258–9
Anti-Atlas Domain, **39**, 41, 164, 249

Apache Group, **73**, 201, *202*
Aphebian Era, *15*, 33
Aphebian provinces (rocks), 35, 135, 136, 144
Appalachian Belt, 33, 229
Appalachian Orogen, 233–4
Apparent polar wander paths (APWP), 8
Appin Group, 227
Arabia, 44, 165
Arabian-Nubian Shield, *18*, 166, 251–254
Arabian Shield, 252, 253–4
Aracuai Fold Belt, 235
Aravalli Craton (Range), **107**, 113–114
Aravalli-Delhi Belt, 45, *113*, 166–7
Aravalli Supergroup (sediments), 18, *19*, 167
Archaean Block, 3, 4, 12, *14*, 32–3, 62–65, *63*, *64*, 134
Archean basement, 46–47
Archean crust, 51–121
   distribution, **51**, *52*
   salient characteristics, 51–4, **53**
Archean domains, 38
Archean gneiss complexes (terrains), 76, 77–9
Archean heat flow, 261–2
Archean Oceanic Crust, 262
Archean provinces, 35
Archean sequences, sedimentation, 70, 271–3
Archean Stage, 275–6
Arctic Platform rifting, 73, 200–1
Arctic Province, 230
Arechchoum Group (Gneiss), *18*, 164
Arequipa-Cuzco Massif, *5*, 36, *37*
Argyll Group, *12*, 227
Arica Elbow, 36, *37*
Armit Lake block, 75, 77
Armorican Massif, 228–229
Arnhem Inlier, 20, 47
Arunta Inlier (Block), 20, 47, 48, 217
Ashburton Fold Belt, 116, *117*, 169
Ashburton Trough, 20, 169
Ashuanipi Complex (Subprovince), 67, 68, 70
Assyntian Orogeny, 12
Atar Group, 248, 249, **250**
Athabasca Basin (Group), *15*, 35, 195–6
Athapuscow Aulacogen, 149, *150*
Atlantic Shield, 38, 82, 87, 152–3
Atlas Belt, 58
Australia, *173*
Australian Craton, *4*, *5*, 19–20, *20*, 46–9, *47*, **112**, 114–20, 168–76, 213–17, 256–8
Avalonian Orogeny, 14, *15*

Bababudan Basin (Group), *19*, 100, *109*
Badaohe Group, **56**, *56*
Badcallion event (metamorphism), *15*, 66, *66*, 135
Baie Comeau Segment, *198*, 199
Baikal Fold Belt, 25, 28, 58–9, 128, *129*
Baikal Region, 223

Baikalian Orogeny, *11*, 11, *12*, 12
Baiyun Obo deposit, 181
Bajkonyr synclinorium, 223
Baker Lake-Thelan basins, *195*, 196
Baltic Shield, *4*, 25, 28–9, 59–62, *61*, 129–33, *133*, 185–8, 222, 224–5
Bambú Group, *82*, 87, 234
Banded Gneissic Complex (BGC), *107*, 113–4, 167
Banded Iron Formation (BIF), 12, 60, 63, 81, 84, 86, 93, 104, 110, 114, 116, 127, 128, 141, 144, 150, 151, 154, 169, 171, 176, 218, 232, 267–9, *268*
Bangemall Basin, *20*, 47, *48*, 216–7
Bangweulu Block (= Zambian Craton), 103, 104, *160*, 162
Baoulé-Mossi (Eburnean) Domain, 163
Baraga Group, *138*, 141, *142*
Barberton Greenstone Belt, 88, 89, 91, 91–3, *92*
Barberton Mountain Land, 91
Barborema Province, 16, 235
Barramundi Orogeny (Igneous Association), *20*, 20, 48, 175–6
Bathurst Inlet Basin, *195*, 196
Bear Province, *4*, *34*, 35
Beardmore Orogeny, 258
Beartooth Mountains (Gneiss Orogeny), **74**, *80*
Beck Spring Formation, 201
Beiyixi Formation, 222
Belcher-Nastapoka Basin (Belt), *143*, 145–6
Belingwe greenstone belt, 88, *101*, 102
Belo Horizonte region, 86, 87
Belomorian Province (Complex, gneiss), 25, 28, 29, 60
Belt-Purcell Supergroup (Basin), *15*, 73, 201–203, *202*, 202, 232
Benin Nigeria Shield, 39, 41, 44, *105*, 164, 251
Berberide Inliers, 39, 41, 249
Berens Plutonic Subprovince, *67*, 70
Bergslagen Region, 132
Bhandara Craton, *45*, 46, 114, 167–8
Bienville Subprovince, *67*, 68
Big Cottonwood Formation, 201
Bighorn Mountains, *80*
Biogenesis, 270–1, *271*
Bimodal Gneiss Suite, 90
Biostratigraphy, Precambrian crust, *271*
Birrimian Subdomain (Supergroup), 41, *96*, 163
Birrindudu Basin, *20*, *47*, 213, 215
Bitter Springs Formation, 258
Blue-Green-Long axis, 233–4
Blueschist assemblages, 57, 167, 222, 227
Bohus Granite, *12*
Boliden ore deposit, 132
Bolivian shield area, 37
Bomu-Kibalian Craton, 39, 40, 103–4
Bou Azzer (-El Graara) ophiolite, 249, **250**
Bouca (Yadé) cratonic fragment, 39, 40, 103
Bourzianian (see Burzyanian) Division
Brasilia Fold Belt, *36*, *82*, 85, 234
Brasíliano Belt (Orogeny), *15*, **82**, 87, 206, 235
British Columbia, 203, 233
British Isles, 12, 14, 225–9, *226*, *227*
Brock and Minto Inliers (Arch), *36*, *231*
Brockman Iron Formation, 116
Broken Hill Block (Subdomain ore deposit), *48*, 174–5, *175*
Bruce River Group, *73*, *76*, 204
Brumado-Anajé area (belts), 86, *87*
Buganda-Toro (-Kibalian) Belt, 160
Bukoban Supergroup (Belt), *241*, *242*, 245
Bulawayan Group (greenstone), 17, 38, 88, *101*, 102, 103
Bulk Earth Heat Production, 261, 272
Bundelkhand Block (Complex Granite), 19, *45*, 46, *107*, 113–14
Bunger Hills, *49*, 218

Burundian Supergroup, *18*, 39, 209–210
Burzyanian (Byrzyan, Bourzianian) Group, *11*, *183*, 183–5
Bushimay (Mbuji Mayi) Supergroup, *241*, **242**, 243
Bushmanland Subprovince, *207*, 207–8
Bushveld Complex, 16, *17*, 153, 156–8, *157*
Bushveld (Transvaal) Sub-basin, 153–4, *154*, 155

Caledonian Fold Belt (Caledonides), 14, 33, 225, 229
Caledonian Orogeny, *12*, *14*, 33, *190*
Callanna Group, 256, *257*
Cameroon, 248
Campo Formoso District, 153
Cana Brava complex, *85*
Canadian Shield, *3*, *4*, 13, 33–36, *149*, 178, *195*, 196
Cape Smith Belt, **74**, *143*, 145
Cape-Botswana (Griqualand West, Ghaap) Sub-basin (Basin), 153, *154*
Capim Belt, *84*, 153
Capricorn Orogen, 168, *169*
Carajás Formation, 81
Carajás Ridge (basement gneiss), 81, *82*
Carolina Slate Belt, *73*, 234
Carolinidian Orogeny, *14*, 33, 190, *190*
Carpentarian Division, 20, 213–16
Cathaysian Craton, 1, 9–10, **10**, 23–4, 26–27, 54–7, *55*, 127–8, *180*, 180–2, 221–2
Caue Formation, 152
Central Africa, 17–18, 103–4, 209–12, *236*, 240–1
Central African (NorthEquatorial) Mobile Belt, 38, *39*, 44, 248
Central Australian Mobile Belt (anorogenic province), *5*, *20*, 46
Central (Midcontinent) Belt (Transcontinental Proterazoic provinces), *3*, *15*, *34*, 36, 191–193
Central Brazil (Province) Shield, *37*, 37, 38, 81–8, *82*, *84*, *85*, 152, 204–5
Central (Equatorial) Africa, 159–62
Central Gneiss Belt, 197, *198*, *199*
Central Granulite Terrain, *198*, 199
Central Karelia zone, 61
Central Metasedimentary Belt, *198*, 199
Central Plains Orogeny, *73*, *127*, 191
Central Rand Group, *90*, 96
Central Siberian Anteclise, 24, 28
Cerro (Ciudad) Bolivar iron ore deposits, 81, *83*
Chaibasa Formation, **108**, 166
Chaillu Massif (Craton), 39, 40, 103
Chang'an Glacial Epoch, 222
Changchengian (Changcheng) system, *180*, 181
Chapada Diamantina Group (platform cover), 16, *82*, 87
Charnian Group, *12*, 227, *228*
Chatisgarh Basin (Supergroup), 46
Chemical Sedimentary Assemblage, 154
Cheyenne Belt, 80
Chitradurga-Gadag Belt, *109*, 110
Chitradurga Group (granite), 19, *109*, 110
Chocolay Group, *138*, 141, *142*
Chotanagpur Block (Craton), *45*, 166
Chotanagpur-Singhbhum Craton, *5*, *45*, 166
Chuar Group, *232*, 233
Chuniespoort Basin, see Bushveld (Transvaal) Sub-basin
Chuos Formation, 238, *238*
Churchill Province, 35, *75*, 136, 146–148
Circum-Superior Fold Belts, 136–46, *143*
(Cis)-Vralian Foredeep (Trough), 12, *29*, 31, 224
Ciudata Pier-Guri fault, 81
Closepet Granite Complex, 19, *45*, 46, **107**, 110
Cobalt Group, *138*
Committee Bay block, *75*, 77
Congo Basin, 18, 41, 103, 240, *241*, 241–5
Congo Craton, 40, *160*, 238
Contendas-Mirante Complex, *87*, 152–3

Continental crust
  Archean stage, 275–6
  biogenesis, 270
  comparative island-arc accretion model, 264
  composition, 264
  development and preservation, 272
  endogenous processes and products, 261–6
  evolution, 261–80
  exogenous processes and products, 266–74
  expanding earth theory, 266
  growth rate, 264, 264–5
  Hadean (formative stage), 274–5
  mineral deposits, 273–4
  preferred evolutionary model, 277–80
  sedimentation, 271
  summary of development by stage, 274–7
Continental reconstructions, 7–8
Copperbelt (Singhbhum) Thrust Zone, 19, *19*, 45, 166
Coppermine Homocline (River area), 190, 195, 201, *230*, 231
Cordilleran Fold Belt (Orogen), 201–3, 230–3, *232*
Coronation Gulf, 196
Coronation Supergroup, **74**, 149–51
Cratonic fragments, 86
Cree Lake Zone, 146–7, *147*
Crustal regeneration, 128–9
Crystal Spring and Beck Spring formations, 201, *202*
Cuddapah basin (Group), 19, 45, 46, 212–13
Curitiba Marginal Massif, 235
Curnamona Craton, 46, 48, 256
Cuyuna Range (BIF), *140*, 141, *142*, 143

Dabie Group (Uplift Mountains), 57
Dahomeyan (-Pharusian) Belt, 39, 41, 249
Dahomeyides, **250**
Dales Gorge Member, 116
Dalradian Supergroup, 12, 227, *227*
Damara-Katanga-Zambezi structure, **3**, 17, 44
Damara Mobile Belt (Orogen, Province), 16, 44, 236
Damara Supergroup, 17, 236–8, *237*, *238*
Darfur Province (Inlier), *18*, 212
Deccan Traps, 44–45, *45*, 113
Delamarian Orogeny, 20, 49
Delhi Belt (Supergroup), 166
Delhi Complex, 167
Dengfeng Group, *10*, 54, 56
Denison Block, 73
Denman Glacier, 258
Dhanjori Group, *108*, 166
Dharwar (Schist) Belts, 19, 45, **107**, *109*, 110
Dharwar Craton, **3**, 45, *45*, 106–10, **107**, *109*
Dharwar-Shimoga Belt, 110
Dharwar Supergroup, 110
Diomiotites (Tillites Mixtites), *15*, 20, 36, 40, 48, 129, 137, 155, *155*, *190*, 219, **221**, 222, 223, 227, 228, 229, 230, *230*, 231, 232, 233, 234, 238, 239, 241, **242**, 243, 244, 245, 249, *250*, 251, 254, 256, 257, *257*, 258, 269–70
Dismal Lakes Group, *73*, 196
Dnieper Complex, **13**, 59, 60
Dnieper-Donets Aulacogen, 29, 30, 133, *187*, 188
Dnieprovian Block, *30*, 59
Dodoman schist belts, 104
Dom (Don) Feliciano Belt, 16, 37
Dominion Group, 90, 96
Don Feliciano Belt, 235
Dongargarh Supergroup, 168
Dorset Orogen, 35, *127*, 148
Doublet Group, 144, *145*
Dravidian (South Indian) Shield, **3**, *5*, *18*, *19*, 106
Dronning Maud Land, 176, 218
Dubawnt Basin (Group), *15*, **73**, 195, 196

Duluth Complex (Gabbro), **73**, 79, 203
Dunnage Zone, *198*

Earaheedy Group (sub-basin), *20*, 170–171
Early Proterozoic crust, 123–76
  distribution, **123**, *123*, *125*
  salient characteristics, 123–6, **125**
Early Proterozoic stage, 276
East Antarctic Metamorphic Shield, 121, 218, 258
East Arm Fold Belt, 69, 72, 149, *149*, 150
East Bolivian Shield, 151
East European Craton, 1, 3, *4*, 11–12, **12**, **13**, 25, *30*, *31*, 59–62, 129–34, 185- 9, *187*, 188, 223–9, *224*
East Finland zone, 61
East Gondwana, 247
East Greenland, 229
East Greenland Caledonian Belt, 13
East Greenland Fold Belt, 190–1
East Karelia zone, 61
East Kootenay Orogeny, *73*, 231
East Qinling Range, 54
East Sayan Fold Belt (Province), *10*, 25, 28, 59, 128
East Sayan Ridge, 128
Eastern Bangemall Basin, 217
Eastern Churchill Province, 148
Eastern Dharwar Domain, 108–10
Eastern Ghats Belt, 45, 46, **107**, 111, 213, 218, *255*
Eastern Ghats Front, *45*, 111
Eastern Ghats Orogeny, 19, *19*
Eastern Goldfields Province (Basin), 118, 119
Eastern Grenville Province, 199
Eastern Hebei District, 56, 128
Eastern Nain Subprovince, 75
Eastern Province, 106
Eastern Tchad Basement, 212
Eburnean cycle (geosyncline), 162, 164, 165
Eburnean Domain, 163
Eburnean (Suggarian) Orogeny, 17, *18*, 41, **97**
Eclogite, 57, 160, 251
Ediacara assemblage, 23, 48, 219, 221, 223, 232, 239, 257–8
Ediacara Range, 257
Ekwi Supergroup, 231
Eleonore Bay Group (sediments), *14*, 33, *190*, 229
El Pao fault, 81
El Graara Ophiolite Suite (see Bou Azzer)
Elliot Lake Group, *138*
Elsonian Orogeny (Disturbance), *14*, *15*, 191, 196–7
Enderby Land, *5*, *21*, 49, 50, 121, 176, 218
Endogenous Processes and Products, 261–6
English River Subprovince (Belt, Superbelt), 67, 70, 71
Ennadai Block, 75, 77
Eriksfjord Formation (sediments), *14*, 189
El-then Group, 150
Evaporite deposits (gypsum, halite etc. pseudomorphs), 115, 129, 171, 172, 189, 201, 214, 215, 229, 230, 256, 257, 269–70
Exogenous Processes and Products, 266–74
Expanding earth theory, 266

Fennoscandia (Fennoscandian Shield), 61, *125*
Fenno-Sarmatia, 25
Fig Tree Group (sediments), 89, 92, *92*, 93
Fiskenaesset Complex, *14*, 64–5
Flin Flon-Snow Lake area, 147
Fold (gneiss) ovals, 57, 141
Fortescue Group, 116, *117*
Fox River Belt, 145
Foxe Orogen, *127*, 148
Francevillian Supergroup (sediments), *18*, 39, 161–2
Franklinian Fold Belt, 230

Frere Formation, 171
Fuping Group, 10, 55, 56
Fupingian Orogeny, 9, 10
Fury and Hecla Basin (strait), 201

Gabon (Ntem) Massif, 39, 40, 103
Gaia hypothesis, 271, 275
Gardar Province (syenites), 13, 14, 189
Gariep Belt (Province), 17, 39, 236, 237, 239
Gascoyne Province (Block), 168, 169, 216
Gawler Craton (Domain), 47, 47, 48, 120, 173, 216
Gawler Range Volcanics, 20, 213, 216
Georgina Basin, 47, 258
Geothermal gradients, 261–2
Glacier National Park, 202
Glacigene deposits, 269–70
Glengarry Group (sub-basin), 169, 171
GLIMPCE (Great Lakes International Multidisciplinary Program on Crustal Evolution), 203
Gneiss domes, see Fold ovals
Gneiss-Granulite Transition Zone, 45, 46, 106, 107–8, 111
Godavari Succession (Rift), 213
Gogebic Range (BIF), 138, 140, 141, 142
Goias Massif, 38, 85, 85, 86, 152
Gondwanaland, 7, 259
Gordonia Subprovince, 208
Gorge Creek Group, 20, **112**, 115, *115*
Goritore Group, 16, 82, 83, 205
Gothian (sub-Jotnian) complexes, 12, 186
Goulburn Group, 78, 150
Gourma Basin (Aulacogen), 39, 105, 250, **251**
Gowganda Formation, 126, 137, 138
Grand Canyon Supergroup, 73, 232, 233
Granitic Zone, 208
Granitoid associations, 262–4
Granitoid basement, 72
Granitoid complexes, 90
Granitoid domes, 91
Granitoid-gneiss batholiths, 116
Granitoid-greenstone subprovinces, 67, 70
Granitoid intrusions, 119–20
Granitoid-rhyolitic complexes, 15, 192, 193
Granitoid terrains, 151–2
Granulite Domain, 46, **107**, 110–11, 168, 265
Granulties: see High-grade metomorphis assemblages
Graó-Para Group, 81, **82**, 85
Great Bear Batholith, 150
Great Bear Lake, 195, 196
Great Dyke, 16, 17, 38, *101*, 102–3
Great Gangetic Plain, 254
Great Lakes Region, 140–2, *142*, *143*
Great Lakes Tectonic Zone (GLTZ), 78, 79, 140, 141
Great Slave Supergroup, 150
Grat Smoky Group (sediments), 15
Green (Pharusian 2) Group (Série Vert), 97, 251
Greenland Shield, 12–13, 14, 31–3, 32, 34, 62–6, 134, 189–91, 229
Greenstone (Schist) belts, 11, 12, 13, 15, 16, 18, 19, 20, 24, 28, 29, 34, 36, 38, 40, 41, 45, *45*, 46, **53**, *53*, 54, 57, 58, 59, 60, 61–2, 70–1, 72–4, **74**, 75, 78, 79, **82**, *83*, 86, **89**, 91–3, **96**, 101–2, *101*, 103, 104, 106, **107**, *107*, 108, *109*, 110, **112**, 114, *115*, 115–6, 119, *125*, 126, 146, 147, 151–2, 153, 163, 271–3, 275
Grenville Front Tectonic Zone, 197, 199
Grenville Province (Belt), 4, 34, 35, 178, 191, *192*, 197–200, *198*, *199*, *200*
Grenville Supergroup, 73, 199
Grenvillian Orogeny, 14, *15*, **173**, 199
Grey Member, 139
Griqualand West Sub-basin (see Cape Botswana Sub-basin)

Grit-shale-carbonate wedge (Windermere supergroup), 233
Guapore Craton, 37, 37, 84
Guebra Osso Group, 86
Guelb el Habib malasse (Group), **96**, *105*, 164
Guiana Shield, 36–7, 81, *82*, *83*, 105, 204–5
Gunflint Range (Formation, BIF), *140*, 141, *143*, 270
Gurian Orogeny, 16

Hackett-River-type volcanic belts, 72
Hadean (Formative) Stage, 274–5
Hadrynian domains (era), *15*, 36, 229–233
Hagen-Fjord Group, 14, 33, 229, 190
Hallandian (-Gothian) Orogeny, 11, *12*, 187
Halls Creek Inlier, 20, 47, *47*, 171, *173*, 216
Hamersley Basin (Group), 20, *47*, 48, 116–18, *117*
Hamersley Range Synclinorium, 116
Hank Supergroup, 41
Hastings Basin, 199
Hay Creek Group, 232
Hebrides (Hebridean Craton), 33, 225, 227
Hecla Hoek Geosyncline, 225
Hedmark Group (Basin), 225
Helikian basins, 195
Helikian provinces (era), 15, 35
High-grade metamorphic assemblages (granulite, charnockite), 51, 52–3, *53*, 54, 56, 57, 58, 60, 62, 65, 66, 67, 68, **74**, 81, *82*, 85, **89**, 91, 99, 100, 103, 104, 106, **107**, 107, 108, 110, 111, 113, 118, 119, 121, 124, 133, 146, 162, 164, 167, 168, 169, 171, 174, 176, 193, 194, 197, 199, 206, 207, 208, 210, 217, 218, 239, 246, 247, 255, *255*, 265
High Lake deposit, 75
Highland Boundary Zone, 225, 226
Highland Complex (Group), 254, *255*
Himalaya Fold Belt, 182
Hogenakal carbonatites, 168
Hornby Bay Group, 73, 196
Hough Lake Group, 138
Hubei Province, 222
Hudsonian Mobile Zone, 146
Hudsonian Orogeny, 13–14, *15*, 146, 148
Huronian Supergroup (Basin), *15*, **74**, 136, 269
Hurwitz Group, **74**, 147
Hutchison Group, **112**, 173–4
Hutuo Group, 9, *10*, 25, **56**

Iberian Massif, 228, 229
Ikorongo Group, *241*, 245
Ilimaussaq Intrusion, 189
Imataca Complex, 14, 16, 81, **82**, *83*
Indian Craton, 3, 4, 5, 18–19, *19*, 44–6, *45*, 106–14, 166–8, 212–13, 254–6, *255*
Indian Ocean Orogeny, 19
In Ouzzal Domain, 39, *105*, 106
Interior Platform, 223–4
Iron formation, see Banded iron formation
Iron Ore Group, 19, **108**, 113, *113*
Iron Quadrangle, 152
Iron River-Crystal Falls Range (BIF), *140*, 141
Irumide Belt (Foreland Zone, Irumides), 18, 41, *160*, 210–12, *211*
Irumide Orogeny, 210
Isua-Akilia (Isukasia) supracrustal rocks, 62, *63*, **64**, 65
Itabira Group, 152
Itacolomi squence, 152
Itombwe Belt (Supergroup), *241*, **242**, 245
IUGS Subcommission on Precambrian Stratigraphy, 23
Ivigtut peralkaline granite stock, 189

Jack Hills Quartzite, 20
Jacobina Belt, 87, 153

Jacobsville sediments, 73
Jamari-Alta Candeias-São Miquel region, 205
Jari-Balsino (Parguazan, Madeira) episode, 15, 16, **82**
Jatulian Group (sediments), 12, 61, 131
Jequié-Matuipe (Jequié) Complex, 86
Jequié metamorphism, 86
Jequié Orogeny, 15, 16
Jiangnan Oldlands region, 182
Ji-Lu (Jilu) Nucleus, 10, 24, 27
Jinningian Orogeny, 10, 221
Jixianian (Jixian) system, 10, 180, 181
Johannesburg Dome, 91
Jotnian sandstone, 12, 187–8

Kaapvaal Craton, 3, 4, 16, 38, 88–98, **90**, 206, 208–9
Kaladgi Succession (Basin), 213
Kalahari Craton, 38–40, 88, 209, 239
Kalevian Group, 12, 28, 61, 129, 131–2
Kalgoorlie Subprovince, 119
Kambui Supergroup, **96**, 104–5
Kaminak (Belt, Group) Lake alkaline complex, **74**, 75, 78
Kaniapiskau Supergroup, 144, *145*
Kanin-Timan Ridge, 224
Kansk Group, 59
Kackoveld Belt (North Coastal Branch), *237*, 236, 239, *238*
Kapuskasing Structural Zone, 66, 67, 70, **74**, 146
Karagwe-Ankolean Belt (Supergroup), see Kibaran Belt
Karatavian Division, 11, 13, 222–3
Karelian-Kalevian relations, 131–2
Karelian Orogeny, 11, 12
Karelian Province, 28, 29, 60–2
Karelian Supergroup (facies, cover), 12, 131
Karrat Group, 135
Kasai (-Angolan) Craton, 4, **13**, 39, 103
Kasila Belt (Group), 104, 105, *105*
Katangan Supergroup (Belt), *39*, 240, 241–3, *242*
Katangan Unconformity, 249
Kavirondian greenstones (schist belt), 18, 104
Kayes Inlier, *39*, *105*
Kemp Coast, 218
Kenema-Man Domain, 104
Kenieba Inlier, *39*, *105*
Kenoran Orogeny, 13, *15*, 68, 70, **74**, 75, 148
Kenya-Tanzania Province, 247
Ketilidian Belt (Orogeny), 13, 14, 32, 33, 135
Ketyet group, 75
Keweenawan Supergroup (sequence), *15*, 202, 203–4
Kham-Yunnan (Dam) Oldlands, 181
Kheis Belt, 17, 159, 207
Khomas Subgroup (Trough, Zone, Belt), *237*, 238, *238*
Khondalite Belt, 111
Kibalian Craton (greenstones), 18, 39, 103–4
Kibaran Belt (Supergroup, Kibarides), 18, 41, *160*, 209–210
Kilohigok Basin, 149, *149*, 150
Kimberley Basin, 213, 215–16, *172*, *173*
Kimberley Region, 258
Kimezian Assemblage (Supergroup), 161, 244, *244*
Kingston Platform, 170
Kisseynew Domain (gneisses), 147
Klipriversberg group, **90**, 97
Knob Lake Group, 144, *145*
Kola Peninsula Province, 25, 60, 61, 129
Kolar Belt (Group), 108, *109*
Kolyma-Omolon-Taigonos block, 3, 4, 59
Konka-Verkhovtsevo (-Oboyan) greenstone belts, **13**, 59–60
Koras-Sinclair-Ghanzi (-Chobe) rift system (redbeds-volcanics), 39, 40, 209
Krivoy Rog Supergroup, 12, **13**, 133
Krivoy Rog-Kremchug Zone, 29, *30*, 133–4
Krummedal sediments, 14, 190, *190*

Kudashian (Kudash) (latest Riphean) Division, 11, 223
Kudremukh-West Coast Belt, 110
Kundelungu sequence, 242, 243
Kunene Anorthosite Complex, 18, 162
Kursk Magnetic Anomaly (supergroup), *13*, 29, 133, 134

Labrador, 76
Labrador Trough, 142–6
Labrador-California Belt, 193
Labradorian Orogeny, 200
Lafaiete District, 86
Lake Huron, *138*
Lake Huron Region, *130*
Lake Superior, 77, 79, *138*
Lake Superior Region, 140, *140*, 142, 191
Lamuyka Complex, 58
Lanzhishan Granite, 56
Lapland Granulite Belt, 25–8, 133
Laporte Group, 144
Lapponian Group, 131
La Ronge-Lynn Lake arc, 147
Late Proterozoic crust, 219–59
    distribution, *219*, 219, *220*
    large interior basins, 219
    mobile belt network, 219–20
    salient characteristics, 219, **221**, *221*
Late Proterozoic platform cover Australian Craton, 48
Late Proterozoic stage, 277
Late Sinian Glacial Epoch, 222
Laurasia, 7
Laurentian Orogney, 15
Lawn Hill Platform, *173*, 213, 215
Letitia Lake Group, 76, 204
Lewisian Complex, 33, 65–6, **66**
Limpopo Mobile Belt, 38, *39*, 98–100, *99*
Lindian Supergroup, 241, 245
Litchfield Complex, 120
Lizard Peninsula, 228
Lochiel Batholith, 91
Lofoten-Vesteralen Islands, 62
Longquanguan Group, 55
Lopian Supergroup (greenstone belts), 61–62
Lower Eleonore Bay Group, *190*, 229
Lower Greenstones Bulawayan Group, 102
Lower Mafic Volcanic and Clastic Assemblage Transvaal Supergroup, 154
Lufilian Arc, 243
Luis Alves Craton, 86, 235
Luliangian episode, 127
Luliangian Orogeny, 24
Luoquan Glacial Epoch, 222
Lurio Belt, *211*, 247
Lutzow-Holm Bay, 176, 258

McArthur Basin, 20, 47, 48, 213, 216–17
Macaubas Group, 82, 87, 234
Mackenzie Fold Belt, 231
Mackenzie Mountains Supergroup, 73, *202*, 231, *232*
Madagascar (Malagasy), 247
Mafic-ultramafic complexes, 56, 59, 61, 64–5, **66**, 68, 79, 80, 85, **89**, **90**, 93, 94, 101–3, *107*, 108, *109*, **112**, 114, 119, 120, 123, **125**, 126, 127, 128, 131, 133, 137–40, 144, 145, 152, 153, 156–8, 164, 166, 167, 171, 201, 203, 211, 247, 249, 251
Mafic dyke swarms, *202*, 265–6
Magondi Mobile Belt, 17, *39*, 158
Makkovik Orogen (Sub–province), 75, 76, 148
Malange Transcurrent Fault, 162, *244*
Malawi Province, 246, *246*
Malene supracrustal rocks, 14, 64, *64*
Malmesbury Group, 239

Man (Guinea, Leo, Liberian) Shield, 41, 104–5, 163
Mantiqueira Province, *37*, *38*, 235
Maquine Group, 86
Marampa Group, *39*, 104, *105*
Marie Byrd Land, 258
Maroni-Itacaiunas Mobile Belt, *16*, *37*, 151–2
Marquette Range Supergroup, *74*, *138*, 141, *142*, *142*
Martin Basins (redbeds), *22*, 195
Marquette Trough, 79
Mashaba Ultramafic Suite, *101*, 103
Matsap Group, *17*, **90**, *158*, *158*
Mawson Coast, 218
Mayumbian Group, *242*, 244, *244*
Mazatzal Orogeny, *127*, 191
Mbuji Mayi Supergroup, *241*, **242**, 243
McArthur Basin (Group), *20*, *47*, *48*, **112**, *173*, 214
McDonald (Wilson) Fault, *149*, 150
Medicine Bow Mountains, *80*, 150
Menominee Range (Group), *138*, *140*, 141, *142*, *143*
Mesabi Range, *140*, 141, *142*
Metamorphic Series, Angolan Craton, 162
Metasedimentary-gneiss-pluton subprovinces, Superior Province, Canaidan Shield, *67*, *67*, *71*
Metazoans, 221, 270–271
Michigan, 77–9
Microfossil types, Adelaide Geosyncline, Australia, 257
Midcontinent Rift System (Gravity High), 141, 203–4
Middleback Ranges, *48*, 174
Midland Craton, *226*, *227*, 227–8
Mid-Proterozoic crust, 177–218
  distribution, *177*, *178*
  salient characteristics, 177–80, **179**
Mid-Proterozoic domains, Australia, *48*
Mid-Proterozoic stage, 276
Mid-Zambezi Province, *246*, *246*
Migmatite and Granite Gneiss Zone, Natal Subprovince, 208
Mille Lacs Group, 140, 141, *142*
Minas Gerais State, 86
Minas Supergroup, **82**, 87, 152
Minnesota River Valley Inlier, *74*, 76–7, *79*
Minto Asch, 230, *231*
Mirelv (Moelv) tillite, 225
Mistassini-Otish Basin, *143*, 144–5
Moelv tillite, see Mirelv tillite
Moine Thrust, 225–6
Moinian (Grampian) assemblage, 227, *227*
Monocyclic Belt Boundary Zone (Grenville Province), 199, *200*
Moodies Group, *89*, *92*, 93
Moralana Supergroup, 256, *257*
Morrissey Metamorphic Suite, 169
Morton gneiss, **74**, 76–7
Mortonian event, 13, *15*
Moscow (-Caspian) Baltic Syneclise, *13*, *29*, 223–4, *224*
Mount Bruce Supergroup, *48*, 116, *117*
Mount Isa Orogen (Inlier, Basin, Group), *20*, *47*, *172*, *173*, 213, 215
Mount Narryer quartzites, *20*, 119
Mount Sones, 121
Mountain River-Redstone River Basin, 232
Moyo event, *18*
Mozaan Group, *90*, *93*, *159*
Mozambique Belt, *44*, *160*, 245–7, *246*
Mozambique Province, 246
Mporokoso Basin (Group), *211*, 211–12
Murchison Belt Africa, 93
Murchison Province Australia, *118*, 119
Murray Fault System, *130*
Musgrave Domain, 217
Musgrave-Fraser Province, 217
Musgravian Division, 216–17

Muskox Instrusion, **73**, 201
Muva Group Supergroup, 211, 212
Muya Group, 129
Mwembeshi dislocation, 240
Mylonite Zone, *186*

Nabberu basin (supergroup), *48*, 116, *169*, 169–71, *170*
Nagssugtoqidian Mobile Belt (Orogeny), 12, *14*, 32–3, *63*, 134
Nain Province, 34, 35, *75*, 148
Nama Group (Basin), *17*, 239
Namaqua Metamorphic Complex (Province, Domain), 239
Namaqua-Natal Mobile Belt, *39*, 206–9
Namaqualand-Rehoboth, 209
Nanambu Complex, 120
Nantuo Glacial Epoch, 222
Napier Complex, 21, 50, 121, 218
Narryer Gneiss Complex, *20*, 119, 120
Natal Province (Complex), 208
Natkusiak lavas (formation), **73**, 230
Naturaliste Block, *47*, 217
Nelson Front, 35
Nellore schist belt, 213
Neohelikian subera, 191, 197–204
New Quebec Orogen, 148
Ngalia Basin, *173*, 258
Ngwane gneiss, 90
Nipissing Diabase, **74**
Nondweni Belt, 93
Non-Metamorphic Caledonides, 227–8, *227*
Norseman-Wiluna Belt, *118*, 119, 120
North American Craton, 1, *4*, 13–14, **14**, *32*, 33–8, 66–81, **73**, 135–51, 191- 204, *202*, 229–34, *230*
North Atlantic Craton, 33
North Australian Craton, *4*, 171–3
North Australian Province, 120
North Baffin Rift Zone, 201
North Equatorial Belt, see Central African Belt
North Greenland Fold Belt, 1, *32*, 229
North Zemlya Massif, 128
Northeast Africa, 44, 165, *165*, 251–4
Northern Africa, 17–18, 212
Northern Australia, *172*
Northern Coastal Branch, 236, *237*, 239
Northern Frontal Zone, Natal Province, 208
Northern Greenland, 229
Northern (Scottish) Highlands, British Isles, 226
Northwest Africa, 41–4, **97**, 163–4, 248–51
North-West Highlands, 225, *226*
Northwest Territories, 231–2
Nosib stage (group), 236–7, *238*
Nova Lima Group (greenstone), 86
Nsuze Group, *17*, 239
Nubian Shield, **3**, *4*, 251–3, *252*
Nûk gneiss, 12, *14*, **64**, 65
Nullarbor Block, 120
Nutak segments (domains), 35, 76
Nyanzian schist (greenstone) belts, *18*, 104

Oates Land, 258
Ocoee Supergroup, 234
Officer Basin, *47*, 258
Ogilvie Mountains, 232, *232*
Ogooué Basin (schists), 161
Okhotsk massif, **3**, *4*, *28*, 59
Okwa Domain (Block), *39*
Older Gneiss-greenstone association, 100
Older Metamorphic Group (OMG), *19*, *46*, *107*, 111, 113, *113*
Olenek Uplift, *28*, 223
Olondo Group, see Subgan Group
Olympic Dam ore deposit (suite, breccia), 216

Omolon Massif, 3, *4*, 59
Onaping Formation, 137, 139. *139*
Onverwacht Group, 92, *92*, 93
Onwatin, Chelmsford formations, 137, *139*
Ophiolites, *12*, **125**, 131–2, 145, 220–1, 228, 236, 239, 247, 249, **250**, 262, 276, 278, 279
Ordos Nucleus, 24, *27*
Orogenies, 9
Otavi Group, 236, 237, *238*
Oumelalen Gneiss
Outokumpu assemblage (nappe), *61*
Owambo Formation, 237, *238*
Oxidation state, 270

Pachelma Trough (Aulaccgen), *29*, *187*, 188–9
Padbury Group (Sub-basin), *170*, 171
Pahrump Group, *73*, 233
Paint River Group, *138*, 141, *142*
Pakairama Nucleus, 81, **82**, *83*
Paleohelikian subera, 191, 191–6
Paleomagnetism, 7–8
Paleosols, 269
Pan-African Belt (Orogeny, Cycle), 5, *18*, **18**, *39*, 44, 236–48, 249–51, **250**, 254
Paraguay-Araguaia Fold Belt, *16*, *38*, *85*, **83**, 234
Parana Basin, *5*, *37*
Parguaza Rapakivi Granite, **82**, 83
Parnaiba Basin, *5*, *37*
Paterson Province (Basin), *47*, 216
Patom Highlands, 223
Pechenga Trough, *61*, 131
Pharusian-Dahomeyan Belt: see Dahomeyan Belt
Pelotas Marginal Massif, 235
Penge Formation, 154, *155*
Peninsular Gneiss Complex, *19*, 45–6, **107**, 110
Penokean Fold Belt (Orogeny), *13*, *15*, 35, 140–1
Pericratonic Downwarps, 223–4
Pericratonic Mobile Belts, 58–9
Phosphorites, 273
Pietersburg Belt, 93
Pikwitonei Subprovince, 66, 67
Pilbara Block, 3, *4*, *20*, 46, 114–116, *169*
Pilbara Supergroup, 115
Pine Creek Inlier (Basin), *20*, **112**, 120, 171–2, *172*, *173*
Platberg group, 97
Pneil group, 97
Point Lake-Contwoyto Lake Basin, 75
Pongola Supergroup (Basin), *17*, *38*, 93, *94*
Pontiac Belt (Subprovince), 71
Precambrian cratons, **3**
  classification scheme, 21–3
  distribution, *5*
  geologic setting, 23–50
Precambrian crust, 1–50
  areal proportions, 5–6
  biostratigraphy, *271*
  global distribution, 2, 5–8
  paleomagnetism and continental reconstructions, 7–8
Pretoria Group, 154, *155*, 155
Prince Albert group, 75
Prince Charles Mountains, 50, 176, 218
Proterozoic sequences, sedimentation, 273
Protogine Zone, 186, *186*
Prydz Bay, 218
Purcell Supergroup, see Bett-Purcell supergroup

Qianxi Group, 56
Qingbaikou system, *10*, *180*, *181*
Qinglonghe Group, 128
Qôrqut Granite Complex, *14*, **64**, 65

Quadrilatero Ferrifero, 86, 152
Queen Maud Land, 176
Quetico Belt, 67, 71
Qyirke Lake Group, *138*

Racklan Orogeny, **73**, 231
Radiometric dating, 8–9
Rae Province (Group), *127*, 148
Rajasthan Block, 113–14
Rapakivi granites (suite), *12*, *13*, *14*, *15*, *16*, *61*, 83, 178, **179**, 187, 193–4, 204, 205, 214, 264
Rapitan Group, *73*, 231, 232
Rare Earth Elements (REE), 56
Rayner Complex, *21*, 176, 218
Redbeds (Red beds), *10*, *11*, *15*, *16*, 35, 40, *128*, 131, 137, 158, **179**, 179, 182, 185, 188, 189, 190, 194–6, 204, 205, 209, 212, **221**, 223, 225, 241, 244, 254, 269, *271*, 272
Reguibat Shield, 3, *4*, *39*, 41, 106
  Eastern Province, 163–4
  Southwestern Province, 106
Rehoboth (Rehobothian) Domain (subprovince), *39*, *207*
Richtersveld Subprovince (Domain), 206
Rifted redbed basins, 194
Rinkian Mobile Belt (Province), *13*, **32**, *32*, 33, 134–5
Rio Apa Massif (Craton), 205
Rio das Velhas Supergroup, *16*, 86
Rio de la Plata Craton (fragment), 37, *38*, 86, 235
Rio Itapicuru greenstone belt, 153
Rio Negro-Juruena Mobile Belt, *16*, *37*, 37, 205
Riphean structures (Division), *11*, 185, *187*, 188
Rondonian (-Sunsas) Belt Complex, 37, 205
Roraima Group, *16*, 205
Ross Orogeny, 21
Rum Jungle Complex, 112, 120, *172*
Russia, *61*, **183**, *184*
Ruwenzori (Buganda-Toro) Fold Belt, *41*, 160–1
Ruzizian Belt see Ubendian Belt

Saglek Bay-Hebron Fjord area, **74**, *75*, 76
St Francois Mountains, 193
Sakoli Group, *45*, 168
Salvador-Juazeiro region, 86, *87*
Sanheming Group, *56*, 128
São Francisco Craton (Province), 86, 152–3, 205–6, 234–5
São Luis Craton, 38, 151, 235
Sargur Belts (Group, enclaves), **107**, 108, *109*
Sariolian (Sariolan) sediments (unit), *12*, *61*, 131
Sausar Group, *45*, 168
Scottish Shield Fragment, 1, *13*, **14**, *33*, 65–6, 135
Sea water composition, 266–7
Seal Lake Group, *73*, 76, 204
Sebakwian Group, *17*, 101, *101*
Sedimentation
  Archean sequences, 271–3
  Proterozoic sequences, 273
Sensitive highmass resolution ion-microprobe (SHRIMP), 8
Serra de Santa Rita, 152
Serra do Jacobina Region (Mn), 153
Serra do Navio District (Amape, Mn), *16*, 151
Serra dos Carajás (BIF), 81, *84*, 152
Serrinha greenstone belt, *16*, 153
Serro do Espinhaço Range, 205
Serwin Ironstone Member, 215
Seychelles Islands, 247–8
Shaba (Aulacogan)-Zambia Copperbelt (Golfe du Katanga), 241
Shackleton Range, 218
Shaler Group, *231*
Shamvaian Group, *17*, *101*, 102
Shanxi Plateau, 127

Shuanshansi Group, 56, 128
Siberian Craton, 1, 10–11, **11**, 24, 57–8, 128–9, 182–5, 222–3
  buried basement, 59
Sierra Madre, 80, 150
Singhbhum Craton, 3, 46, **107**, **108**, 111–13, *113*
Singhbhum Granite Complex (granite), 111, 113
Sinian strata (system), 10, 222
Sino-Korean Craton, 3, 24, 54, **55**, **56**, 127–8
Sioux-Baraboo-Barron quartzites, 73, 194–5
Slave Province (Craton), 35, 68, 72–5, 148
Sleaford Complexes, 120
Snake River Basin, 232
South America, *236*
South American Craton, 3, 4, 5, 14–15, *16*, 36, 81–8, *84*, 151–3, 204–6, 234–5
South Australia, 120, 173–5, *174*
Southeast Africa, *160*
Southeast Rae Province, 148
Southern Africa, 15–17, 88–103, *89*, 153–9, 208–11, *236*, 236
Southern Coastal Branch, 236, 239
Southern Cross Province (greenstone, belts), *118*, 119
Southern Granulite Zone, 208
Southern Province, *34*, 35
Southern Urals, 188–9
Southern West Greenland, 62, **64**
Southern-Western Superior Province, 70–1
Southwest Scandinavian Domain (Province), 28–9, 185–7
Sparagmite Formation, 225
Spitzbergen Archipelago, 32, 225
Sri Lanka (Craton), 3, 4, 253, 254–6
Stanley Fold Belt, 170
Stanovoy Fault, 128
Stanovoy Fold Belt, 24, 28, 57, 58, 128
Stillwater Complex, **74**, 80
Stromatolites, 256
Strontium in sea water, 267
Sub-Andean Foredeep, 36
Subgan (Olondo) Group, 57
Sudbury Igneous Complex (Irruptive), **73**, 137, 139
Sudbury Structure, *130*, 137–40
Suisaarian Group, 131
Sumian-sariolan succession, *12*, 131
Superior Province, *34*, 35, 66–72, *67*, **68**
Svecofennian (flysch) facies (supergroup), 132–3
Svecofennian Orogeny, *12*, 133, 185
Svecofennian Province (Domain), 28, *29*, *61*, 131, 185
Svecokarelian rocks (complex, terrain), 131
Swakop Group (stage, subprovince), 237
Swaziland Supergroup (stratigonphy), **17**, 91, 92, *92*

Taigonos massif (Block), 3, 4, 59
Taihua Complex, 54, **56**
Tampere Region, 132
Tanshan Mountains, 181
Tanzania Craton, 3, 4, 39, 40, 104, *160*
Taoudeni Basin, 3, *18*, 39, 41, 248–9
Tapajos (Guapore) Craton, *16*, 37, *37*
Tarim Craton (Basin), 3, 4, 9, 24, 27, 221
Tarim-Tianshan Province, 222
Tasmania, *4*, 47, 258
Taymyr fold belt, *4*, 25, 128, 183
Tectonic cycles, 9
Tectonic development, 22
Tectonic framework, 9–21
Terre Adélie/King George V Land, 4, 49, *50*
Tertiary Rift Volcanics, 247
Thelon Orogen, 148
Thelon Formation, **73**, 196
Thompson Belt, 145
Tillite Group, *190*, 229

Tillites, see Diamictites
Timan-Pechora Extension (Geosyncline, Syneclise), 29, 30, 224
Tindir Group, **73**. 232
Tocantins Province, 37, 38, 85, *85*, 151, 152, 206, 234
Togo Discontinuity, 249
Tonalite Trondhjemite Granodiorite (TTG) (Na-tonalite trend), 56, 262–3
Torngat Orogen, *127*, 148
Torrens Hinge Zone, 256
Torridonian Supergroup (Group), *12*, 225–6
Transamazonian Orogeny, 15, *16*, **82**
Transantarctic Mountains (Fold Belt), 21, 258
Trans-Aravalli Vindhyan Basin, 254
Trans-Hudson Orogen, 70, 136, 146–8, *147*
Trans-Saharan (Pharusian-Dahomeyan) Mobile Belt, 39, 41–4, 97, 104–6, 249
Transscandinavian (Smaland-Varmland) Belt, *12*, 28, 185, 186
Transvaal Basin (Supergroup), **17**, 153–156, *154*, *155*, *157*
Tuareg Shield, 3, 44, 97, *105*, 106, 164, 212, 249–51, **250**
Turee Creek Group, 48, 116, *117*
Turukhansk Fold Belt (Uplift), 3, 25, 128

Uatuma Volcanic-Plutonic Complex (Supergroup), *16*, 204–5
Ubendian (-Ruzizian) Belt, *18*, 39, 41, 159, *160*, *211*
Uchur-Maya Region, 223
Udokan (Oudokan) Group (Trough), 11, 129
Uivak Gneiss, *15*, 74
Uivakian Orogeny, 13
Ukrainian Shield, 1, 11, 29–30, *30*, 59–60, 133–4, 185, 188
Ultramafic-anorthosite bodies, 64–5
Ultramafic complexes, see Mafic-ultramafic complexes
Umberatana tillites, 256, **257**
Umkondo Belt (Group), 38–9, *39*, 246
Ungava Craton (Domain), 35, 66, **68**, 68–70, *143*, 144
Ungava Orogen, *127*, 148
Uppernavik association (supracrustals), **74**, 75
Upper Clastic Assemblage, 155, *155*
Upper Eleonore Bay Group, 229
Upper Greenstones Bulawayan Group, 102
Upper Vendian-Riphean sequence, *184*
Uralian (Ural) Fold Belt, 29, 31
Uraniferous conglomerates, 95, 137, 150–151, 153, 171–2
Uruao Belt (assemblage), *85*, 206
Usagaran Belt, 39, 41, 159–60, *160*
Usushwana Intrusive Suite, **89**, 93, *94*
Uweinat Inlier, 165–6

Varanger Fjord, 225
Varangian Glacial Period, 225
Variscan Fold Belt (Variscides), *228*, 228–9
Variscan-Hercynian fold, 1
Variscan massifs, 228, *228*
Variscan Orogeny, 228
Varmland-Smaland Belt, see Transscandinavian Belt
Vendian Division (deposits), 188, 223, *224*
Ventersdorp Supergroup (Basin), 97–8, *98*
Vepsian (Subjotnian) Group, 131
Vestfold Hills, 218
Victoria River Basin, 47, *173*, 216
Vijyan Complex, 255, *255*
Vindhyan Basin (Supergroup), 46, 107, 113, 254, *255*
Volga (-Vrals)-Kama Anteclise (massif), 1, 29, *187*
Volta Basin, *105*, *250*, 251
Volyn (Volynian) Block, *30*, 188
Voronezh Anteclise (Massif, uplift), 1, 30, 133–4, *187*
Vredefort cryptoexplosion event, 91
Vredefort Dome, **17**, 91, 95, 139, *157*

Wabigoon Subprovince (Belt), 67, 70
Wakeham Bay Supergroup, 199

Walsh Lake Fault, 144, *145*
Warrawoona Group, *20*, 115, *115*, 270, *271*
Wanni Complex, 254, 255, 256
Waterberg-Soutpansberg-Matsap-Umkondo Basins, *17*, 94, *158*, 158–9
West African Craton, **3**, 41, **96**, 104–6, *105*, 163, **250**
West Australian Shield (Craton), **3**, 115–120, 168–71, 216
West Congo Fold Belt, **3**, 44, 243–5, *244*
West Congolian Supergroup, 244–5
West Gondwana, 247
West Karelia zone, 61
Western Bangemall Basin, 216–19
Western Churchill Province, **77**, *78*, 146, 148
Western Dharwar Domain, 110
Western Gneiss Terrain, 46, 47, 118–119, *118*
Western Hubei region, *180*, 181
Willyama Domain (Block supergroup), *20*, 174–5, 256
Wilson Cycle, 7, 200
Windermere Supergroup, **73**, 203, 230–1, *232*, 233
Windmill Islands, 218
Wisconsin Magmatic Terranes (Volcanic Belt), **74**, *140*, 141–2
Witwatersrand Basin (Supergroup, Group), *17*, **89**, **90**, *95*, 97
Woodburn (-Ketyet) group, 75, *78*
Wopmay Orogen (Fold Belt), *15*, 149–50

Wutai Group, **56**, 127, 128
Wutai-Taihang District, 127–8
Wutai-Taihang-Luliang District, **55**, *55*
Wyloo Group, *112*, 169
Wyoming Uplift, 1, 36, 79–81, *80*

Xingu Nucleus, 37, 81, **82**, *83*

Yangtze Craton, **3**, 24, 27, 180, 181–2, 221
Yangtze Gorges, 222
Yanshan Mountains (District), 26, *55*, **56**, 181
Yellowknife Supergroup, 72, **74**
Yellowknife-type volcanic successions, 72
Yenisei Fold Belt (Ridge), **3**, 25, 59, 128, 223
Yilgarn Block, **3**, 19, 46, 47, *118*, 118–20, 119, 120, *169*, 170
Yinshan District, *55*, **56**, 128
Yukon, 231–2
Yurmatinian Group, *11*, *13*, **183**, 185

Zadinian Group (Supergroup), 244
Zambezi Belt, **39**, 239–40, *240*
Zambian Craton, 4, 104, 162
Zimbabwe Craton, 38, 100, 100–3, *101*, 160, 239, 240
Zinjian Province, 26, 222